FLORA ZAMBESIACA

Flora terrarum Zambesii aquis conjunctarum

PTERIDOPHYTA

PLATYCERIUM ELEPHANTOTIS   PLATYCERIUM ALCICORNE

FLORA ZAMBESIACA

MOZAMBIQUE

MALAWI, ZAMBIA, RHODESIA

BOTSWANA

PTERIDOPHYTA

by E. A. C. L. E. SCHELPE
Bolus Herbarium, University of Cape Town

Edited by
A. W. EXELL and E. LAUNERT

on behalf of the Editorial Board:

J. P. M. BRENAN
Royal Botanic Gardens, Kew

A. W. EXELL
Commonwealth Forestry Institute, Oxford

A. FERNANDES
Junta de Investigações do Ultramar, Lisbon

H. WILD
University College of Rhodesia

Published by the Managing Committee on behalf of
the contributors to Flora Zambesiaca
Portugal, Malawi, Zambia
and Southern Rhodesia
Crown Agents for Oversea Governments and Administrations
4 Millbank, London S.W.1.
October 23, 1970

LIST OF FAMILIES AND GENERA

1. *PSILOTACEAE:*
   Psilotum

2. *LYCOPODIACEAE:*
   Lycopodium

3. *SELAGINELLACEAE:*
   Selaginella

4. *ISOETACEAE:*
   Isoetes

5. *EQUISETACEAE:*
   Equisetum

6. *OPHIOGLOSSACEAE:*
   Ophioglossum

7. *MARATTIACEAE:*
   Marattia

8. *OSMUNDACEAE:*
   Osmunda
   Todea

9. *GLEICHENIACEAE:*
   Gleichenia
   Dicranopteris

10. *SCHIZAEACEAE:*
    Schizaea
    Anemia
    Mohria
    Lygodium

11. *MARSILEACEAE:*
    Marsilea

12. *SALVINIACEAE:*
    Salvinia

13. *AZOLLACEAE:*
    Azolla

14. *CYATHEACEAE:*
    Cyathea

15. *HYMENOPHYLLACEAE:*
    Trichomanes
    Hymenophyllum

16. *DENNSTAEDTIACEAE:*
    Blotiella
    Histiopteris
    Lonchitis
    Pteridium
    Microlepia
    Hypolepis

17. *VITTARIACEAE:*
    Vittaria
    Antrophyum

18. *ADIANTACEAE:*
    Acrostichum
    Anogramma
    Coniogramme
    Ceratopteris
    Pityrogramma
    Adiantum
    Aspidotis
    Pteris
    Doryopteris
    Cheilanthes
    Pellaea
    Actiniopteris

19. *LINDSAEACEAE:*
    Lindsaea

20. *GRAMMITIDACEAE:*
    Grammitis
    Xiphopteris

21. *POLYPODIACEAE:*
    Platycerium
    Pyrrosia
    Drynaria
    Loxogramme
    Pleopeltis
    Phymatodes
    Microgramma
    Microsorium
    Polypodium
    Belvisia

22. *DAVALLIACEAE:*
    Nephrolepis
    Arthropteris
    Oleandra
    Davallia

23. *ASPLENIACEAE:*
    Asplenium
    Ceterach

24. *THELYPTERIDACEAE:*
    Thelypteris
    Ampelopteris

25. *ATHYRIACEAE:*
    Athyrium
    Diplazium
    Dryoathyrium

26. *LOMARIOPSIDACEAE:*
    Elaphoglossum
    Lomariopsis
    Bolbitis

27. *ASPIDIACEAE*:
Didymochlaena
Dryopteris
Polystichum
Arachniodes
Hypodematium
Ctenitis
Tectaria

28. *BLECHNACEAE*:
Blechnum
Stenochlaena

A SELECTED BIBLIOGRAPHY CONCERNING THE PTERIDOPHYTA OF THE FLORA ZAMBESIACA AREA

ALSTON, A. H. G., Enumeração de Criptogâmicas vasculares de Moçambique in Estudos, Ensaios e Documentos **12** (Contribuições para conhecimento da Flora de Moçambique, **2**: 3–38 (1954)).

BALLARD, F., Pteridophyta in BRENAN, J. P. M. *et al.* Plants collected by the Vernay Nyasaland Expedition of 1946 in *Mem. N. York Bot. Gard.* **8**, 3: 197–210 (1953).

SCHELPE, E. A. C. L. E., A revision of the African species of *Blechnum* in *Journ. Linn. Soc. Lond., Bot.* **53**: 487–510 (1952).

— The genus *Pyrrosia* (Polypodiaceae) in Africa in *Journ. S. Afr. Bot.* **18**: 123–134 (1952).

— The ecology of *Salvinia auriculata* and associated vegetation of Lake Kariba in *Journ. S. Afr. Bot.* **27**: 181–187 (1961).

— Pteridophyta collected on an expedition to northern Moçambique in *Journ. S. Afr. Bot.* **30**: 177–200 (1964).

— A review of the Southern African species of *Thelypteris* in *Journ. S. Afr. Bot.* **31**: 259–269 (1965).

— The Pteridophyta of Gorongosa Mountain, southern Moçambique in *Bol. Soc. Brot.*, Sér. 2, **40**: 149–179 (1966).

— New Taxa of *Pteridophyta* from South East Tropical Africa in *Bol. Soc. Brot.*, Sér. 2, **41**: 203–217 (1967).

— Three new species of ferns from Southern Africa in *Journ. S. Afr. Bot.* **34**: 235–241 (1968).

— Contr. Fl. Rhod. XI, Pteridophyta in *Bot. Notis.* **121**: 361–382 (1968).

— A revised Check List of the Pteridophyta of Southern Africa in *Journ. S. Afr. Bot.* **35**: 127–140 (1969).

SIM, T. R., The Ferns of South Africa, 2nd ed., Cambridge (1915).

A BIBLIOGRAPHY OF OTHER WORKS RELEVANT TO AFRICAN PTERIDOLOGY

ADAMS, C. D. & ALSTON, A. H. G., A List of the Gold Coast Pteridophyta in *Bull. Brit. Mus. (Nat. Hist.)* **1**, 6: 145–185 (1955).

ALSTON, A. H. G., Mr. John Gossweiler's plants from Angola and Portuguese Congo in *Journ. of Bot.* **72**, Suppl. Pterid.: 1–11 (1934).

— Notes on *Selaginella*. IX. The South African species in *Journ. of Bot.* **77**: 221–224 (1939).

— Pteridophyta in EXELL, A. W., Catalogue of the Vascular Plants of São Tomé (with Principe and Annobon). London (1944) and Supplement (1956).

— New African Ferns in *Bol. Soc. Brot.*, Sér. 2, **30**: 5–27 (1956).

— Les Ptéridophytes d'Afrique Intertropicale Française (Fin) in *Mém. Inst. Fr. Afr. Noire*, **50**: 10–49 (1957).

— The Ferns and Fern Allies of West Tropical Africa. (Suppl. to 2nd ed., F.W.T.A. London (1959)).

BAKER, J. G., Handbook of the Fern Allies. London (1887).

CHRISTENSEN, C., Index Filicum. Copenhagen (1905–6) and Supplements (1913, 1917, 1934, 1965).

— The Pteridophyta of Madagascar in *Dansk Bot. Ark.* **7**: 1–253, t. 1–80 (1932).

COPELAND, E. B., Genera Filicum. Waltham, Mass. (1947).

HARLEY, W. J., The ferns of Liberia in *Contr. Gray Herb.* **177**: 58–103 (1955).

HOOKER, W. J. & BAKER, J. G., Synopsis Filicum, 2nd ed. London (1874).

KUHN, M., Filices Africanae. Leipzig (1868).

LAUNERT, E., A monographic survey of the genus *Marsilea* L. I. The species of Africa and Madagascar in *Senckenb. Biol.* **49**: 273–315 (1968).

— Pteridophyta in *Prodr. Fl. SW. Afr.*, Fam. 1–12 (1969).

MACLEAY, K. N. G., The Ferns and Fern Allies of the Sudan in *Sudan Notes and Records*, **34**: 286–298 (1953).

PICHI-SERMOLLI, R. E. G., Adumbratio Florae Aethiopicae:

    3. Ophioglossaceae, Osmundaceae, Schizaeaceae in *Webbia*, **9**: 623–660 (1954).

    4. Hymenophyllaceae, Negripteridaceae, Cyatheaceae in *Webbia*, **12**: 121–146 (1956).

    5. Parkeriaceae, Adiantaceae, Vittariaceae in *Webbia*, **12**: 645–704 (1957).

    8. Gleicheniaceae in *Webbia*, **17**: 33–43 (1962).

    9. Cryptogrammaceae in *Webbia*, **17**: 299–315 (1963).

    10. Actiniopteridaceae in *Webbia*, **17**: 317–328 (1963).

    11. Oleandraceae in *Webbia*, **20**: 745–769 (1965).

    13. Hemionitidaceae in *Webbia*, **21**: 487–505 (1966).

    15. Elaphoglossaceae in *Webbia* **23**: 209–246 (1968).

— On the Fern Genus *Actiniopteris* Link in *Webbia*, **17**: 1–32 (1962).

— Taxonomic notes on the fern genus *Arthropteris* in *Webbia*, **21**: 507–516 (1966).

SCHELPE, E. A. C. L. E., The Pteridophyta of Mount Kenya in *Amer. Fern Journ.* **41**: 65–74 (1951).

— Ferns in Central Ethiopia in *Brit. Fern Gaz.* **8**: 61–64 (1953).

— Distributional, ecological and phytogeographical observations on the ferns of South-West Africa in *Journ. S. Afr. Bot.* **22**: 5–22 (1956).

Taton, A., Revision des Hymenophyllacées au Congo Belge in *Bull. Soc. Roy. Bot. Belg.* **78**: 5–42 (1946).

Tardieu-Blot, M. L., in Humbert, H., Flore de Madagascar et des Comores: Marattiacées, Ophioglossacées, Hymenophyllacées, Cyatheacées. Paris (1951).

Polypodiacées I. Paris (1958).

Polypodiacées II. Paris (1960).

— Les Ptéridophytes de l'Afrique Intertropicale Française in *Mém. Inst. Fr. Afr. Noire*, **28**: 1–241 (1953).

— Les fougères des Mascareignes et des Seychelles in *Notul. Syst.* **16**: 151–201 (1960).

— in Aubréville, A. *Flore du Cameroun*, **3**: Ptéridophytes. Paris (1964).

— in Aubréville, A. *Flore du Gabon*, **8**: Ptéridophytes. Paris (1964).

GLOSSARY

Most of the terms used in this work correspond largely to Tryon's proposals in Taxon **9**: 104–109 (1960). However, since the present author's concept of certain terms differs to a minor extent it is necessary that these differences be made clear in a glossary not covered by the introduction to Flora Zambesiaca.

acroscopic (acroscópico), on the side towards the apex.

acrostichoid (acrosticóide), with the sporangia spread over the lower surface of the fertile lamina (as in *Acrostichum* and *Elaphoglossum*).

annulus (anel), the hygroscopic thickened cells causing the dehiscence of the sporangium.

basipetal (basípeto), arising and maturing in succession from apex to base (e.g. sporangia in *Hymenophyllaceae*).

basiscopic (basiscópico), on the side towards the base.

caudex (caudex), stem surrounded by leaf-bases and adventitious roots, as in arborescent ferns (e.g. *Cyathea*).

clathrate (clatrado), of rhizome and other scales, with a latticed appearance, due to the cells having thickened lateral walls and transparent surface walls.

costa (costa), the midrib of the pinna.

costule (cóstula), the midrib of a pinnule or pinna segment.

cristo-reticulate (cristado-reticulado), crests anastomosing to form a network.

dimidiate (dimidiado), of pinnae or pinnules in which the midrib forms the basiscopic margin for a significant distance.

dorsal (dorsal), synonymous with abaxial or lower in regard to lamina surfaces.

echinate (equinado, espinhoso), spinose, with projections tapering from a broad base to a $\pm$ sharp apex.

exindusiate (sem indúsio), without an indusium.

gemma (gema), an adventitious bud arising on the frond which can produce a new plant.

granulate (granuloso), with more or less isodiametric projections not less than 1μ.

heterosporous (heterospórico), producing spores of 2 sizes, the larger giving rise to a female megagametophyte, the smaller giving rise to a male microgametophyte.

homosporous (homospórico ou isospórico), producing spores of the same size.

indusium (indúsio), a thin flap of tissue covering at least the young sorus; pseudo-indusia are formed by modification of the lamina margin.

laesura (lesura), the dehiscence fissure of a spore and its margin.

lamina (lâmina), the blade of a frond.

ligule (lígula), a small membranous triangular organ on the adaxial side of the fertile leaf-base in *Isoetes*.

lophate (lofado), ridged with simple flange-like ridges, seldom much shorter than the shortest diameter of the spore.

mammillate (mamilado), having small nipple-like projections.

megagametophyte (megagametófito ou macrogametófito), the female gametophyte produced by the megaspore which bears the female sex organs.

megaspores (megásporos ou macrósporos), in heterosporous pteridophytes the large spores which give rise to the female gametophyte.

megasporangium (megasporângio ou macrosporângio), the sporangium containing megaspores.

microgametophyte (microgametófito), the male gametophyte produced by the microspore.

microspores (micrósporos), the small spores which give rise to the male gametophyte in heterosporous pteridophytes.

microsporangium (microsporângio), the sporangium containing microspores.

midrib (nervura média), the main vascular supply of a simple lamina.

monolete (monolete), with the dehiscence line unbranched (as in bilateral spores).

paraphyses (paráfises), sterile hairs, sometimes clavate or with an enlarged apical cell, occurring among sporangia in a sorus.

perispore (perispório), a layer outside the exine in certain spores.

pinna (pina), the first order division of a dissected lamina.

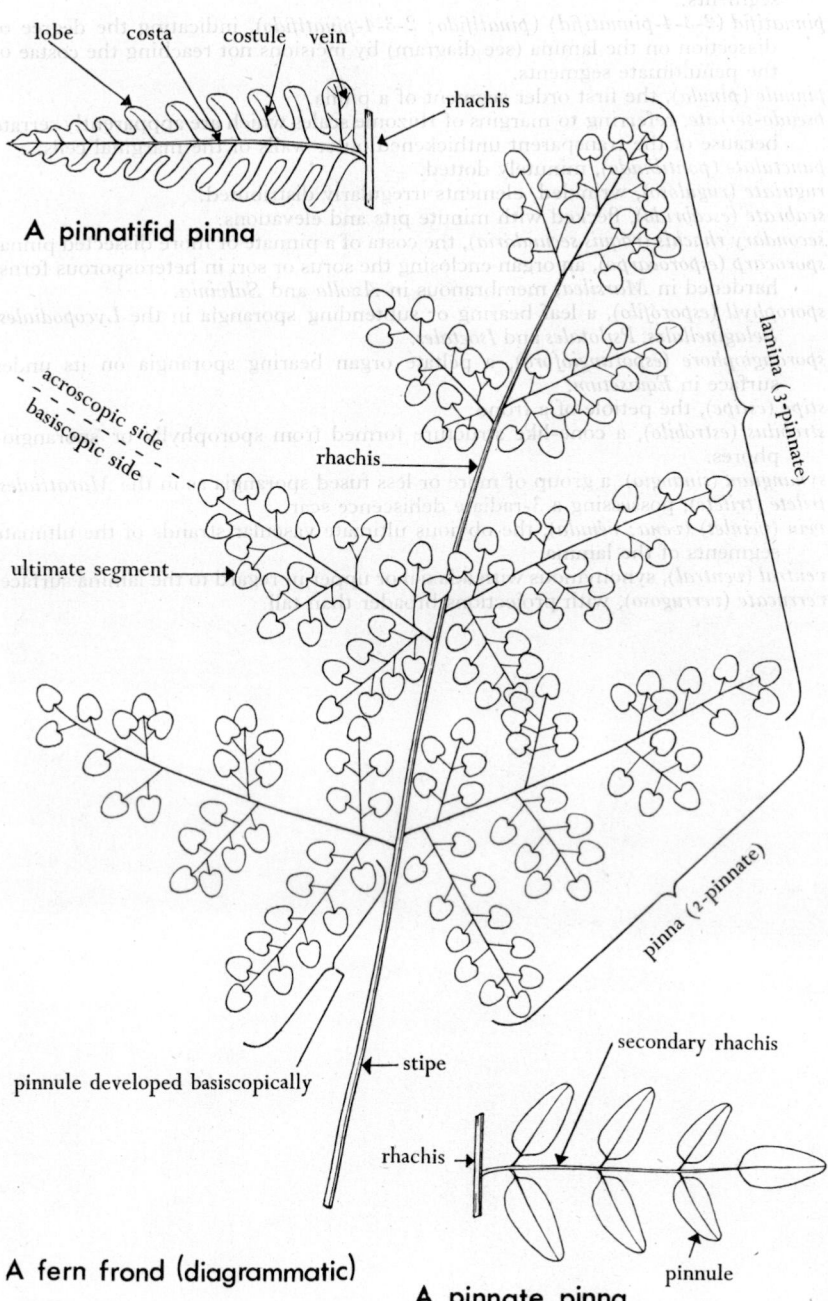

lobe   costa   costule   vein

rhachis

A pinnatifid pinna

lamina (3-pinnate)

acroscopic side

basiscopic side

rhachis

ultimate segment

pinna (2-pinnate)

pinnule developed basiscopically

stipe

rhachis

secondary rhachis

pinnule

A fern frond (diagrammatic)

A pinnate pinna

Tab. 1. Illustration of terms in the Glossary.

pinnate (2-3-4-pinnate) (*pinado; 2-3-4-pinado*), indicating the degree of dissection of the lamina (see diagram) by incisions to the costae of the penultimate segments.

pinnatifid (2-3-4-pinnatifid) (*pinatífido; 2-3-4-pinatífido*), indicating the degree of dissection on the lamina (see diagram) by incisions not reaching the costae of the penultimate segments.

pinnule (pínula), the first order segment of a pinna.

pseudo-serrate, referring to margins of rhizome scales which are appparently serrate because of the transparent unthickened outer walls of the marginal cells.

punctulate (pontilhado), minutely dotted.

rugulate (ruguloso), wrinkled, elements irregularly distributed.

scabrate (escábrido), flecked with minute pits and elevations.

secondary rhachis (ráquis secundária), the costa of a pinnate or more dissected pinna.

sporocarp (esporocarpo), an organ enclosing the sorus or sori in heterosporous ferns, hardened in *Marsilea,* membranous in *Azolla* and *Salvinia.*

sporophyll (esporófilo), a leaf bearing or subtending sporangia in the *Lycopodiales, Selaginellales, Psilotales* and *Isoetales.*

sporangiophore (esporangióforo), a peltate organ bearing sporangia on its under surface in *Equisetum.*

stipe (estipe), the petiole of a frond.

strobilus (estróbilo), a cone-like structure formed from sporophylls or sporangiophores.

synangium (sinângio), a group of more or less fused sporangia as in the *Marattiales.*

trilete (trilete), possessing a 3-radiate dehiscence scar.

vein (veinlet) (*vena; vénula*), the obvious ultimate vascular strands of the ultimate segments of the lamina.

ventral (ventral), synonymous with adaxial or upper in regard to the lamina surface.

verrucate (verrugoso), with projections broader than tall.

KEY TO THE ORDERS

Leaves all narrow and simple, entire (usually small) with unbranched veins; the sporophylls either with a sporangium at the base or subtending a sporangium or with the sporangia borne within a cone-like structure:
Sporangia borne in the axils of the sporophylls or on their bases:
Plants homosporous; leaves without ligules:
Sporangia 3-lobed; stems leafless except for minute much reduced sporophylls - - - - - - - **Psilotales** (p. 15)
Sporangia not lobed; stems with numerous leaves **Lycopodiales** (p. 15)
Plants heterosporous; leaves with ligules:
Plants aquatic, wholly or partially submerged during the wet season, sedge-like; sporangia borne on the leaf bases; stem a short lobed rootstock
Isoetales (p. 29)

Plants not aquatic, often moss-like with elongated erect or creeping stems; sporangia borne in the axils of differentiated sporophylls
Selaginellales (p. 22)

Sporangia borne on peltate sporangiophores arranged in a cone-like structure; leaves reduced to a short toothed sheath about the nodes **Equisetales** (p. 32)
Leaves usually broad, simple or dissected, the lamina with a branched vascular supply:
Sporangia thick-walled, without an annulus; homosporous:
Sporangia borne in 2 rows on a distinct slender fertile segment; sterile lamina entire (or lobed) - - - - - - **Ophioglossales** (p. 34)
Sporangia fused in small groups on the dorsal surface on undifferentiated lamina segments - - - - - - - **Marattiales** (p. 38)
Sporangia thin-walled, with an annulus; homosporous or heterosporous
Filicales (p. 40)

PSILOTALES

1. PSILOTACEAE

Epiphytic or saxicolous plants. Aerial stems chlorophyllose, simple or dichotomously branched produced from non-chlorophyllous mycorrhizal rootless rhizomelike horizontal axes. Sporangia 2-locular or 3-locular, subtended by a bifurcate bract.

Only the genus *Psilotum* is known from our area.

PSILOTUM Sw.

Psilotum Sw., Syn. Fil.: 187 (1807).

Epiphytic or saxicolous plants; aerial stems repeatedly dichotomously branched with much reduced leaves. Sporangia 3-locular, each subtended by a small bifurcate bract and borne on the upper parts of the aerial branches.

Psilotum nudum (L.) Beauv., Prodr. L'Aethéog.: 106, 112 (1805). TAB. **2**.
Type probably from Asia.
  Lycopodium nudum L., Sp. Pl. **2**: 110 (1753). Type as above.
  Psilotum triquetrum Sw., Syn. Fil.: 187 (1806) *nom. illegit.*—Sim, Ferns S. Afr.: 342, t. 181 (1915). Type as for *Psilotum nudum*.
  Bernhardia capensis K. Müll. in Bot. Zeit. **1858**: 239 (1858). Type from S. Africa.
  Bernhardia mascarenica K. Müll., loc. cit. Type from Mascarene Is.
  Psilotum natalense Gandog. in Bull. Soc. Bot. Fr. **66**: 306 (1920). Type from S. Africa.

Rhizome c. 1·5 mm. in diam., short, rootless. Aerial stems up to 24 × 0·2 cm., dichotomously branched, triangular in cross-section, glabrous, with widely spaced lanceolate scale-leaves up to 1·5 mm. long. Sporangia c. 2·5 mm. in diam., 3-locular, each subtended by a bifurcate bract c. 1·3 mm. long.

Zambia. N: Ntumbacushi Falls, 22.vi.1957, *Robinson* 2344 (K; SRGH). W: Ichimpi, Kitwe, 18.v.1963, *Mutimushi* 312 (K; SRGH). **Rhodesia.** N: Masene R. Gorge, 10.viii.1951, *Whellan* 540 (SRGH). W: Victoria Falls, Rain Forest, v.1904, *Eyles* 115 (BM; BOL; SRGH). C: Rumani Distr., Enterprise, 25.vi.1947, *Wild* 1967 (SRGH). E: Melsetter Distr., Haroni R., 3.xii.1964, *Wild* 6601 (BOL; SRGH). **Mozambique.** MS: Moribane, xi.1911, *Dawe* 489 (K). LM: Inhaca I., vii.1959, *Mogg* 30800 (SRGH).

Also in eastern S. Africa and widespread in tropical Africa and elsewhere in the tropics; Epiphytic and saxicolous in forests, 910–1830 m.

LYCOPODIALES

2. LYCOPODIACEAE

Epiphytic, lithophytic or terrestrial plants. Stems erect, pendulous or prostrate, unbranched or dichotomously branched. Leaves small, simple with a single vein, without ligules. Sporophylls uniform, usually restricted to distinct or indistinct

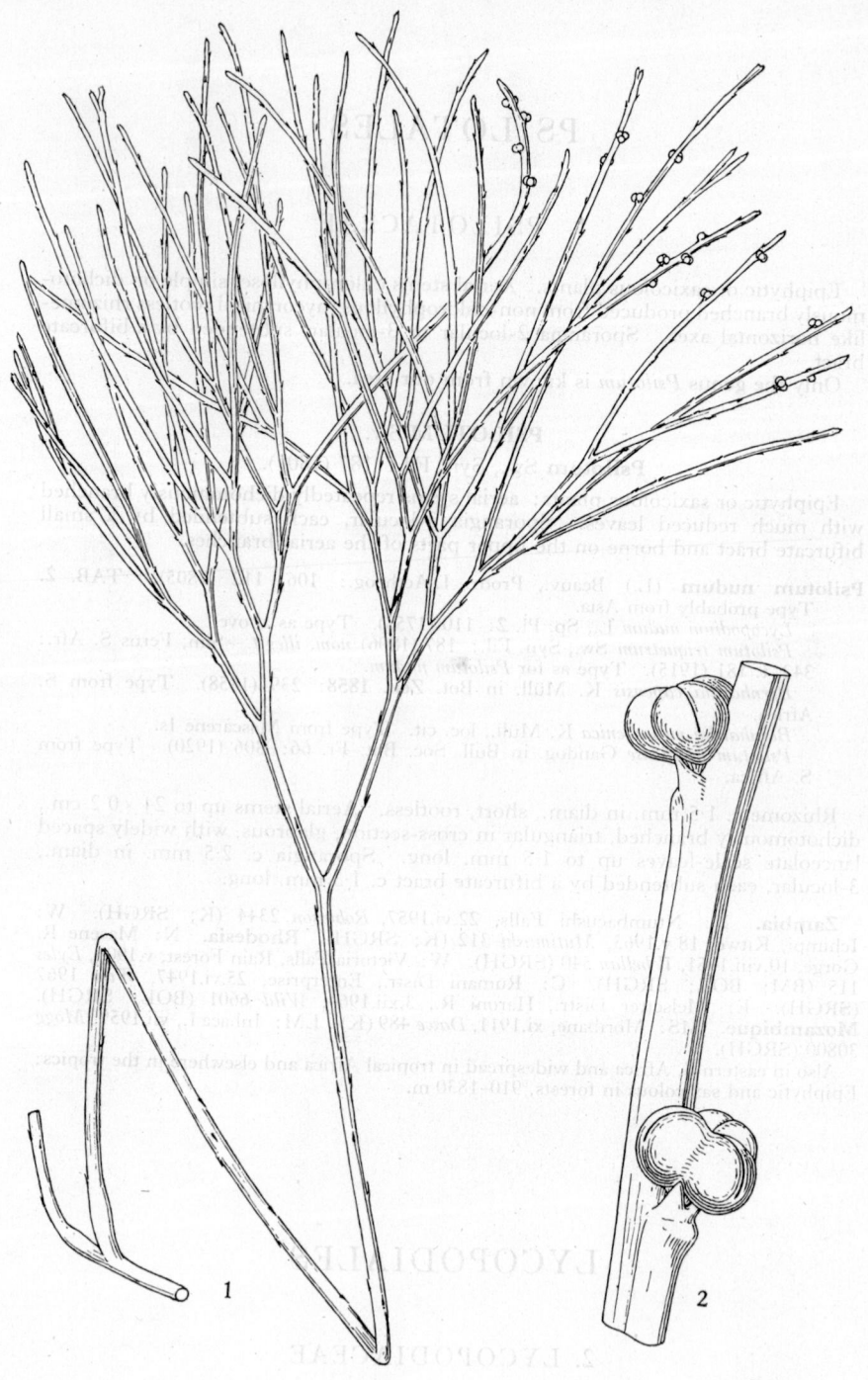

D.E

Tab. 2. PSILOTUM NUDUM. 1, habit (×⅘); 2, enlargement of fertile branch (×8) both from Kew Ferneries.

fertile regions arranged in strobili; sporangia borne in the axils of the sporophylls, solitary, unilocular, reniform to globose, homosporous. Gametophytes chlorophyllose, or mycorrhizal and colourless.

LYCOPODIUM L.

Lycopodium L., Sp. Pl. **2**: 110 (1753): Gen. Pl. ed. 5: 486 (1754).

Description as for the family.

The only living genus of the family, with a world-wide distribution and comprising over 450 spp. The species which do not form distinct strobili have been referred to the genus *Urostachys* by some authors.

Sporophylls of the same shape and size as the foliage leaves:
  Leaves narrowly oblong-lanceolate; spore-bearing stems erect, unbranched
      1. *saururus*
  Leaves acicular; spore-bearing stems pendulous, dichotomously branched
      2. *verticillatum*
Sporophylls different from the foliage leaves:
  Sporophylls not grouped into pedunculate strobili:
    Sporophylls 6·5–7 mm. long - - - - - - - 3. *dacrydioides*
    Sporophylls less than 4 mm. long:
      Fertile region more than 2 cm. long; sporophylls not lacerate; plants lithophytic or epiphytic:
        Sporophylls 2·5–3 mm. long; leaves coriaceous imbricate - 4. *gnidioides*
        Sporophylls 1–2 mm. long:
          Leaves firmly herbaceous, tapering to the base, very loosely imbricate; sporophylls c. 2 mm. long, covering the sporangia and grading into sterile leaves - - - - - - - 5. *ophioglossoides*
          Leaves coriaceous, abruptly rounded at the base, squarrose; sporophylls c. 1 mm. long, incompletely covering the sporangia and sharply differentiated from the sterile leaves - - - - - - 6. *phlegmaria*
      Fertile region less than 1·5 cm. long; sporophylls lacerate; plants terrestrial
      7. *cernuum*
  Sporophylls grouped into pedunculate strobili; plants terrestrial:
    Strobili 2–5 on a branched peduncle; sporophylls finely lacerate - 8. *clavatum*
    Strobili solitary on an unbranched peduncle; sporophylls erose - 9. *carolinianum*

1. **Lycopodium saururus** Lam., Encycl. Méth., Bot. **3**: 653 (1789).—Sim, Ferns S. Afr ed. 2: 324, t. 175 (1915). Type from Réunion.
    Plananthus saururus (Lam.) Beauv., Prodr. Aethéog.: 100 (1805). Type as above.
    Urostachys saururus (Lam.) Herter in Fedde, Repert. **19**: 162 (1923). Type as above.
    Huperzia saururus (Lam.) Rothm. in Fedde, Repert. **54**: 60 (1944). Type as above.

Terrestrial or lithophytic. Aerial stems up to 20 × c. 0·3 cm., usually unbranched, erect, crowded, produced from compact branching horizontal stems. Leaves c. 1 × 0·2 cm., erect, closely imbricate, narrowly oblong-lanceolate. Sporophylls not distinguishable from the foliage leaves, 8–10 × 1·8–2 mm.; sporangia hidden.

Rhodesia. E: Inyanga Distr., 8 km. N. of Troutbeck, 31.vii.1951, *Taylor* 3223 (BOL).
Malawi. S: Mt. Mlanje, 11.v.1963, *Wild* 6203 (BOL; SRGH).
Also in S. Africa and on the tropical African mountains and in S. America. Montane grasslands above 1900 m.

2. **Lycopodium verticillatum** L.f., Suppl. Pl.: 448 (1781).—Sim, Ferns S. Afr. ed. 2: 325, t. 178 (1915). Type from Réunion.
    Lycopodium setaceum Lam., Encycl. Méth., Bot. **3**: 653 (1806). Type from tropical America.
    Lycopodium acerosum Sw., Fl. Ind. Occ. **3**: 1575 (1806). Type from tropical America.
    Urostachys acerosus (Sw.) Herter ex Nessel & Hoehne in Arch. Bot. S. Paulo **4**: 45, t. 18 (1927). Type as above.
    Urostachys verticillatus (L.f.) Herter ex Nessel, Bärlappgew.: 121 (1939). Type as for *Lycopodium verticillatum*.
    Huperzia verticillata (L.f.) Rothm. in Fedde, Repert. **54**; 60 (1944). Type as above.

Epiphytic or lithophytic. Stems up to 50 × c. 1 cm., pendulous, repeatedly dichotomously branched. Leaves c. 5 × 0·2 mm., acicular, suberect, membranous, loosely imbricate. Sporophylls c. 5 × 0·2 mm., almost indistinguishable from the foliage leaves; sporangia usually evident.

Rhodesia. E: Chimanimani Mts., 6.v.1959, *Mitchell* 502 (BOL; SRGH). **Malawi.** S: Zomba Plateau, 7.vi.1946, *Brass* 16307 (K; SRGH). **Mozambique.** Z: Namuli Mt. (NE. sector), 27.vii.1962, *Schelpe & Leach* 7055 (BOL). MS: Gorongosa Mt., Gogogo Peak, 6.vii.1955, *Schelpe* 5528 (B; BM; BOL; K; P; US).

Also in S. Africa, Fernando Po, Madagascar, Comoro and Mascarene Is. and tropical America. Moist forest in shade, 1500–1980 m.

3. **Lycopodium dacrydioides** Bak., Fern Allies: 17 (1887).—Sim, Ferns S. Afr. ed. 2: 327, t. 176 (1915). Syntypes from S. Africa, Cameroon, Fernando Po and S. Tomé.
  Urostachys dacrydioides (Bak.) Herter ex Nessel, Bärlappgew.: 188 (1939). Syntypes as above.

Epiphytic or infrequently lithophytic. Stems up to 90 × 0·3 cm., pendulous, dichotomously branched. Leaves 1·5–1·8 × 0·15–0·2 cm. at the base, suberect, loosely imbricate, subulate. Sporophylls 6·5–7 × c. 2mm., narrowly oblong-lanceolate, imbricate.

Rhodesia. E: Umtali Distr., 1500 m., ix. 1953, *Whellan* 753 (BM; SRGH). **Malawi.** N: Rumpi Distr., E. Nyika escarpment, 6.v.1963, *Chapman* 2007 (BOL; SRGH). C: Nchisi Mt., 21.iii.1961, *Chapman* 1818 (SRGH). S: Mlanje Mt., 28.vi.1946, *Brass* 16527 (K). **Mozambique.** Z: Serra de Gúruè, Marrequelo, 20.ix.1944, *Mendonça* 2145 (BM; LISC). MS: Gorongosa Mt., Gogogo Peak, 5.vii.1955, *Schelpe* 5517 (B; BM; BOL; K; P; US).

Also in S. Africa, tropical Africa and the Comoro Is. Moist forest in shade, 1500–1800 m.

4. **Lycopodium gnidioides** L.f., Suppl. Pl.: 448 (1781).—Sim, Ferns S. Afr. ed. 2: 326, t. 177 (1915). Type from Mauritius.
  Lycopodium funiculosum Lam., Encycl. Méth., Bot. **3**: 649 (1789). Type from S. Africa.
  Plananthus gnidioides (L.f.) Beauv., Prodr. Aethéog.: 110 (1805). Type as for *Lycopodium gnidioides*.
  Lepidotis funiculosa (Lam.) Beauv., tom. cit.: 108 (1805). Type as for *Lycopodium funiculosum*.
  Lycopodium flagelliforme Schrad. in Gött. Gel. Anz. **1818**: 920 (1818). Type from S. Africa.
  Lycopodium ambiguum Schrad., loc. cit. Type from S. Africa.
  Lycopodium pinifolium Kaulf., Enum. Fil.: 7 (1824) non Bl. (1828). Type from S. Africa.
  Lycopodium strictum Bak. in Journ. of Bot. **20**: 271 (1882). Type from Madagascar.
  Lycopodium gnidioides var. *strictum* (Bak.) C. Chr. in Dansk Bot. Ark. **7**: 190 (1932). Type from Madagascar.
  Urostachys gnidioides (L.f.) Herter ex Nessel, Bärlappgew.: 187 (1944). Type as for *Lycopodium gnidioides*.
  Huperzia gnidioides (L.f.) Rothm. in Fedde, Repert. **54**: 61 (1944). Type as above.

Epiphytic or lithophytic. Stems up to 50 × 0·4 cm., erect, arching or pendulous, dichotomously branched. Leaves up to 1·4 × 0·3 cm., very narrowly oblong, coriaceous, acute to broadly acute, imbricate. Sporophylls up to 3 × 2 mm., much shorter than the foliage leaves, broadly ovate, acute, closely imbricate.

Rhodesia. E: Chimanimani Mts., 23.ii.1957, *Goodier* 170 (BM; BOL; SRGH). **Malawi.** N: S. Vipya, Kawendama, 29.xi.1961, *Chapman* 1495 (SRGH). **Mozambique.** MS: Gorongosa Mt., Gogogo Peak, 5.vii.1955, *Schelpe* 5515 (BM; BOL).

Also in S. Africa, Madagascar, the Comoro and Mascarene Is. Montane forest in shade above 1750 m.

5. **Lycopodium ophioglossoides** Lam., Encycl. Méth., Bot. **3**: 646 (1789). TAB. 3. Type from Mauritius.
  Lepidotis longifolia Beauv., Prodr. Aethéog.: 109 (1805). Type from Mascarene Is.
  Lycopodium longifolium (Beauv.) Sw., Syn. Fil.: 177 (1806). Type as above.
  Urostachys ophioglossoides (Lam.) Herter ex Nessel, Bärlappgew.: 239 (1939). Type as for *Lycopodium ophioglossoides*.
  Huperzia ophioglossoides (Lam.) Rothm. in Fedde, Repert. **54**: 62 (1944). Type as above.

Tab. 3. LYCOPODIUM OPHIOGLOSSOIDES. 1, habit (× ⅔); 2, detail of sporophylls (× 9), both from *Brass* 16417.

Epiphytic or lithophytic. Stem up to 30 × c. 0·1 cm., pendulous, dichotomously branched. Leaves up to 1·1 × 0·12 cm., narrowly lanceolate, acute, narrowed towards the base, very loosely imbricate. Sporophylls c. 2 × 1·5 mm., broadly ovate, entire, closely imbricate.

Rhodesia. E: Inyanga Distr., Pungwe R., Rest Huts, 11.viii.1950, *Chase* 3182 (BM; SRGH). **Malawi.** S: Mlanje Mt., 12.vi.1962, *Richards* 16649 (K; SRGH). **Mozambique.** MS: Gorongosa Mt., Gogogo Peak, 5.vii.1955, *Schelpe* 5516 (B; BM; BOL; K; P; US).

Also in S. Africa, tropical Africa, Fernando Po, Madagascar, Réunion and the Comoro Is. Moist forest in shade, above 1600 m.

6. **Lycopodium phlegmaria** L., Sp. Pl. **2**: 1101 (1753). Type from Ceylon.
 Urostachys phlegmaria (L.) Herter in Bot. Arch., **3**: 17 (1923). Type as above.
 Huperzia phlegmaria (L.) Rothm. in Fedde, Repert. **54**: 62 (1944). Type as above.

Epiphytic. Stems up to 40 × 0·3 cm., pendulous, dichotomously branched, with the fertile region sharply differentiated from the sterile region. Leaves up to 3·3 × 0·4 cm., narrowly lanceolate-acuminate, abruptly rounded at the base, coriaceous, squarrose. Sporophylls 1·2 × 1·5 mm., oblate acute-acuminate, only partially covering the sporangia.

Malawi. S: Mlanje Distr., Ruo Gorge near Lujeri Power House, 26.iii.1967, *Berrie* 123 (BOL).

Widespread in tropical Africa and Asia.

7. **Lycopodium cernuum** L., Sp. Pl. **2**: 1103 (1753).—Sim, Ferns S. Afr. ed. 2: 327, t. 179 (1915). Type from India or E. Indies.
 Lepidotis cernua (L.) Beauv., Prodr. Aethéog.: 101 (1805). Type as above.
 Lycopodium marianum Willd. in L., Sp. Pl., ed. 4, **5**: 31 (1810). Type from Philippine Is.
 Lycopodium boryanum A. Rich., Sert. Astrol.: 42 (1834). Type from Mascarene Is.
 Lycopodium heeschii K. Müll. in Bot. Zeit. **19**: 164 (1861). Type from Sierra Leone.
 Lycopodium hupeanum K. Müll., tom. cit.: 165 (1861). Type from Borneo.
 Lycopodium moritzii K. Müll., loc. cit. Type from Java.
 Lycopodium secundum K. Müll., tom. cit.: 164 (1861). Type from S. Africa.
 Lycopodium sikkimense K. Müll., loc. cit. Type from Sikkim.
 Lycopodium cymosum L'Herm. ex Hieron. in Engl., Bot. Jahrb. **34**: 574 (1905). Type from Colombia.
 Lycopodium lehmannii Hieron., loc. cit. Type from Colombia.
 Lycopodium trianae Hieron., loc. cit. Type from Colombia.

Terrestrial. Stems up to 85 × 0·3 cm., erect, leafy, with much-branched leafy lateral branches bearing strobili at their apices, rooting at the base and producing arching leafy stolons from near the base. Leaves up to 4 × 0·3 mm. at the base, subulate, patent, curved forward. Strobili up to 1 cm. long, not pedunculate, solitary at the ends of the branches. Sporophylls up to 1·8 × 1·2 mm., broadly ovate-acuminate, margins lacerate.

Zambia. B: c. 5 km. S. of Kalabo, 16.xi.1959, *Drummond & Cookson* 6545 (SRGH). N: Saisi R., 23.x.1957, *Richards* 2114 (K). W: Mwinilunga Distr., Matonchi R., 24.x.1937, *Milne-Redhead* 2928 (K). C: Serenje, 14.vi.1960, *Leach & Brunton* 10038 (K; SRGH). **Rhodesia.** W: Matobo Distr., Chesterfield, *Miller* 5895 (BOL; SRGH). C: Goromonzi Distr., Ngomakurira Pass, 1.vii.1963, *Loveridge* 1009 (SRGH). E: Umtali Distr., Mt. Nuza, 9.vi.1934, *Gilliland* 225 (BM; SRGH). S: Victoria Distr., E. of Zimbabwe, 1.vii.1930, *Hutchinson & Gillett* 3393 (K). **Malawi.** N: Nchena-chena, 21.viii.1946, *Brass* 17373 (K; SRGH). S: Zomba, 15.xi.1957, *Banda* 357 (SRGH). **Mozambique.** N: Lualeze R., x.1964, *Magalhães* 64 (COI). Z: Serra de Gúruè, Marrequelo, 20.ix.1944, *Mendonça* 2146 (BM; LISC). MS: Vila Gouveia, Serra Chôa, R. Nhacangara, 24.viii.1949, *Pedro & Pedrógão* 8039 (LMJ). SS: Gaza, Macia, S. Martinho de Bilene, 1947, *Pedro* 3536 (LMJ).

Cosmopolitan. Continually moist streambanks and ditches, 1000–1980 m.

8. **Lycopodium clavatum** L., Sp. Pl. **2**: 1101 (1753). Type from Europe.

Var. **inflexum** (Beauv.) Spring, Monogr. Lycopod., **1**: 90 (1842). Type from Mauritius.
 Lepidotis inflexa Beauv., Prodr. Aethéog.: 109 (1805). Type as above.
 Lycopodium inflexum (Beauv.) Sw., Syn. Fil.: 179 (1806). Type as above.
 Lycopodium clavatum sensu Sim, Ferns S. Afr. ed. 2; 328, t. 180 (1915).

Lycopodium clavatum var. *natalense* Nessel in Fedde, Repert. **36**: 191 (1934)
("nataliense"). Type from S. Africa.
    Lycopodium aberdaricum Chiov. in Lav. Ist. Bot. Univ. Modena, **6**: 147 (1935).
Type from Kenya.

Terrestrial. Main stem up to 3 mm. in diam., creeping, producing dichotomously branched erect stems 5–10 cm. apart. Leaves up to 5 × 0·5 mm., narrowly
lanceolate, patent to suberect or imbricate, with a translucent hair-point up to
3 mm. long at the apex, margin usually entire but sometimes ciliate in leaves on the
main stem. Strobili pedunculate in groups of 2–5 at the apex of a sparsely leafy
branched peduncle up to 19 cm. long. Sporophylls up to 3 × 2 mm., broadly ovate-
acuminate, finely lacerate.

Zambia. N: Nyika, 30.xii.1962, *Fanshawe* 7325 (K; SRGH). **Rhodesia.** E:
Stapleford, Inyamquarara Valley, 29.iv.1952, *Chase* 4507 (BM; SRGH). **Malawi.**
Mlanje Mt., Luchenya Plateau, 27.vi.1946, *Brass* 16482 (BM; K; SRGH). **Mozam-
bique.** MS: Chimanimani Mts., NW. of Martins Falls, *Mitchell* 348 (BOL; SRGH).
Also in S. Africa, tropical African mountains, Madagascar and the Mascarene Is. Con-
tinually moist margins of montane forest, 1600–2300 m.

9. **Lycopodium carolinianum** L., Sp. Pl. **2**: 1567 (1753). Type from N. America.
    Lycopodium ericetorum Schrad. in Gött. Gel. Anz. **1818**: 920 (1818). Type from
S. Africa.

Terrestrial. Main stem up to 6 mm. in diam., creeping on the ground, somewhat
dorsiventrally flattened. Leaves up to 1·5 × 0·4 cm., lanceolate to oblong, often
falcate, the lateral leaves spreading horizontally, the dorsal leaves smaller than the
lateral leaves, appressed or curving erect. Strobili up to 7 × 0·6 cm., solitary, born
at the apex of an unbranched sparsely leafy peduncle up to 30 cm. long. Sporophylls up to 5·5 × 2 mm., broadly ovate-acuminate, erose.

L. carolinianum var. *carolinianum* differs from var. *tuberosum* in not producing tubers
at the end of the growing season and from var. *affine* in having crowded broader lateral
leaves 3 times as long as broad.

Dorsal leaves similar in shape and slightly shorter than the lateral leaves:
  Plants not producing tubers - - - - - - - - - var. *affine*
  Plants producing tubers - - - - - - - - - var. *tuberosum*
Dorsal leaves much reduced and appressed - - - - - var. *grandifolium*

Var. **affine** (Bory) Schelpe in Bol. Soc. Brot., Sér. 2, **41**: 214 (1967). Type from Mascarene Is.
    Lycopodium affine Bory, Voy. Quatre Princ. Iles **2**: 204, 262 (1804). Type from
Mascarene Is.

Zambia. B: Mongu, 10.xi.1959, *Drummond & Cookson* 6322 (K; LISC; SRGH).
N: Chinsali Distr., Lake Young, Shiwa Ngandu, 15.i.1959, *Richards* 10665 (K). W:
Mwinilunga, Mugesa R., 30.xi.1952, *White* 3432 (COI; K). **Rhodesia.** E: Chimanimani Mts., 11.ix.1945, *Granelli* (K). **Malawi.** N: Nyika Plateau, Lake Kaulime,
15.xi.1958, *Robson & Fanshawe* 634 (K; LISC; SRGH). **Mozambique.** SS: Inhambane, Panda, 3.xii.1935, *Moss* (K; SRGH). MS: Cheringoma, Dondo, near R.
Inhansal, *Mendonça* 2483 (LISC).
Widespread in tropical Africa and Mascarene Is. Swampy ground and swamp forest,
1350–2150 m.

Var. **tuberosum** (Welw. & A. Braun ex Kuhn) Nessel, Bärlappgew.: 274 (1939). Type
from Angola.
    Lycopodium tuberosum Welw. & A. Braun ex Kuhn, Fil. Afr.: 211 (1868). Type
as above.

Zambia. W: Mwinilunga Distr., 9.xii.1937, *Milne-Redhead* 3580 (K).
Also in Angola.

Var. **grandifolium** Spring in Mém. Acad. Roy. Brux. **24**: 46 (1849–1850). Syntypes
from S. Africa and Angola.
    Lycopodium sarcocaulon Welw. & A. Braun ex Kuhn, Fil. Afr.: 210 (1868). Type
from Angola.

Zambia. N: Shiwa Ngandu, 24.vii.1938, *Greenway* 5488 (K). W: Mwinilunga
Distr., Zambezi R. Falls, 10.xi.1962, *Richards* 17134 (K). **Rhodesia.** W: Matobo
Distr., Besna Kobila, ii.1953, *Miller* 1600 (SRGH). E: Inyanga Distr., Troutbeck,

17.xi.1956, *Robinson* 1932 (K; SRGH). **Malawi.** S: Mlanje Mt., Tuchila Plateau, v.1901, *Purves* 20 (K). **Mozambique.** MS: Gorongosa Mt., Gogogo Peak, 6.vii.1955, *Schelpe* 5538 (B; BM; BOL; K; P; US).
Also in S. Africa and Angola. Marshy ground, 1200–1800 m.

SELAGINELLALES

3. SELAGINELLACEAE

By A. C. Jermy

Terrestrial herbs, annual or perennial. Stems erect, decumbent or creeping, bearing aerial roots (rhizophores); primary and secondary branch systems often regular in shape forming a pseudofrond. Leaves ligulate (i.e. with microscopic tongue of tissue on adaxial base), small, thin, either spirally arranged and ± similar in shape and size or in two ranks, those in the centre facing uppermost (the median leaves) and those below on either side (the lateral leaves). Sporangia dehiscent with 2 valves, borne singly in axils of specialised leaves (sporophylls) arranged in strobili (which may or may not be dorsiventral); sporophylls may be similar or dimorphic as in the leaves. Spores of 2 kinds, borne in separate sporangia. Megaspores, of which there are usually 4 (rarely 1 or 8) in each sporangium, trilete, laesurae usually forming a 3-radiate ridge reaching an equatorial ridge; on germination give rise to prothalli bearing female organs (archegonia) only. Microspores trilete or more rarely monolete, laesurae usually less distinct, produced in large numbers in each sporangium, giving rise to prothalli bearing male organs (antheridia) only.
Usually 1 genus only, *Selaginella*, is recognized

SELAGINELLA Beauv.

Selaginella Beauv., Mag. Encycl., Par. **5**: 478 (1804) *nom. conserv.*

Description as for the family. Those species with uniform leaves are placed in subgenus *Selaginella*; heterophyllous species in subgenus *Stachygynandrum* Beauv. The latter has been given generic rank by some authors.

Leaves of all branch systems ± similar:
  Seta on leaf-apex ⅓ or more the length of the blade; sporophylls in 2 ranks   1. *dregei*
  Seta on leaf-apex ¼ or less the length of the blade; sporophylls in 4 ranks:
    Leaves arranged dorsiventrally; plant green, older leaves persistent, brown
                                          2. *njam-njamensis*
    Leaves not dorsiventrally arranged; plant bluish-green at short apices otherwise
      whitish, older leaves white, soon breaking away -   -   -   -   3. *nivea*
Leaves of at least the secondary branch systems markedly dimorphic:
  Primary branch system erect, without branches at least in lower ½:
    Upper stems hairy below; branchlets not curling when dry; leaves uniform on both
    sides - - - - - - - - - - 8. *eublepharis*
    Upper stems not hairy below; branchlets curling in when dry; leaves dark green
    above, pale beneath - - - - - - - 5. *imbricata*
  Primary branch system creeping or if erect then branched from base:
    Primary branches creeping although secondary branches may be suberect, not truly
    suffruticose; rhizomorphs arising along greater length of stem unless plants less
    than 5 cm.; sporophylls either all similar or dimorphic:
      Lateral leaves 3–4 mm. long, not curled beneath stem when dry; plant robust, often
      30 cm. or more across - - - - - - 7. *kraussiana*
      Lateral leaves 2 mm. or less; plant forming small mats rarely more than 10 cm.
      across:
        Lateral leaves curling beneath stem when dry, bright green, ciliate at base with
        narrow hyaline border; sporophylls of one kind -   - - 6. *mittenii*

Lateral leaves not curling beneath stem when dry, pale green, denticulate at
    base, not bordered; sporophylls dimorphic   -   -   -   10. *perpusilla*
Primary branches semi-prostrate or ± erect, suffruticose; rhizomorphs arising at base
    only; sporophylls dimorphic:
    Median leaves long-acuminate, margins ciliate-dentate:
        Lower part of stem bearing filiform runners; plant usually more than 10 cm.
        high; megaspores mammillate   -   -   -   -   9. *abyssinica*
        Plant without filiform runners, less than 10 cm. high; megaspores lophate
                11. *tenerrima*
    Median leaves acute, margins spinose   -   -   -   -   -   4. *subisophylla*

1. **Selaginella dregei** (C. Presl) Hieron. in Hedwigia, **39**: 315, fig. 36 map 38 (1900).
    TAB. 4. Type from S. Africa.
    Lycopodium dregei C. Presl in Abh. Böhm. Ges. Wiss. **5**, 3: 583 (1845); reimpr.
in C. Presl, Bot. Bemerk.: 153 (1846). Type as above.
    Selaginella rupestris forma *dregei* (C. Presl) Milde, Fil. Europ. Atlant.: 262 (1867).
Type as above.
    Selaginella rupestris var. *recurva* forma *dregeana* A. Braun in Kuhn, Fil. Afr.: 214
(1868). Type as above.
    Selaginella grisea Alston in Journ. of Bot. **77**: 222 (1939). Type from East Africa.

A creeping plant often forming loose mats; stems prostrate, primary branches
often irregularly ascendent, 1–3-furcate; plants often forming further primary
branches from continuation of growth of lower secondary branch upon cessation of
growth or death of upper secondary branches; leafy stem radially symmetrical or
occasionally slightly dorsiventral in position of leaves. Leaves 1·5–2·25 × 0·25 mm.,
herbaceous, often brown and papyraceous on the under side, narrowly triangular
to linear or subulate, adnate at the base, margins with piliform cilia, acuminate at
the apex tapering into a usually curved seta ⅓–½ as long as the blade, white end
opaque when young often soon becoming tawny. Coning irregular; strobili
4–8 mm. long, suberect, dorsiventral; sporophylls in 2 ranks on the under side,
broader (up to 1 mm.) and more lanceolate than leaves, apex not setose. Megaspores 350–440 μ, subgranulate; 3-radiate ridges indistinct. Microspores 40–50 μ,
scabrate.

Zambia. N: Abercorn Distr., Lufubu R., Iyendwe Valley, 780 m., 8.xii.1959,
Richards 11923 (K). **Rhodesia.** N: Karoi, K 34. Experimental Farm, 4.x.1946, *Wild*
1315 (BM; SRGH). W: 8 km. E. of Filabusi, Insiza Distr., 1070 m., 8.v.1953, *Schelpe*
3974 (BM). C: Salisbury, Twentydales, 1370 m., 9.ix.1946, *Wild* 1218 (BM; SRGH).
E: Umtali, 1220 m., 31.i.1954, *Chase* 5188 (SRGH). S: Nuanetsi, 610 m., xii.1955,
Davies 1845 (BM; SRGH). **Malawi.** C: Bunda Hill, Lilongwe, 1250 m., 7.ii.1959,
Robson 1497 (K; LISC; SRGH). **Mozambique.** T: Tete 1858, *Kirk* (K). MS:
Chimanimani Mts., around overhang near Martin Falls, 1520 m., 9.ii.1958, *Mitchell* 342
(BM; SRGH). LM: Namaacha, ix.1930, *Gomes & Sousa* 85 (BM; COI; LMJ).
    Also in Uganda, Kenya, Angola and S. Africa. Growing in the open and under scrub
in rocky situations, especially over igneous (acid) rocks, 610–1520 m.

2. **Selaginella njam-njamensis** Hieron. in Hedwigia, **39**: 312, fig. 28 map 33 (1900).
    Type from Central Africa (Njam-Njam).

A creeping plant with a short prostrate branch system, 1- or 2-furcate; leafy stem
dorsiventral regarding position and length of leaves. Upper leaves 2·0 × 0·25 mm.,
herbaceous, lanceolate to ligulate-triangular, base adnate, margins with short
dentiform cilia above, longer piliform cilia towards the base, apex acuminate
tapering into stout tawny or subopaque white seta ¼ or less the length of the blade;
under leaves up to 2·5 mm., similar but more linear-lanceolate and slightly longer.
Strobili 3–10 mm. long, sporophylls in 4 ranks, those on upper side of strobili
lanceolate, acuminate, margins with short cilia, apex acuminate, those on the
lower side slightly longer and broader. Megaspores 200–325 μ, granulate; 3-
radiate ridges indistinct. Microspores 35–40 μ, narrowly winged, scabrate with
firm ridges.

Zambia. E: Chipiri Hill near Chadiza, 900 m., 29.xi.1958, *Robson* 784 (K; LISC;
SRGH). **Malawi.** S: Slopes of Mt. Mlanje, 910 m., 9.xii.1957, *Chapman* 473 (BM; K;
SRGH). **Mozambique.** N: Nampula, 11.vii.1948, *Pedro & Pedrógão* 4396 (LMA).
Z: Montes de Ile, 22.vi.1943, *Torre* 5566 (LISC).
    Also in the Sudan, Dahomey, Cameroon, Nigeria and Central Africa (Njam-Njam).
A plant of rocky places, 900–910 m.

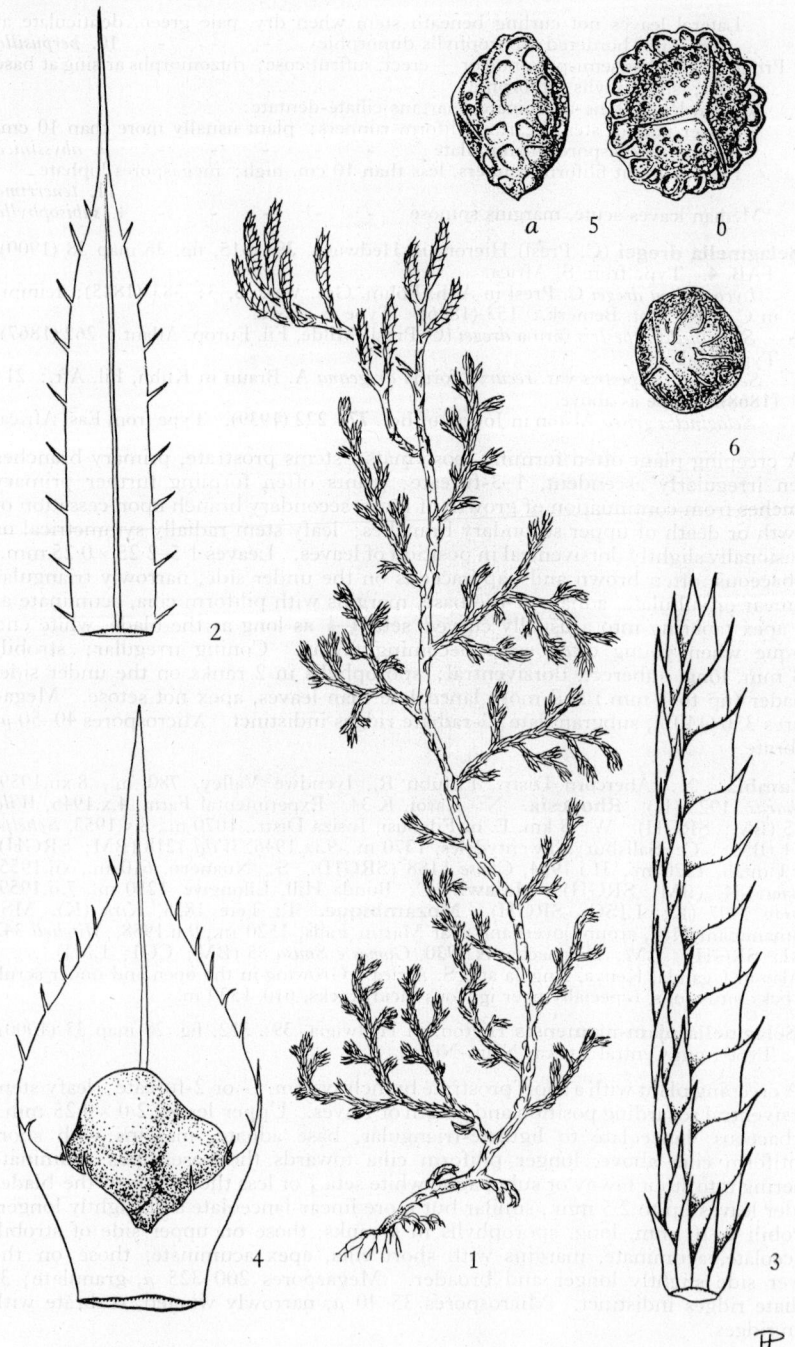

Tab 4. SELAGINELLA DREGEI. 1, habit (× ⅔); 2, stem leaf (× 20); 3, strobilus (× 6);
4, macrosporophyll (× 35); 5, megaspores (a) lateral view (b) proximal view (× 70);
6, microspore (× 350), all from *Monro* 1021.

3. **Selaginella nivea** Alston [in Perrier, Acad. Malgache, Cat. Pl. Madag., Ptérid.: 71 (1932) *nom. nud.*] in Dansk Bot. Ark. **7**: 194 (1932). Type from Madagascar.

Creeping plants with primary branches 4–7 mm. long, 1–4 furcate; leafy stems not obviously dorsiventral. Leaves 1·5–2·0 × 0·25 mm., bluish-green, soon becoming opaque-white, linear-lanceolate, adnate at the base, apex acute or acuminate ending in an opaque white straight seta ¼ as long as the blade, margins with short cilia above becoming piliform towards the base, rather squarrose at the apex forming club-shaped branch tips; lower leaves soon becoming papyraceous and whitish (or colourless) lying close to the stem and soon breaking away. Strobilus 2·5 mm., forming at top of lowermost secondary branch; sporophylls in 4 ranks, those on the upper side lanceolate-ovate, on the under side broadly ovate and acuminate, margins long ciliate towards the base. Megaspores 125–200 μ, subscabrate to almost smooth, 3-radiate ridges absent. Microspores 45–50 μ, rugulate, finely scabrate, ± winged.

Mozambique. SS: Guijá, between Caniçado and Mabalane, 5.xi.1944, *Mendonça* 2767 (LISC).
Occurs also in Madagascar. Growing in open dry sandy places in *Colophospermum mopane* woodland savanna.

4. **Selaginella subisophylla** Jermy, Brit. Fern Gaz. **10**: 30, fig. 1–9 (1968). Type: Zambia, Kawambwa Distr., Kawambwa, Ntumbachushi Falls, *Richards* 9325 (BM, holotype; K).

Plant erect, 9–15 cm. high, rooting at base or from axils of lowest branches only. Lower primary branches up to 3 cm., once or twice furcate; upper branches decreasing to branchlets 3 mm. long, overall shape of flattened branch system lanceolate or narrowly triangular. Leaves of the main stem and primary branches similar, those of the ultimate branches of 2 kinds; stem leaves 2·0 × 0·5–0·75 mm., oblong-lanceolate, bordered with narrow hyaline cells, spinose with occasionally long fine teeth at base, denticulate towards apex; median leaves 1·25–1·5 × 0·5 mm., lanceolate with small pointed auricles, margins long-spinose below, sharply dentate above; lateral leaves of ultimate branches similar to stem leaves. Strobili on ultimate segments of all branch systems maturing ± simultaneously, with single female sporophyll at base and 3–4 male sporophylls and several sterile leaves above. Female sporophylls 1·75–2·0 × 1·0 mm., broadly ovate, acuminate, margins entire below, long-ciliate in middle becoming dentate at apex; male sporophylls 1·75 × 0·75 mm., ovate-lanceolate ± acuminate, margins sparsely ciliate except at base. Megaspores 375–425 μ, lophate, 3-radiate ridge indistinct. Microspores 35–38 μ, trilete, gemmate or mammillate.

Zambia. N: Kawambwa Distr., Kawambwa, Ntumbachushi Falls, 19.iv.1957, *Richards* 9325 (BM; K).
Endemic. In wet sands, 1260 m.

5. **Selaginella imbricata** (Forsk.) Spring ex Decne. in Arch. Mus. Par. **2**: 193, t. 7 (1841–2?); Monogr. Lycopod. **2**: 70 (1850).—Bak., Fern Allies: 87 (1887).—Sim, Ferns S. Afr., ed. 2: 336, t. 184 fig. 3 (1915). Type from Arabia.
  Lycopodium imbricatum Forsk., Fl. Aegypt.-Arab.: CXXV, 187 (1775). Type as above.
  Lycopodium sanguinolentum sensu Forsk., tom. cit.: CXXV (1775) ("sanguin.") non L. (1753).
  Lycopodium yemense sensu Grev. & Hook. in Hook., Bot. Misc. **2**: 382 (1831) non Sw. (1806). Type from Arabia.

A caulescent plant; stems 5–30 cm., 1–3 arising from a stout creeping stolon, branched in the upper ½. Primary branch system ovate to lanceolate in outline when moist but branches and leaves inrolling when dry; secondary branches with 3–8 ultimate branches which may be further forked, ultimate branches 0·4–0·8 cm. Leaves heteromorphic, upper surfaces bright green, pale below, ± fleshy, often drying and turning grey-brown in situ; median leaves c. 1·25 mm. long, oblong-lanceolate, adnate at the base, tapered to an obtuse subfalcate apex, margins with a broad band of hyaline cells dentate except at the apex; lateral leaves contiguous or overlapping, c. 1·5 × 0·5 mm., oblong-elliptic, adnate at the base, apex obtuse, margins entire. Strobili 0·5–0·8 cm. long, male below and female above, at the tips of ultimate branches, forming simultaneously throughout a branch system; sporophylls of 1 kind, similar to the median leaves but with a broader marginal band of colourless cells and ± irregularly toothed. Megaspores of 2 sizes, 150 and 300 μ, ± smooth or punctulate with a faint 3-radial ridge. Microspores 70–80 μ, minutely foveo-reticulate.

Zambia. S: Livingstone Distr., 31.viii.1947, *Greenway & Brenan* 8039 (BM; K). **Rhodesia.** N: Lake Kariba, Sanyati Gorge, 1750 m., 9.xi.1961, *Mitchell* 600 (K; SRGH). W: Wankie, Victoria Falls, 2990 m., 2.ix.1955, *Chase* 5763 (BM; SRGH). E: Near Freshwater Farm, Vumba Mts., Umtali, 1.viii.1956, *Chase* (BM). **Mozambique.** T: Tete, *Kirk* (K). MS: Lupata, vi.1859, *Kirk* (D; K).

Also in Arabia, Sudan, Ethiopia, Kenya, Natal and Madagascar. Rocky places, in moist crevices of rock walls etc., 490–2990 m.

6. **Selaginella mittenii** Bak. in Journ. of Bot. **21**: 81 (1883). Type from Usagura Mts., Central Africa.
    Selaginella welwitschii Bak., loc. cit. Type from Angola.
    Selaginella mackenii Bak., op. cit. **22**: 89 (1884). Type from S. Africa.
    Selaginella cooperi Bak., loc. cit. Type from S. Africa.
    Selaginella tectissima Bak., loc. cit. Type from S. Africa.
    Selaginella depressa sensu Sim, Ferns S. Afr., ed. 2: 337 (1915) non Spring (1843).

A terrestrial creeping plant often forming interwoven mats; stems 1–12 cm. long, branches all ± distant, those of secondary order 3–8 furcate, divaricate; ultimate branches 0·3–0·8 mm. long. Leaves heteromorphic, distinctly mid-green, minutely punctulate, remaining intact on the stem when dead; median 0·5–0·75 mm. long, ovate-lanceolate, acute or acuminate, margins bordered with hyaline cells, dentate; lateral leaves 1·25–1·75 × 0·75–1 mm., ovate-elliptic, apex acute, rounded, base unequal, uppermost lobe amplexicaul, margins dentate, becoming ciliate at the base with a narrow hyaline border, the whole, when dry, characteristically curled ensheathing the stem. Strobili 0·4–0·8 cm. long, erect at the apex of the ultimate branches, maturing simultaneously throughout the branch system, female at the base, male above; sporophylls of 1 kind, similar to the median leaves. Megaspores 250–300 μ, subrugulose, scabrate; 3-radiate ridge shallow. Microspores 34–50 μ, verrucate.

Rhodesia. N: Vicinity of Umvukwes Mts., 24–27.iv.1948, *Rodin* 4459 (K). W: Matobo, Besna Kobila Farm, 1220 m., v.1954, *Miller* 2391 (SRGH). C: Makoni, Forest Hill Kop, 1520 m., vii.1917, *Eyles* 828 (BM; K; SAM; SRGH). E: Umtali, Stapleford Forest Res., 1520 m., 29.iv.1952, *Chase* 4502 (BM; SRGH). **Malawi.** S: Likubula Gorge, Mlanje Mt., 840 m., 20.vi.1946, *Brass* 16372 (K). **Mozambique.** N: Serra de Ribáuè, c. 540 m., 28.i.1964, *Correia* 126 (LISC).

From Tanzania to Angola and the Transvaal. On peaty or inorganic soils, 540–1600 m.
Very close in leaf and spore characters to *S. cathedrifolia* Spring from S. and W. Africa but differs in habit, texture and shape of lateral leaf.

7. **Selaginella kraussiana** (Kunze) A. Braun, App. Ind. Sem. Hort. Berol.: 22 (1860). Kuhn, Fil. Afr.: 190 (1868).—Bak., Fern Allies: 65 (1887). Type from S. Africa.
    Lycopodium kraussianum Kunze in Linnaea, **18**: 114 (1844). Type as above.
    Selaginella hortensis Mett., Fil. Hort. Lips.: 125, 128 (1856). Origin unknown.

A wide-creeping terrestrial plant; stems up to 50 cm., with swollen " joints " occasionally below the furcation of the branch. Primary branch-systems ovate to broadly elliptic in mature specimens but branches straggly, interwoven and outline obscured; secondary branches elliptic-lanceolate; tertiary branches 1–3-furcate, 1–1·5 cm. long. Leaves of 2 kinds; median 2·5 mm. long, lanceolate or ovate-lanceolate, apex acute, base cordate unequal, margins sparsely toothed; lateral leaves 3–4 × 0·75–1·5 mm., subsessile, linear-elliptic, subdimidiate with a pale median line, apex acuminate, margins serrate; axillary leaves similar to the lateral leaves but wider. Strobili 0·5–0·8 mm. long, infrequently formed at apex of ultimate branches, simultaneously maturing throughout a secondary branch, production appearing to be seasonal, possibly requiring light; sporophylls of 1 kind only, similar to the median leaves but narrower. Megaspores c. 750 μ, reticulate with thin narrow wings. Microspores 26–32 μ, echinate, bases of spines joined to form ridges.

Rhodesia. E: Melsetter, 1520 m., 10.x.1950, *Wild* 3563 (BM; SRGH). **Malawi.** N: Karonga Distr., Musissi, 30.iv.1947, *Benson* 1219 (BM). S: Mlanje Distr., Luchenya

Plateau, 2100 m., 5.vi.1962, *Richards* 16523a (K). **Mozambique.** Z: Serra de Gúruè, Marrequelo, 18.ix.1944, *Mendonça* 2097 (BM; LISC). MS: Mt. Vumba near Vila de Manica, 1200 m., 25.ix.1961, *Gomes e Sousa* 4740 (COI; K).

Also in Cameroon, Congo, Kenya, Uganda, Tanzania, S. Africa and the Azores. Shady situations, from open habitats such as grassland to more closed bush, often as a weed on shady paths, 1200–2500 m.

8. **Selaginella eublepharis** A. Braun apud Hieron. in Engl. & Prantl, Nat. Pflanzenfam. **1**, 4: 677 (1902). Type from Tanzania.

Plants terrestrial, caulescent from a robust, wide-creeping rhizome; stems 20–50 cm., branched and sparsely hairy beneath in the upper ½. Outline of branch-system broadly ovate to lanceolate, acute or rarely flagelliform at the apex, second order branch-systems broadly lanceolate to linear in outline dichotomising 2–6 times, ultimate branchlets 0·6–1·5 cm. long; the lowermost basiscopic tertiary branch conspicuously larger than the acroscopic one. Leaves heteromorphic; stem-leaves 2–3 mm. long, ovate-lanceolate, acuminate, finely serrate, becoming ciliate at the cordate clasping base; median leaves similar, 3 mm. long, apex abruptly tapered into whitish seta ¼ as long as leaf; lateral leaves (of secondary branches) 4 × 1 mm., linear with an acute apex or linear-lanceolate, adnate at the base, margin sinuous, sparsely toothed becoming ciliate towards the base. Strobili 1·0–1·5 cm., at tips of ultimate branches, maturing ± simultaneously, bearing sporangia at the base and towards the top; sporophylls of 1 kind, broadly ovate tapering to a setose apex, margins dentate becoming ciliate at the base. Megaspores of 2 sizes c. 200 and 300–375 μ, finely papillate to subverrucose with 3-radiate ridges. Microspores 25–32 μ, clavate-verrucose.

Malawi. N: Kavali (?Karali), 1891, *Carson* (K). **Mozambique.** Z: between Mocuba and Maganja da Costa, 43 km. from Mocuba, margin of R. Metolude, 17.v.1949, *Barbosa & Carvalho* 2699 (LISC). LM: Lourenço Marques, 21.v.1955, *Pedro* 5051 (LMJ).

Also in Tanzania. In rocky ground amongst grasses in transition of forest-grassland zones, 300–800 m.

This species is closely related to *S. pervillei* Spring from Madagascar and *S. vogelii* Spring from W. Africa. Both these species have numerous short stiff hairs on the under-part of the stem; the former has median leaves ± entire with only an occasional cilium at the base; *S. vogelii* is without cilia on both median and lateral leaves.

9. **Selaginella abyssinica** Spring, Monogr. Lycopod. **2**: 99 (1850). TAB. **5**. Type from Ethiopia.

    Selaginella kirkii Bak., Journ. of Bot. **23**: 76 (1885). Type from Tanzania.
    Selaginella whytei Hieron. in Engl. & Prantl, Nat. Pflanzenfam., **1**, 4: 697 (1901). Type: Malawi, *Whyte* (K, holotype).
    Selaginella preussii Hieron. in Engl. & Prantl, tom. cit.: 686 (1901). Syntypes from Cameroon.
    Selaginella bueensis Hieron. in Hedwigia, **43**: 20 (1904). Type from Cameroon.
    Selaginella goetzei Hieron. in Engl., Bot. Jahrb. **30**: 265 (1901). Type from Tanzania.

A semi-prostrate plant, sometimes becoming suffruticose with a ± erect branch-system, 10–40 cm. high, ill-defined in outline. Primary branches narrowly tri-angular to lanceolate, the longest below the middle, 7–12 cm.; secondary branches with 2–4 branchlets, 0·5–1·0 cm. long. Leaves heteromorphic; median 2–3 mm. long, broadly lanceolate or ovate drawn into a fine serrated tip c. ½ the length of the leaf, margins ciliate-dentate of thick-walled hyaline cells; lateral leaves 1·75–2·5 × 0·5–0·75 mm., linear-lanceolate, semi-dimidiate, with a pale median band, upper branch lobe rounded, lightly toothed, amplexicaul on leaves of primary stem, apex acute. Strobili 0·5–1·2 mm., terminal on ultimate branchlets, maturing simultaneously on each branch system, sporophylls of 2 kinds, lateral and median, those in a median position situated on the ventral not the dorsal side as with the branch leaves; lateral sporophylls similar to lateral leaves, usually bearing micro-sporangia only; median sporophylls similar to median leaves but wider, ovate and attenuate or lanceolate, bearing megasporangia below and microsporangia above. Megaspores c. 250 μ, evenly subgranulate with shallow 3-radiate ridges. Micro-spores 26–32 μ, covered with broad flat verrucae.

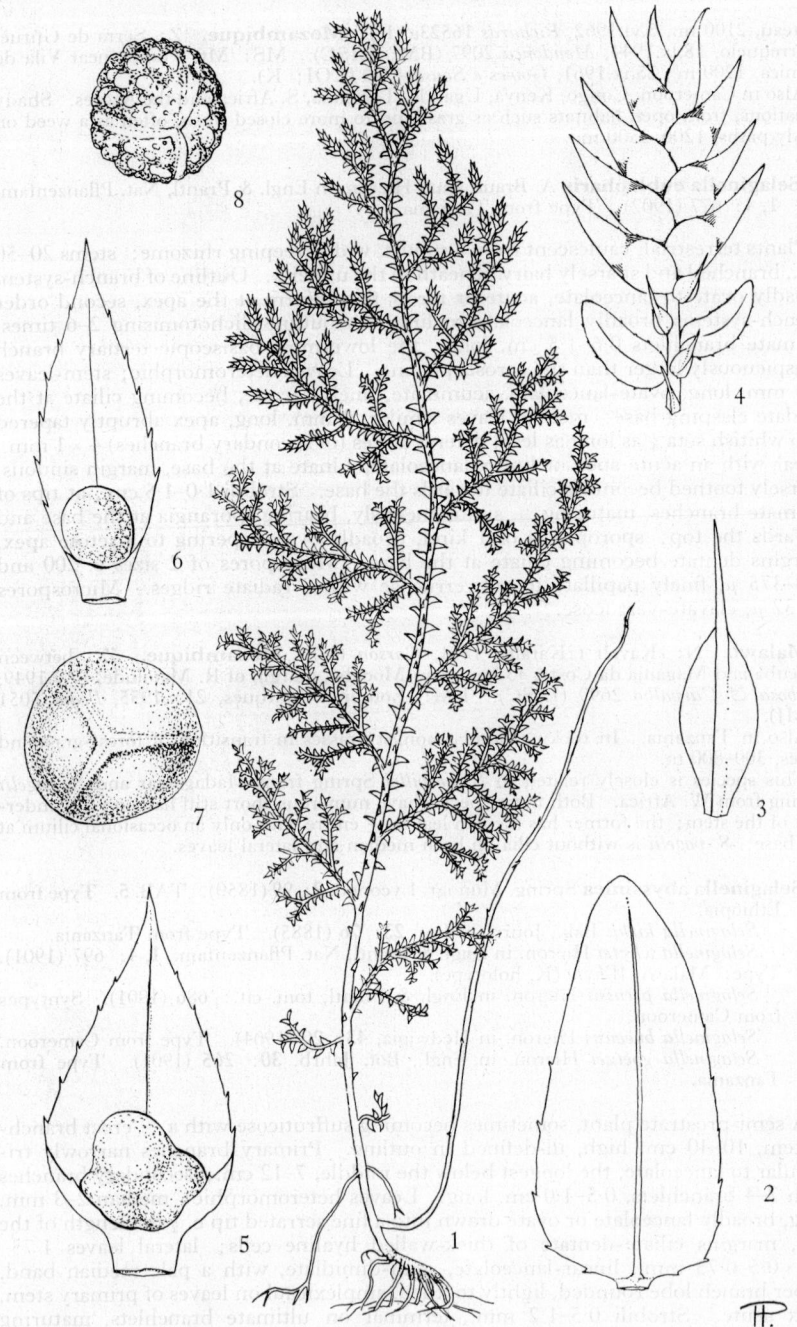

Tab. 5. SELAGINELLA ABYSSINICA. 1, habit (×⅔); 2, lateral leaf (×25); 3, median leaf (×20); 4, strobilus (ventral view) (×3); 5, dorsal megasporophyll (×20); 6, dorsal microsporophyll (×20); 7, megaspore (×100); 8, microspore (×350), all from *E.M. & W.* 1271.

Zambia. N: Abercorn Distr., Lunzua R., 1260 m., 6.vi.1961, *Richards* 15206 (K; SRGH). W: Solwezi, 28.ii.1964, *Fanshawe* 8370 (SRGH). S: Mazabuka, Mabwingombe Hills, 5.xi.1960, *White* 6823 (K). **Rhodesia.** E: Umtali, Vumba, " Cloudlands ", vii.1948, *Fisher & Schweickerdt* 221 (K; SRGH). **Malawi.** N: Karonga Distr., Misuku, 1520 m., 11.v.1947, *Benson* 1221 (BM). S: Mlanje Mt., path to Luchenya Plateau, 2100 m., 5.vi.1962, *Richards* 16523b (K). **Mozambique.** N: Vila Cabral, iv.1934, *Torre* 101 (BM; COI; LISC). T: near Munguzi in damp bed of stream, iv.1859, *Kirk* (K). MS: Manica, Mavita, Rotanda, 29.i.1948, *Barbosa* 903 (BM).

Also in Ethiopia, Congo, Fernando Po, Kenya and Tanzania. Along stream banks and in moist situations; in evergreen forest amongst rocks and mosses, 1200–2100 m.

10. **Selaginella perpusilla** Bak. in Journ. of Bot. **23**: 292 (1885). Type from Tanzania.

A small creeping or prostrate moss-like plant. Stems 1–2·5 cm., rarely branched in upper part. Leaves of 2 kinds: median leaves 0·8–1·0 × 0·35 mm., ovate, tapered to an acute tip; lateral leaves 1·5–1·75 × 0·75–1·0 mm., ovate-elliptic, apex acute, margin denticulate. Strobili 1·5–3·0 mm. at the tips of branchlets, maturing simultaneously; sporophylls 5–9, slightly larger than the leaves, of 2 kinds, those similar to the median leaves on the under side, i.e. opposite the median leaves on the stem, sporangia mostly female with occasional male at top of strobilus. Megaspores 250 μ, lophate, ridges narrow. Microspores 20–38 μ, echinate or more rarely gemmate.

Rhodesia. S: Chibi Distr. near Lundi R. bridge, 2.v.1962, *Drummond* 7851 (BM; SRGH).

Also in the Congo, E. Africa and Madagascar. In damp places.

11. **Selaginella tenerrima** A. Braun ex Kuhn, Fil. Afr.: 193 (1868). Type from Angola.

A ± erect plant, 2–8 cm. high, rarely taller with 2–3 main stems. Stem soft, flexuous, loosely branched, rooting from axils of lower branches, outline of system ovate. Leaves of 2 kinds: the median leaves 1·25–1·75 × 0·25–0·4 mm., lanceolate-elliptic, margins denticulate, apex acuminate into a fine setose point; the lateral leaves 1·25–1·5 × 0·75 mm., ovate-elliptic, margins denticulate, apex obtuse or rounded, base unequal, that on the acroscopic side enlarged. Strobilus 3–4 mm. long, at the tips of branches, maturing ± simultaneously, with male sporangia only at tips; sporophylls of 2 kinds, similar to, but narrower than their respective leaves and more spiny on margins, those similar to the median leaves on the underside, i.e. on the opposite side to those on the stem. Megaspores 200–225 μ, with a thin lophate perispore with narrow ridges. Microspores 30–48 μ, sparsely granular.

Zambia. E: Muchinga foothills, Luangwa Valley, 5.vi.1957, *Savory* 181 (SRGH). **Mozambique.** MS: Gorongosa, Serra de Gorongosa, Quedas do R. Malombosi, 23.x.1945, *Pedro* 435 (LMA).

Also in Angola, Congo and Nigeria. Overhanging pools amongst moss on vertical rock; usually in wet places on rocks.

ISOETALES

4. ISOETACEAE

By A. C. Jermy

Aquatic plants or plants of seasonally flooded or boggy ground. Stem short, 2- or 3-lobed. Leaves in a rosette; leaf-base spathulate, imbricate with membranous margin and a delicate triangular ligule on the adaxial surface at the point where the leaf narrows. Heterosporous; megaspores tetrahedral, with conspicuous 3-radial and equatorial ridges; microspores bilateral, frequently with a conspicuous apical ridge, in septate sporangia sunken into separate leaf-bases, with covering tissue (velum) entire, with a large or small aperture or absent.

Two genera only; *Stylites* (S. America) and *Isoetes* (widespread).

ISOETES L.

Isoetes L., Sp. Pl. **2**: 1100 (1753); Gen. Pl. ed. 5: 486 (1754).

Leaves terete or subterete often becoming flattened below. Megaspores often of 2 sizes. Megagametophytes developing within the spore-wall and often within the sporangium; sporophytic stage of life-cycle frequently achieved without fertilisation by male gamete (apogamy).

There are 3 spp. in our area.

Sporophylls c. 60 × 0·5–0·7 cm., conspicuously winged in the basal half - 2. *rhodesiana*
Sporophylls less than 50 × 0·3–0·5 cm., conspicuously winged only at the extreme base:
  Stem 2-lobed, with 30–40 leaves; sporophyll tissue ± thick at base, sporangia not pronounced on abaxial side; microspores cristo-reticulate, winged - - 1. *alstonii*
  Stem 3-lobed, with 10–20 leaves; sporophyll tissue thin at the base, sporangia pronounced on the abaxial side; microspores scabrate - - - 3. *aequinoctialis*

1. **Isoetes alstonii** Reed & Verdcourt in Kirkia, **5**: 19 (1965).—Launert in Prodr. Fl. SW. Afr. **1**: 1 (1969). TAB. **6**. Type: Rhodesia, Victoria Falls, south bank in front of main falls, *Greenway & Brenan* 8012 (BM, isotype; EA, holotype; K, isotype; PRE, isotype; SRGH, isotype).

Stem 1–2 cm. in diam., 2-lobed, with 30–40 leaves. Leaves 15–40 × c. 0·2 cm., yellow or mid-green, semiterete, with a hyaline wing in lower ¼, stiff. Sporangium 5–10 × 3·5–6 mm., oblong, not protruding on the abaxial side of the sporophyll; velum reduced to a thin marginal ring. Megaspores of 2 sizes, c. 350 μ and c. 500 μ, dark-grey, drying white; proximal faces with single large or cluster of smaller verrucae in the centre or granular with indistinct subsurface verrucae, distal faces papillate-verrucate, verrucae often anastomosing. Microspores 25–30 μ, winged, reticulate.

Zambia. N: Kasama Distr., 13 km. E. of Chisera, 28.iii.1961, *Drummond & Rutherford-Smith* 7473 (BM; BOL; EA; SRGH). S: Mumbwa Distr., in open woodland country, 28.v.1948, *Bunting* (K). **Rhodesia.** W: Victoria Falls, Danger Point, 3.vi.1962, *Schweickerdt* 2613 (BOL). S: Nuanetsi R., gorge upstream from Buffalo Bend, 28.iv.1961, *Drummond & Rutherford-Smith* 7557 & 7557a (BM; EA; SRGH). **Mozambique.** N: 51·2 km. E. of Camuana, 21.v.1961, *Leach & Rutherford-Smith* 10952 (BM; BOL; SRGH).
Occurs also in SW. Africa, Madagascar and Tanzania. Aquatic plant of pool edges or more rarely on wet soil in spray of waterfalls, 225–460 m.

2. **Isoetes rhodesiana** Alston in Bol. Soc. Brot., Sér. 2, **30**: 17 (1956). Type: Rhodesia, Nyamandhlovu Distr., Bongolo, *West* 3075 (BM, holotype; K, isotype; SRGH, isotype).

Leaves c. 60 × 0·5–0·7 cm., fewer (<10), flaccid, winged conspicuously in the lower ½. Megaspores as in *I. alstonii;* microspores unknown.

Rhodesia. W: Nyamandhlovu Distr., Bongolo, 24.ii.1949, *West* 3075 (BM; K; SRGH).
Unknown from elsewhere. A plant introduced into deeper water from open pans in vlei with clay puddled in bottom of storage tank; possibly a habitat form of *I. alstonii.*

3. **Isoetes aequinoctialis** A. Braun in Kuhn, Fil. Afr.: 195 (1868). Type from Angola.
Isoetes nigritiana A. Braun, tom. cit.: 196 (1868). Type from Nigeria.

Stem 0·5–1·0 cm. in diam., 3-lobed, with 10–20 leaves. Leaves 4–35 × 0·1–0·15 mm., ± trigonous, pale-green, stiff, apex abruptly pointed; sporophylls with a broad membranous margin (soon disappearing) c. 3 mm. wide at the base, above sporangia. Sporangia 3–7 × 1·5–3 mm., protruding either side of the mid-rib on the adaxial side of the sporophyll; velum covering lower ½ of sporangium or reduced to a marginal ring less than 1 mm. wide. Megaspores of 2 sizes, c. 350 μ and 500 μ,

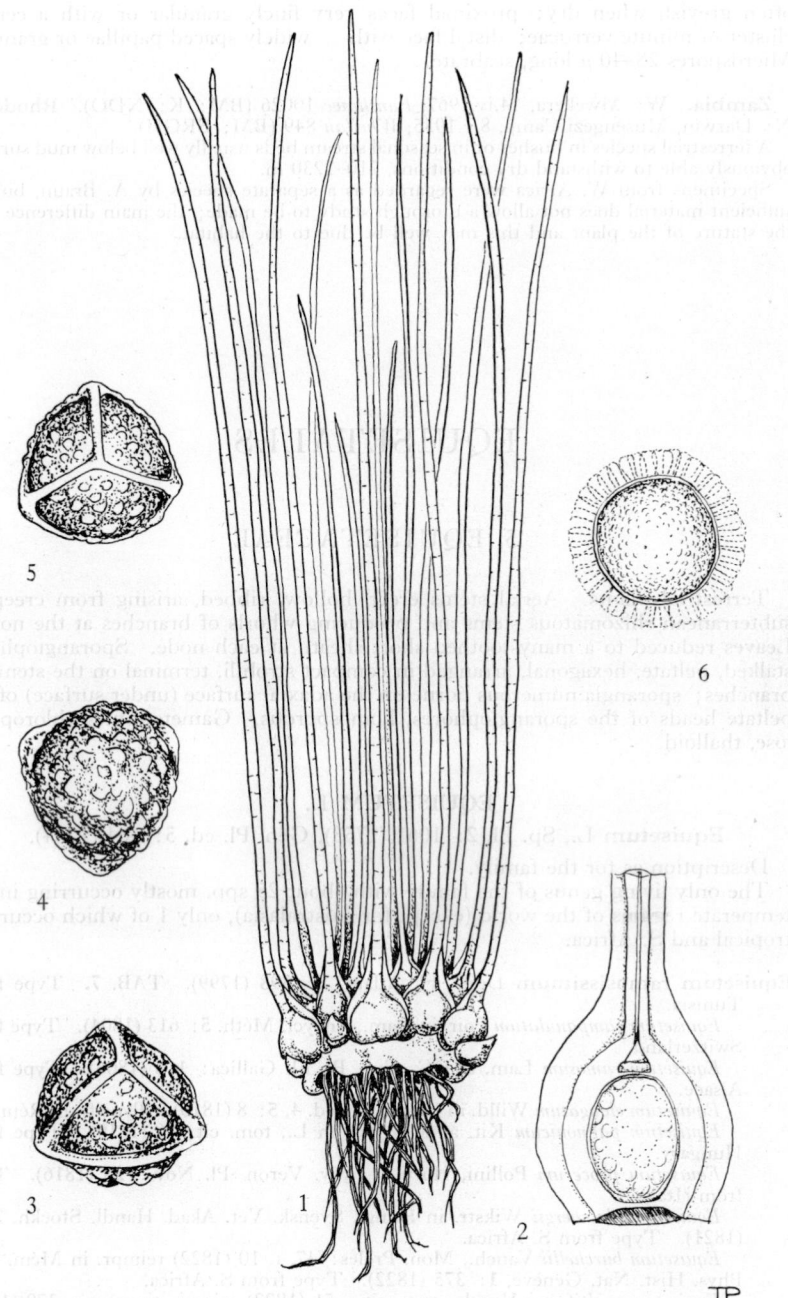

Tab. 6. ISOETES ALSTONII. 1, habit (×⅔); 2, inner face of sporophyll showing mega-sporangium (×48); 3, proximal view of megaspore (×48); 4, distal view of mega-spore (×48); 5, oblique side of megaspore (×48); 6, proximal view of microspore (×900), all from *Leach & Rutherford-Smith* 10952.

often greyish when dry; proximal faces very finely granular or with a central cluster of minute verrucae; distal face with ± widely spaced papillae or granules. Microspores 28–40 μ long, scabrate.

Zambia. W: Mwekera, 14.iv.1967, *Fanshawe* 10026 (BM; K; NDO). **Rhodesia.** N: Darwin, Musengezi Camp, 8.v.1955, *Whellan* 849 (BM; SRGH).

A terrestrial species in flushes or in seasonal stream beds usually well below mud surface, obviously able to withstand dry conditions, 210–1230 m.

Specimens from W. Africa were regarded as a separate species by A. Braun, but insufficient material does not allow a thorough study to be made; the main difference is in the stature of the plant and this may well be due to the habitat.

EQUISETALES

5. EQUISETACEAE

Terrestrial plants. Aerial stems erect, hollow, ribbed, arising from creeping subterranean rhizomatous stems and producing whorls of branches at the nodes. Leaves reduced to a many-toothed short sheath at each node. Sporangiophores stalked, peltate, hexagonal, arranged in compact strobili, terminal on the stems or branches; sporangia numerous borne on the adaxial surface (under surface) of the peltate heads of the sporangiophores, homosporous. Gametophytes chlorophyllose, thalloid.

EQUISETUM L.

Equisetum L., Sp. Pl. **2**: 1061 (1753), Gen. Pl. ed. 5: 484 (1754).

Description as for the family.

The only living genus of the family with about 25 spp. mostly occurring in the temperate regions of the world (except for Australasia), only 1 of which occurs in tropical and S. Africa.

Equisetum ramosissimum Desf., Fl. Atlant. **2**: 398 (1799). TAB. **7**. Type from Tunisia.

  Equisetum campanulatum Poir. in Lam., Encycl. Méth. **5**: 613 (1804). Type from Switzerland.

  Equisetum ramosum Lam. & DC., Syn. Pl. Fl. Gallica: 118 (1806). Type from Alsace.

  Equisetum elongatum Willd. in L., Sp. Pl., ed. 4, **5**: 8 (1810). Type from Réunion.

  Equisetum pannonicum Kit. apud Willd. in L., tom. cit.: 6 (1810). Type from Hungary.

  Equisetum procerum Pollini, Horti et Prov. Veron. Pl. Nov.: 28 (1816). Type from ?Italy.

  Equisetum thunbergii Wikstr. in Kungl. Svensk. Vet. Akad. Handl. Stockh. **2**: 4 (1821). Type from S. Africa.

  Equisetum burchellii Vauch., Mon. Prêles: 47, t. 10 (1822) reimpr. in Mém. Soc. Phys. Hist. Nat. Genève, **1**: 375 (1822). Type from S. Africa.

  Equisetum multiforme Vauch., tom. cit.: 51 (1822) reimpr. tom. cit.: 379 (1822) *nom. illegit.*

  Equisetum incanum Vauch., tom. cit.: 54, t. 12 (1822) reimpr. tom. cit.: 382 (1822). Type from Grand Canary.

  Equisetum ephedroides Bory, Nouv. Fl. Peloponn.: 66 (1838) *nom. illegit.*

  Equisetum hungaricum Sandor & Kit. in Verh. Zool.-Bot. Ges. Wien, **13**: 574 (1863). Type as for *E. pannonicum.*

  Equisetum sieboldii Milde in Ann. Mus. Bot. Lugd.-Bat. **1**: 62 (1864). Type from Japan.

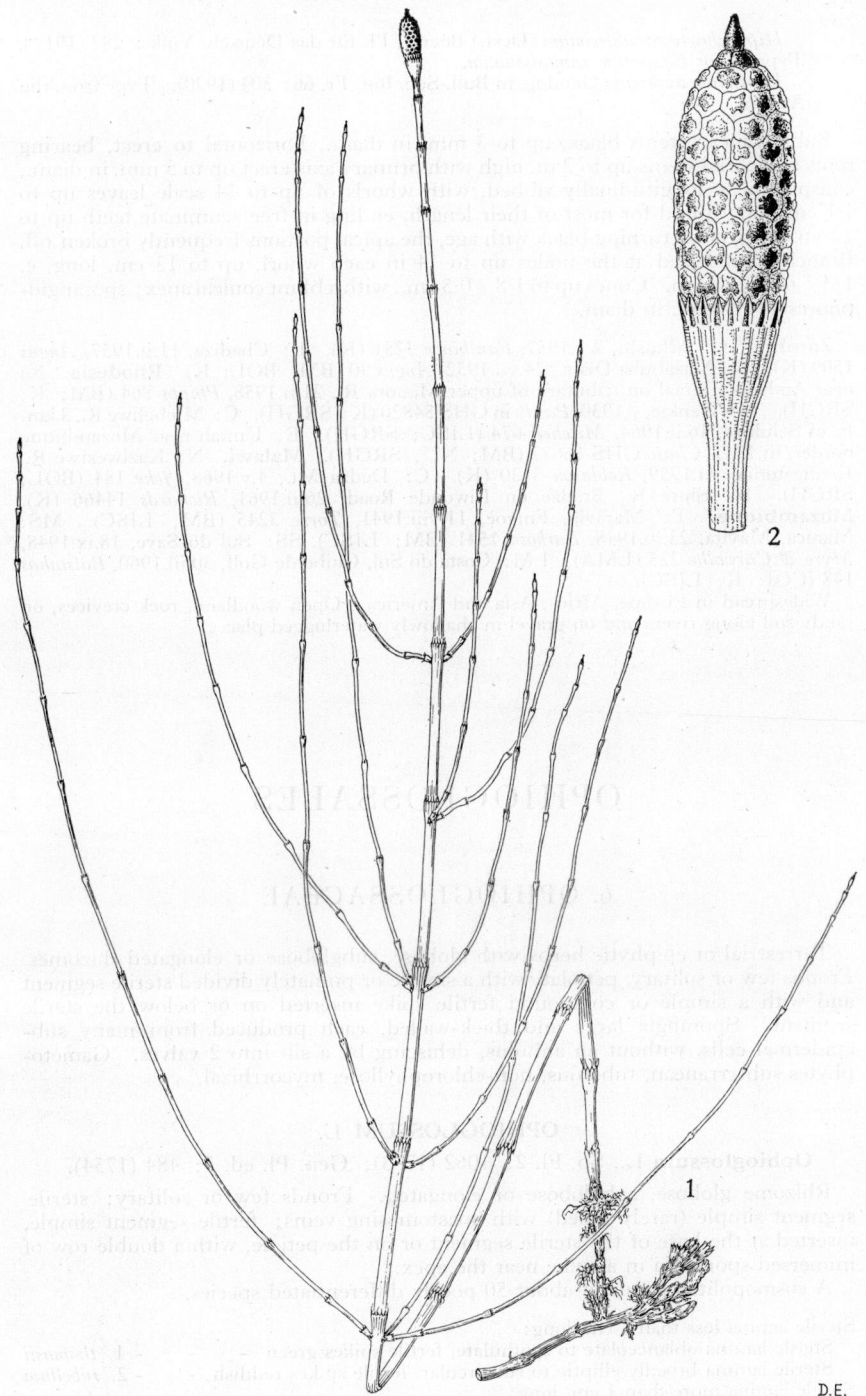

D.E.

Tab. 7. EQUISETUM RAMOSISSIMUM. 1, habit (× ½) from *Lusaka Natural History Club* 98;
2, cone (× 5) *Newbold* 909.

Hippochaete ramosissimum (Desf.) Boern., Fl. für das Deutsch. Volk.: 282 (1912). Type as for *Equisetum ramosissimum*.

Equisetum azoricum Gandog. in Bull. Soc. Bot. Fr. **66**: 304 (1920). Type from the Azores.

Subterranean stems black, up to 3 mm. in diam., horizontal to erect, bearing roots and aerial stems up to 2 m. high with primary axis erect up to 5 mm. in diam., conspicuously longitudinally ribbed, with whorls of up to 14 scale leaves up to 1·1 cm. long, fused for most of their length, ending in free acuminate teeth up to 2 mm. long, often turning black with age, the apical portions frequently broken off. Branches produced at the nodes up to 14 in each whorl, up to 13 cm. long, c. 1–1·5 mm. in diam. Cones up to 1·8 × 0·5 cm., with a blunt conical apex; sporangiophores 1–1·5 mm. in diam.

Zambia. C: Mkushi, 2.v.1957, *Fanshawe* 3251 (K). E: Chadiza, 11.ii.1957, *Angus* 1509 (K). S: Mazabuka Distr., 24.vii.1952, *Angus* 30 (BM; BOL; K). **Rhodesia.** N: near Andrew's Kraal on tributary of upper Mauora R., 21.ii.1958, *Phipps* 864 (BM; K; SRGH). W: Wankie, v.1930, *Pardy* in GHS 54820 (K; SRGH). C: Mtebehwe R., 3 km. E. of Selukwe, 16.ii.1964, *Mitchell* 674 (LISC; SRGH). E: Umtali near Mozambique border, iii.1947, *Chase* GHS 16622 (BM; NU; SRGH). **Malawi.** N: Kaziweziwe R. Livingstonia, 9.i.1959, *Robinson* 3130 (K). C: Dedza Mt., 4.v.1968, *Jeke* 184 (BOL; SRGH). S: Shire R., Bradze on Liwonde Road, 26.ii.1961, *Richards* 14466 (K). **Mozambique.** T: Marávia, Fingoè, 11.viii.1941, *Torre* 3245 (BM; LISC). MS: Manica, Mavita, 23.iv.1948, *Barbosa* 1541 (BM; LISC). SS: Sul do Save, 18.ix.1948, *Myre & Carvalho* 225 (LMA). LM: Costa do Sol, Clube de Golf, 30.iii.1960, *Balsinhas* 148 (COI; K; LISC).

Widespread in Europe, Africa, Asia and America. Open woodland, rock crevices, on sandy soil along rivers and on gravel in shallowly waterlogged places.

OPHIOGLOSSALES

6. OPHIOGLOSSACEAE

Terrestrial or epiphytic herbs with globose, subglobose or elongated rhizomes. Fronds few or solitary, petiolate with a simple or pinnately divided sterile segment and with a simple or compound fertile spike inserted on or below the sterile segment. Sporangia large and thick-walled, each produced from many subepidermal cells, without an annulus, dehiscing by a slit into 2 valves. Gametophytes subterranean, tuberous, non-chlorophyllose, mycorrhizal.

OPHIOGLOSSUM L.

Ophioglossum L., Sp. Pl. **2**: 1062 (1753); Gen. Pl. ed. 5: 484 (1754).

Rhizome globose, subglobose or elongated. Fronds few or solitary; sterile segment simple (rarely lobed) with anastomosing veins; fertile segment simple, inserted at the base of the sterile segment or on the petiole, with a double row of immersed sporangia in a spike near the apex.

A cosmopolitan genus of about 50 poorly differentiated species.

Sterile lamina less than 1 cm. long:
  Sterile lamina oblanceolate to spathulate, fertile spikes green - - - 1. *thomasii*
  Sterile lamina broadly elliptic to subcircular, fertile spikes reddish - - 2. *rubellum*
Sterile lamina more than 1 cm. long:
  Rhizome subglobose - - - - - - - - - 8. *costatum*
  Rhizome cylindric:
    Old leaf-sheaths persistent:
      Sterile lamina less than 3 cm. long - - - - - - 3. *gomezianum*
      Sterile lamina more than 3 cm. long - - - - - 5. *polyphyllum*

Old leaf-sheaths not persistent:
Sterile lamina cordate - - - - - - - - 7. *reticulatum*
Sterile lamina narrowly elliptic to ovate:
Sterile lamina very narrowly elliptic to oblanceolate, less than 1·2 cm. broad
4. *lancifolium*

Sterile lamina oblong-elliptic, more than 1·2 cm. broad:
Sterile lamina base narrowly cuneate; veins obscure - - 6. *vulgatum*
Sterile lamina base broadly cuneate; veins evident - - 7. *reticulatum*

1. **Ophioglossum thomasii** Clausen in Mem. Torrey Bot. Club, **19**, 2: 152 (1938). TAB. 8 fig. B. Type from Uganda.

Rhizome very small, with a few horizontal roots. Petiole subterranean, up to 5 mm. long. Sterile lamina up to 8 × 1·5 mm., oblanceolate, apex rounded. Fertile segment up to 5·6 cm. long, inserted at the base of the sterile lamina, with 2–7 pairs of sporangia, filiform.

Zambia. N: Abercorn Distr., Chilongowelo Escarpment, 20.ii.1959, *Richards* 1049 (K).
Also in Uganda. Wet peaty sand on laterite plateau, 1500–1650 m.

2. **Ophioglossum rubellum** Welw. ex A. Braun in Kuhn, Fil. Afr.: 179 (1868). TAB. 8 fig. E. Type from Angola.

Rhizome very short with a single leaf; old leaf-bases not persistent. Petiole up to 0·5–1 cm. long, subterranean for most of its length. Sterile lamina prostrate, broadly ovate to almost circular, broadly obtuse, rounded, base almost truncate. Fertile spike up to 7·5 cm. long, inserted at the base of the lamina, with up to 13 pairs of sporangia, apex acute.

Zambia. N: Abercorn Distr., c. 2·4 km. above Sansia Falls, Kalambo R., 29.xii.1958, *Richards* 10371 (K).
Also in Angola and Uganda. Recorded in our area only from damp peat at 1740 m.

3. **Ophioglossum gomezianum** Welw. ex A. Braun in Kuhn, Fil. Afr.: 176 (1868). TAB. 8 fig. A. Type from Angola.

Rhizome short, with 1–3 leaves with persistent leaf-bases. Petiole 0·7–2·5 cm. long, subterranean for most of its length. Sterile lamina up to 2·6 × 0·8 cm., lanceolate-elliptic to narrowly elliptic, acute without a prominent mucro, base narrowly cuneate. Fertile spike 3–9 cm. long, inserted at the base of the lamina, with up to 12 pairs of sporangia, apex acute.

Zambia. N: Kasama Distr., 19.i.1961, *Robinson* 4279 (K; SRGH). W: 11 km. N. of Chingola, 3.i.1961, *Robinson* 4218 (K; SRGH). C: Lusaka, 17.i.1965, *Robinson* 6363 (K; SRGH). S: Mazabuka Distr., 5.ii.1958, *Drummond* 5481 (BOL; LISC; SRGH). **Rhodesia.** N: Mtoko Distr., Mudzi Dam, 16.ii.1962, *Wild* 5672 (BM; BOL; K; SRGH). C: Salisbury, southern slopes of Ngomakurira, Chindamora Reserve, 27.i.1956, *Drummond* 5102 (BM; SRGH). **Malawi.** S: Shire R. near Liwonde Ferry, 13.iii.1955, *E.M. & W.* 843 (LISC; SRGH).
Also in S. Africa and widespread in tropical Africa. Damp soil on rock outcrops, 470–2400 m.

4. **Ophioglossum lancifolium** C. Presl, Suppl. Tent. Pterid.: 50 (1845) reimpr. in Abh. Königl. Böhm. Ges. Wiss., Ser. 5, **4**: 310 (1847). TAB. 8 fig. C. Type from Mauritius.
Ophioglossum lusoafricanum Welw. ex Prantl in Ber. Deutsch. Bot. Ges. **1**: 351 (1883). Type from Angola.

Rhizome cylindric, short to long, with usually a single leaf; leaf-bases not persistent; petiole 1·6–10 cm. long, subterranean for c. ⅓ of its length. Sterile lamina 2–6 × 0·2–1 cm., narrowly to very narrowly elliptic, sharply acute, sometimes shortly mucronate. Fertile spike up to 14·4 cm. long, inserted at the base of the lamina, with up to 30 pairs of sporangia, apex shortly or long-apiculate.

Zambia. N: Abercorn Distr., Sansia Falls, Kalambo R., 1.i.1957, *Richards* 7449a (K). W: Kitwe, 24.viii.1964, *Mutimushi* 965 (SRGH). C: Lusaka, 17.i.1965, *Robinson* 6363 (SRGH). S: Mapanza, 15.i.1956, *Robinson* 1322 (K). **Rhodesia.** W: Bulalima Mangwe, Embakwe, 10.v.1942, *Feiertag* (SRGH). E: W. bank of Haroni Gorge,

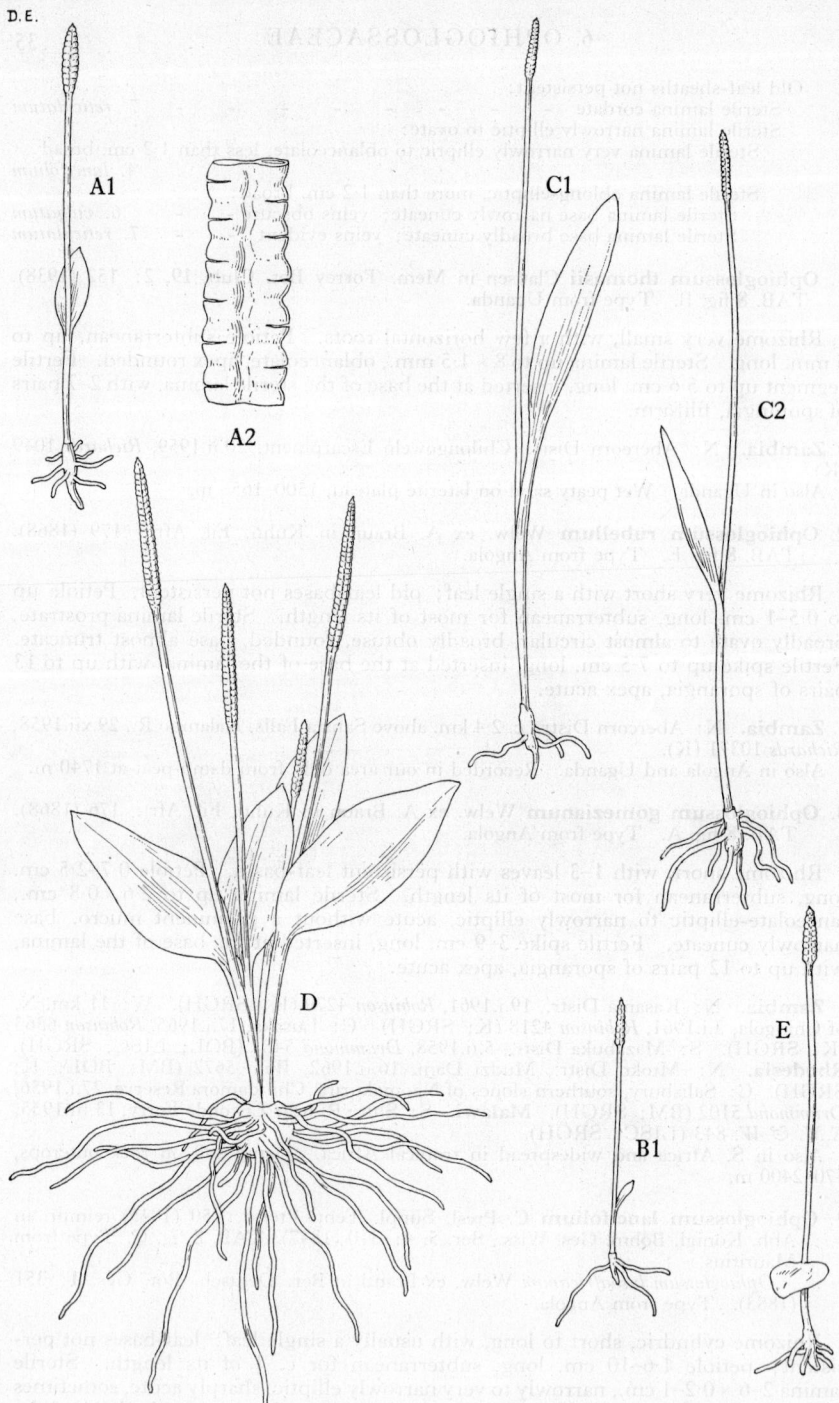

Tab. 8. A.—OPHIOGLOSSUM GOMEZIANUM. A1, habit (× ⅔); A2, enlargement of fertile spike (× 4). both from *Robinson 4518*. B.—OPHIOGLOSSUM THOMASII. B1, habit (× ⅔) *Richards 10949*. C.—OPHIOGLOSSUM LANCIFOLIUM. C1, habit (× ⅔); C2, habit (× ⅔). C1–C2 (showing variation) *Robinson 1322*. D.—OPHIOGLOSSUM COSTATUM. D1, habit (× ⅔) *Robinson 4219*. E.—OPHIOGLOSSUM RUBELLUM. E1, habit (× ⅔) *Richards 10371*.

16.ii.1958, *Mitchell* 406 (SRGH). **Malawi.** S: Mlanje Mt., Luchenya Forest Hut, 27.iii.1960, *Phipps* 2758 (BM; SRGH).
Widespread in tropical Africa. Seasonally moist soil, 1060–1770 m.

5. **Ophioglossum polyphyllum** A. Braun in Seub., Fl. Azor.: 17 (1844) emend. P.-Sermolli in Webbia, **9**: 632, t. 2a (1954).—Launert in Prodr. Fl. SW. Afr. **2**: 1 (1969). Type from S. Africa.

Ophioglossum capense var. *regulare* Schlechtend., Adumbr. **1**: 9, t. 1 fig. 2 (1825) non *O. capense* Sw. (1803).—Sim, Ferns S. Afr. ed. 2: 321, t. 166 fig. 2 (1915). Type from S. Africa.

Ophioglossum cuspidatum Milde in Bot. Zeit. **22**: 107 (1864). Type from Ethiopia.

Ophioglossum vulgatum var. *polyphyllum* Milde, Fil. Europ. Atlant.: 188 (1867) pro parte.

Ophioglossum vulgatum var. *aitchisonii* C.B.Cl. in Trans. Linn. Soc. Lond., Bot. **1**: 586 (1880). Type from India.

Ophioglossum regulare (Schlechtend.) C. Chr., Ind. Fil.: 472 (1906). Type from S. Africa.

Ophioglossum aitchisonii (C.B.Cl.) D'Almeida in Journ. Ind. Bot. Soc. **3**: 63, figs. 12–13 (1922). Type from India.

Ophioglossum tapinum A. Peter in Fedde, Repert. Beih. **40**: 86, t. 4 fig. 3 (1929). Type from Tanzania.

Rhizome short to long, with 1–2 leaves with persistent leaf-bases. Petiole up to 10 cm. long, subterranean for about ½ its length, often with a reddish tinge. Sterile lamina up to 5·5 × 1·7 cm., elliptic to oblong-lanceolate, acute to obtuse, mucronate, base broadly cuneate. Fertile spike up to 10 cm. long, inserted at the base of the sterile lamina, up to 30 pairs of sporangia, apex sharply acute.

Botswana. N: 73·6 km. W. of Francistown, 3.ii.1966, *Drummond* 8495 (SRGH). SW: Takatshwane Pan, 19.ii.1960, *Wild* 5077 (BM; SRGH). **Rhodesia.** N: Sebungwe, 15.xi.1958, *Phipps* 1463 (BM; BOL; SRGH). C: Salisbury, Domboshawa, 16.ii.1947, *Wild* 1664 (BM; SRGH). E: Inyanga, fort, 31.i.1948, *Fisher* 1453 (K). **Malawi.** C: Lilongwe Distr., 24.i.1954, *Jackson* 1195 (BM). **Mozambique.** SS: Inhambane, Zavala, Quissico, 11.xii.1944, *Mendonça* 3392 (BM; LISC).
Also in S. Africa, SW. Africa and in E. tropical Africa and outside Africa in Arabia, Afghanistan, Northern and Western India. Damp flushes on rock outcrops and seasonally moist soil, 850–1830 m.

6. **Ophioglossum vulgatum** L., Sp. Pl. **2**: 1062 (1753). Type from Europe.

Ophioglossum vulgatum var. *kilimandscharicum* Hieron. in Engl., Pflanzenw. Ost-Afr. **C**: 89 (1895). Type from Tanzania.

Rhizome short, erect, usually with only a single leaf, leaf-bases not persistent. Petiole 6·5–14 cm. long, subterranean for less than ½ its length. Sterile lamina narrowly oblong-acute; mucronate, base narrowly cuneate, veins obscure. Sterile spike up to 9 cm. long inserted at the base of the lamina, or just below, with up to 20 pairs of sporangia, apex acute.

Zambia. C: Fiwila, Mkushi, 7.i.1958, *Robinson* 2667 (BOL; SRGH). S: Mazabuka Distr., 0·8 km. from Chirundu Bridge, 5.ii.1958, *Drummond* 5480 (LISC; SRGH). **Rhodesia.** E: Inyanga, 18.xii.1955, *Chase* 5913 (BOL; SRGH). **Malawi.** C: Kasungu Distr., Chamama, 16.i.1959, *Robson* 1219 (K). **Mozambique.** N: Nampula. 21 km. from Nampula towards Nametil, 1.iv.1964, *Torre & Paiva* 11571 (LISC).
Also in S. Africa and sporadic in E. tropical Africa. Damp flushes on rock outcrops and in grassland, 820–1830 m.

7. **Ophioglossum reticulatum** L., Sp. Pl. **2**: 1063 (1753).—Sim, Ferns S. Afr. ed. 2: t. 167 fig. 2 (1915).—Launert in Prodr. Fl. SW. Afr. **2**: 2 (1969). Type from Mauritius.

Rhizome short, with 1–2 leaves; leaf-bases not persistent. Petiole up to 15 cm, long, subterranean for less than a ¼ of its length. Sterile lamina up to 8 × 7·5 cm., cordate or sometimes almost reniform to broadly ovate, rarely broadly elliptic-oblong or broadly ovate, apex rounded with or without a small mucro, veins evident, base cordate to broadly cuneate. Fertile spike up to 17 cm. long, inserted at the base of the lamina (or up to 1 cm. below the apparent base), with up to 45 pairs of sporangia, apex narrowly to broadly acute.

Zambia. S: Livingstone Distr., Victoria Falls, Livingstone I., 21.xii.1961, *Whellan* 1908 (SRGH). **Rhodesia.** N: Mazoe, i.1909, *Eyles* 559 (BM; SRGH). C: Selukwe,

Ferny Creek, 8.xii.1953, *Wild* 4296 (BM; BOL; SRGH). E: Umtali, 28.xii.1956, *Chase* 6273 (K; SRGH). **Malawi.** C: Dedza, Kachere near Mphunzi, 22.i.1959, *Robson* 1303 (K; LISC; SRGH). S: Chilwa I., 11.iii.1955, *E.M. & W.* 798 (LISC; SRGH). **Mozambique.** Z: Lugela, 31.i.1948, *Faulkner* 193 (K). ?T: between Tete and the coast, 1860, *Kirk* (K).

Also in S. Africa and widespread in tropical Africa and in Mauritius. Seasonally wet soils, 1060–1550 m.

Plants in which the fertile spike is inserted rather lower than the apparent base of the fertile lamina occur together with plants in which the insertion is normal for the species (*Chase* 6273); the former have sometimes been referred to the tropical American *O. petiolatum* Hook. which has an acute lanceolate sterile lamina.

8. **Ophioglossum costatum** R. Br., Prodr. Fl. Nov. Holl.: 163 (1810).—P.-Sermolli in Webbia, **9**: 627, t. 1 (1954). TAB. **8** fig. D Type from Australia.
    Ophioglossum pedunculosum Desv. in Mag. Ges. Naturf. Fr. Berl. **5**: 306 (1811).— Tardieu in Fl. Madag., Ophiogloss.: 4, t. 1 fig. 1–2 (1951). Type of unknown origin.
    Ophioglossum fibrosum Schumach., Beskr. Guin. Pl.: 452 (1827); in Kongel. Dansk. Vid. Selsk. Afh. **4**: 226 (1829). Type from W. Africa.
    Ophioglossum wightii Grev. & Hook., Bot. Misc. **3**: 218 (1833). Type from India.
    Ophioglossum brevipes Bedd., Ferns S. India: 23, t. 72 (1863). Type from India.
    Ophioglossum aphrodisiacum Welw. ex. Hook., Syn. Fil.: 446 (1868). Type from Angola.
    Ophioglossum felixii Tardieu in Notul. Syst. **13**: 169 (1948). Type from Guinea.

Rhizome up to 1 cm. in diam., subglobose with 1–4 leaves, the leaf-bases not persistent. Petiole 1–5·5 cm. long, subterranean for most of its length. Sterile lamina up to 11 × 1·8 cm., narrowly elliptic, oblong-lanceolate to broadly elliptic, acute to obtuse with or without a short mucro, base narrowly cuneate to rounded. Fertile spike 5–25 cm. long, inserted at the base of the lamina, with up to 60 pairs of sporangia, apex acute.

    Zambia. N: Abercorn Distr., Sansia Falls, 1.i.1957, *Richards* 7449a (K). W: Luanshya, 13.i.1965, *Fanshawe* 2706 (K). S: Mapanza, 18.ii.1956, *Robinson* 1348 (BM; K; SRGH). **Rhodesia.** N: Urungwe, 3.ii.1958, *Drummond* 5452 (BOL; K; LISC; SRGH). W: Bulalima Mangwe, Embakwe, 10.v.1942, *Feiertag* GH. 45451 (BM; SRGH). C: Salisbury, Gilnockie, 20.i.1958, *Mitchell* 215 (BOL). **Malawi.** N: 5 km. SE. of Fort Hill, 16.iii.1961, *Robinson* 4532 (K). **Mozambique.** T: R. Zambesi, ii.1859, *Kirk* (K).
    Widespread in tropical Africa, Asia and Australasia. Damp flushes on rock outcrops and damp depressions in *Brachystegia* woodland, 450–1130 m.

MARATTIALES

7. MARATTIACEAE

Large to very large terrestrial plants with often very large fronds and stipular outgrowths at the base of the stipe. Rhizome short and massive (in African species) or creeping. Fronds 2-pinnate (in African species) or simply pinnate or palmate; veins free (in African species) or rarely reticulate. Sporangia in clusters on the veins, each derived from several initial cells, variously fused into synangia. Gametophytes thalloid, dark-green.
    Of this family only the genus *Marattia* occurs in Africa.

MARATTIA Sw.

Marattia Sw., Prodr. Fl. Ind. Occ.: 128 (1788).

    Rhizome erect, massive, short, covered with fleshy persistent stipular outgrowths and fleshy mycorrhizal roots. Lamina very large 2-pinnate, veins free. Sporangia

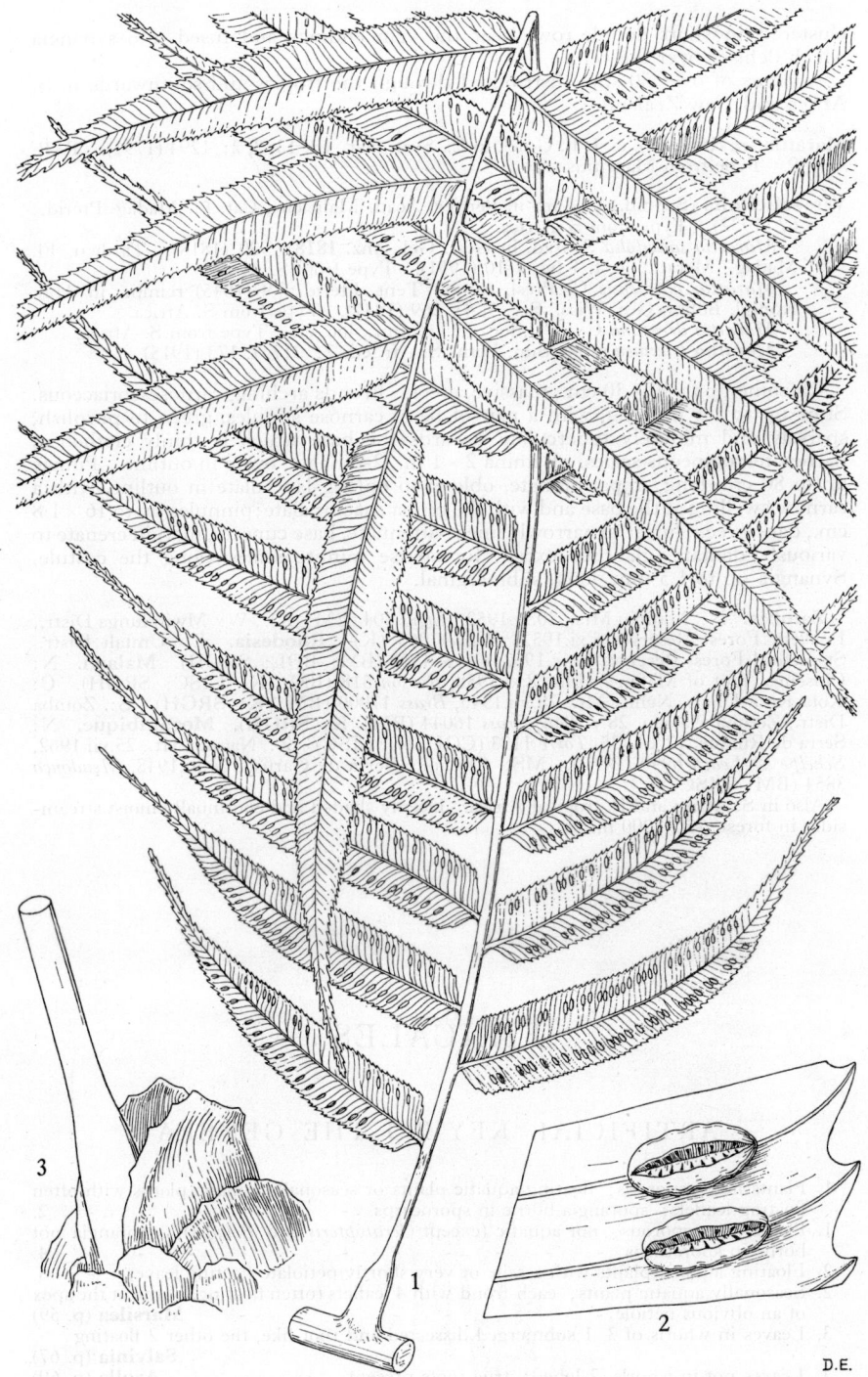

D.E.

Tab. 9. MARATTIA FRAXINEA. 1, pinna (×⅔) *Fanshawe* 5602; 2, synangia (×9); 3, stipular base of stipe (×⅔), 2–3 from *Brass* 17761.

clustered in short double rows near the vein endings and fused into synangia which dehisce over the vein.

A genus of over 50 spp. distributed through the tropics and southwards to S. Africa and New Zealand.

Marattia fraxinea Sm. ex J. F. Gmel. in L., Syst. Nat., ed. 13, **2**, 2: 1294 (1791). TAB. 9. Type from Mauritius.

Var. **salicifolia** (Schrad.) C. Chr. in Perrier, Acad. Malgache, Cat. Pl. Madag. Ptérid.: 67 (1932). Type from S. Africa.
     Marattia salicifolia Schrad. in Gött. Gel. Anz. **1818**: 920 (1818).—Tardieu, Fl. Madag., Maratt.: 6, t. 1 fig. 9–10 (1951). Type from S. Africa.
     Marattia natalensis C. Presl, Suppl. Tent. Pterid.: 9 (1845) reimpr. in Abh. Königl. Böhm. Ges. Wiss., Ser. 5, **4**: 269 (1847). Type from S. Africa.
     Marattia dregeana C. Presl, loc. cit., reimpr., loc. cit. Type from S. Africa.
     Marattia fraxinea sensu Sim, Ferns S. Afr. ed. 2: 317, t. 173 (1915).

Rhizome up to 40×30 cm., massive, erect. Fronds arching, carnose-coriaceous. Stipe up to 1·5 m. long, with a pair of basal carnose stipules, green to purplish, sparsely and minutely tuberculate towards the base which is thinly clothed in narrow ferrugineous scales. Lamina 2×1 m., 2-pinnate, ovate in outline. Pinnae up to 80 cm. long, imparipinnate, oblong to oblong-lanceolate in outline, with a carnose swelling at the base and with the pinna rhachis alate; pinnules up to $16 \times 1·8$ cm., dark-green, linear to narrowly linear-attenuate, base cuneate, margin crenate to variously dentate, glabrous except for sparse minute scales along the costule. Synangia up to 1·5 mm. long, submarginal.

     Zambia. N: Mafingi Mts., 20.xi.1952, *Angus* 804 (BM; K). W: Mwinilunga Distr., Luakera Forest Reserve, 5.xi.1952, *Holmes* 965 (K). **Rhodesia.** E: Umtali Distr., Stapleford Forest Reserve, 29.iv.1952, *Chase* 4508 (BM; BOL; SRGH). **Malawi.** N: Chisenga, foot of Mafinga Hills, 8.xi.1958, *Robson* 518 (BM; K; LISC; SRGH). C: Kota Kota Distr., Nchisi Mt., 28.vii.1946, *Brass* 17000 (BM; K; SRGH). S: Zomba Distr., Zomba Plateau, 28.v.1946, *Brass* 16044 (BM; K; SRGH). **Mozambique.** N: Serra de Ribáuè, 12.x.1935, *Torre* 1123 (COI; K; LISC). Z: Namúli Mt., 25.vii.1962, *Schelpe & Leach* 6999 (BOL). MS: Chimoio, Serra de Garuso, 24.iii.1948, *Mendonça* 3854 (BM; LISC).

Also in S. Africa and E. tropical Africa. Deeply shaded and continually moist stream-sides in forest, 750–2000 m.

FILICALES

ARTIFICIAL KEY TO THE GENERA

1. Plants heterosporous; floating aquatic plants or seasonally aquatic plants with often floating leaflets; sporangia borne in sporocarps - - - - - - 2.
1. Plants homosporous; not aquatic (except *Ceratopteris* and *Bolbitis*); sporangia not borne in sporocarps - - - - - - - - - - 4.
2. Floating aquatic plants with sessile or very shortly petiolate floating leaves - 3.
2. Seasonally aquatic plants; each frond with 4 leaflets (often floating) borne at the apex of an obvious petiole - - - - - - - - **Marsilea** (p. 59)
3. Leaves in whorls of 3, 1 submerged dissected and root-like, the other 2 floating **Salvinia** (p. 67)
3. Leaves not in whorls, 2-lobed; true roots present - - **Azolla** (p. 69)
4. Sporangial annulus an uninterrupted ring of thickened cells, or in the form of a lateral group of thickened cells; sporangia maturing simultaneously, or in basipetal succession - - - - - - - - - - - - - 5.
4. Sporangial annulus vertical and interrupted by a few thin-walled cells at the point of

dehiscence; sporangia maturing at different times (*Polypodiaceae* sensu Christensen)
      15.
5. Sporangial annulus in the form of a group of thickened cells on the side of the
   sporangium (*Osmundaceae*) -   -   -   -   -   -   -   -   -   -   -   6.
5. Sporangial annulus an apical, lateral, oblique, or sometimes nearly vertical, uninterrupted ring of thickened cells   -   -   -   -   -   -   -   -   -   7.
6. Fertile pinnules much narrower than the sterile pinnules   -   **Osmunda** (p. 44)
6. Fertile pinnules not differentiated from the sterile pinnules -   -   **Todea** (p. 46)
7. Fronds semitransparent, 1 cell thick, small; sori subtended by 2 indusial lobes;
   sporangia maturing in basipetal succession (*Hymenophyllaceae*)   -   -   -   8.
7. Fronds opaque, more than 1 cell thick   -   -   -   -   -   -   -   9.
8. Indusial lobes mostly free   -   -   -   -   -   **Hymenophyllum** (p. 78)
8. Indusial lobes partly fused into a tube   -   -   -   **Trichomanes** (p. 75)
9. Fronds spaced on a creeping rhizome, falsely dichotomously branched due to the
   abortion of apical buds (*Gleicheniaceae*)   -   -   -   -   -   -   10.
9. Fronds tufted, or if spaced, then climbing -   -   -   -   -   -   -   11.
10. Stipule-like lobes present at the junction of the frond branches
      Dicranopteris (p. 50)
10. Stipule-like lobes not present at the junction of the frond branches
      Gleichenia (p. 48)
11. Sporangial annulus apical (*Schizaeaceae*)   -   -   -   -   -   -   12.
11. Sporangial annulus oblique or nearly vertical; mostly arborescent ferns (*Cyatheaceae*)
      Cyathea (p. 70)
12. Frond not climbing -   -   -   -   -   -   -   -   -   -   -   13.
12. Frond climbing -   -   -   -   -   -   -   -   -   **Lygodium** (p. 57)
13. Sporangia borne on small fertile pinnae at the apex of a very narrowly simple linear
   frond   -   -   -   -   -   -   -   -   -   **Schizaea** (p. 52)
13. Sporangia borne on modified dissected basal pinnae or on unmodified pinnae   14.
14. Sporangia borne on modified much dissected basal pinnae   -   -   **Anemia** (p. 52)
14. Sporangia borne on unmodified pinnae   -   -   -   -   **Mohria** (p. 54)
15. Fronds simple, or shallowly to deeply lobed   -   -   -   -   -   -   16.
15. Fronds distinctly pinnate, 2-pinnatifid or more dissected   -   -   -   -   38.
16. Fronds simple, entire   -   -   -   -   -   -   -   -   -   -   17.
16. Fronds pinnatifid (usually deeply lobed, rarely shallowly lobed) or dichotomously
   lobed or flabellately divided   -   -   -   -   -   -   -   -   30.
17. Fertile fronds with sporangia covering the whole or part of the lamina or a linear
   apical segment (acrostichoid)   -   -   -   -   -   -   -   -   18.
17. Fertile fronds with sporangia in discrete sori or in linear marginal grooves or forming
   a reticulate pattern along the veins; nest leaves absent -   -   -   -   20.
18. Sporangia covering the whole or part of the under surface of the fertile lamina   19.
18. Sporangia covering the under surface of a linear apical segment of th ʋ lamina
      Belvisia (p. 159)
19. Sporangia covering nearly the whole of the under surface; nest leaves absent
      Elaphoglossum (p. 209)
19. Sporangia covering an oval area in the upper ½; nest leaves present
      Platycerium (p. 145)
20. Sori in 2 grooves or borne along the veins -   -   -   -   -   -   21.
20. Sori discrete, round or linear, not in grooves   -   -   -   -   -   22.
21. Sori in 2 narrow grooves   -   -   -   -   -   -   **Vittaria** (p. 94)
21. Sori forming a reticulate pattern along the veins -   -   -   **Antrophyum** (p. 96)
22. Under surface of fronds densely clothed with matted stellate hairs
      Pyrrosia (p. 146)
22. Fronds glabrous or with scattered hairs or scales -   -   -   -   -   -   23.
23. Sori linear and oblique to the midrib -   -   -   -   -   -   -   24.
23. Sori round or oval   -   -   -   -   -   -   -   -   -   -   25.
24. Sori with lateral linear indusia   -   -   -   -   -   **Asplenium** (p. 167)
24. Sori without indusia   -   -   -   -   -   -   **Loxogramme** (p. 149)
25. Veins free (or if veins obscure, the plants with tufted linear fronds less than 5 cm.
   long)   -   -   -   -   -   -   -   -   -   -   -   -   26.
25. Veins anastomosing   -   -   -   -   -   -   -   -   -   -   27.
26. Sori with indusium; stipes articulated -   -   -   -   -   **Oleandra** (p. 165)
26. Sori without indusia; stipes not articulated -   -   -   -   **Grammitis** (p. 141)
27. Fronds large, coriaceous with copious scattered minute sori c. 1 mm. in diam.
      Microsorium (p. 156)
27. Fronds membranous to coriaceous with sori more than 2 mm. in diam. in a row or
   scattered irregularly on either side of the midrib   -   -   -   -   28.
28. Rhizome very long, scandent; included free veinlets only formed in the areoles
   nearest the costa   -   -   -   -   -   -   -   **Microgramma** (p. 155
28. Rhizome short creeping; included free veinlets not confined to the areoles nearest the
   costa   -   -   -   -   -   -   -   -   -   -   -   -   29.

29. Immature sori with peltate paraphyses; sori in a regular single row on either side of
the midrib   -   -   -   -   -   -   -   -   -   **Pleopeltis** (p. 151)
29. Immature sori without peltate paraphyses; sori scattered irregularly on either side
of the midrib   -   -   -   -   -   -   -   -   **Microsorium** (p. 156)
30. Fertile fronds dichotomously lobed or flabellately divided   -   -   -   -   31.
30. Fronds pinnately lobed (pinnatifid)   -   -   -   -   -   -   -   -   32.
31. Fertile fronds dichotomously lobed; sterile fronds (nest-leaves) subentire or shallowly
lobed; large epiphytic plants   -   -   -   -   -   -   -   **Platycerium** (p. 145)
31. Fertile and sterile fronds deeply flabellately divided; small terrestrial plants
Actiniopteris (p. 136)
32. Sori linear   -   -   -   -   -   -   -   -   -   -   -   -   -   33.
32. Sori round   -   -   -   -   -   -   -   -   -   -   -   -   -   34.
33. Under surface of lamina densely paleaceous; frond uniform   **Ceterach** (p. 188)
33. Under surface of lamina glabrous to pubescent; fronds dimorphic
Blechnum (p. 235)
34. Veins free, not anastomosing   -   -   -   -   -   -   -   -   -   35.
34. Veins anastomosing   -   -   -   -   -   -   -   -   -   -   -   36.
35. Under surface of lamina paleaceous   -   -   -   -   -   **Polypodium** (p. 158)
35. Under surface of lamina not paleaceous, glabrous or hairy   - **Xiphopteris** (p. 141)
36. Lobes of the fertile lamina articulated to the midrib; lobed sterile nest-leaves also
produced   -   -   -   -   -   -   -   -   -   -   -   **Drynaria** (p. 149)
36. Lobes of the fertile and sterile laminae not articulated to the midrib; fronds uniform
37.
37. Immature sori with peltate paraphyses, not sunk into the lamina
Pleopeltis (p. 151)
37. Immature sori with non-peltate paraphyses, sunk into the lamina
Phymatodes (p. 153)
38. Fertile and sterile fronds pinnate, with or without only the basal pinnae lobed
basiscopically (except for 2-pinnate acrostichoid fertile fronds in *Stenochlaena*)   39.
38. Fronds (fertile) distinctly 2-pinnatifid or more dissected   -   -   -   -   53.
39. Pinnae of sterile fronds minutely serrate   -   -   -   -   -   -   40.
39. Pinnae of sterile fronds entire to crenate   -   -   -   -   -   -   42.
40. Basal pinnae lobed basiscopically   -   -   -   -   -   -   -   41.
40. Basal pinnae not lobed basiscopically   -   -   -   -   **Stenochlaena** (p. 240)
41. Pinnae caudate; sori superficial on the veins, without indusia
Coniogramme (p. 102)
41. Pinnae attenuate; sori marginal with linear indusia   -   -   -   **Pteris** (p. 115)
42. Under surface of the fertile pinnae more or less evenly covered with sporangia   43.
42. Sporangia in individual round or linear sori   -   -   -   -   -   -   45.
43. Veins anastomosing   -   -   -   -   -   -   -   -   -   -   -   44.
43. Veins free, not anastomosing   -   -   -   -   -   **Lomariopsis** (p. 216)
44. Pinnae thickly coriaceous   -   -   -   -   -   -   **Acrostichum** (p. 98)
44. Pinnae membranous   -   -   -   -   -   -   -   -   - **Bolbitis** (p. 218)
45. Sori marginal or submarginal, or borne on fertile pinnae much narrower than the
sterile pinnae   -   -   -   -   -   -   -   -   -   -   -   46.
45. Sori superficial on the veins   -   -   -   -   -   -   -   -   -   51.
46. Fronds dimorphic, with the fertile pinnae much narrower than the sterile pinnae;
indusium linear -   -   -   -   -   -   -   -   -   **Blechnum** (p. 235)
46. Fronds not dimorphic   -   -   -   -   -   -   -   -   -   -   47.
47. Indusia opening outwards towards the margin or reniform   -   -   -   48.
47. Indusia or indusial flaps opening inwards from the margin   -   -   -   49.
48. Pinnae articulated to the rhachis -   -   -   -   -   **Nephrolepis** (p. 159)
48. Pinnae not articulated to the rhachis   -   -   -   -   **Lindsaea** (p. 139)
49. Sporangia borne on the under surface of an indusial flap   -   **Adiantum** (p. 108)
49. Sporangia borne in a submarginal linear sorus covered by an indusium opening
inwards from the margin   -   -   -   -   -   -   -   -   -   50.
50. Fertile pinnae with a distinct sterile apex   -   -   -   -   -   **Pteris** (p. 115)
50. Fertile pinnae without a distinct sterile apex   -   -   -   -   **Pellaea** (p. 128)
51. Sori round, or if oblong then exindusiate   -   -   -   -   -   -   52.
51. Sori linear to oblong, indusiate   -   -   -   -   -   -   **Asplenium** (p. 167)
52. Soral paraphyses absent; fronds not proliferous or only proliferous near the apex
Thelypteris (p. 189)
52. Soral paraphyses present; frond proliferous anywhere on the rhachis
Ampelopteris (p. 200)
53. Plant aquatic with succulent fronds   -   -   -   -   **Ceratopteris** (p. 102)
53. Plant not aquatic   -   -   -   -   -   -   -   -   -   -   -   54.
54. Sori without indusia, or indusia minute and obscured in the mature sori   -   55.
54. Sori with obvious indusia   -   -   -   -   -   -   -   -   -   62.
55. Sori acrostichoid, covering the under surface of the fertile pinnules
Stenochlaena (p. 240)

55. Sori not acrostichoid - - - - - - - - - 56.
56. Under surface of fronds with white, yellow or orange coloured powder - -
           Pityrogramma (p. 105)
56. Under surface of fronds without powder - - - - - - 57.
57. Under surface of fronds villous or densely paleaceous - - - - - 58.
57. Under surface of frond pubescent, subglabrous or glabrous - - - - 59.
58. Under surface of frond villous - - - - - - **Cheilanthes** (p. 122)
58. Under surface of frond densely paleaceous - - - **Ceterach** (p. 188)
59. Lamina glabrous; fronds annual, up to 15 cm. high; sporangia borne along the veins
           Anogramma (p. 99)
59. Lamina pubescent or with minute short or long hairs along the veins, costules or
    costae; fronds persistent - - - - - - - - 60.
60. Frond proliferous, glabrous at maturity - - - - **Dryopteris** (p. 220)
60. Frond not proliferous, with hairs on the costules - - - - - 61.
61. Lamina deeply 3-pinnatifid, 0·6–1·0 m. long with minute blunt multicellular hairs on
    the costules and veins - - - - - - **Dryoathyrium** (p. 207)
61. Lamina 2-pinnatifid, less than 50 cm. long, with sharp unicellular hairs at least on
    the costae and costules - - - - - - **Thelypteris** (p. 189)
62. Sori marginal or submarginal - - - - - - - - 63.
62. Sori superficial on the lamina - - - - - - - - 77.
63. Veins obvious and anastomosing freely to form numerous areoles - - 64.
63. Veins free or only forming a series or areoles along the costa, or obscure 65.
64. Frond pubescent to hirsute - - - - - - **Blotiella** (p. 81)
64. Frond glabrous and glaucous - - - - - **Histiopteris** (p. 84)
65. Indusium opening towards the margin - - - - - - 66.
65. Indusium or indusial flaps opening inwards from the margin - - 68.
66. Stipe articulated - - - - - - - - **Davallia** (p. 167)
66. Stipe not articulated - - - - - - - - - 67.
67. Rhizome with hairs - - - - - - - **Microlepia** (p. 89)
67. Rhizome with scales - - - - - - - **Asplenium** (p. 167)
68. Sori borne on an indusial flap opening inwards - - - **Adiantum** (p. 108)
68. Sori not borne on an indusial flap - - - - - - - 69.
69. Sori linear to narrowly oblong - - - - - - - - 70.
69. Sori round - - - - - - - - - - - 74.
70. Rhizome with scales - - - - - - - - - - 71.
70. Rhizome hairy - - - - - - - - - - - 73.
71. Ultimate fertile segments with a distinct sterile apex - - - **Pteris** (p. 115)
71. Ultimate fertile segments without a distinct sterile apex - - - - 72.
72. Fronds pedate, the upper ½ of the rhachis alate - - **Doryopteris** (p. 121)
72. Fronds pinnate to 4-pinnate, the rhachis not alate - - - **Pellaea** (p. 128)
73. Frond coriaceous; soral paraphyses absent - - - **Pteridium** (p. 88)
73. Frond membranous; soral paraphyses present - - - **Lonchitis** (p. 86)
74. Sori borne only in the sinuses between teeth or lobes - - - - 75.
74. Sori not borne only in sinuses between segment lobes - - - - 76.
75. Frond less than 0·5 m. tall; rhizome scaly - - - - **Aspidotis** (p. 113)
75. Frond 1–3 m. tall; rhizome hairy - - - - - **Hypolepis** (p. 92)
76. Fronds pedate, glabrous, with the upper ½ of the rhachis alate
           Doryopteris (p. 121)
76. Fronds 3-4-pinnatifid or, if pedate, with the upper ½ of the rhachis alate, then pilose
           Cheilanthes (p. 122)
77. Indusium lateral, linear-oblong or J-shaped - - - - - - 78.
77. Indusium peltate or reniform - - - - - - - - 80.
78. Rhizome and stipe scales clathrate - - - - - **Asplenium** (p. 167)
78. Rhizome scales not clathrate - - - - - - - - 79.
79. Frond lamina less than 0·8 m. long - - - - - **Athyrium** (p. 202)
79. Frond lamina more than 1 m. long - - - - - **Diplazium** (p. 204)
80. Indusium peltate - - - - - - - - - - 81.
80. Indusium reniform - - - - - - - - - - 82.
81. Pinnules dimidiate articulated to the costa; sori oblong
           Didymochlaena (p. 220)
81. Pinnae neither dimidiate nor articulated to the costa; sori round
           Polystichum (p. 226)
82. Basal pinnae developed basiscopically - - - - - - 83.
82. Basal pinnae not developed basiscopically - - - - - - 87.
83. Veins free; fronds usually not proliferous, but if proliferous then only on the apical
    half of the rhachis - - - - - - - - - 84.
83. Veins anastomosing; fronds proliferous - - - - **Tectaria** (p. 234)
84. Pinna costa with raised edges on the ventral surface - - - - 85.
84. Pinna costa without raised edges on the ventral surface - - - - 86.

85. Pinnule margins crenate to serrate or if aristate-dentate then the lamina not triangular in outline - - - - - - - - - - - **Dryopteris** (p. 220)
85. Pinnule margins aristate-dentate; lamina triangular in outline
                                                 Arachniodes (p. 228)
86. Hairs on lamina unicellular, needle-like - - - - **Hypodematium** (p. 230)
86. Hairs on lamina multicellular - - - - - - - **Ctenitis** (p. 230)
87. Stipe articulated - - - - - - - - **Arthropteris** (p. 162)
87. Stipe not articulated - - - - - - - - - - - 88.
88. Veins from costules forked - - - - - - - - **Dryopteris** (p. 220)
88. Veins from costules not forked - - - - - - - - - 89.
89. Indusia ciliate - - - - - - - - - - **Ctenitis** (p. 230)
89. Indusia entire to erose, pilose or glabrous - - - **Thelypteris** (p. 189)

8. OSMUNDACEAE

Terrestrial plants with erect or procumbent creeping rhizomes enclosed by a mass of persistent stipe-bases which are winged at the base. Fronds deeply 2-pinnatifid to 3-pinnatifid often clothed in woolly simple hairs during development, glabrous at maturity. Sporangia large, eusporangiate in development, maturing simultaneously, with a rudimentary annulus of a group of cells near the equator and borne either on both surfaces of narrow fertile pinnules or on the under surface of undifferentiated pinnae. Gametophytes cordate or elongate, thick, dark-green and with a discernable midrib on the under surface.

Fertile pinnules much narrower than sterile pinnules - - - - **1. Osmunda**
Fertile pinnules not differentiated from sterile pinnules - - - - **2. Todea**

1. OSMUNDA L.

Osmunda L., Sp. Pl. **2**: 1063 (1753); Gen. Pl. ed. 5: 484 (1754).

Rhizome erect, without scales, clothed with a mass of persistent winged leaf-bases and embedded in a mass of black roots. Lamina 2-pinnate, veins free. Sporangia borne in dense clusters on both surfaces of narrow fertile pinnules with a much reduced laminar surface.

A genus of about 13 spp. distributed through the more temperate parts of the world. One species represented in our area.

Osmunda regalis L., Sp. Pl. **2**: 1065 (1753).—Sim, Ferns S. Afr. ed. 2: 310, t. 170 (1915). TAB. **10**. Type from Europe.
    Osmunda capensis C. Presl, Suppl. Tent. Pterid.: 63 (1845) reimpr. in Abh. Königl. Böhm. Ges. Wiss., Ser. 5, **4**: 323 (1847) non L. (1771). Type from S. Africa.
    Struthiopteris regalis (L.) Bernh. in Schrad., Journ. Bot. **1800**, 2: 126 (1801). Type as for *Osmunda regalis*.
    Osmunda regalis var. *capensis* Milde, Fil. Europ. Atlant.: 179 (1867). Type from S. Africa.

Rhizome erect to suberect, covered with the persistent winged bases of the stipes, usually embedded in a mass of black fibrous roots and with tufted fronds. Fronds erect, with a rufous tomentum when young, becoming glabrous at maturity. Lamina 2-pinnate up to 1 m. long, with the fertile pinnae borne in the apical portion, oblong to narrowly oblong in outline, with subopposite pinnae. Sterile pinnules up to 6 × 1·5 cm., herbaceous to thinly coriaceous, very narrowly to narrowly oblong, unequally truncate at the base, obtuse to broadly acute, minutely crenulate, petiolate except for adnate pinnules towards the apices of the pinnae. Fertile pinnules up to c. 2·5 × 0·2 cm., linear, petiolate or adnate, bearing groups of sporangia at intervals.

Zambia. N: Abercorn Distr., Inono Stream, 6.v.1942, *Richards* 1594 (K). W: Mwinilunga Distr., Kabompo R., 17.ix.1952, *Angus* 473 (K). **Rhodesia.** N: Mazoe

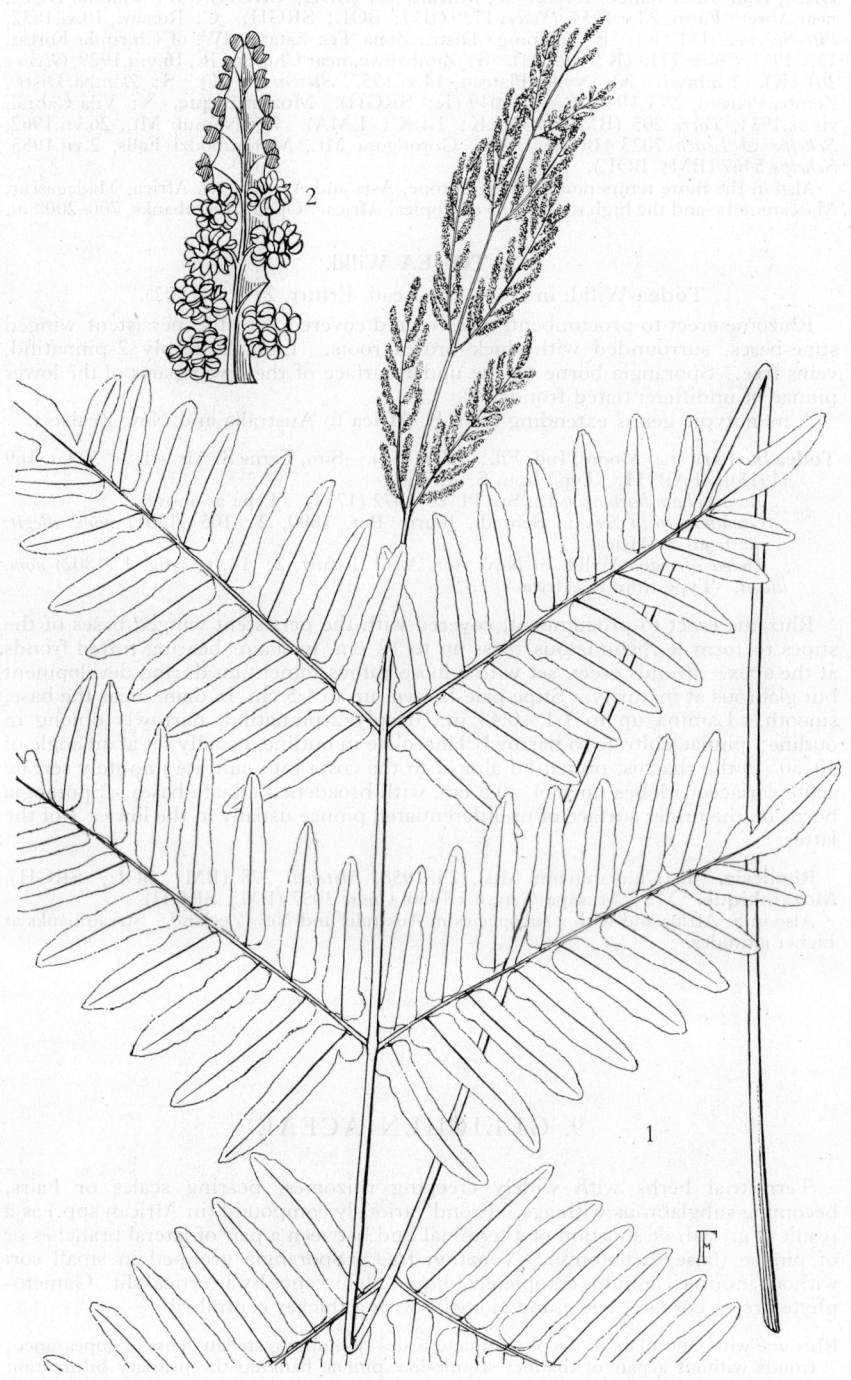

Tab. 10. OSMUNDA REGALIS. 1, fertile frond ($\times \frac{2}{3}$); 2, enlargement of fertile pinnule ($\times 2$), both from *Richards* 1594.

Distr., Iron Mask Range, 15.vii.1958, *Mitchell* 451 (BOL; SRGH). W: Matobo Distr., near Absent Farm, 23.v.1954, *Plowes* 1739 (BM; BOL; SRGH). C: Ruzawi, 10.xi.1932, *Pitt-Schenkel* 111 (K). E: Chipinga Distr., Zona Tea Estate, SW. of Chirinda Forest, 15.v.1962, *Chase* 7713 (K; SRGH). S: Zimbabwe, near Chipopo R., 16.viii.1929, *Glynne* 261 (K). **Malawi.** N: Nyika Plateau, 14.xi.1957, *Stewart* 1 (K). S: Zomba Distr., Zomba Plateau, 28.v.1946, *Brass* 16049 (K; SRGH). **Mozambique.** N: Vila Cabral, vii–ix.1934, *Torre* 205 (BM; COI; K; LISC; LMA). Z: Namúli Mt., 26.vii.1962, *Schelpe & Leach* 7023 (BOL). MS: Gorongosa Mt., Morambodzi Falls, 2.vii.1955, *Schelpe* 5467 (BM; BOL).

Also in the more temperate parts of Europe, Asia and America, S. Africa, Madagascar, Mascarene Is. and the high mountains of tropical Africa. Open streambanks, 700–2000 m.

2. TODEA Willd.

Todea Willd. in Nov. Act. Acad. Erfurt, **2**: 14 (1802).

Rhizome erect to procumbent, massive and covered with the persistent winged stipe-bases, surrounded with thick brown roots. Lamina deeply 2-pinnatifid, veins free. Sporangia borne on the under surface of the lower parts of the lower pinnae of undifferentiated fronds.

A monotypic genus extending from S. Africa to Australia and New Zealand.

Todea barbara (L.) Moore, Ind. Fil.: 119 (1857).—Sim, Ferns S. Afr. ed. 2: 309, t. 169 (1915). TAB. **11.** Type from S. Africa.
  Acrostichum barbarum L., Sp. Pl. **2**: 1072 (1753). Type as above.
  Osmunda totta Sw. in Schrad., Journ. Bot. **1800**, 2: 105 (1801) *nom. illegit.* Type from S. Africa.
  Todea africana Willd. in Nov. Act. Acad. Erfurt, **2**: 14, t. 3 fig. 1 (1802) *nom. illegit.* Type from S. Africa

Rhizome erect to procumbent, covered with the persistent winged bases of the stipes to form a rhizomatous mass up to 12 cm. in diam. bearing tufted fronds at the apex. Fronds erect, set with a loose rufous tomentum during development but glabrous at maturity. Stipe pale-brown, up to 1·5 cm. in diam. near the base, smooth. Lamina up to 1·1 × 0·44 m., deeply 2-pinnatifid, narrowly oblong in outline; pinnae cultrate to narrowly lanceolate in outline, usually set at an angle of 40–50° to the rhachis, pinnatifid almost to the costa into cultrate minutely serrate acute coriaceous lobes up to 4 × 0·5 cm. with broadened adnate bases. Sporangia borne on the under surface of undifferentiated pinnae usually in the lowest ⅓ of the latter.

Rhodesia. E. Chimanimani Mts., 2.ii.1958, *Mitchell* 278 (BM; BOL; SRGH). **Mozambique.** MS: Musapa Gap, 6.x.1950, *Chase* 3057 (BM; SRGH).

Also in S. Africa and with a subspecies in Australia and New Zealand. Streambanks at higher altitudes.

9. GLEICHENIACEAE

Terrestrial herbs with widely creeping rhizomes, bearing scales or hairs, becoming subglabrous with age. Frond variously compound (in African spp.) as a result of growth or abortion of a terminal bud between a pair of lateral branches or of pinnae (false dichotomy). Venation free. Sporangia grouped in small sori without indusia; annulus complete, oblique; dehiscence by a vertical slit. Gametophyte green, cordate, becoming elongate with a thicker central rib.

Rhizome with lanceolate or deeply laciniate scales resembling stellate hairs in appearance; fronds without a pair of distinct stipule-like pinnae flanking the primary bifurcation of the frond - - - - - - - - - - **1. Gleichenia**
Rhizome with multicellular hairs but without scales; fronds with a pair of stipule-like pinnae flanking primary and often secondary bifurcations of the frond
                                                              2. Dicranopteris

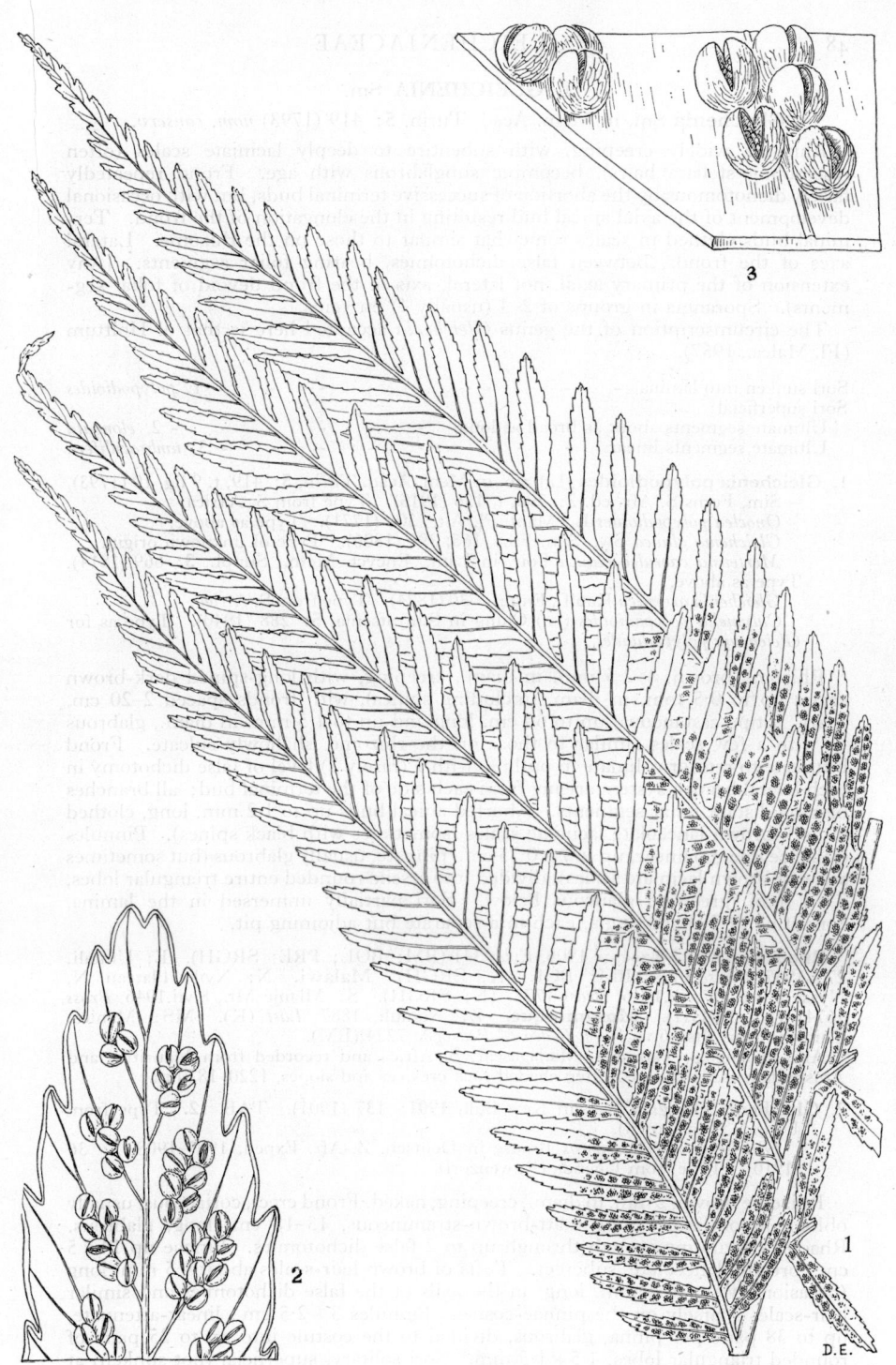

Tab. 11. TODEA BARBARA. 1, habit (× ⅔); 2, enlargement of pinna (×6); 3, enlargement of pinna showing sporangia (×18), all from *Johnson* 222.

1. GLEICHENIA Sm.

Gleichenia Sm. in Mem. Acad. Turin, **5**: 419 (1793) *nom. conserv.*

Rhizome widely creeping, with subentire to deeply laciniate scales (often resembling stellate hairs), becoming subglabrous with age. Frond repeatedly falsely dichotomous by the abortion of successive terminal buds, but with occasional development of the axial apical bud resulting in the elongation of the frond. Terminal buds clothed in scales somewhat similar to those on the rhizome. Lateral axes of the fronds, between false dichotomies, bearing foliar segments. (Any extension of the primary axial, not lateral, axis of the frond devoid of foliar segments). Sporangia in groups of 2–4 (usually 4) on veins.

The circumscription of the genus *Gleichenia* accepted here is that of Holttum (Fl. Males., 1957).

Sori sunken into lamina - - - - - - - - - 1. *polypodioides*
Sori superficial:
 Ultimate segments about as broad as long - - - - - - 2. *elongata*
 Ultimate segments linear - - - - - - - - 3. *umbraculifera*

1. **Gleichenia polypodioides** (L.) Sm. in Mem. Acad. Turin, **5**: 419, t. 9 fig. 10 (1793).
 —Sim, Ferns S. Afr. ed. 2: 296, t. 156 (1915). Type from S. Africa.
  Onoclea polypodioides L., Mant. Pl. Alt.: 306 (1771). Type as above.
  Gleichenia glauca Sw., Syn. Fil.: 165, 393 (1806). Type of unknown origin.
  Mertensia caeruleo-glauca Poir. in Lam., Encycl. Méth., Suppl., **3**: 669 (1814). Type as above.
  Gleichenia argentea Kaulf., Enum.: 36 (1824). Type from S. Africa.
  Calymella polypodioides (L.) Ching in Sunyatsenia, **5**: 288 (1940). Type as for *Gleichenia polypodioides*.

Rhizome brown, 1–2·5 mm. in diam., creeping, with long-spined dark-brown scales up to 0·5 mm. in diam. (including spines), with fronds spaced 2–20 cm. apart. Stipe castaneous, up to 60 cm. long and up to 1·5 mm. in diam., glabrous or with a few scales similar to those on the rhizome, shallowly sulcate. Frond bifurcate to reniform-lunate in outline, with 1 (rarely 2) level of false dichotomy in each lateral branch system arising from each side of the terminal bud; all branches bearing distant foliar segments. Aborted apical buds up to 1·2 mm. long, clothed in dark-brown lanceolate laciniate scales (sometimes with black spines). Pinnules (foliar segments) linear, up to 7 × 0·75 cm., pinnate, usually glabrous (but sometimes set with brown laciniate scales), divided into sessile rounded entire triangular lobes, 3 × 2 mm., green to glaucous below. Sori partially immersed in the lamina, consisting of 2–4 sporangia, each in a separate but adjoining pit.

Rhodesia. N: Mazoe, iv.1906, *Eyles* 341 (BM; BOL; PRE; SRGH). E: Umtali. 27.x.1957, *Chase* 6734 (BM; BOL; K; SRGH). **Malawi.** N: Nyika Plateau, N, Nyasa Distr., 17.viii.1946, *Brass* 17277 (K; SRGH). S: Mlanje Mt., 8.vii.1946, *Brass* 16741 (K; SRGH). **Mozambique.** Z: Namúli, 1887, *Last* (K). MS: Manica, Chimanimani Mts., 6.vii.1949, *Pedro & Pedrógão* 7224 (BM).
 Also in Angola, Tanzania, Madagascar, S. Africa and recorded from Mauritius and Amsterdam Is. Sheltered often shaded rock crevices and slopes, 1220–1870 m.

2. **Gleichenia elongata** Bak. in Kew Bull. **1901**: 137 (1901). TAB. **12.** Type from Uganda (Ruwenzori).
  Gleichenia ruwenzoriensis Brause in Deutsch. Z.-Afr. Exped. 1907–1908, **2**: 36 (1910). Type from Uganda (Ruwenzori).

Rhizome brown, 2 mm. in diam., creeping, naked. Frond erect, coriaceous, usually oblong in outline. Stipe matt-brown-stramineous, 13–18 cm. long, glabrous. Rhachis continuing growth through up to 2 false dichotomies. Pinnae up 14 × 5 cm., broadly lanceolate, suberect. Tufts of brown hair-scales about 1·5 mm. long (occasionally up to 4 mm. long) in the axils of the false dichotomies and similar hair-scales sparsely on the pinnae-costae. Pinnules 3 × 2·5 cm., linear-attenuate, up to 38 pairs per pinna, glabrous, divided to the costule into up to 25 pairs of rounded triangular lobes, 1·5 × 1·5 mm. Sori solitary, superficial (not sunken) at the end of simple veins, consisting of 6–14 sporangia.

Malawi. N: outside Nyika Juniper Reserve, xii.1952, *Chapman* 44 (BM).
 Also from Ruwenzori. The only specimen from our area is recorded from along the banks of a small stream, at high altitude.

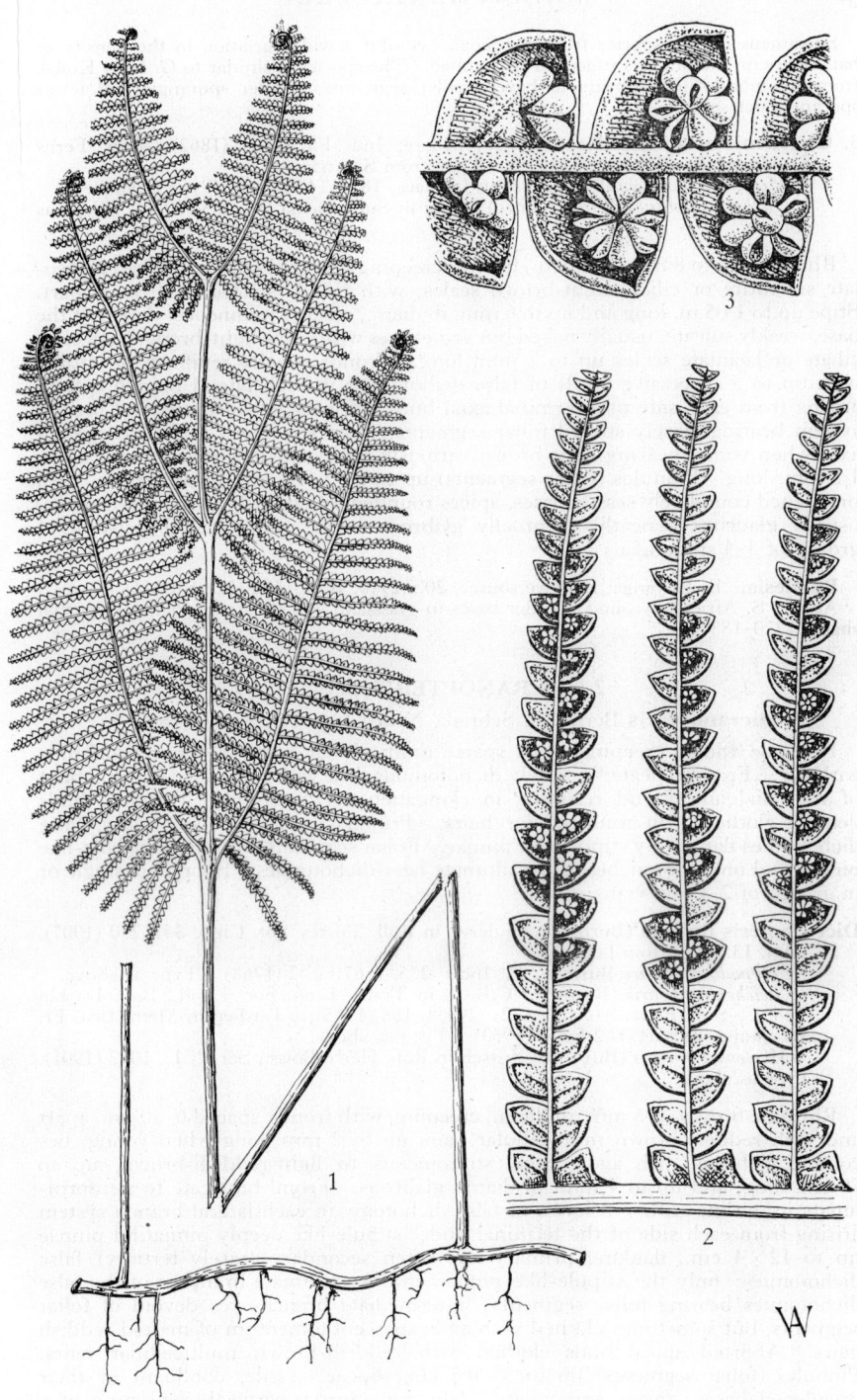

Tab. 12. GLEICHENIA ELONGATA. 1, habit ($\times \frac{2}{3}$); 2, pinnules ($\times 4$); 3, sporangia ($\times 14$), all from *Chapman* 44.

Specimens of this species from Ruwenzori exhibit a wide variation in the density of hair-scales on the under surfaces of the pinnae. The species is similar to *G. boryi* Kunze from Réunion which appears to have smaller segments, smaller sporangia and fewer sporangia per sorus.

3. **Gleichenia umbraculifera** (Kunze) Moore, Ind. Fil.: 384 (1862).—Sim, Ferns S. Afr. ed. 2: 298, t. 157 (1915). Type from S. Africa.
   Mertensia umbraculifera Kunze in Linnaea, **18**: 114 (1844). Type as above.
   Sticherus umbraculiferus (Kunze) Ching in Sunyatsenia, **5**: 285 (1940). Type as above.

Rhizome up to 8 mm. in diam., brown, creeping, covered with appressed lanceolate subentire or ciliate light-brown scales, with fronds spaced 2–10 cm. apart. Stipe up to 1·05 m. long and up to 6 mm. in diam., erect, stramineous, darker at the base, weakly sulcate, usually naked but sometimes with a few light-brown lanceolate ciliate or laciniate scales up to 2 mm. long. Fronds usually reniform in outline with up to 5 successive levels of false dichotomies in each lateral branch system arising from each side of a terminal axial bud; all branches of the lateral branch system bearing closely spaced foliar segments. Aborted apical buds and branch axes when young bearing light-brown variously laciniated lanceolate scales up to 1·5 mm. long. Pinnules (foliar segments) up to 3·5 × 0·3 cm., linear, entire, with broadened completely sessile bases, apices rounded to narrowly acute, green above, usually glaucous beneath, eventually glabrous. Sori superficial, consisting of groups of 3–4 sporangia.

Rhodesia. E: Inyanga, Pungwe source, 20.x.1946, *Wild* 1475 (K; SRGH).
Also in S. Africa. Around boulder bases in grassland and on stream-banks in partial shade, 1220–1830 m.

2. DICRANOPTERIS Bernh.

Dicranopteris Bernh. in Schrad., Neues Journ. **1**, 2: 26 (1806).

Rhizome widely creeping, with sparse multicellular hairs, becoming glabrous with age. Frond repeatedly falsely dichotomous, but with occasional development of the axial apical bud resulting in elongation of the frond. Terminal buds densely clothed with multicellular hairs. Primary, and often secondary, false dichotomies flanked by stipule-like pinnae. Foliar segments only on the stipule-like pinnae and on the branches of the ultimate false dichotomies. Sporangia single or in groups of 2–15 on veins.

Dicranopteris linearis (Burm. f.) Underw. in Bull. Torrey Bot. Club, **34**: 250 (1907). TAB. **13**. Type from Java.
   Polypodium lineare Burm. f., Fl. Ind.: 235, t. 67 fig. 2 (1768). Type as above.
   Gleichenia linearis (Burm. f.) C.B.Cl. in Trans. Linn. Soc. Lond., Bot., **1**: 428 (1880).—Sim, Ferns S. Afr. ed. 2: 299, t. 158 (1915);—Tardieu in Mém. Inst. Fr. Afr. Noire, **28**: 34, t. 2 fig. 6 (1953). Type as above.
   Mertensia linearis (Burm. f.) Fritsch in Bull. Herb. Boiss., Sér. 2, **1**: 1092 (1901). Type as above.

Rhizome brown, 1–5 mm. in diam., creeping, with fronds spaced 6–20 cm. apart and with reddish-brown multicellular hairs up to 2 mm. long when young, becoming glabrous with age. Stipe stramineous to light-reddish-brown, up to 70 cm. long, and up to 4 mm. in diam., glabrous. Frond bifurcate to reniform-lunate in outline with 1–3 levels of false dichotomy in each lateral branch system arising from each side of the terminal bud; stipule-like deeply pinnatifid pinnae up to 12 × 4 cm., flanking primary and often secondary (rarely tertiary) false dichotomies; only the stipule-like pinnae and the ultimate branches of the false dichotomies bearing foliar segments; intermediate branch-axes devoid of foliar segments, but sometimes clothed with an evanescent tomentum of matted reddish hairs. Aborted apical buds clothed with reddish-brown multicellular hairs. Pinnules (foliar segments) up to 3 × 0·5 cm., linear, sessile, confluent at their broadened bases, apices emarginate, glabrous. Sori superficial, consisting of a single sporangium or groups of 2–10 sporangia.

Zambia. N: Luapula Distr., Mbereshi, 15.i.1960, *Richards* 12361 (K). W: Mwinilunga, Zambezi R., 22.ix.1952, *Angus* 520 (K). **Rhodesia.** E: Inyanga, Aberfoyle Tea

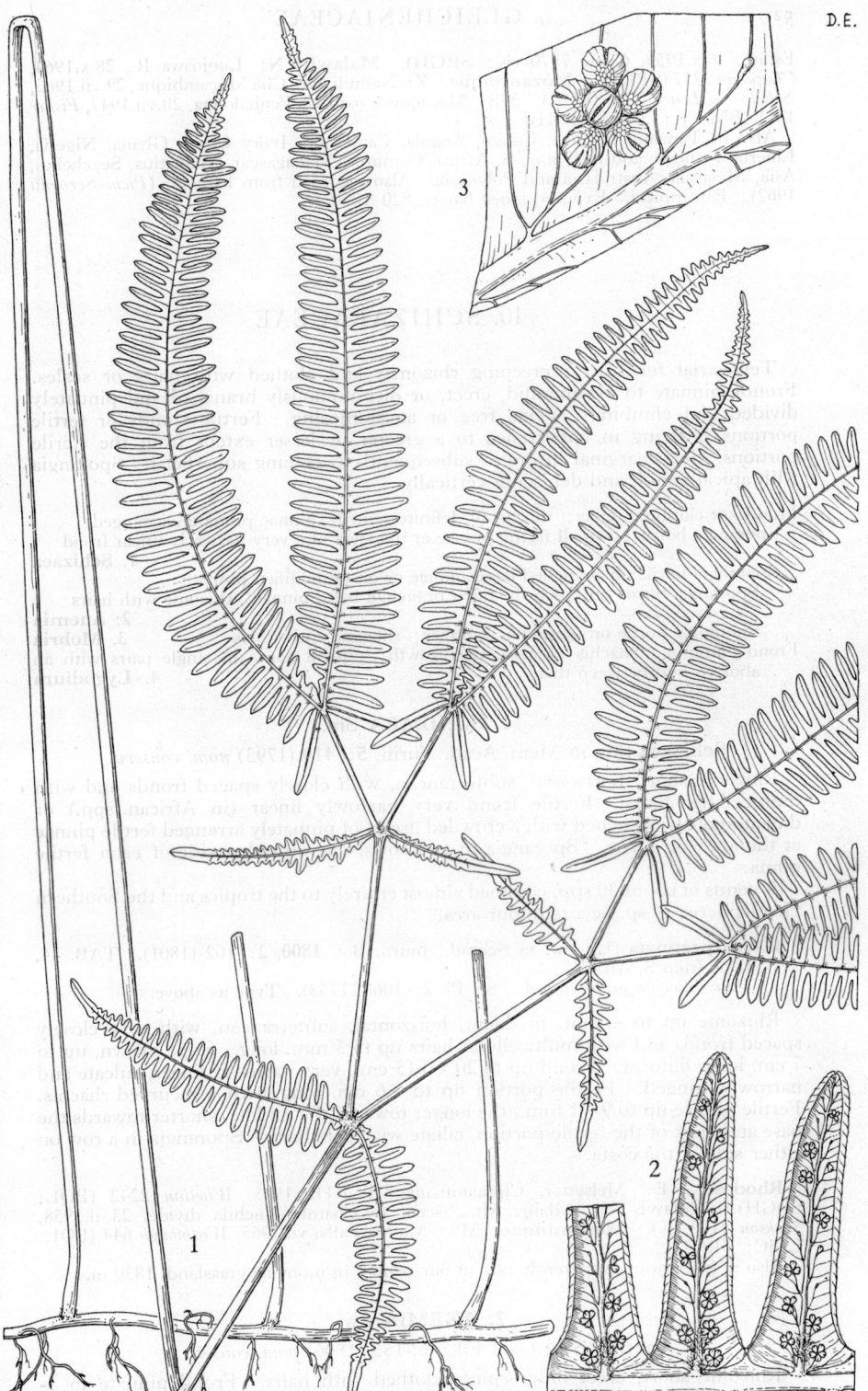

D.E.

Tab. 13. DICRANOPTERIS LINEARIS. 1, habit (×½); 2, section of pinnules (×3); 3, sori
(×15), all from *Angus* 812.

Estate, 26.x.1959, *Chase* 7870 (K; SRGH). **Malawi.** N: Lonjoswa R., 28.x.1964, *Chapman* 2277 (SRGH). **Mozambique.** Z: Namúli Mt., Chà Moçambique, 29.vii.1962, *Schelpe & Leach* 7078 (BOL). MS: Macequece road via Penhalonga, 20.vii.1947, *Fisher* 1351 (BM; K; NU; SRGH).

Also in Tanzania, Kenya, Congo, Angola, Cameroon, Ivory Coast, Ghana, Nigeria, Liberia, Príncipe and S. Tomé, S. Africa, Comoros, Madagascar, Mauritius, Seychelles, Asia, Malaysia, Australasia and Polynesia. Also recorded from Ethiopia (*Pichi-Sermolli* 1962). Pantropical. Exposed moist banks, 820–1050 m.

10. SCHIZAEACEAE

Terrestrial ferns with creeping rhizomes and clothed with hairs or scales. Fronds pinnate to 3-pinnatifid, erect, or dichotomously branched and pinnately divided and climbing. Veins free or anastomosing. Fertile fronds or fertile portions differing in appearance to a greater or lesser extent from the sterile portions. Sori marginal in origin subsequently becoming superficial. Sporangia with apical annuli and dehiscing vertically.

Frond not climbing, erect; rhachis of definite growth; pinnae pinnately arranged:
Sporangia borne on small fertile pinnae at the apex of a very narrowly linear frond
1. **Schizaea**
Sporangia borne on branched basal pinnae or on unmodified pinnae:
Sporangia borne on much modified branched basal pinnae; rhizome with hairs
2. **Anemia**
Sporangia borne on unmodified pinnae; rhizome with scales   -   3. **Mohria**
Frond climbing; rhachis of indefinite growth; pinnae borne in single pairs with an aborted bud between them   -   -   -   -   -   -   -   - 4. **Lygodium**

1. SCHIZAEA Sm.

Schizaea Sm. in Mem. Acad. Turin, **5**: 419 (1793) *nom. conserv.*

Rhizome short, horizontal, subterranean, with closely spaced fronds and with multicellular hairs. Fertile frond very narrowly linear (in African spp.) or dichotomously branched with a crowded group of pinnately arranged fertile pinnae at the apex or apices. Sporangia in two rows, one on either side of each fertile pinna.

A genus of about 30 spp. confined almost entirely to the tropics and the Southern Hemisphere. 1 sp. occurs in our area.

Schizaea pectinata (L.) Sw. in Schrad., Journ. Bot. **1800**, 2: 102 (1801). TAB. **14**.
Type from S. Africa.
Acrostichum pectinatum L., Sp. Pl. **2**: 1068 (1753). Type as above.

Rhizome up to 4 mm. in diam., horizontal, subterranean, with very closely spaced fronds and with multicellular hairs up to 5 mm. long. Stipe brown, up to 7 cm. long, filiform. Frond up to 20 × 0·15 cm., very narrowly linear, sulcate and narrowly winged. Fertile portion up to 1·6 cm. long, with a recurved rhachis. Fertile pinnae up to 9 × 1 mm., the longer towards the middle, shorter towards the base and apex of the fertile portion, ciliate with pale hairs. Sporangia in a row on either side of the costa.

Rhodesia. E: Melsetter, Chimanimani Mts., 11.v.1965, *Whellan* 2242 (BOL; SRGH). **Malawi.** S: Mlanje Mt., ascent to Chambe-Tuchila divide, 23.iii.1958, *Jackson* 2178 (K). **Mozambique.** MS: Martin Falls, vii.1965, *Watmough* 644 (BOL; LISC; SRGH).
Also in S. Africa. Apparently rare in our region, in montane grassland, 1830 m.

2. ANEMIA Sw.

Anemia Sw., Gen. Fil.: 6, 155 (1806) *nom. conserv.*

Rhizome short, erect or creeping, clothed with hairs. Frond pinnate to 2-pinnate, with the basal pair of pinnae usually fertile. Fertile pinnae longl-

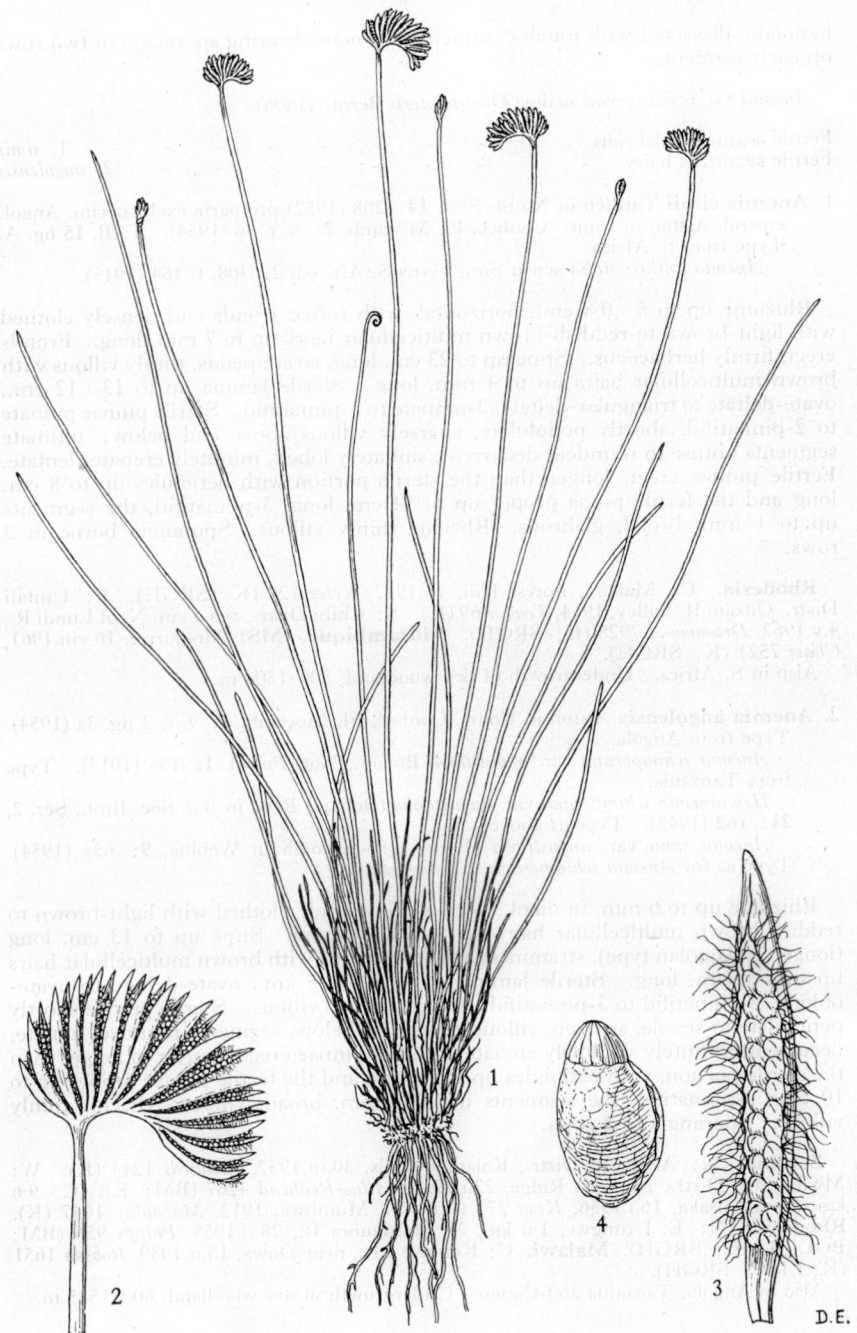

D.E.

Tab. 14. SCHIZAEA PECTINATA. 1, habit (×⅘); 2, fertile lamina (×2); 3, fertile pinna (×10); 4, sporangium (×50), all from *Jackson* 2178.

petiolate, dissected with much contracted segments bearing sporangia in two rows on each segment.

Anemia Sw. is conserved against *Ornithopteris* Bernh. (1806).

Fertile segments glabrous - - - - - - - - - 1. *simii*
Fertile segments hairy - - - - - - - - - 2. *angolensis*

1. **Anemia simii** Tardieu in Notul. Syst. **14**: 208 (1952) pro parte excl. specim. Angol. emend. Alston in Contr. Conhec. Fl. Moçamb. **2**: 8, t. 36 (1954). TAB. **15** fig. A. Type from S. Africa.
    Anemia anthriscifolia sensu Sim, Ferns S. Afr. ed. 2: 308, t. 164 (1915).

Rhizome up to 5 × 0·8 cm., horizontal, with tufted fronds and densely clothed with light-brown to reddish-brown multicellular hairs up to 7 mm. long. Fronds erect, firmly herbaceous. Stipe up to 23 cm. long, stramineous, thinly villous with brown multicellular hairs up to 4 mm. long. Sterile lamina up to 13 × 12 cm., ovate-deltate to triangular-deltate, 2-pinnate to 3-pinnatifid. Sterile pinnae pinnate to 2-pinnatifid, shortly petiolulate, sparsely villous above and below; ultimate segments obtuse to rounded, decurrent, sinuately lobed, minutely crenate-dentate. Fertile pinnae erect, longer than the sterile portion with petiolules up to 8 cm. long and the fertile pinna proper up to 14 cm. long, 3-pinnatifid, the segments up to 1 mm. broad, glabrous. Rhachis thinly villous. Sporangia borne in 2 rows.

Rhodesia. C: Makoni, Forest Hill, vii.1917, *Eyles* 726 (K; SRGH). E: Umtali Distr., Odzani R. Valley, 1914, *Teague* 69 (K). S: Chibi Distr., c. 6·4 km. N. of Lundi R., 4.v.1962, *Drummond* 7928 (K; SRGH). **Mozambique**. MS: Mossurize, 16.viii.1961, *Chase* 7521 (K; SRGH).
Also in S. Africa. Undergrowth of dry woodland, 500–1500 m.

2. **Anemia angolensis** Alston in Contr. Conhec. Fl. Moçamb. **2**: 9, t. 2 fig. 3a (1954). Type from Angola.
    Anemia schimperana var. *angustiloba* Bonap., Not. Ptérid. **1**: 133 (1915). Type from Tanzania.
    Hemianemia schimperana var. *angustiloba* (Bonap.) Reed in Bol. Soc. Brot., Sér. 2, **21**: 162 (1948). Type as above.
    Anemia simii var. *angustiloba* (Bonap.) P.-Sermolli in Webbia, **9**: 654 (1954). Type as for *Anemia schimperana* var. *angustiloba*.

Rhizome up to 6 mm. in diam., horizontal, densely clothed with light-brown to reddish-brown multicellular hairs up to 5 mm. long. Stipe up to 13 cm. long (longer in Angolan type), stramineous, thinly villous with brown multicellular hairs up to 2·5 mm. long. Sterile lamina up to 11 × 7·5 cm., ovate-deltate to ovate-oblong, 2-pinnatifid to 3-pinnatifid; rhachis thinly villous. Sterile pinnae shortly petiolulate to sessile, sparsely villous above and below, segments rounded, adnate, decurrent, minutely shallowly crenate. Fertile pinnae erect, shorter to longer than the sterile portion, with petiolules up to 5·5 cm. and the fertile pinnae proper up to 10 cm., 3-pinnatifid, the segments up to 1 mm. broad, villous; rhachis thinly villous. Sporangia in 2 rows.

Zambia. N: Abercorn Distr., Kalambo Falls, 30.iii.1952, *Richards* 1244 (K). W: Mwinilunga Distr., Katunda Ridge, 22.i.1938, *Milne-Redhead* 4267 (BM; K). C: 9·6 km. E. of Lusaka, 16.i.1956, *King* 271 (K). S: Mumbwa, 1912, *Macaulay* 1047 (K). **Rhodesia.** N: E. Urungwe, 1·6 km. N. of Mauora R., 28.ii.1958, *Phipps* 956 (BM; BOL; LISC; SRGH). **Malawi.** C: Kongwe Mt., near Dowa, 18.ii.1959, *Robson* 1651 (K; LISC; SRGH).
Also in Angola, Tanzania and Congo. Undergrowth of dry woodland, 600–1525 m.

3. MOHRIA Sw.
Mohria Sw., Syn. Fil.: 5, 159 (1806).

Rhizome short, horizontal, clothed with brown scales. Frond 2-pinnatifid to 3-pinnatifid, not or slightly dimorphic. Sori of few sessile submarginal sporangia partly covered by the reflexed margins of the ultimate segments.

A genus of 2 spp. confined to southern and SE. tropical Africa, Madagascar and

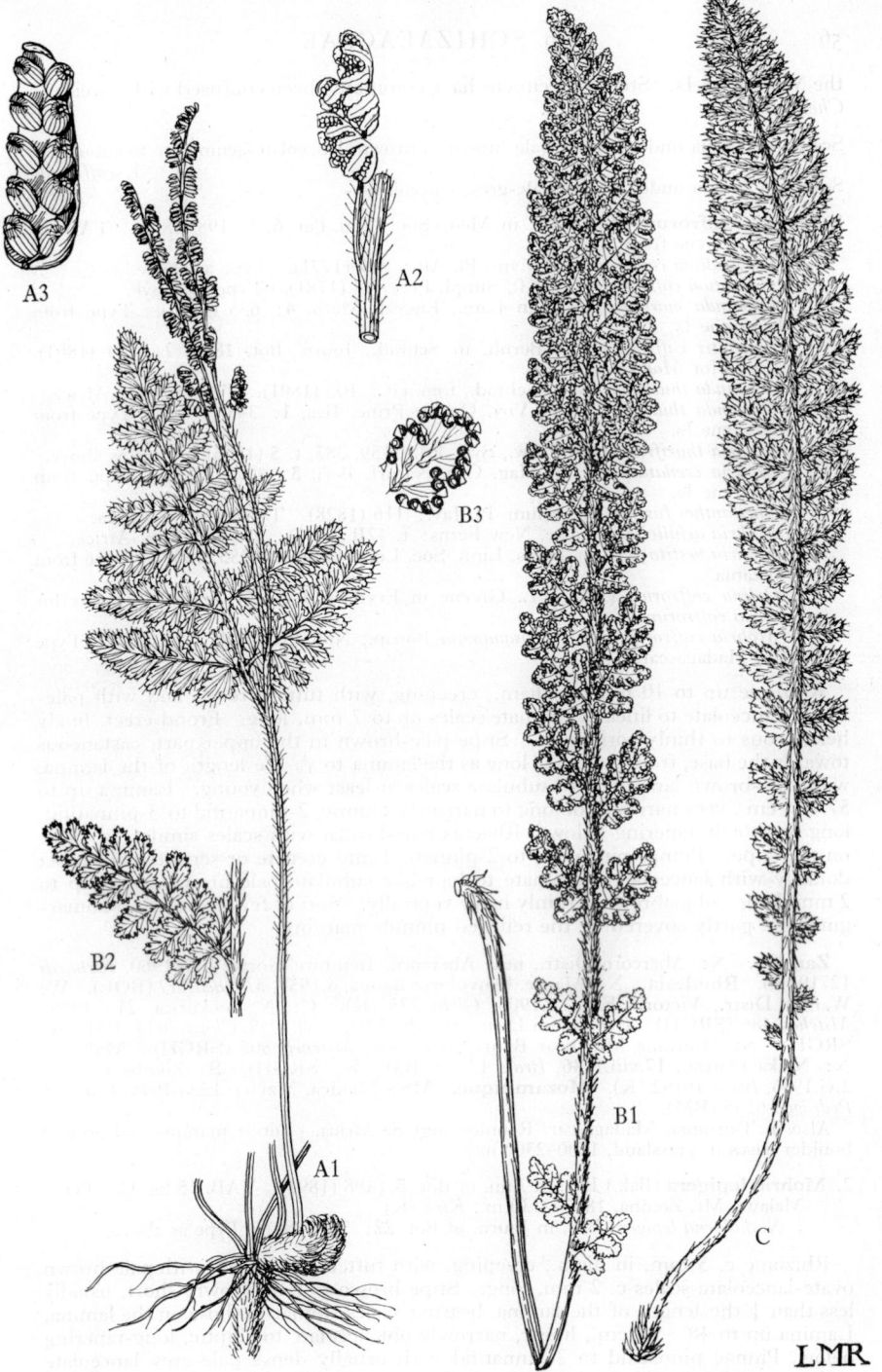

Tab. 15. A.—ANEMIA SIMII. A1, habit (×⅔) *Eyles* 726. A2, fertile pinna (×2); A3, sporangia (×10). A2–A3 from *Pedro & Pedrógão* 6668. B.—MOHRIA CAFFRORUM. B1, fertile frond (×⅔) *Schelpe* 4010. B2, fertile pinna (×⅔); B3, segment of fertile pinna (×2). B2–B3 both from *Schelpe* 5654. C.—MOHRIA LEPIGERA. Sterile frond (×⅔) *Chase* 4917.

the Mascarene Is.   Sterile specimens have commonly been confused with species of *Cheilanthes*.

Scales on lamina under surface pale-brown to brown, lanceolate-acuminate to subulate
                                                                              1. *caffrorum*
Scales on lamina under surface pale-grey, lanceolate -   -   -   -   2. *lepigera*

1. **Mohria caffrorum** (L.) Desv. in Mém. Soc. Linn. Par. **6**, 2: 198 (1827).   TAB. **15** fig. B.   Type from S. Africa.
    Polypodium caffrorum L., Mant. Pl. Alt.: 307 (1771).   Type as above.
    Adiantum caffrorum (L.) L.f., Suppl. Pl.: 447 (1781).   Type as above.
    Osmunda marginalis Sav. in Lam., Encycl. Méth. **4**: 655 (1797).   Type from Mascarene Is.
    Lonchitis caffrorum (L.) Bernh. in Schrad., Journ. Bot. **1800**, 2: 124 (1801). Type as for *Mohria caffrorum*.
    Osmunda thurifera Sw. in Schrad., tom. cit.: 105 (1801).   Type from S. Africa.
    Osmunda thurifraga Bory, Voy. Quatre Princ. Iles, **1**: 348 (1804).   Type from Mascarene Is.
    Mohria thurifraga (Bory) Sw., Syn. Fil.: 159, 385, t. 5 (1806).   Type as above.
    Mohria crenata Desv. in Mag. Ges. Naturf. Berl. **5**: 307 (1811).   Type from Mascarene Is.
    Cheilanthes fuscata Bl., Enum. Pl. Jav.: 116 (1828).   Type from S. Africa.
    Mohria achilleifolia Lowe, New Ferns: t. 42B (1862).   Type from S. Africa.
    Mohria vestita Bak. in Trans. Linn. Soc. Lond., Bot., **2**: 355 (1887).   Type from Tanzania.
    Colina caffrorum (L.) E. L. Greene in Erythea, **1**: 247 (1893).   Type as for *Mohria caffrorum*.
    Mohria caffrorum var. *multisquamosa* Bonap., Not. Ptérid. **4**: 85 (1917).   Type from Madagascar.

Rhizome up to 10 mm. in diam., creeping, with tufted fronds and with pale-brown lanceolate to linear acuminate scales up to 7 mm. long.   Frond erect, finely herbaceous to thinly coriaceous.   Stipe pale-brown in the upper part, castaneous towards the base, from almost as long as the lamina to $\frac{1}{10}$ the length of the lamina, with pale-brown lanceolate to subulate scales at least when young.   Lamina up to 57×11 cm., very narrowly oblong to narrowly elliptic, 2-pinnatifid to 3-pinnatifid, long or shortly tapering below.   Rhachis pale-brown with scales similar to those on the stipe.   Pinnae pinnatifid to 2-pinnatifid into crenate or serrate lobes beset dorsally with lanceolate-acuminate to hair-like subulate pale-brown scales up to 2 mm. long, subglabrous or thinly hairy ventrally.   Sori of few sporangia, submarginal and partly covered by the reflexed pinnule margins.

Zambia.   N: Abercorn Distr., near Abercorn, Itembwe Gorge, 6.iii.1960, *Richards* 12719 (K).   **Rhodesia.**   N: Mazoe, Umvukwes Range, ii.1957, *Mitchell* 117 (BOL).   W: Wankie Distr., Victoria Falls, x.1905, *Gibbs* 235 (K).   C: Ngomokurira, 21.ii.1959, *Mitchell* 486 (SRGH).   E: Umtali Distr., Vumba Mts., 1.vi.1958, *Chase* 7033 (BM; K; SRGH).   S: Hunyana Mt., near Bikita, 22.ii.1964, *Mitchell* 805 (SRGH).   **Malawi.**, N: Nyika Plateau, 17.viii.1946, *Brass* 17286 (BM; K; SRGH).   S: Zomba Plateau, 2.vi.1946, *Brass* 16162 (K).   **Mozambique.**   MS: Manica, Mavita, 15.vi.1949, *Pedro & Pedrógão* 6555 (BM).
    Also in Tanzania, Madagascar, Réunion and S. Africa.   Forest margins and around boulder bases in grassland, 1500–2300 m.

2. **Mohria lepigera** (Bak.) Bak. in Ann. of Bot. **5**: 498 (1891).   TAB. **15** fig. C.   Type: Malawi, Mt. Zomba, 1830–2130 m., *Kirk* (K).
    Notholaena lepigera Bak. in Journ. of Bot. **22**: 53 (1884).   Type as above.

Rhizome c. 5 mm. in diam., creeping, with tufted fronds and with pale-brown ovate-lanceolate scales c. 2 mm. long.   Stipe brown to pale-brown, short, usually less than $\frac{1}{4}$ the length of the lamina, bearing scales similar to those on the lamina. Lamina up to 48×5.5 cm., linear, narrowly oblanceolate to elliptic, long-tapering below. Pinnae pinnatifid to 2-pinnatifid with usually dense pale-grey lanceolate scales up to 2.5 mm. long dorsally, subglabrous and with yellow glands ventrally. Sori of few sporangia, submarginal and partly covered by the reflexed pinnule margins.

Zambia.   N: 20 km. E. of Kasama, 23.iii.1961, *Robinson* 4553 (SRGH).   C: Mkushi Distr., Fiwila, 7.i.1958, *Robinson* 2673 (K; SRGH).   **Rhodesia.**   N: Mt. Binga, 29.i.1966, *Pereira, Sarmento & Marques* 23 (LMU).   W: Bulawayo, iii.1902, *Eyles* 1042 (GRA;

SRGH). C: Chindamora, iv.1922, *Eyles* 3417 (SRGH). E: Vumba Mts., Castle Beacon, 8.viii.1948, *Fisher* 1639 (K). S: Victoria, near Birchenough Bridge. 2.viii.1951, *Taylor* 3411 (BOL). **Malawi.** S: Mlanje Mt., Luchenya Plateau, 27.vi.1946 *Brass* 16484 (BM; K; SRGH). **Mozambique.** N: Chiradula summit, 1905, *Cameron* 178 (K). Z: Namúli, 1887, *Last* (K). T: Macanga, Furancungo, 15.iii.1966, *Pereira, Sarmento & Marques* 1693 (LMU). MS: Gorongosa Mt., Gogogo Peak, 4.vii.1955, *Schelpe* 5493 (B; BM; BOL; K; P; US).

Endemic to SE. tropical Africa. Boulder bases and rock ledges in grassland, 1370–1890 m.

4. LYGODIUM Sw.

Lygodium Sw. in Schrad., Journ. Bot., **1800,** 2: 106 (1801) *nom. conserv.*

Rhizome horizontal, dichotomous, covered with dark multicellular hairs. Frond climbing, with a slender twisting rhachis. Pinnae (secondary rhachis-branches) borne in pairs along the rhachis on short secondary rhachises ending in an aborted bud between the pinnae. Pinnae pinnate to 3-pinnatifid. Sporangia borne along the margins of pinnule lobes, each sporangium subtended by an indusium.

Lygodium Sw. has been conserved against *Ramondia* Mirb. (1801) and *Ugena* Cav. (1801).

Pinnae (secondary rhachis-branches) pinnate, oblong, pinnules articulated; veins glabrous
1. *microphyllum*
Pinnae 2-pinnate to 3-pinnatifid, triangular, pinnules not articulated; veins hairy
2. *kersteni*

1. **Lygodium microphyllum** (Cav.) R. Br., Prodr. Fl. Nov. Holl. **1**: 162 (1810). TAB. **16** fig. B. Type from Philippine Is.
    Ugena microphylla Cav., Icon. Descr. Pl. **6**: 76, t. 595 fig. 2 (1801). Type as above.
    Lygodium scandens sensu Sim, Ferns S. Afr. ed. 2: 302, t. 161–162 (1915) non (L.) Sw.

Rhizome 3 mm. in diam., subterranean, producing fronds 4–13 cm. apart and with short black multicellular hairs c. 1·5 mm. long. Fronds up to 10 × 0·30 m., climbing, herbaceous. Rhachis up to 2 mm. in diam., twining, matt grey-green, producing secondary rhachises; aborted apical bud densely clothed with brown multicellular hairs 1–4 mm. long. Pinnae (secondary rhachis branches) oblong, pinnate. Sterile pinnules up to 6·2 × 1·8 cm., petiolate, articulated, lanceolate to oblong-lanceolate, base cordate, apex acute to acuminate, glabrous, margin minutely crenate, veins free. Fertile pinnules 1·5–4·5 × 1–1·8 cm. (excluding fertile lobes), broadly lanceolate to oblong, base cordate, apex acute to rounded, glabrous, margin minutely crenate with fertile linear lobes up to 8 × 1 mm., produced at irregular intervals. Fertile lobes bearing up to 25 sporangia in 2 rows.

Zambia. B: Kalabo, 14.x.1963, *Fanshawe* 8075 (K). N: Fort Rosebery Distr., Samfya Mission, 26.viii.1952, *White* 3152 (FHO; K). W: Mwinilunga Distr., Zambezi R., 20.ix.1952, *Holmes* 876 (FHO; K). **Mozambique.** SS: Vila de João Belo, near R. Lumane, 25.viii.1947, *Barbosa* 363 (BOL; LMA).

Also in S. Africa and tropical Africa, Asia and America. Twining climber in moist forest, up to 1035 m.

2. **Lygodium kerstenii** Kuhn, Fil. Deck.: 28 (1867).—Sim, Ferns S. Afr. ed. 2: t. 163 (1915). TAB. **16** fig. A. Syntypes from Madagascar and Kenya.
    Lygodium subalatum Boj. ex Kuhn, Fil. Afr.: 170 (1868). Type from Comoro Is.
    Lygodium brycei Bak. in Kew Bull. **1901**: 138 (1901). Type: Rhodesia, Mashonaland, near the Portuguese boundary, drift of Renio R., *Bryce* (K).

Rhizome up to 4 mm. in diam., creeping, with dark-brown to black multicellular hairs c. 1·5 cm. long. Fronds up to 20 × 0·48 m., climbing. Rhachis up to 3 mm. in diam., twining, matt-pale-brown to greyish-green, producing secondary rhachises up to 10 mm. long, at intervals of 7–20 cm. each with an aborted apical bud densely clothed with brown multicellular hairs, 1·5 mm. long. Pinnae (secondary rhachis-branches) triangular, 2-pinnate to 3-pinnatifid. Secondary and tertiary rhachises with narrowly winged, petiolate, not articulated, lanceolate-oblong to lanceolate, simple to 2-pinnatifid sterile pinnules up to 11 cm. long, the

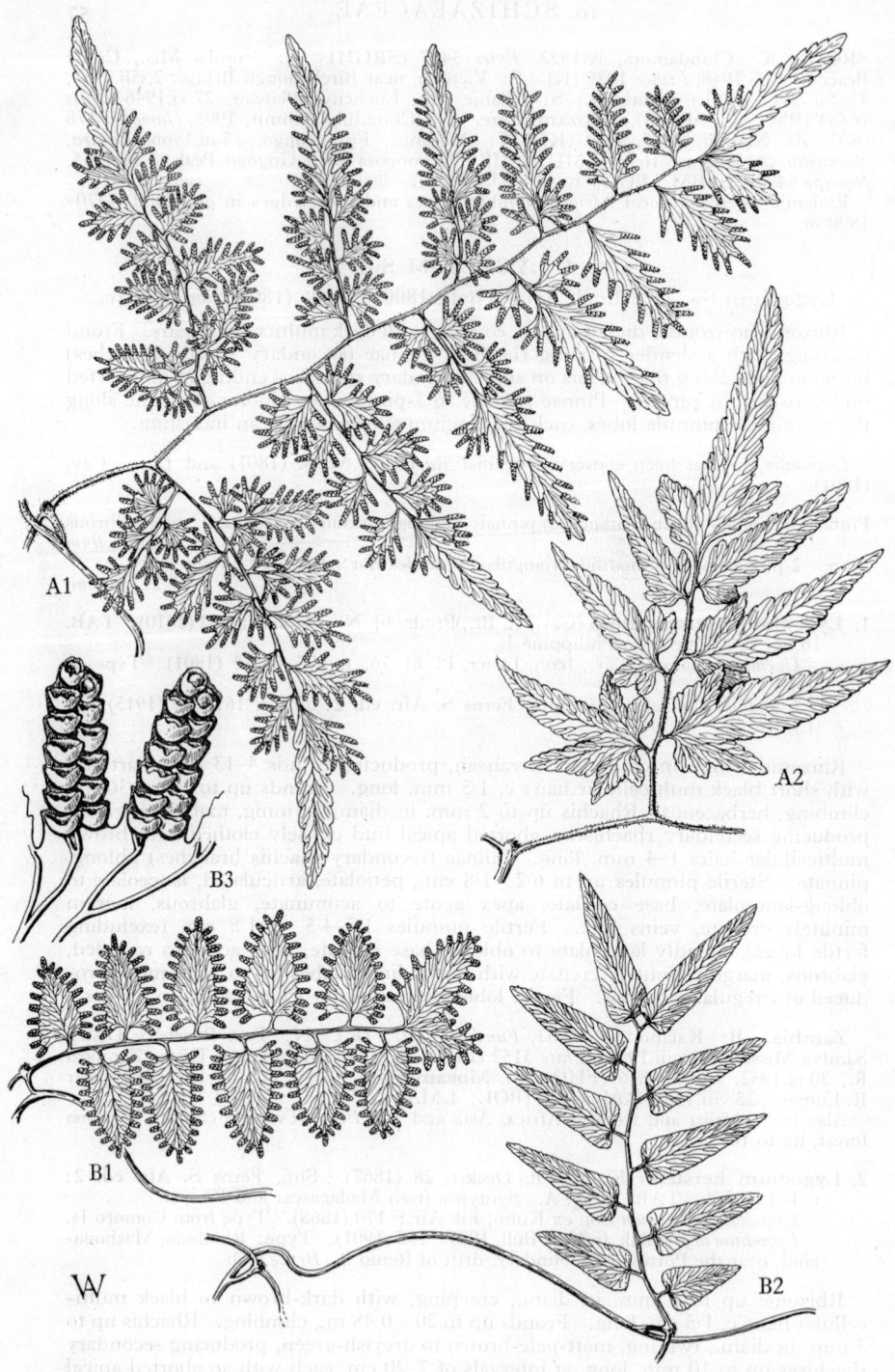

Tab. 16. A.—LYGODIUM KERSTENII. A1, fertile pinna (×⅔), *Mendonça* 1415; A2, sterile pinnule (×⅔) *Fisher & Schweickerdt* 225. B.—LYGODIUM MICROPHYLLUM. B1, fertile pinna (×⅔); B2, sterile pinna (×⅔); B3, sporangia (×6), all from *White* 3152.

ultimate segments usually with prominent basal lobes, apices broadly acute, margins both coarsely crenate and with crenations finely crenate, veins hairy on both surfaces. Fertile pinnules up to 9 cm. long, lanceolate, simple to 2-pinnate, usually with a long apical segment and shorter basal segments, veins hairy on both surfaces, margin crenate but produced into numerous linear fertile lobes up to 6 × 1 mm., often at irregular intervals. Fertile lobes bearing up to 20 sporangia in 2 rows.

Zambia. W: Mwinilunga, Lisombo R., 9.vi.1963, *Loveridge* 877 (BM; LISC). **Rhodesia.** E: Vumba Mts., Hoboken, 8.viii.1958, *Chase* 6975 (BM; K; LISC; SRGH). **Mozambique.** MS: Lower R. Zona, Jihu, xi.1905, *Swynnerton* 208 (BM; K; PRE).
Also in S. Africa, Madagascar and Comoro Is. Twining climber in gallery forest, up to 1280 m. alt.

11. MARSILEACEAE

By E. Launert

Small aquatic or semi-aquatic ferns with a creeping, branched, solenostelic rhizome. Fronds circinate when young, simple or with 2 or 4 opposite leaflets borne terminally on a long stipe; veins dichotomously branched, anastomosing. Sporangia contained in closed sporocarps which are inserted on short pedicels on the stipe (usually at the very base). Spores of 2 kinds: solitary megaspores and numerous microspores (contained in megasporangia and microsporangia respectively).

MARSILEA L.

Marsilea L., Sp. Pl. **2**: 1099 (1753); Gen. Pl., ed. 5: 485 (1754).

Fronds with 4 leaflets in a terminal cluster, arranged symmetrically cross-wise at the apex of the stipe, floating in submerged plants, otherwise erect or decumbent. Leaflets herbaceous, obdeltate to obovate, rarely narrowly deltate, with the outer margin entire, sinuate, crenate or lobate; veins dichotomously branched, anastomosing, often with interstitial suberous streaks on the lower surface (in submerged plants), sometimes with sclerenchymatous interstitial pellucid streaks (*M. coromandelina*). Sori numerous on a gelatinous string-like receptacle attached to the wall of the sporocarp and released in the form of a ring when moistened.

Sporocarps (3) 5–25 in branched clusters; pedicels adnate, dichotomously branched from a common peduncle-like base - - - - - - - - 2. *ephippiocarpa*
Sporocarps solitary or in groups of 2–3 (very rarely 5); pedicels free from each other or just fused at the very base, never branched dichotomously:
  Sporocarps with hairs of 2 kinds, erect cylindric hairs and appressed flattened ones, the cells of the erect hairs collapsed in dried specimens thus giving the sporocarp a farinose appearance - - - - - - - - 7. *farinosa*
  Sporocarps only with laterally attached flattened hairs:
    Sporocarps almost square in lateral view, distally truncate, with a distinct dorsal and frontal furrow and a broader lateral dorsiventral one - - 3. *aegyptiaca*
    Sporocarps longer than high, or if as long as high then never with a broad lateral dorsiventral furrow:
      Pedicels growing horizontally or more often downwards, burying the sporocarps in the ground - - - - - - - - - - 6. *vera*
      Pedicels growing upright or ascending (very rarely a few in a plant may grow horizontally):
        Pedicels 2–3 (rarely 4), united at their base to a varying degree but not branched:
          Sporocarp with both teeth distinct, the superior one usually acute - 1. *minuta*
          Sporocarp with only the superior tooth present and this subobtuse to acute, the inferior one absent or rarely appearing as a shallow hump 8. *apposita*

Pedicels always free from each other (sometimes crowded):
  Sporocarps with both teeth present, often different in size - - 1. *minuta*
  Sporocarps with only the superior tooth developed and this obtuse or acute, the
    inferior one absent or represented as a very shallow hump:
      Sporocarps narrowly elliptic to elliptic or subcircular in dorsiventral cross-
        section (see t. 18, fig. C2 & D2), without dorsal or frontal furrow:
          Sporocarps very small, inflated, pyriform to subglobose, 1·25–2·5 (very
            rarely 3) mm. long, usually crowded - - - 9. *burchellii*
          Sporocarps larger, 2·7–7 mm. long (if slightly smaller than 2·7 mm. then
            not inflated and different in shape from above), usually not crowded:
              Superior tooth of the sporocarps always obtuse, usually very short:
                Sporocarps almost sessile, often deflexed against the 1·4 mm. long
                  pedicel; leaflets very densely villous or tomentose when young, later
                  thinly tomentose, rarely partly glabrous; stipes very robust, up to
                  2·5 mm. wide - - - - - - - - 5. *villifolia*
                Sporocarps distinctly pedicelled (pedicels (3·5) 6–14 mm. long), rarely
                  deflexed; leaflets pilose, usually becoming glabrous with age; stipes
                  slender - - - - - - - - 4. *macrocarpa*
              Superior tooth of the sporocarp acute, rarely subacute, often prominent
                10. *capensis*
      Sporocarps narrowly rectangular or biscoctiform in dorsiventral cross-section
        (see t. 17, fig. D2), usually with a dorsal, frontal and ventral furrow:
          Superior tooth of sporocarp obtuse (if subacute or acute then sporocarps
            1·5–2 times longer than high or broadly elliptic to subcircular in lateral
            view), rather inconspicuous (t. 18, fig. B2):
              Sporocarps almost sessile, often deflexed against the 1·4 mm. long pedicel;
                leaflets very densely villous or tomentose when young, later thinly
                tomentose, rarely partly glabrous; stipes very robust, up to 2·5 mm.
                wide - - - - - - - - - 5. *villifolia*
              Sporocarps distinctly pedicelled (pedicels (3·5) 6–14 mm. long), rarely
                deflexed; leaflets pilose, usually becoming glabrous with age; stipes
                slender - - - - - - - - - 4. *macrocarpa*
          Superior tooth of sporocarp acute, rarely subacute, sporocarps obtusely
            rectangular in lateral view - - - - - 8. *apposita*

1. **Marsilea minuta** L., Mant. Pl. Alt.: 308 (1771).—Launert in Senckenb. Biol. **49**:
291, fig. 32–34, 69 (1968). TAB. **17** fig. B. Type from India.
    Marsilea crenulata Desv., Prodr.: 179 (1827).—Tardieu & Alston in Mém. Inst.
Fr. Afr. Noire, **50**: 12, t. 1 fig. 6 (1957). Type from Mascarene Is.
    Marsilea diffusa Leprieur ex A. Braun in Monatsber. Königl. Akad. Wiss. Berl.
1863: 419 (1864).—Tardieu, Fl. Madag., Ptérid. **10**: 2 (1952).—Alston in
F.W.T.A., ed. 2 Suppl.: 24 (1959). Type from Senegal.
    Marsilea rotundata sensu Wild in Pl. Aquat. Nuis. Afr. et Madag., Proj. con.
CCTA/CSA: 14, t. 11 fig. D (1964) non L. (1771).
    Marsilea tenax A. Peter in Abh. Ges. Wiss. Gött., Math.-phys. Kl. n. F. **13**: 38
(1929); in Fedde, Repert. Beih. **40**: 91 (1929). Type from Tanzania.

Stipes of leaves usually 2–8 (15) cm. long, in floating leaves up to 25 cm. long,
slender, usually glabrous. Leaflets 0·4–2·5 (3) × 0·3–2·25 (2·75) cm., obdeltate to
broadly obdeltate, usually quite glabrous (sparsely pilose in terrestrial forms), with
brownish suberous streaks between the veins of the lower surface of floating leaves;
flanks varying from straight to convex, rarely concave; outer margin round, entire
in floating leaves, entire or sinuate or crenulate in aerial or subaerial ones. Sporocarps (2·8) 3–4 (6) mm. long, 2·4–3 (4·75) mm. high, 0·8–1·4 (2) mm. thick,
usually crowded, more rarely in groups of 2 or 3, very rarely solitary (see also
description of pedicel) extremely variable in size (see tab.), broadly oblong,
broadly elliptic or very rarely subcircular in lateral view, distally always rounded,
elliptic in dorsiventral cross-section, without dorsal or frontal furrow or suture,
densely appressed-pilose when young, usually quite glabrous when mature, dark-
brown to almost black at maturity; lateral ribs usually invisible; sori 8–12; teeth
usually prominent; the inferior shorter than or occasionally as long as or sometimes
even longer than the superior, obtuse to subacute, erect or rarely slightly recurved;
the superior narrowly conical, acute, rarely subacute or obtuse, erect or recurved.
Pedicels (2) 3–7 mm. long, fairly stout, terete, appressed-pilose at first, later becoming glabrous, erect or gently curved upwards, free or united to some extent
with 2 or 3 (rarely 4) others (see tab.), usually inserted at the very base of the
stipe but not infrequently arising separately (or in groups of 2) one above the other
above the base of the stipe (as shown in tab.).

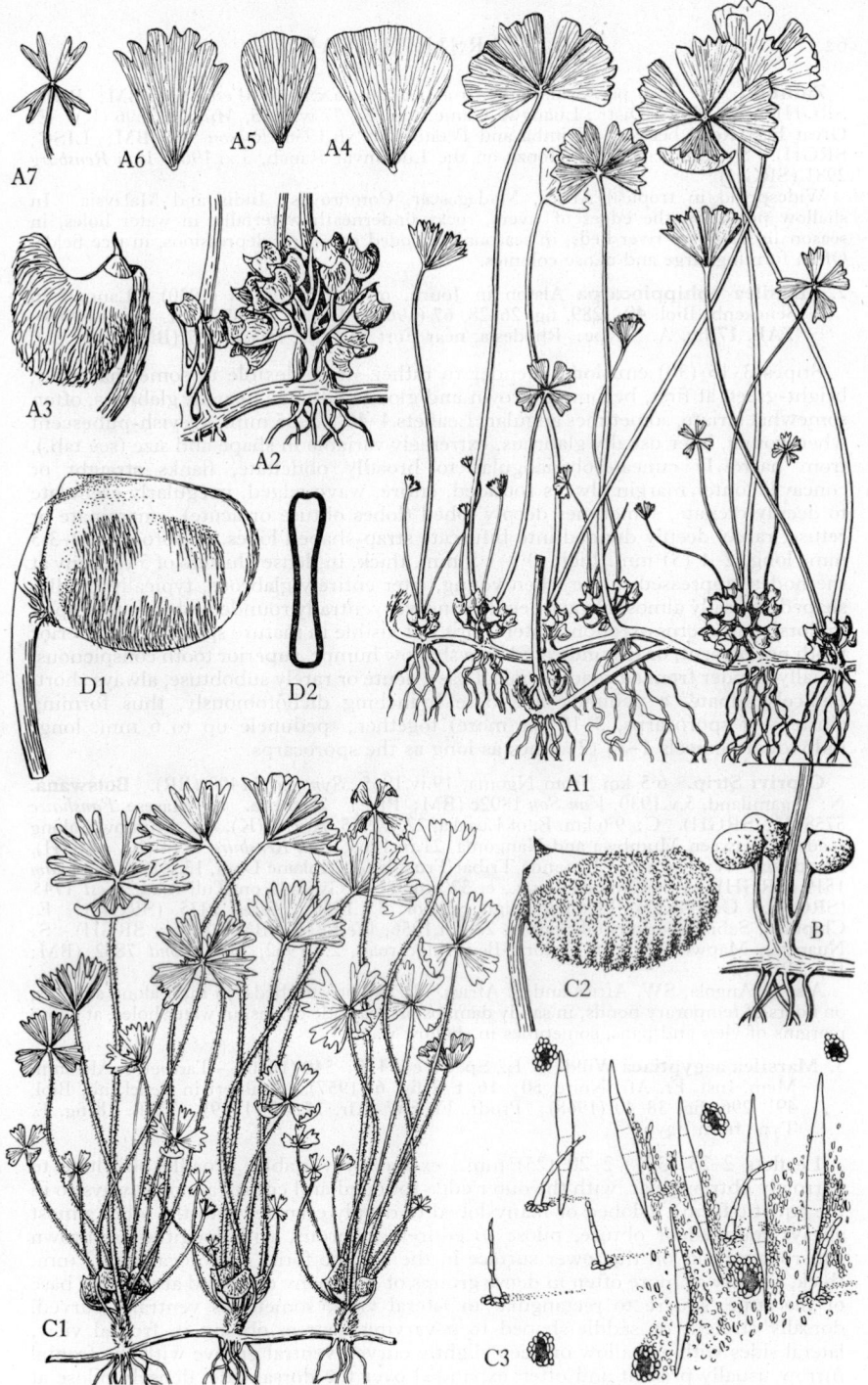

Tab. 17. A.—MARSILEA EPHIPPIOCARPA. A1, habit (× ½); A2, base of stipes showing
arrangement of sporocarps (× 2); A3, sporocarp in lateral view (× 6); A4–A6,
leaflets showing range of variability (× ⅔). A1–A6 from *West 3450*, A7 from *Walter
2211*. B.—MARSILEA MINUTA, base of stipe showing sporocarps (× 2), *Robson
947*. C.—MARSILEA FARINOSA. C1, habit (× ⅔); C2, sporocarps in lateral view (× 6);
C3, section of sporocarp surface showing 2 different types of hair as well as scars left
by fallen-off hairs (× 20), all from *Schönfelder 7688*. D.—MARSILEA MACROCARPA.
D1, sporocarp in lateral view (× 6); D2, same in dorsi-ventral cross-section (× 6),
both from *Kinges 2200*.

Zambia. B: small pan near Zambezi at Lukulu, 6.x.1957, *West* 3520 (BM; BOL; SRGH). N: Mpika Distr., Luangwa Game Reserve, 27.iv.1965, *Mitchell* 2696 (K). E: Great East Road between Nyimba and Petauke, 14.xii.1958, *Robson* 947 (BM; LISC; SRGH). S: Choma Distr., Monze on the Lochinvar Ranch, 3.xi.1964, *Van Rensburg* 2981 (SRGH).

Widespread in tropical Africa, Madagascar, Comoro Is., India and Malaysia. In shallow pools, on the edges of rivers, rocks underneath waterfalls, in water holes, in seasonally dried up river beds, in seasonally flooded grassland depressions, in rice fields. Often forming large and dense colonies.

2. **Marsilea ephippiocarpa** Alston in Journ. of Bot. **68**: 118 (1930).—Launert in Senckenb. Biol. **49**: 289, fig. 26–28, 67 (1968); Prodr. Fl. SW. Afr. **11**: 3 (1969). TAB. **17** fig. A. Type: Rhodesia, near Fort Victoria, *Rendle* 307 (BM, holotype).

Stipes 3–15 (30) cm. long, slender to rather stout, flexible to somewhat rigid, bright-green at first, becoming brown and glossy with age, usually glabrous, often somewhat striate, sometimes angular. Leaflets 4–40 × 3–35 mm., greyish-pubescent when young, later usually glabrous, extremely variable in shape and size (see tab.), from narrowly cuneate-obtriangular to broadly obdeltate, flanks straight or concave; outer margin always rounded, entire, wavy-edged, irregularly crenulate to deeply crenate, sometimes deeply lobed (lobes obtuse or acute), emarginate or retuse, rarely deeply divided into bifurcate strap-shaped lobes. Sporocarps 2–3·5 mm. long, 2–4 (5) mm. high, 0·5–1·5 mm. thick, in dense clusters of 3 to many at the nodes, appressed-pilose when young, later entirely glabrous, typically saddle-shaped, dorsally almost always deeply concave, ventrally rounded, oblong to elliptic in dorsiventral cross-section; lateral ribs not visible in mature specimens; inferior tooth not present, or just indicated by a shallow hump; superior tooth conspicuous, usually slender from a broadly conical base, acute or rarely subobtuse, always short. Pedicels adnate, a peduncle-like base branching dichotomously, thus forming clusters of sporocarps, 3–10 (or more) together; peduncle up to 6 mm. long; individual pedicels 1–1½ (2) times as long as the sporocarps.

Caprivi Strip. 6·5 km. from Ngoma, 19.iv.1965, *Symoens* 11490 (BR). **Botswana.** N: Ngamiland, 5.v.1930, *Van Son* 1802c (BM; PRE). **Zambia.** B: Masese, *Fanshawe* 5758 (K; SRGH). C: 9·6 km. E. of Lusaka, 22.v.1955, *King* 6 (K). S: Mumbwa, along roadside between Mumbwa and Nangoma, 21.iii.1963, *Van Rensburg* 1774 (K; SRGH). **Rhodesia.** N: Darwin, Chimanda Tribal Trust land, Mukane Dam, 13.v.1965, *Bingham* 1519 (SRGH). W: Gwanda Distr., c. 32 km. S. of Gwanda on Tuli road, *West* 1745 (SRGH). C: Gwelo, Fletcher High School, 3.x.1966, *Biegel* 1335 (SRGH). E: Chipinga, Sabi Experimental Station, 23.viii.1956, *West* 3450 (BM; BOL; SRGH). S: Nuanetsi, Mapwe R., Fort Victoria-Beitbridge road, 2.v.1962, *Drummond* 7862 (BM; K; SRGH).

Also in Angola, SW. Africa and S. Africa. In dry stream beds, in mud along streams, on floors of temporary ponds, in sandy dambos, in roadside drains, in water holes, at damp margins of vleis and pans, sometimes in shallow water.

3. **Marsilea aegyptiaca** Willd. in L., Sp. Pl. ed. 4, **5**: 540 (1810).—Tardieu & Alston in Mém. Inst. Fr. Afr. Noire, **50**: 16, t. 2 fig. 6 (1957).—Launert in Senckenb. Biol. **49**: 296, fig. 38–40 (1968); Prodr. Fl. SW. Afr. **11**: 3 (1969). TAB. **18** fig. E. Type from Egypt.

Leaflets 2–25 (30) × 2–20 (25) mm., extremely variable, broadly obdeltate to narrowly obtriangular, with the outer edge rounded and entire (almost always so in the aquatic form), 2-lobed or many-lobed to deeply crenate, with the lobes almost always rounded or obtuse, pilose to entirely glabrous, with longitudinal brown suberous streaks on the lower surface in the aquatic form. Sporocarps 1–2 mm. thick, solitary or more often in dense groups of 2 to many clustered at the very base of the stipe, square to rectangular in lateral view, sometimes ventrally curved, dorsally straight or saddle-shaped to a varying degree, oblong in frontal view, lateral sides with a shallow or deep slightly curved ventral groove with the frontal furrow usually present and often expanded over the dorsal side, densely pilose at first (the hairs always appressed and pointed downwards), later usually glabrous; raphe present but often indistinct; sori 4–6; inferior tooth not developed, the superior one always distinct, varying from obtusely conical to (more often) acute, prominent, straight or sometimes slightly recurved. Pedicels always free, (2) 3–8 (12) mm. long, somewhat stout, curved or straight, densely appressed-pilose at first, later becoming glabrous.

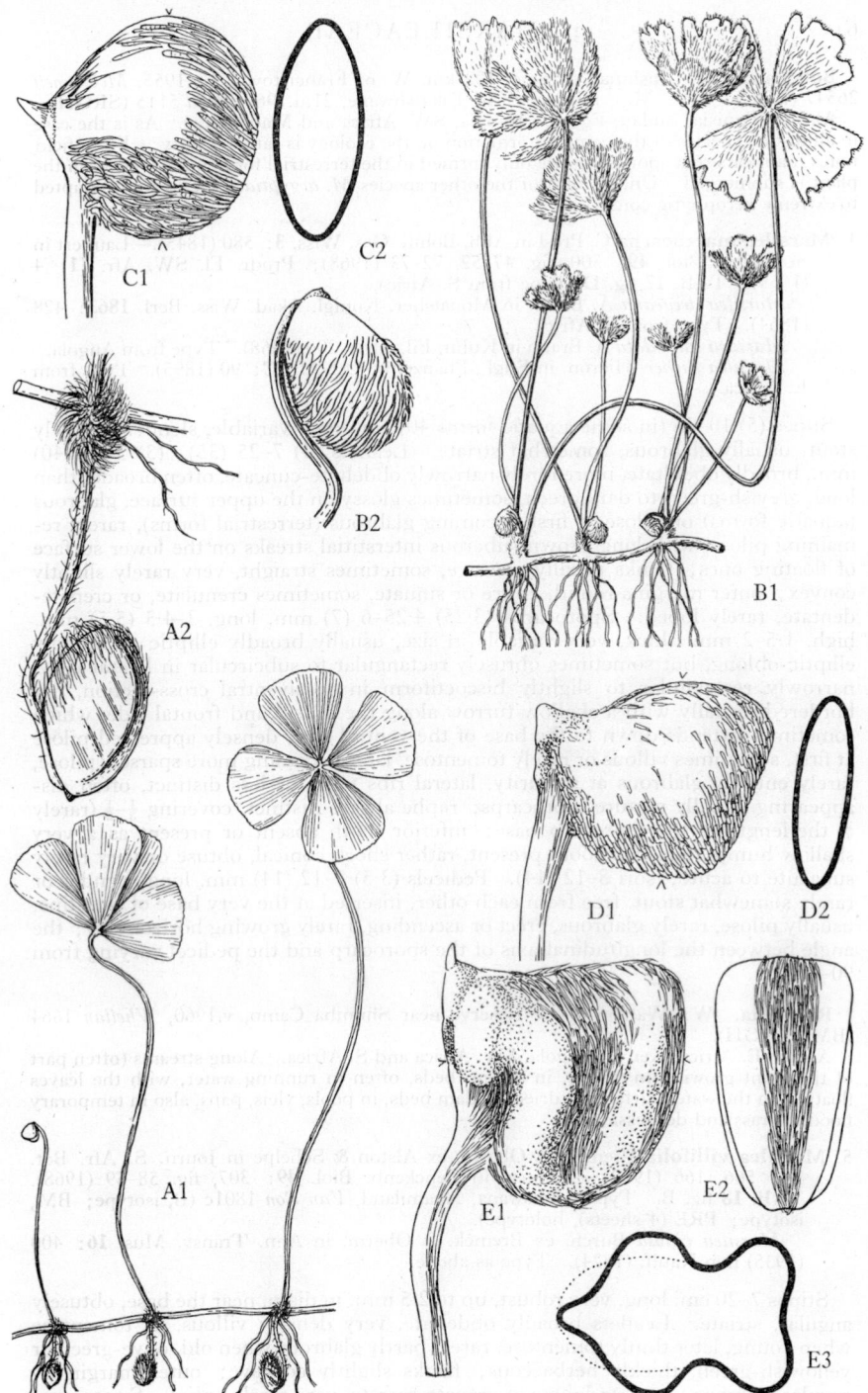

Tab. 18. A.—MARSILEA VERA. A1, habit (×⅔); A2, sporocarp in lateral view (×4), both from *Schinz* s.n. B.—MARSILEA VILLIFOLIA. B1, habit (×⅔); B2, sporocarp in lateral view (×4), both from *Van Son* 1801C. C.—MARSILEA BURCHELLII. C1, sporocarp in lateral view (×12); C2, same in dorsiventral cross-section (×12), both from *Burchell* 1625. D.—MARSILEA CAPENSIS. D1, sporocarp in lateral view (×10); D2, same in dorsiventral cross-section (×10), both from *Drège* s.n. E.—MARSILEA AEGYP-TIACA. E1, sporocarp in lateral view (×10); E2, same in frontal view (×10); E3, horizontal-longitudinal cross-section (×10), all from *Schimper & Wiest* 33.

Botswana. N: Bushman Mine c. 102 km. W. of Francistown, 4.v.1955, *McConnell* 26547 (BOL). SW: Minyani Pan, S. of Takatshwane, 21.ii.1960, *Wild* 5115 (SRGH).

Also in Tunisia, Sudan, Egypt, Ethiopia, SW. Africa and Madagascar. As is the case in almost all species of the genus information on the ecology is rather scanty. From field notes it appears that sporocarps are only formed in the terrestrial form and never when the plant is submerged. Unlike most of the other species *M. aegyptiaca* seems to be adapted to extreme xerophytic conditions.

4. **Marsilea macrocarpa** C. Presl in Abh. Böhm. Ges. Wiss. **3**: 580 (1845).—Launert in Senckenb. Biol. **49**: 300, fig. 47–52, 72–73 (1968); Prodr. Fl. SW. Afr. **11**: 4 (1969). TAB. **17** fig. D. Type from S. Africa.

    Marsilea dregeana A. Braun in Monatsber. Königl. Akad. Wiss. Berl. **1863**: 428 (1864). Type from S. Africa.

    Marsilea rotundata A. Braun in Kuhn, Fil. Afr.: 200 (1868). Type from Angola.

    Marsilea fischeri Hieron. in Engl., Pflanzenw. Ost-Afr. **C**: 90 (1895). Type from E. Africa.

Stipes (5) 10–25 (in some aquatic forms 40) cm. long, variable, slender or rarely stout, usually glabrous, somewhat striate. Leaflets (4) 7–25 (35) × (3) 5–28 (40) mm., broadly obdeltate, more rarely narrowly obdeltate-cuneate, often broader than long, greyish-green to dark green, sometimes glossy on the upper surface, glabrous (aquatic forms) or pilose at first becoming glabrous (terrestrial forms), rarely remaining pilose, with long brown suberous interstitial streaks on the lower surface of floating ones; flanks usually concave, sometimes straight, very rarely slightly convex; outer margins round, entire or sinuate, sometimes crenulate, or crenate-dentate, rarely lobed. Sporocarps (3·25) 4·25–6 (7) mm. long, 3–4·5 (5·5) mm. high, 1·5–2 mm. thick, very variable in size, usually broadly elliptic or broadly elliptic-oblong, but sometimes obtusely rectangular to subcircular in lateral view, narrowly rectangular to slightly biscoctiform in dorsiventral cross-section, not bordered, usually with a shallow furrow along the dorsal and frontal side, which sometimes extends down to the base of the ventral side, densely appressed-pilose at first, sometimes villous or rarely tomentose, later becoming more sparsely pilose, rarely entirely glabrous at maturity, lateral ribs more or less distinct, often disappearing in fully mature sporocarps; raphe always distinct, covering $\frac{1}{3}$–$\frac{1}{2}$ (rarely $\frac{2}{3}$) the length of the sporocarp base; inferior tooth absent or present as a very shallow hump; superior tooth present, rather short, conical, obtuse or very rarely subacute to acute; sori 8–12 (14). Pedicels (3·5) 7–12 (14) mm. long, slender or rarely somewhat stout, free from each other, inserted at the very base of the stipe, usually pilose, rarely glabrous, erect or ascending, rarely growing horizontally; the angle between the longitudinal axis of the sporocarp and the pedicel varying from 90–180°.

Rhodesia. W: Wankie Game Reserve near Shumba Camp, v.1960, *Whellan* 1664 (BM; SRGH).

Also in E. Africa, Kenya, Angola, SW. Africa and S. Africa. Along streams (often part of the plant growing in water), in stream beds, often in running water, with the leaves floating on the water surface, in dried up dam beds, in pools, vleis, pans, also in temporary flooded grassland depressions.

5. **Marsilea villifolia** Bremek. & Oberm. ex Alston & Schelpe in Journ. S. Afr. Bot. **18**: 566, 166 (1952).—Launert in Senckenb. Biol. **49**: 307, fig. 58–59 (1968). TAB. **18** fig. B. Type: Botswana, Ngamiland, *Van Son* 1801c (B, isotype; BM, isotype; PRE (4 sheets), holotype).

    Marsilea villosa Burch. ex Bremek. & Oberm. in Ann. Transv. Mus. **16**: 400 (1935) non Kaulf. (1824). Type as above.

Stipes 7–20 cm. long, very robust, up to 2·5 mm. in diam. near the base, obtusely angular, striate. Leaflets broadly obdeltate, very densely villous, or tomentose when young, later thinly tomentose, rarely partly glabrous when old, olive-green or yellowish-green, thickly herbaceous; flanks slightly concave; outer margin irregularly crenate, crenate-lobate or crenate-serrate, very rarely entire. Sporocarps 5–6 mm. long, 4–5 mm. high, 1–1·5 mm. thick, solitary, square to obtusely rectangular (rarely circular) in lateral view, rectangular in dorsiventral cross-section, with or without a shallow dorsal furrow extending over the frontal surface, densely appressed-pilose, gradually becoming glabrous; lateral ribs not visible in fully mature specimens; lateral veins (as seen on the interior surface) not anastomosing;

sori 8–12; raphe not very distinct, covering $\frac{1}{3}$–$\frac{1}{2}$ the sporocarp base; inferior tooth absent or just visible as a very shallow hump; superior tooth present but not very conspicuous, rather short, broadly conical, obtuse. Pedicels up to 4 mm. long, very short, appressed-pilose.

Botswana. N: Ngamiland, pan S. of Kopjes, 7.v.1950, *Van Son* 1801c (B; BM; PRE). SE: Gaberones, on Norris's farm, 30.ix.1932, *Giffen* (BM).
Also in S. Africa. In (seasonally?) dry pans, dry river-beds and along rivers.

6. **Marsilea vera** Launert in Mitt. Bot. Staatssamml. Münch. **3**: 505 (1960); Senckenb. Biol. **49**: 308, fig. 60–62 (1968); Prodr. Fl. SW. Afr. **11**: 4 (1969). TAB. **18** fig. A Type from SW. Africa.

Stipes 4–20 (30) cm. long, slender, rarely stout, very sparsely pilose to glabrous. Leaflets 9–20 (25) × 4–15 mm., broadly triangular-obovate to obdeltate, sometimes narrowly obdeltate, very densely greyish-villous when young, becoming almost entirely glabrous later; flanks straight or slightly outwardly curved; outer margin entire, retuse, 2-lobed, crenate, irregularly dentate or rarely deeply bifurcate with narrow strap-shaped lobes, which vary from rounded to acute. Sporocarps 5–7 mm. long, 4–5 mm. high, up to 3 mm. thick, always distinctly pedicelled, deflexed against the pedicel, sub-rectangular to obliquely broad-elliptic in lateral view, elliptic to obtusely rectangular in dorsiventral cross-section, not bordered when fully mature, usually without dorsal or frontal furrow, densely appressed-pilose, villous or rarely tomentose; raphe distinct, usually attached to the whole base of the sporocarp. Sori 6–8 (10); inferior tooth absent, superior tooth very short, broadly conical, always obtuse, dark-brown. Pedicel 4–10 (25) mm. long, stout, or somewhat slender, straight or curved (sometimes with the upper part curved around the sporocarp in a semi-circle), usually growing downwards, thus burying the sporocarp in the soil, rarely spreading, usually glabrous.

Botswana. N: Between Totem and Maun, 17.v.1930, *Van Son* 1803c (PRE).
Also in SW. Africa. In seasonally dry pans, at the edges of vleis or river lagoons or in similar habitats.

7. **Marsilea farinosa** Launert in Senckenb. Biol. **49**: 298, fig. 41–46, 70–71 (1968); Prodr. Fl. SW. Afr. **11**: 4 (1969). TAB. **17** fig. C. Type from SW. Africa.
    Marsilea biloba sensu Eyles in Trans. Roy. Soc. S. Afr. **5**, 4: 290 (1916) non Willd. (1810).
    Marsilea macrocarpa sensu Eyles, loc. cit. non C. Presl (1845).

Stipes 2–15 (20) cm. long, slender, usually hispid, rarely glabrous, erect. Sporocarps 4·25–7 mm. long, 3·5–4·5 (4·75) mm. high, 1·5–2·25 mm. thick, densely crowded at the base of the stipe, bean-shaped, usually horizontal, with a dorsal furrow which extends over the front towards the base on the ventral side, laterally slightly bulging, flat or rarely concave; lateral ribs 8–11, not conspicuous in fully mature specimens; covered with multicellular uniseriate hairs of 2 kinds; basally attached, erect ordinary cylindrical hairs and flattened hairs, which are laterally attached to the surface of the sporocarp by an usually 1-celled funicle-like stalk (as illustrated in TAB. **17** fig. C3); raphe present, extending $\frac{1}{2}$–$\frac{2}{3}$ the length of the base of the sporocarp; sori 8–11; inferior tooth absent or very weakly developed (only indicated as a shallow hump); superior tooth present but usually rather inconspicuous, conical, obtuse or very rarely subacute. Pedicels (4) 8–15 (25) mm. long, usually curved, rarely straight, relatively slender, flexible, almost always hispid, with the angle between the pedicel and the longitudinal axis of the sporocarp varying from 90° to almost 180°.

Botswana. SE: 8 km., E. of Lothlekane, 24.iii.1965, *Wild & Drummond* 7263 (SRGH). **Rhodesia.** W: N. of Bulawayo near Mt. Pasipas, vi.1903, *Eyles* 26 (BM; K; PRE; SRGH). C: Gwelo, 28.i.?, *Gardner* 28 (K).
Also in Ethiopia, Tanzania, Angola, SW. Africa and S. Africa. In dry sandy river beds, in dried-out vleis, dried mud pans, on moist banks of dry river beds, usually forming large mats.

8. **Marsilea apposita** Launert in Senckenb. Biol. **49**: 306, fig. 75 (1968). Type: Rhodesia, Matopo Hills, *Gibbs* 289 (BM, holotype; BOL, isotype).
    Marsilea capensis sensu Eyles in Trans. Roy. Soc. S. Afr. **5**, 4: 290 (1916).

C

Stipes (3) 7–30 (40) cm. long, slender in juvenile, rather stout in older specimens, obtusely angular, striate, red-brown in the lower, green in the upper part, scattered-pilose or glabrous. Leaflets 6–26 × 6–25 mm., obdeltate, appressed-pilose at first, later glabrous or scattered-pilose towards the base; flanks shallowly concave or straight; outer margins rounded, irregularly crenate or crenate-dentate. Sporocarps (2·4) 3–4 mm. long, (1·5) 2–3·25 mm. high, 1·5–2 mm. thick, obtusely rectangular in lateral view, dorsally usually concave to a varying degree, ventrally slightly convex, rectangular in dorsiventral cross-section (fully mature sporocarps sometimes almost bean-shaped), with a continuous shallow furrow from the dorsal side over the frontal side towards the ventral base, densely appressed-pilose when young, becoming gradually glabrous; lateral ribs hardly visible in mature specimens; raphe distinct, covering almost the entire base of the sporocarp; sori 6–8, inferior tooth absent or present as a very shallow hump; superior tooth present, rather short, subobtuse to acute. Pedicels 8 mm. long, terete, slender, wiry, erect or arching, 2 or 3 (or more), connate at the base, rarely solitary, pilose or rarely glabrous.

Rhodesia. W: Bulawayo, Hillside, 2.i.1954, *Eccles* in *Schelpe* 3915a (BM; BOL).
Also in S. Africa. On sandy wet soil in river beds, in puddles of seasonally flooded areas (" in and at edge and out of water,—some out of water had leaves erect or semi-erect, others prostrate " (*Schelpe* 3915a)).

9. **Marsilea burchellii** A. Braun in Monatsber. Königl. Preuss. Akad. Wiss. Berl. **1863**: 429 (1864).—Launert in Senckenb. Biol. **49**: 311, fig. 65–66, 77 (1968); Prodr. Fl. SW. Afr. **11**: 3 (1969). TAB. **18** fig. C. Type from S. Africa.
Marsilea biloba Willd. in L., Sp. Pl., ed. 4, **5**: 540 (1810). Type from S. Africa.

Stipes 0·5–6 (8) cm. long (very rarely longer), rather slender, glabrous, rarely scattered-pilose. Leaflets 2–3·5 (7) × 0·5–2 (8) mm., variable, narrowly cuneate-spathulate or very narrowly obdeltate to obdeltate, appressed-pilose or glabrous; flanks straight or slightly convex; outer margin entire, emarginate, retuse or 2-lobed (the lobes up to $\frac{1}{3}$ the length of the entire leaflet), very rarely crenate. Sporocarps 1·25–2·5 (3) mm. long, 1·5–2 (2·8) mm. high, 0·6–1·5 (1·9) mm. thick, usually crowded, very small, subcircular, pyriform, very rarely obtusely square or shortly rectangular in lateral view, the dorsiventral cross-section varying from circular to elliptic, never bordered, densely covered with appressed or spreading hairs (all pointing downwards) when young, later often glabrous; lateral ribs not apparent in fully mature specimens; raphe present, not very distinct, covering $\frac{1}{3}$–$\frac{1}{2}$ the length of the sporocarp base; sporangia 3–4 (6); inferior tooth completely absent, superior always developed, conical, short and obtuse or long (up to 0·3 mm.) and acute. Pedicels (2) 3–8 (12) mm. long, always free from each other, rather slender, flexible, erect or ascending, pilose at first.

Botswana. N: Nata R., Makarikari Basin, 13.viii.1930, *Van Son* 1804c (BM).
Also in SW. Africa and S. Africa. Forming dense carpets in vleis, at edges of ponds, pools or along water courses, usually in muddy or black turfy soil. Submerged plants produce larger leaflets with entire margins.
For discussion of *M. biloba* Willd. see Launert in Senckenb. Biol. **49**: 313 (1968).

10. **Marsilea capensis** A. Braun in Monatsber. Königl. Akad. Wiss. Berl. **1863**: 428 (1864).—Launert in Senckenb. Biol. **49**: 310, fig. 63–64, 76 (1968). TAB. **18** fig. D. Type from S. Africa.

Stipes (4) 7–20 (25) cm. long, usually crowded, rarely solitary, slender, flexible, sparsely pilose to glabrous. Leaflets narrowly obdeltate to obdeltate-obovate, flanks slightly curved outwards, outer margins 2-lobate, crenate, or rarely entire, at first appressed-pilose, later glabrous. Sporocarps (2·25) 2·75–3·75 (4·25) mm. long, 2·25–3·5 (very rarely 4) mm. high, 0·8–1·25 mm. thick, obliquely broad-oblong or irregularly rhombic (rarely rectangular or square) in lateral view, often typically axe-shaped, distally obliquely truncate or more rarely rounded, narrowly elliptic in dorsiventral cross-section, always unbordered, laterally compressed or slightly inflated only, dark-brown to almost black at maturity, densely appressed-pilose at first, later becoming glabrous, very rarely remaining pilose; lateral ribs always absent; lateral veins (as seen on the interior surface) not anastomosing; sori 8–10 (12); raphe distinct, covering $\frac{4}{5}$ to whole of the length of the sporocarp base; inferior tooth absent or very rarely present as a very shallow hump; superior tooth distinct, short, conical, acute or subacute, rarely obtuse, erect or somewhat recurved. Pedicels (2) 4–7 (12) mm. long, solitary, free from each other, slender, erect or ascending, pilose at first, later usually glabrous.

Zambia. S: Kafue, 2.xii.1919, *Shantz* 461 (P).
Also in Egypt, SW. Africa and S. Africa. Along rivers, in river back-waters, in seasonally flooded grassland, in water holes.

12. SALVINIACEAE

Floating aquatic plants with slender branched siphonostelic rootless rhizomes. Leaves dimorphous, borne in whorls of 3, 2 floating, oblong to orbicular, entire, variously papillate on the aerial surface, the 3rd submerged, finely dissected, hairy and root-like. Sori in thin walled sporocarps borne on the dissected submerged leaf; sporangia borne on branched receptacles; microspores germinating within the microsporangium.

SALVINIA Adans.

Salvinia Adans., Fam. Pl. **2**: 15 (1763).

Description as for the family.
The only genus of the family, comprising about 10 spp., mostly in tropical America, but with 2 indigenous spp. in continental Africa.

Apex of each leaf-papilla with an open basket of 4 curved cells - - 1. *auriculata*
Apex of each leaf-papilla without an open basket of cells - - - - 2. *hastata*

1. **Salvinia auriculata** Aubl., Hist. Pl. Guian. **2**: 969, t. 367 (1775). TAB. **19** fig. B.
   Type from tropical America.
   Salvinia adnata Desv. in Mém. Soc. Linn. Par. **6**, 2: 177 (1827). Type erroneously given as W. Africa, probably America.

Stem up to 1·2 mm. in diam., horizontal, set on the under surface with brown multicellular hairs up to 2 mm. long. Floating leaves up to 2·5 × 2·4 cm., emarginate, base cordate, flat in weak specimens, folded along the midrib in vigorous specimens; ventral surface with dense multicellular papillae c. 2 mm. high, with 4 separated curved segments adjoining at their apices to form an open basket structure; dorsal surface thinly beset with pale multicellular hairs up to 1 mm. long. Submerged leaves up to 12 cm. long, the lobes with dense brown multicellular hairs up to 2 mm. long. Sporocarps spherical, hairy, up to 33 arranged in 2 rows along one lobe of the dissected submerged leaf.
   Botswana. N: Chobe Distr., Kasane, Chobe R., viii.1962, *Yalala* 227 (SRGH).
   Zambia. W: Ndola, Kafubu, 6.vi.1964, *Mutimushi* 745 (K; SRGH). S: Zambezi R., Kasungula, 11.iv.1955, *E.M. & W.* 1457 (BM; LISC). **Rhodesia.** N: Chirundu, 16.vii..1959, *Bates* (K; LISC). W: Victoria Falls, 22.x.1958, *Wild* 4851 (K; LISC; SRGH). C: Makoni Distr., Harleigh Farm, iii.1957, *Boay* (SRGH). E: Umtali Distr., c. 17 km. W. of Odzi, 20.viii.1959, *Phipps* 2171 (SRGH).
   Introduced from tropical America, now widespread on Lake Kariba (Schelpe in Journ. S. Afr. Bot. **27**: 169 (1961)).

2. **Salvinia hastata** Desv. in Mém. Linn. Soc. Par. **6**: 177 (1827). TAB. **19** fig. A.
   Type from Mascarene Is.
   Salvinia mollis Mett. ex Kuhn, Fil. Afr.: 201(1868). Type from Madagascar.
   Salvinia hildebrandtii Bak. in Journ. of Bot. **24**: 98 (1886). Type from Madagascar.

Stem c. 1 mm. in diam., horizontal, beset on the dorsal surface with dark-brown multicellular hairs up to 2 mm. long. Floating leaves up to 2 × 1·3 cm., opposite, oblong, sometimes slightly broadened towards the base, emarginate, base cordate to

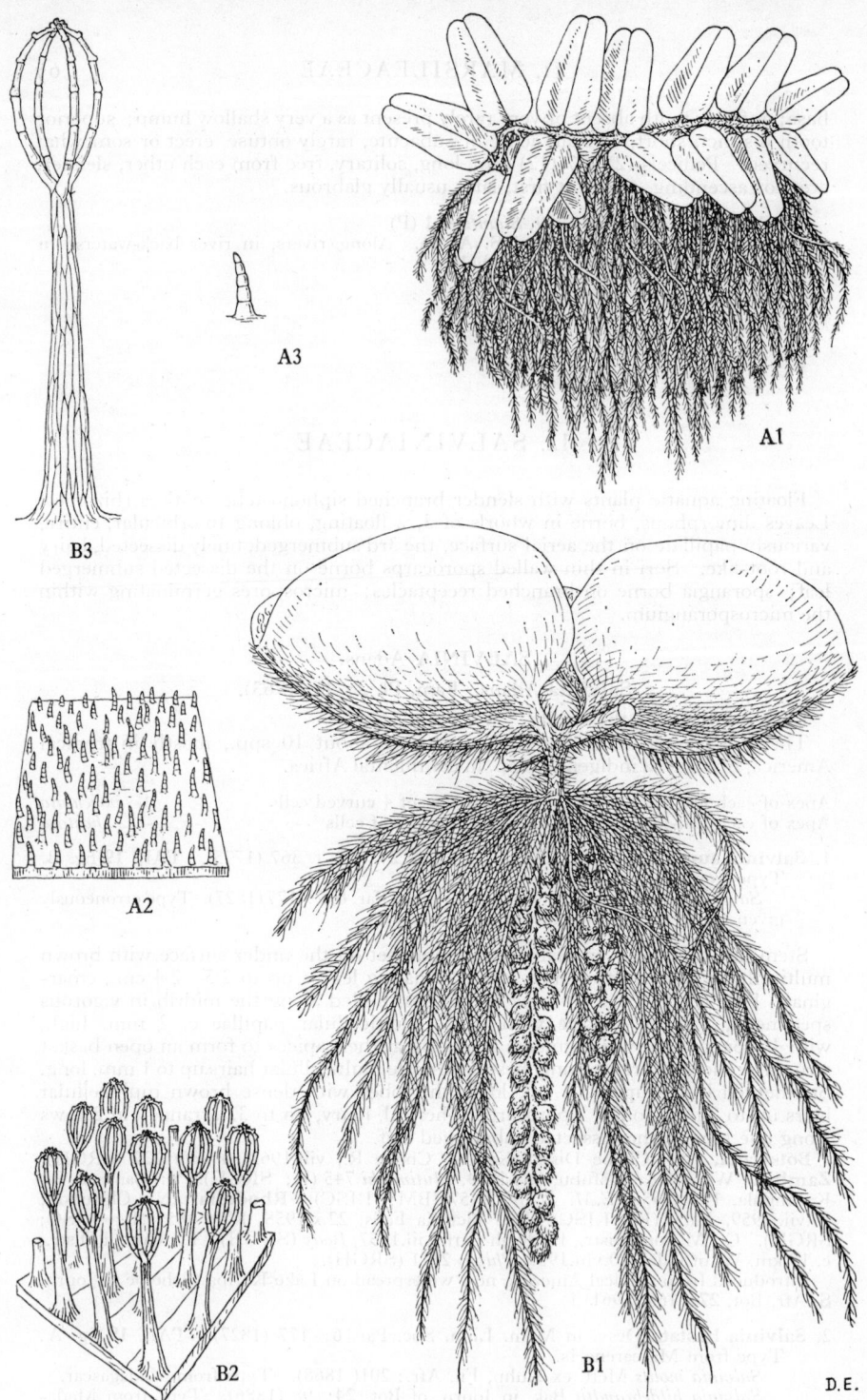

Tab. 19. A —SALVINIA HASTATA. A1, habit (× 4/5); A2, leaf surface (× 20); A3, hair from leaf surface (× 30), all from *Torre & Paiva* 11474. B.—SALVINIA AURICULATA. B1, pair of floating leaves with submerged leaf and sporocarps (× 4/5); B2, leaf surface (× 2); B3, hair from leaf surface (× 30), all from Kew Ferneries.

almost truncate, flat (or infolded along the midrib when leaves closely spaced), ventral surface papillate with minute whitish simple multicellular hairs, dorsal surface pilose with brownish multicellular hairs; dissected submerged leaf up to 6·5 cm. long with lobes densely beset with dark-brown multicellular hairs up to 3 mm. long. Sporocarps spherical, hairy, up to 6, arranged in 2 rows along the midrib of one dissected leaf-lobe.

Malawi. C: Kota Kota Distr., Chia Lagoon, 13.iv.1960, *Eccles* (BOL). S: Port Herald Distr., Chiromo, 22.iii.1960, *Phipps* 2604 (K; SRGH). **Mozambique.** N: Mogincual, 30.iii.1964, *Torre & Paiva* 11474 (LISC). Z: R. Macuse, 23.viii.1962, *Wild & Pedro* 5891 (BOL; K; SRGH). MS: R. Sungué, Gorongosa Game Reserve, 9.vii.1955, *Schelpe* 5639 (BOL).
E. tropical Africa, Madagascar and Mascarene Is. Sluggish streams and backwaters.

13. AZOLLACEAE

Floating aquatic plants with slender pinnately branched siphonostelic rhizomes and with roots borne singly or in fascicles. Leaves alternate, imbricate, 2-lobed, with an aerial chlorophyllous lobe and a thin colourless submerged lobe. Sori in thin-walled sporocarps borne on the basal leaf of a branch, heterosporous; microsporocarp with numerous microsporangia; megasporocarp with single megaspore surmounted by apical massulae. Female gametophyte submerged; microspores germinating in a massula derived from the microsporangial contents.

AZOLLA Lam.

Azolla Lam., Encycl. Méth., Bot. **1**: 343 (1783).

Description as for the family.
The only genus of the family, comprising 5 spp., 2 of which occur in continental Africa.

Root fascicles of 2–3 roots at nodes, or roots single; rhizome up to 0·2 mm. in diam.
1. *pinnata*
Root fascicles of more than 5 roots at nodes; rhizome of up to 2 mm. in diam.
2. *nilotica*

1. **Azolla pinnata** R. Br., Prodr. Fl .Nov. Holl.: 167 (1810). Type from Australia.

Var. **africana** (Desv.) Bak., Fern Allies: 138 (1887). Type from W. Africa.
Azolla africana Desv. in Mém. Soc. Linn. Par. **6**: 178 (1827). Type as above.
Azolla guineensis Schumach. in Kongel. Dansk. Vid. Selsk. **4**: 236 (1829). Type from W. Africa.

Main stem up to 2 cm. long and 0·2 mm. in diam., horizontal, minutely papillate, repeatedly alternately branched and with a few fascicles of 2–3 roots at intervals. Roots up to 3·5 cm. long, hairy with a long conspicuous root-cap. Leaves 2-lobed, the upper lobe up to 1·1 mm. long, ovate to broadly elliptic, with a papillate chlorophyllous central portion surrounded with a hyaline border, the lower lobe almost of similar size but hyaline. Microsporocarps borne singly or subtended by a megasporocarp, only partly covered by a hyaline lower leaflet. Megasporocarps up to 0·8 mm. long with a prominent dark apex and containing a single granular megaspore surmounted by numerous " flotation bodies ". Microsporocarps spherical, up to 1·7 mm. in diam. with a minute dark apex and containing numerous long-stalked microsporangia; massulae of microspores with few or no weak outgrowths.

Botswana. N: Linyanti R. between Dunangara and Simanja, c. 910 m., 31.viii.1930, *Stephens* 60 (BM). **Zambia.** B: Mongu, 9.xi.1959, *Drummond & Cookson* 6272 (K; LISC; SRGH). N: Shiwa Ngandu, 22.ix.1938, *Greenway & Trapnell* 5760 (K). C:

Kafue Gorge, 9.vii.1955, *King* 51 (K). S: Livingstone, 25.vii.1958, *Van Someren* 87308 (BOL; K; SRGH). **Mozambique.** Z: between Quelimane and Marral, 10.x.1941, *Torre* 3618 (LISC). MS: Gorongosa Game Reserve, Lago Nhaurine, *Chase* 5080 (BM; SRGH). SS: near Vila João Belo, 13.vii.1944, *Torre* 6737 (LISC). LM: Incanhini, 14.i.1898, *Schlechter* 12033 (BOL).

Widespread in tropical Africa. Ponds and backwaters of rivers, 3–1000 m.

2. **Azolla nilotica** Decne. ex Mett. in Kotschy & Peyr., Pl. Tinn.: 54, t. 25 (1867).
    TAB. 20. Type from the Sudan.

Main stem up to 32 cm. long and 2 mm. in diam., horizontal to slightly ascending, pubescent with pale multicellular hairs and producing branches alternately, and with fascicles of numerous roots at the nodes. Roots up to 15 cm. long, hairy, with short root-caps. Leaves 2-lobed, set along the branches and branchlets, the upper lobe up to 2 mm. long, broadly elliptic with a papillate chlorophyllous central portion surrounded by a broad hyaline margin; the lower lobe smaller and thinner. Microsporocarps and megasporocarps produced on the older parts of the plant in groups consisting of up to 4 megasporocarps or of up to 3 microsporocarps, or more usually of a mixture of 1 microsporocarp and 1–3 megasporocarps, the groups of sporocarps being enclosed at first in a hyaline ovoid membrane. The mega-sporocarps up to 0·8 mm. in diam., spherical, with a small dark apex and con-taining numerous long-stalked microsporangia; massulae of microspores with few or without weak outgrowths.

Rhodesia. W: Wankie Game Reserve, 10.viii.1960, *Weir* (SRGH). **Malawi.** S: Lower Shire, 19.vii.1958, *Seagrief* 3163 (K; SRGH). **Mozambique.** Z: Morrumbala, near Vila Bocage, Chire R., 3.x.1944, *Mendonça* 2338a (LISC). T: R. Zambese, Sisitso Station, 8.vii.1950, *Chase* 2628 (BOL; SRGH). MS: Gorongosa Game Reserve, R. Sungué, *Schelpe* 5640 (BM; BOL).

Widespread in E. tropical Africa. In sluggish streams and backwaters.

14. CYATHEACEAE

Arborescent plants with a caudex (trunk) made up of a dictyostelic axis enveloped by numerous short adventitious roots and old stipe-bases. Fronds usually large, deeply 2-pinnatifid to 3-pinnate or 4-pinnatifid, borne at the apex of the caudex. Sori superficial on the veins; sporangia developing mostly in basipetal succession, short-stalked, dehiscing horizontally; annulus complete, oblique; indusium cup-shaped, 1-sided or absent. Gametophyte thick, cordate.

The only genus of the Cyatheaceae occurring in our area is *Cyathea* Sm.

CYATHEA Sm.

Cyathea Sm. in Mem. Acad. Turin, **5**: 416 (1793).

Caudex erect, short or tall, with scales produced at the apex and on the stipe bases. Frond membranous to coriaceous, deeply 2-pinnatifid to 3-pinnate or 4-pinnatifid, glabrous to villous or tomentose on the dorsal surface; veins free. Sori borne on the veins; receptacle elongated or hemispherical, indusium basal, cup-shaped or one-sided.

A genus of over 800 spp. throughout the tropics and the southern temperate regions. Species with asymmetric indusia (*C. capensis*) were previously referred to the genus *Hemitelia*.

Frond 2-pinnate, rarely 3-pinnatifid; caudex short, usually less than 0·5 m. high
                                                                1. *mossambicensis*
Frond 3-pinnate; caudex tall, up to 4 m. high:
    Stipe and rhachis smooth or slightly muricate or tuberculate; pinnules green below
        or with pale-brown bullate or linear scales or with pale hairs on the veins:

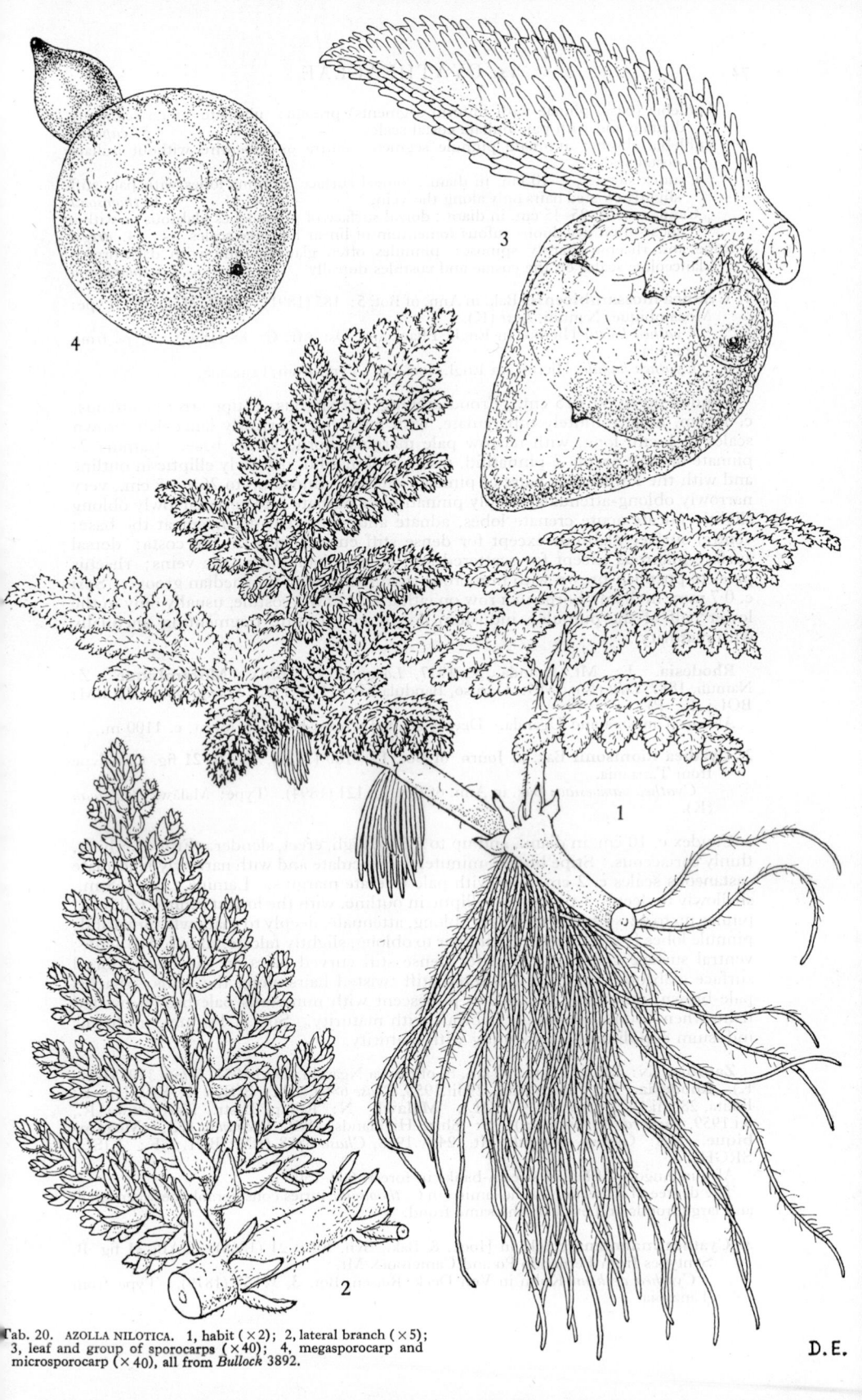

Tab. 20. AZOLLA NILOTICA. 1, habit (×2); 2, lateral branch (×5);
3, leaf and group of sporocarps (×40); 4, megasporocarp and
microsporocarp (×40), all from *Bullock* 3892.

D.E.

Aphlebia (basal pinnae with filiform segments) present; ultimate pinna segments
    dentate and with pale bullate costal scales   -   -   -   -   5. *capensis*
Aphlebia absent; ultimate pinnule segments entire or subentire without bullate
    costal scales:
    Caudex slender, c. 10 cm. in diam.; dorsal surface of pinna lobes with pale stiff
        slightly twisted hairs only along the veins   -   -   -   -   2. *thomsonii*
    Caudex stout, 25–45 cm. in diam.; dorsal surface of pinna lobes glabrous or with a
        loose often deciduous rufous tomentum of linear hair-like scales   -   4. *dregei*
Stipe and rhachis sharply spinose; pinnules often glaucous and with dark-brown
    lanceolate scales on the costae and costules dorsally   -   -   -   3. *manniana*

1. **Cyathea mossambicensis** Bak. in Ann. of Bot. **5**: 185 (1891). TAB. **21** fig. A. Type:
    Mozambique, Namúli, *Last* (K).
    Cyathea holstii Hieron. in Engl., Pflanzenw. Ost-Afr. **C**: 88 (1895). Type from
    Tanzania.
    Cyathea humilis Hieron. in Engl., loc. cit. Type from Tanzania.

Caudex up to 80 × 15 cm. Frond arching, herbaceous. Stipe atropurpureous,
c. 17 cm. long, minutely tuberculate, with numerous narrowly lanceolate brown
scales c. 1 cm. long, with narrow pale margins, borne at the base. Lamina 2-
pinnate or very deeply 2-pinnatifid, rarely 3-pinnatifid, narrowly elliptic in outline
and with the lowest four pairs of pinnae reduced; pinnae up to 26 × 4·5 cm., very
narrowly oblong-attenuate, deeply pinnatifid or pinnate into very narrowly oblong
slightly falcate acute crenate lobes, adnate and somewhat decurrent at the base;
ventral surface glabrous except for dense stiff curved hairs on the costa; dorsal
surface glabrous except for scattered stiff twisted hairs along the veins; rhachis
pale- to purplish-brown with numerous rigid hairs along the median groove. Sori
c. 0·7 mm. in diam., borne in a row on either side of the costule, usually only in the
lower half of the pinnule; indusium deeply cupuliform, fragmenting unequally
with maturity.

    Rhodesia. E: Mt. Selinda, 5.x.1937, *Longfield* 11 (BM). **Mozambique.** Z:
Namúli, 1887, *Last* (K). MS: Garuso, Bandula Mt., 11.vii.1955, *Schelpe* 5620 (B; BM;
BOL; K; P).
    Also in Tanzania and Uganda. Deeply shaded stream-banks in forest, c. 1100 m.

2. **Cyathea thomsonii** Bak. in Journ. of Bot. **19**: 180 (1881). TAB. **21** fig. C. Type
    from Tanzania.
    Cyathea zambesiaca Bak. in Ann. of Bot. **8**: 121 (1894). Type: Malawi, *Buchanan*
    (K).

Caudex c. 10 cm. in diam., and up to 2·5 m. high, erect, slender. Frond arching,
thinly coriaceous. Stipe brown, minutely tuberculate and with narrowly lanceolate
castaneous scales c. 2 cm. long with pale lacerate margins. Lamina c. 2 × 0·8 m.,
shallowly to deeply 3-pinnatifid, elliptic in outline, with the lowest pinnae reduced;
pinnae up to 42 × 14 cm., narrowly oblong, attenuate, deeply to shallowly pinnatifid;
pinnule lobes from almost semicircular to oblong, slightly falcate, rounded to acute;
ventral surface glabrous except for dense stiff curved hairs on the costa; dorsal
surface glabrous except for scattered stiff twisted hairs along the veins; rhachis
pale-brown, minutely tuberculate, pubescent with numerous pale-brown twisted
hairs when young, becoming glabrous with maturity. Sori c. 1·5 mm. in diam.;
indusium cupuliform fragmenting with maturity.

    Zambia. N: Mpika Distr., 60 km. S. of Shiwa Ngandu, 29.xi.1952, *White* 3792 (BM;
K). **Rhodesia.** E: Vumba Mts.. 29.iii.1958, *Chase* 6864 (BM; BOL; K; SRGH). S:
Bikita, 28.viii.1922, *Eyles* 3644 (BOL). **Malawi.** N: Rumpi Distr., Kaziweziwe R.,
8.i.1959, *Richards* 10564 a (K). S: Shire Highlands, 1881, *Buchanan* (K). **Mozambique.** MS: Garuso, Bandula Mt., 24.iv.1958, *Chase* 6878 (BM; BOL; K; LISC;
SRGH).
    Also in Angola. Shaded stream-banks in forest, 600–1500 m.
    The degree of dissection of the lamina in *C. thomsonii* varies considerably between small
and large fronds and even on the same frond.

3. **Cyathea manniana** Hook. in Hook. & Bak., Syn. Fil.: 21 (1868). TAB. **21** fig. B.
    Syntypes from Fernando Po and Cameroons Mt.
    Cyathea deckenii Kuhn in Von Deck. Reisen, Bot. **3**, 3: 57 (1879). Type from
    Tanzania.

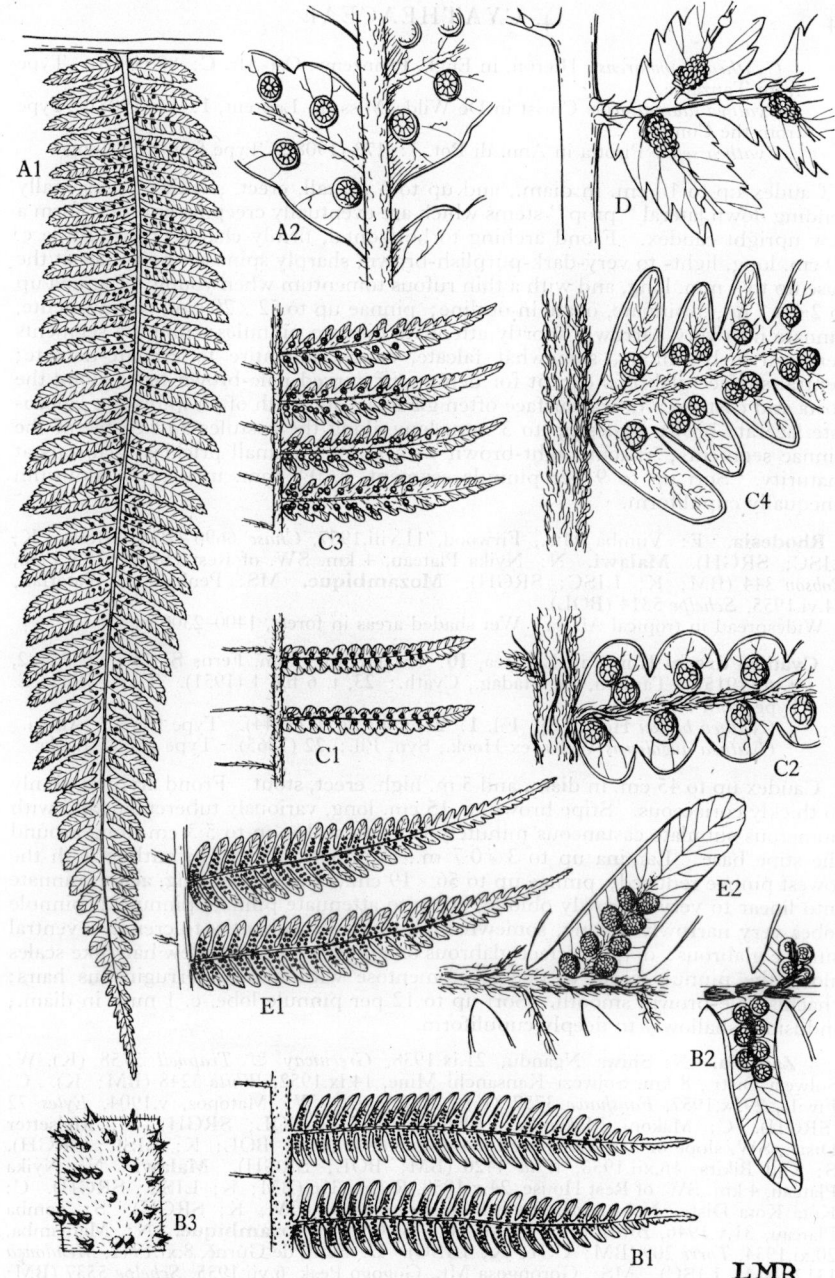

LMR

Tab. 21. A.—CYATHEA MOSSAMBICENSIS. A1, pinna (×⅔); A2, pinnule base (×4), both
from *Chase* 6587. B.—CYATHEA MANNIANA. B1, pinnules (×⅔); B2, pinnule lobe
(×4); B3, surface of stipe (×10), all from *Fisher & Schweickerdt* 312. C.—
CYATHEA THOMSONII. C1, part of small pinna (×⅔); C2, pinnule base (×4), both
from *Chapman* 300; C3, part of large pinna (×⅔); C4, pinnule base (×4), both
from *Chase* 6886. D.—CYATHEA CAPENSIS. Pinnule base (×4), *Schelpe* 6181.
E.—CYATHEA DREGEI. E1, pinnules (×⅔); E2, base of pinnule lobe (×4), both from
Swynnerton 6030.

Cyathea usambarensis Hieron. in Engl., Pflanzenw. Ost-Afr. **C**: 88 (1895). Type from Tanzania.

Cyathea laurentiorum Christ in De Wild., Miss. E. Laurent, **1**: 14 (1905). Type from the Congo.

Cyathea sellae Pirotta in Ann. di Bot. **7**: 173 (1908). Type from the Congo.

Caudex up to 10 cm. in diam., and up to 6 m. tall, erect, slender, occasionally sending down lateral " prop " stems which are eventually creeping and may form a new upright caudex. Frond arching to horizontal, firmly chartaceous. Stipe c. 30 cm. long, light- to very-dark-purplish-brown, sharply spinose, the spines at the base up to 4 mm. long, and with a thin rufous tomentum when young. Lamina up to 2·4 × 1 m., 3-pinnate, ovate in outline; pinnae up to 52 × 20 cm., oblong, acute, pinnate into very narrowly shortly attenuate pinnate pinnules; pinnule segments very narrowly oblong, somewhat falcate, acute, subentire to crenate-dentate; ventral surface glabrous except for dense stiff curved pale-brown hairs along the costa and costules; dorsal surface often glaucous and with often imbricate lanceolate lacerate brown scales up to 3 mm. long along the costules and costae of the pinnae segments; rhachis light-brown with scattered small prickles, glabrous at maturity. Sori up to 9 per pinnule segment, c. 0·8 mm. in diam.; indusium unequally cupuliform.

Rhodesia. E: Vumba Mts., Firwood, 11.viii.1957, *Chase* 6696 (BM; BOL; K; LISC; SRGH). **Malawi.** N: Nyika Plateau, 4 km. SW. of Rest House, 25.x.1958, *Robson* 344 (BM; K; LISC; SRGH). **Mozambique.** MS: Penhalonga Waterfall, 24.vi.1955, *Schelpe* 5314 (BOL).

Widespread in tropical Africa. Wet shaded areas in forest, 1400–2300 m.

4. **Cyathea dregei** Kunze in Linnaea, **10**: 551 (1836).—Sim, Ferns S. Afr. ed. 2: 82, t. 6 (1915).—Tardieu, Fl. Madag., Cyath.: 23, t. 6 fig. 1 (1951). TAB. **21** fig. E. Type from S. Africa.

Cyathea burkei Hook., Sp. Fil. **1**: 23, t. 17 fig. b (1844). Type from S. Africa.

Cyathea angolensis Welw. ex Hook., Syn. Fil.: 22 (1865). Type from Angola.

Caudex up to 45 cm. in diam. and 5 m. high, erect, stout. Frond arching, thinly to thickly coriaceous. Stipe brown, c. 15 cm. long, variously tuberculate and with numerous subulate castaneous minutely lacerate scales up to 5·3 cm. long around the stipe base. Lamina up to 3 × 0·7 m., 3-pinnate, elliptic in outline with the lowest pinnae reduced; pinnae up to 56 × 19 cm., narrowly oblong, acute, pinnate into linear to very narrowly oblong, acute to attenuate pinnate pinnules; pinnule lobes very narrowly oblong, somewhat falcate, acute, subentire to crenate; ventral surface glabrous; dorsal surface glabrous or subglabrous with a few hair-like scales along the pinnule costa to densely tomentose with matted ferrugineous hairs; rhachis pale-brown, smooth. Sori up to 12 per pinnule lobe, c. 1 mm. in diam.; indusium shallowly to deeply cupuliform.

Zambia. N: Shiwa Ngandu, 21.ix.1938, *Greenway & Trapnell* 5758 (K). W: Solwezi Distr., 8 km. Solwezi–Kansanchi Mine, 14.ix.1952, *White* 3248 (BM; K). C: Fiwila, 29.ix.1957, *Fanshawe* 3757 (K). **Rhodesia.** W. Matopos, v.1904, *Eyles* 72 (SRGH). C: Makoni, Makamba, i.1956, *Mitchell* 33 (BOL; SRGH). E: Melsetter Distr., SW. slope of Mt. Pene, 6.v.1958, *Chase* 6894 (BM; BOL; K; LISC; SRGH). S: Old Bikita, 16.xii.1953, *Wild* 4420 (BM; BOL; SRGH). **Malawi.** N: Nyika Plateau, 4 km. SW. of Rest House, 24.x.1958, *Robson* 316 (BM; K; LISC; SRGH). C: Kota Kota Distr., Nchisi Mt., 2.viii.1946, *Brass* 17102 (BM; K; SRGH). S: Zomba Plateau, 31.v.1946, *Brass* 16137 (BM; K; SRGH). **Mozambique.** N: Maniamba, 20.xi.1934, *Torre* 209 (BM; COI; K; LISC). Z: Serra de Gúruè, 8.xi.1942, *Mendonça* 1312 (BM; LISC). MS: Gorongosa Mt., Gogogo Peak, 6.vii.1955, *Schelpe* 5537 (BM; BOL).

Also in S. Africa and widespread in tropical Africa. Stream-banks in grassland and on forest margins, 900–1800 m.

Many of the specimens from our area have smaller indusia than in the typical South African and other forms.

5. **Cyathea capensis** (L.f.) Sm. in Mem. Acad. Turin, **5**: 417 (1793).—Sim, Ferns S. Afr. ed. 2: 85, t. 7 (1915). TAB. **21** fig. D. Type from S. Africa.

Polypodium capense L.f., Suppl. Pl.: 445 (1781). Type as above.

Trichomanes incisum Thunb., Prodr. Pl. Cap.: 173 (1800). Type from S. Africa.

Aspidium capense (L.f.) Sw. in Schrad., Journ. Bot. **1800**, 2: 42 (1801). Type as for *Cyathea capensis*.

Cyathea riparia Willd. in L., Sp. Pl., ed. 4, **5**: 493 (1810). Type from S. Africa.
Hemitelia capensis (L.f.) Kaulf., Enum. Fil.: 253 (1824). Type as for *Cyathea capensis*.
Trichomanes cormophyllum Kaulf., tom. cit.: 266 (1824). Type from S. Africa.
Hemitelia riparia (Willd.) Desv. in Mém. Soc. Linn. Par. **6**, 2: 322 (1827). Type as for *Cyathea riparia*.
Alsophila capensis (L.f.) J. Sm. in Hook., Lond. Journ. Bot. **1**: 666 (1842). Type as for *Cyathea capensis*.
Amphicosmia riparia (Willd.) Gardn. in Hook., Lond. Journ. Bot. **1**: 441, t. 12 (1842). Type as for *Cyathea riparia*.
Polystichum capense (L.f.) J. Sm. in Curt. Bot. Mag. **72** Comp.: 35 (1846). Type as for *Cyathea capensis*.
Cormophyllum capense (L.f.) Newm. in Phytol. **5**: 238 (1854). Type as above.
Amphicosmia capensis (L.f.) Klotzsch in Allg. Gartenzeit. **1856**: 107 (1856). Type as above.

Caudex c. 10 cm. in diam. and up to 4·5 m. high, erect, slender. Frond arching, herbaceous. Stipe pale-brown ventrally, dark purplish-brown dorsally, with a pair of depauperate pinnae (aphlebia) at the base and sparsely clothed when young with narrowly lanceolate castaneous scales c. 1 cm. long with paler lacerate margins. Lamina up to 2 × 0·8 m., deeply 3-pinnatifid, elliptic in outline; pinnae up to 46 × 16 cm., oblong, attenuate, pinnate into very narrowly oblong attenuate deeply pinnatifid pinnules; pinnule segments dentate, narrowly oblong, slightly falcate, acute; ventral surface glabrous; dorsal surface with pale bullate scales along the costules; rhachis pale- to dark-brown, smooth, glabrous. Sori 1–2 at the base of each pinnule segment, c. 1 mm. in diam., with paraphyses; indusium asymmetric.

Rhodesia. E: Inyanga Distr., Pungwe Gorge, 23.vii.1957, *Chase* 6644 (BM; BOL; K; SRGH). **Malawi.** S: Mlanje Mt., Luchenya Plateau, 2.vii.1946, *Brass* 16600 (K). **Mozambique.** MS: Chimanimani Mts., Martin Falls, 8.ii.1958, *Mitchell* 328 (SRGH). Also in S. Africa and Tanzania. Shaded moist forested ravines, 1370–1800 m.

15. HYMENOPHYLLACEAE

Delicate epiphytic or terrestrial small herbs with slender creeping, or erect, protostelic rhizomes. Fronds simple or variously divided, lamina 1 cell thick, without stomata, with free veins or the veins joining a submarginal vein; false veins, unconnected with the vascular system of the lamina, sometimes present. Sori marginal with sporangia borne in basipetal succession on an elongated receptacle within a tubular, obconic or bivalved indusium; annulus complete, oblique. Prothallus filamentous or thalloid and narrow.

Copeland (1938, 1947) divided the *Hymenophyllaceae* into 33 genera, many of which the present author prefers to treat as subgenera of *Hymenophyllum* and *Trichomanes* as far as the African species are concerned.

Indusium tubular to obconic with 2 small lobes at the apex; receptacle often elongated beyond the sorus; rhizome usually thickly clothed in black or brown hairs
1. **Trichomanes**
Indusium of 2 separate valves, sometimes with a small obconic base; receptacle shorter than the valves; rhizome subglabrous - - .. .. - 2. **Hymenophyllum**

1. TRICHOMANES L.

Trichomanes L., Sp. Pl. **2**: 1095 (1753); Gen. Pl. ed. 5: 485 (1754).

Rhizome filiform, widely creeping or erect, densely clothed with black or brown hairs, or shorter and thicker (1–3 mm. thick). Fronds simple, or up to 4-pinnatifid, glabrous or subglabrous or with marginal hairs; veins free or connected to a submarginal vein; false veins sometimes present. Sori marginal, turbinate to obconic, with 2 rounded apical lobes; filamentous receptacle often exserted beyond the sorus.

The genus *Trichomanes*, as construed here, consists of over 300 predominantly tropical species.

Frond simple or lobed less than ½ way to the midrib - - - - **1**. *erosum*
Frond 2-pinnatifid to 4-pinnatifid:
    Rhizome stout, 2–3 mm. in diam.; fronds 3–4 pinnatifid - - - **4**. *rigidum*
    Rhizome filiform; frond 2–3-pinnatifid:
        Rhizome set with black hairs; sorus conical, almost as deep as broad   **3**. *pyxidiferum*
        Rhizome set with brown hairs; sorus turbinate, about twice as deep as broad
                                                  2. *borbonicum*

1. **Trichomanes erosum** Willd. in L., Sp. Pl. ed. 4, **5**: 501 (1810).—Tardieu in Mém. Inst. Fr. Afr. Noire, **28**: 41, t. 4 fig. 3–4 (1955). Type from W. Africa.

Rhizome filiform, covered with dark-brown hairs, bearing fronds up to 3 mm. apart. Stipe short, bearing dark-brown hairs when young. Lamina up to 4 × 1·5 cm., narrowly ovate-oblong, simple to variously pinnatifid, with occasional brown hairs on the costae of the young fronds, submarginal vein present. Sori borne in the upper ½ of the frond, c. 1 mm. long and 0·4 mm. in diam., with rounded entire valves and winged by the lamina throughout its length.

Lamina entire to irregularly crenate - - - - - - - - - var. *erosum*
Lamina pinnatifid - - - - - - - - - - - - var. *aerugineum*

Var. **erosum**. TAB. **22** fig. B.

    Rhodesia. E: Vumba Mts., Elephant Forest, 10.xi.1955, *Chase* 5851 (BM; BOL; SRGH). **Mozambique.** Z: Namúli, 1887, *Last* (K).
    Also in Tanzania, Congo, Angola, Cameroon, Nigeria, Ghana, Liberia, Sierra Leone, Réunion, Comoro Is. and Seychelles. Very deeply shaded and continually moist habitats in forest, sometimes epiphytic on *Cyathea manniana*, 1100–1830 m.

Var. **aerugineum** (Van d. Bosch) Bonap., Not. Ptérid. **13**: 165 (1929). Type from Fernando Po.
    Trichomanes aerugineum Van d. Bosch in Ned. Kruid. Archf., Ser. 1, **5**, 3: 201 (1863). Type as above.
    Trichomanes palmicola Van d. Bosch ex Goddijn in Med. Rijks-Herb. No. **17**: 32, fig. 19 (1913). Type from West Africa.
    Trichomanes chamaedrys Taton in Bull. Soc. Roy. Bot. Belg. **78**: 29, t. s. fig. K, L (1946). Type from Congo.
    Trichomanes erosum var. *chamaedrys* (Taton) Tardieu in Mém. Inst. Fr. Afr. Noire, **28**: 42, t. 4 fig. 5 (1955). Type as above.

    Mozambique. MS: Garuso, Jaegersberg, 11.vii.1955, *Schelpe* 5612 (BM; BOL).
    Also in S. Africa, Tanzania, Congo, Cameroon, Ivory Coast, Liberia, Sierra Leone and Fernando Po. Deeply shaded boulder faces in forest, 1500–1600 m.

2. **Trichomanes borbonicum** Van d. Bosch in Ned. Kruid. Archf., Ser. 1, **5**, 2: 158 (1861).—Tardieu in Mém. Inst. Fr. Afr. Noire, **28**: 44, t. 4 fig. 11–12 (1953). TAB. **22** fig. D. Type from Réunion.
    Trichomanes goetzei Hieron. in Engl., Bot. Jahrb. **28**: 339 (1900).—Taton in Bull. Soc. Roy. Bot. Belg. **78**: 34, t. 3 fig. A–B (1946). Type from Tanzania.

Rhizome filiform, covered with brown hairs, bearing fronds 1–4 cm. apart. Stipe filiform, narrowly winged in the upper ½. Rhachis winged. Lamina up to 10 × 3·5 cm., oblong-lanceolate to narrowly elliptic. Pinnae bilaterally pinnatifid to 2-pinnatifid into up to 12 linear entire rounded lobes c. 0·8 mm. broad. Sori c. 2 mm. deep and c. 0·6 mm. in diam., usually borne in the upper ½ of the frond, cylindric to turbinate, winged by the lamina for its whole length, with entire rounded valves and with a persistent prominent filiform receptacle.

    Rhodesia. E: Slopes of Vumba Mts., 1500 m. 3.vii. 1957, *chase* 6573 (BM; SRGH). **Malawai.** S: Mlanje Mt., Lower Ruo Plateau, 17.viii.1956, *Newman & Whitmore*, 485 (SRGH). **Mozambique.** MS: Gorongosa Mt., near Gogogo Peak, 6.vii.1955, *Schelpe* 5540a (BM; BOL).
    Also in S. Africa, Tanzania, Kenya, Congo, Cameroon, Ghana, Fernando Po, S. Tomé, Madagascar and the Mascarene Is. Moist shaded boulder faces in forest, 1500–1900 m.

3. **Trichomanes pyxidiferum** L., Sp. Pl. **2**: 1098 (1753). Type from S. Domingo.

G.P.A.

Tab. 22. A.—TRICHOMANES RIGIDUM. Habit (×⅔) *Callens* 3998. B.—TRICHOMANES EROSUM var. EROSUM. B1, habit (×⅔); B2, apex of frond (×4), both from *Chase* 5851. C.—TRICHOMANES PYXIDIFERUM var. MELANOTRICHUM. C1, habit (×⅔); C2, segment of fertile pinna (×4), both from *White* 2728. D.—TRICHOMANES BORBONICUM. Segment of fertile pinna (×4) *Schweickerdt* 2397. E.—HYMENOPHYLLUM TUNBRIDGENSE. E1, habit (×⅔); E2, segment of fertile pinna (×8), both from *Mitchell* 287. F.—HYMENOPHYLLUM POLYANTHOS var. MOSSAMBICENSE. F1, habit (×⅔); F2, segment of fertile pinna (×8), both from *Chase* 6679. G.—HYMENOPHYLLUM POLYANTHOS var. KUHNII, Habit (×⅔) *Chase* 6088. H.—HYMENOPHYLLUM CAPILLARE. Habit (×⅔) from *Phipps* 1265. I.—HYMENOPHYLLUM SIBTHORPIOIDES, habit (×⅔), *Whellan* 1271.

Var. **melanotrichum** (Schlechtend.) Schelpe in Journ. S. Afr. Bot. **30**: 181 (1964).
TAB. **22** fig. C.   Type from S. Africa.
  Trichomanes melanotrichum Schlechtend., Adumbr.: 56 (1825); in Linnaea, **10**:
553 (1836).—Tardieu, Fl. Madag., Hymenophyll.: 26, t. 3 fig. 3–4 (1951); in
Mém. Inst. Fr. Afr. Noire, **28**: 46, t. 5 fig. 1–2 (1953).   Type as above.
  Trichomanes frappieri Cordem. in Bull. Soc. Sci. Art Réunion 1890–1891 et Fl.
Réunion: 37 (1895).   Type from Réunion.
  Trichomanes pyxidiferum sensu Sim, Ferns S. Afr. ed. 2: 69, t. 3 fig. 3 (1915).
  Trichomanes pyxidiferum forma *major* Taton in Bull. Soc. Roy. Bot. Belg. **78**: 34,
t. 3 fig. E–F (1946).   Type from Congo.
  Vandenboschia melanotricha (Schlechtend.) P.-Sermolli in Webbia, **12**: 127, t. 1
(1955).   Type as for *Trichomanes pyxidiferum* var. *melanotrichum*.

Rhizome filiform, covered with black hairs, bearing fronds up to 3 cm. apart.
Stipe up to 2·5 cm. long, narrowly winged in the upper ½.   Lamina up to 7×2.5
cm., narrowly oblong to lanceolate or narrowly ovate, 2–3-pinnatifid.   Pinnae
pinnatifid to 2-pinnatifid into rounded entire lobes c. 0·7 mm. broad;   pinnae
dark-green when fresh but the lobes folded longitudinally on drying.   Sori c.
1·5 mm. long and c. 1 mm. in diam., conical, very narrowly winged by the lamina
for part of its length, with entire rounded valves.

Zambia.   E: Lundazi Distr., Nyika Plateau, 6.v.1952, *White* 2728 (BM; K).   **Rho-
desia.**   E: Umtali, 17.ii.1926, *Eyles* 4451 (K; SRGH).   **Malawi.**   N: Misuku Distr.,
Misuku Hills, 11.i.1959, *Richards* 10607 (K).   C: Nchisi Mt., 19.ii.1959, *Robson* 1659 (K;
LISC; SRGH).   S: Mlanje Mt., 13.vii.1946, *Brass* 16823 (K; SRGH).   **Mozambique.**
N: Ribáuè Mt., 19.vii.1962, *Schelpe & Leach* 6940 (BOL).   Z: Namúli Mt., 26.vii.1962,
Schelpe & Leach 7038 (BOL).   MS: Penhalonga, Mahunguè Falls beyond border in
ravine, 8.ix.1957, *Chase* 6708 (BM; K; LISC; SRGH).
  Also in S. Africa, Tanzania, Kenya, Sudan, Ethiopia, Congo, Angola, Nigeria and
Madagascar.   Lithophyte and low-level epiphyte in forest, 1450–2160 m.

4. Trichomanes rigidum Sw., Prodr.: 137 (1788).—Sim, Ferns S. Afr. ed. 2: 68, t. 1
(1915). TAB. **22** fig. A.   Type from Jamaica.
  Trichomanes mandiocanum Raddi, Pl. Bras. **1**: 64, t. 79 fig. 2 (1825).   Type from
Brazil.
  Trichomanes cupressoides Desv. in Mém. Soc. Linn. Par. **6**, 2: 330)1827).—Tardieu
in Mém. Inst. Fr. Afr. Noire, **28**: 46, t. 5 fig. 7–8 (1953).   Type from Seychelles.
  Trichomanes dregei Van d. Bosch in Ned. Kruid. Archf. **4**: 372 (1859).   Type from
S. Africa.
  Selenodesmium rigidum (Sw.) Copel. in Philipp. Journ. Sci. **67**: 81 (1938).   Type
as for *Trichomanes rigidum*.
  Selenodesmium cupressoides (Desv.) Copel., loc. cit.   Type from Seychelles.

Rhizome c. 3 mm. in diam., shortly creeping to suberect, covered with dark-
brown hairs and bearing tufted fronds.   Stipe grey-brown when dried, up to 16
cm. long, with a few dark-brown hairs at the base.   Lamina up to 20×10 cm., 3–4
pinnatifid, narrowly ovate or lanceolate to oblong-lanceolate.   Rhachis narrowly
winged.   Pinnae 2–3-pinnatifid into numerous acute to rounded lobes c. 0·2 mm.
broad.   Sori c. 1·5 mm. long and c. 0·6 mm. in diam., borne near the costae of the
pinnae, turbinate with rounded entire valves, often with a long persistent receptacle.

Zambia.   N: Kasama Distr., 33 km. ESE. of Kasama, 18.ii.1961, *Robinson* 4394 (K
SRGH).   **Rhodesia.**   E: Melsetter, Chimanimani Mts., 16.x.1950, *Wild* 3635 (BM;
SRGH).   **Malawi.**   N: Nkata Bay, Makwadzi Forest, 12.v.1966, *Whellan* 2277 (SRGH).
S: Mt. Mlanje, Luchenya Plateau, 13.vii.1946, *Brass* 16819 (SRGH).   **Mozambique.**
Z: Gúruè, Serra de Gúruè, 20.ix.1944, *Mendonça* 2161 (LISC).   MS: Chimanimani
Mts., R. Mevumozi, 7.vii.1949, *Pedro & Pedrógão* 7307 (BOL).
  Also in S. Africa and widespread in tropical Africa and America.   Shaded stream-banks
in forest, 1370–1890 m.

2. HYMENOPHYLLUM Sm.

Hymenophyllum Sm. in Mem. Acad. Turin, **5**: 418 (1793).

Rhizome filiform, widely creeping, subglabrous.   Fronds pinnately divided to
3-pinnatifid, or flabellate, glabrous or set with stellate hairs, segments entire
or serrate; veins free.   Sori marginal with 2 indusial valves which enclose the
receptacle.

The genus *Hymenophyllum*, as construed here, comprises some 300 spp. and is largely confined to the tropics and the temperate and subantarctic zones of the southern Hemisphere.

Fronds with stellate hairs - - - - - - - - - 5. *capillare*
Fronds glabrous:
    Ultimate segments of the frond dentate; pinnae divided on both sides; indusial lobes
    dentate - - - - - - - - - - 4. *tunbridgense*
    Ultimate segments of the frond entire:
        Frond flabellate; indusial lobes dentate - - - - - 1. *sibthorpioides*
        Frond linear, oblong or triangular; indusial lobes entire:
            Basal pinnae with less than 5 lobes - - - - - - - 2. *capense*
            Basal pinnae with more than 10 lobes - - - - - 3. *polyanthos*

1. **Hymenophyllum sibthorpioides** (Bory ex Willd.) Mett. ex Kuhn, Fil. Afr.: 41 (1868). TAB. **22** fig. I. Type from Réunion.
    Trichomanes parvulum Poir. in Lam., Encycl. Méth. Bot. **8**: 64 (1808). Type from Madagascar.
    Trichomanes sibthorpioides Bory ex Willd. in L., Sp. Pl., ed. 4, **5**: 498 (1810). Type as for *Hymenophyllum sibthorpioides*.
    Trichomanes thouarsianum C. Presl, Hymenophyll.: 16, 40 (1844) (1845). Type from Réunion.
    Microtrichomanes parvulum (Poir.) Copel. in Philipp. Journ. Sci. **67**: 37 (1938). Type as for *Trichomanes parvulum*.

Rhizome filiform, bearing fronds up to 2·5 cm. apart. Stipe 0·6–3·2 cm. long, filiform, not winged. Frond up to 1·5 cm. in radius, flabellate, palmately divided into up to 30 linear entire rounded lobes, up to 4 × 1·5 mm. Sori at the apices of the lobes, obconic with erose-dentate ovate indusial valves c. 1 mm. in diam.; lamina flanking lower part of sorus.

    Rhodesia. E: Melsetter, Chimanimani Mts., 23.iv.1957, *Whellan* 1271 (BM; BOL; LISC; SRGH). **Malawi.** S: Mlanje Mt., Ruo Gorge, 14.v.1963, *Wild* 6251 (SRGH). **Mozambique.** Z: Namúli, 1887, *Last* (K). MS: Chimanimani Mts., 8.ii.1958, *Mitchell* 321 (BM; BOL; COI; K; SRGH).
    Also in Tanzania, Madagascar, Comoros, Mauritius and Réunion. Epiphytic or lithophytic on moist shaded rocks in forest, 1370–1830 m.

2. **Hymenophyllum capense** Schrad. in Gött. Gel. Anz. **1818**: 919 (1818). Type from S. Africa.
    Hymenophyllum natalense Van d. Bosch in Ned. Kruid. Archf. **4**: 386 (1859). Type from S. Africa.
    Hymenophyllum zeyheri Van d. Bosch, tom cit.: 388 (1859). Type from S. Africa.
    Hymenophyllum tabulare Van d. Bosch, tom. cit.: 397 (1859). Type from S. Africa.
    Hymenophyllum fumarioides sensu Sim, Ferns. S. Afr. ed. 2: 74, t. 3 fig. 2 (1915).
    Hymenophyllum parvum C. Chr. in Acad. Malgache, Pl. Madag., Ptérid.: 18 (1932); in Dansk Bot. Ark. **7**: 8, t. 2 fig. 1–3 (1932). Type from Madagascar.

Rhizome filiform, creeping, with fronds 0·5–7 cm. apart. Stipe up to 5 cm. long, filiform, narrowly winged for part of its length. Lamina up to 10 × 1·6 cm., usually 2-pinnatifid, lanceolate to narrowly oblong with the lowest pinnae often reduced. Pinnae up to 1 cm. long, entire, bifurcate to pinnatifid into 3–7 entire closely spaced lobes up to 1·5 mm. broad with rounded apices. Sori on acroscopic segments anywhere in the upper ½ of the frond, broadly obconic with rounded entire indusial valves 2 mm. broad and with the lamina segments usually dilated about the sorus.

    Malawi. S: Mlanje Distr., 30.vi.1946, *Brass* 16547 (K). **Mozambique.** Z: Gúruè, Serra de Gúruè, 1.x.1941, *Torre* 3569 (LISC). MS: Gorongosa Mt., Gogogo Peak, 4.vii.1955, *Schelpe* 5495 (BM; BOL).
    Also in S. Africa, Tanzania and Madagascar. Lithophytic in sheltered rock crevices. 1500–1890 m.

3. **Hymenophyllum polyanthos** Sw. in Schrad., Journ. Bot. **1800**, 2: 102 (1801). Type from Jamaica.

Rhizome filiform, bearing fronds up to 4 cm. apart. Stipe up to 2·5 cm. long, filiform, narrowly winged at least in the upper ½. Lamina up to 45 × 4·5 cm,.

narrowly elliptic to lanceolate, triangular or narrowly oblong. Rhachis narrowly winged. Pinnae up to 5 × 1·3 cm., pinnatifid to 2-pinnatifid into up to 180 linear entire rounded lobes each up to 6 × 2·5 mm. Sori on acroscopic basal segments of pinnae, very broadly obconic with ovate-acute entire indusial valves up to 1·8 mm. broad.

Ultimate lobes set widely apart; lamina usually lanceolate -   -   var. *mossambicense*
Ultimate lobes set close together or overlapping; lamina usually oblong to elliptic in large
plants -   -   -   -   -   -   -   -   -   -   -   -   -   - var. *kuhnii*

Var. **mossambicense** Schelpe in Bol. Soc. Brot., Sér. 2, **40**: 157 (1966).   TAB. **22** fig. F·
Type: Mozambique, Gorongosa Mt., *Schelpe* 5540 (BM, isotype; BOL, holotype)·

Rhodesia. E: Inyanga, Pungwe Gorge, 15.vii.1955, *Schelpe* 5716 (BM; BOL; K;
Malawi. S: Mlanje Mt., 25.vi.1946, *Brass* 16413 (K; SRGH).   **Mozambique.** MS:
Gorongosa Mt., 6.vii.1955, *Schelpe* 5540 (BM; BOL).
Known only from our area.   Shaded boulders on forest stream banks, 1520–1890 m.

Var. **kuhnii** (C. Chr.) Schelpe in Bol. Soc. Brot., Sér. 2, **40**: 156 (1966).   TAB. **22** fig. G.
Type from Tanzania.
    Hymenophyllum meyeri Kuhn in Engl., Hochgebirgsfl. Trop. Afr.: 95 (1892) non
C. Presl (1843).   Type from Tanzania.
    Hymenophyllum kuhnii C. Chr., Ind. Fil.: 363 (1905).—Taton in Bull. Soc. Roy.
Bot. Belg. **78**: 18, t. 1 fig. K–L (1946).   Type from Tanzania.
    Hymenophyllum henkelii Sim in S. Afr. Journ. Sci. **20**: 309, t. 9 (1923).   Type:
Rhodesia, Inyanga, *Eyles* 2559 (K; PRE).
    Mecodium kuhnii (C. Chr.) Copel. in Philipp. Journ. Sci. **67**: 19 (1938).   Type as
for *Hymenophyllum polyanthos* var. *kuhnii*.

Rhodesia. E: Umtali, Pioneer Farm, vii.1948, *Fisher & Schweickerdt* 321 (BM; K;
NU; SRGH).   **Malawi.** N: Nyika Plateau, 16.viii.1946, *Brass* 17260 (K).   S: Mlanje
Mt., 13.vii.1946, *Brass* 16822 (K; SRGH).   **Mozambique.** Z: Namúli, 1887, *Last* (K).
MS: Gorongosa Mt., Gogogo Peak, 6.vii.1955, *Schelpe* 5541 (B; BM; BOL; K; P;
US).
Also in Tanzania, Kenya, Congo, Uganda, Cameroon, Nigeria, Sierra Leone, Liberia,
Fernando Po and S. Tomé.   Common epiphyte in montane forest, 1700–2250 m.

4. **Hymenophyllum tunbridgense** (L.) Sm. in Sowerby, English Bot. **3**: t. 162 (1794).—
    Sim, Ferns S. Afr. ed. 2: t. 3 fig. 1 (1915).—Tardieu in Mém. Inst. Fr. Afr. Noire,
    28: t. 3 fig. 10–11 (1953).   TAB. **22** fig. E.   Syntypes from England and Italy.
    Trichomanes tunbridgense L., Sp. Pl. **2**: 1098 (1753).   Syntypes as above.
    Hymenophyllum dregeanum C. Presl, Hymenophyll.: 32, 52 (1844) reimpr. Abh.
Königl. Böhm. Ges. Wiss. **5**, 3: 124, 144 (1845).   Type from S. Africa.
    Hymenophyllum thomassetii C. H. Wright in Kew Bull. **1906**: 170 (1906).   Type:
Malawi, Mt. Mlanje, *Thomasset* (K).

Rhizome filiform, bearing fronds up to 7 cm. apart. Stipe filiform, narrowly winged in upper part. Rhachis narrowly winged. Lamina up to 10 × 3 cm., lanceolate, narrowly oblong or narrowly elliptic. Pinnae bilaterally pinnatifid to 2-pinnatifid into up to 25 linear serrate rounded lobes c. 1 mm. broad. Sori usually borne on the lowest acroscopic pinnules with ovate serrate indusial valves c. 1.1 mm. broad, the base of the sorus not winged by the lamina lobes.

Rhodesia. E: Inyanga, 25.vii.1957, *Chase* 6678 (BM; BOL; K; SRGH).   **Malawi.**
S: Mlanje, *Thomasset* (K).   **Mozambique.** MS: Gorongosa Mt., Gogogo Peak,
4.vii.1955, *Schelpe* 5494 (BM; BOL).
Also in Tanzania, Kenya, central Madagascar, Madeira, S. Africa, W. and S. Europe
and also in Gabon (*fide* Tardieu).   Sheltered rock faces, 1700–2440 m.

5. **Hymenophyllum capillare** Desv. in Mém. Soc. Linn. Par. **6**, 2: 333 (1827).—Tar-
    dieu in Mém. Inst. Fr. Afr. Noire, **28**: 36, t. 3 fig. 1 (1953).   TAB. **22** fig. H.   Type
    probably from Mascarene Is.
    Hymenophyllum pendulum Bory in Bélanger, Voy. Bot. **2**: 81, t. 8 fig. 2 (1833).
Type from Mascarene Is.
    Sphaerocionium pendulum (Bory) C. Presl, Hymenophyll.: 34 (1844) reimpr. in
Abh. Königl. Böhm. Ges. Wiss., ser. 5, **3**: 126 (1845).   Type as above.
    Hymenophyllum fulvum Van d. Bosch in Ned. Kruid. Archf. **5**: 3, 196 (1863).
Type from Madagascar.
    Hymenophyllum lineare sensu Sim, Ferns S. Afr. ed. 2: 78, t. 5 fig. 2 (1915).
    Hymenophyllum holotrichum A. Peter in Fedde, Repert. Beih. **40**: 16, t. 1 fig. 3–4
(1929).   Type from Tanzania.

Rhizome filiform, bearing fronds 1–5 cm. apart. Stipe 1–3·5 cm. long, filiform, not winged but clothed with stellate hairs. Rhachis not winged, bearing stellate hairs. Lamina up to 15 × 2·3 cm., linear. Pinnae pinnatifid (with frequent 2-lobing of acroscopic basal segment) into up to 7 linear entire, broadly acute to rounded lobes 9 × 1·5 mm., set with stalked stellate hairs on the margins and veins. Sori 1–1·2 mm. in diam., at the apices of the lobes, obconic with rounded indusial valves set with stellate hairs.

Rhodesia. E: Inyanga Distr., 11.viii.1950, *Chase* 3187 (BM; SRGH). **Malawi.** S: Mlanje Mt., 25.vi.1946, *Brass* 16420 (K; SRGH). **Mozambique.** Z: Serra de Gúruè, R. Napissóquè, 19.x.1942, *Mendonça* (LISC). MS: Manica, Chimanimani Mts., R. Mevumozi, 7.vii.1949, *Pedro & Pedrógão* 7308 (BOL).

Also from Tanzania, Kenya, Uganda, Congo, Cameroon, Ghana, S. Tomé, Madagascar, Réunion, Comoro Is. and S. Africa. Sheltered rock faces in forest, 1700–2140 m.

16. DENNSTAEDTIACEAE

Large terrestrial ferns with hairy creeping or erect rhizomes with single or complex double solenosteles. Stipes with undivided or dissected U-shaped vascular strands. Fronds large, 3-pinnate to 4-pinnate, with continuous apical growth in *Hypolepis*. Veins free or anastomosing; sori marginal, submarginal or superficial near the margin, small, subcircular to elongate. Indusium absent, or ovate and shallowly cupped (*Microlepia*), or linear or with the leaf margin modified to form a pseudo-indusium (*Hypolepis*), or both pseudo-indusia and true indusia present; soral paraphyses present or absent.

Veins anastomosing freely:
  Fronds pubescent to hirsute, at least on the under surface -   -   1. **Blotiella**
  Fronds glabrous and glaucous -   -   -   -   -   -   - 2. **Histiopteris**
Veins mostly free, rarely anastomosing:
  Sori elongate, marginal or apparently marginal:
    Soral paraphyses present; frond with very short flattened squamulose hairs on the upper surface -   -   -   -   -   -   -   -   - 3. **Lonchitis**
    Soral paraphyses absent; upper surface of frond glabrous or with a few cylindric long hairs -   -   -   -   -   -   -   -   - 4. **Pteridium**
  Sori small, subcircular:
    Sori opening towards the margin; spores trilete -   -   -   - 5. **Microlepia**
    Sori opening inwards; spores monolete -   -   -   -   - 6. **Hypolepis**

1. BLOTIELLA Tryon

Blotiella Tryon in Contr. Gray Herb. **191**: 96 (1962).

Rhizome erect or creeping, often massive, bearing tufted large fronds and with reddish-brown hairs. Lamina pinnate to 3-pinnate, usually thinly pubescent at least on the dorsal surface, the ultimate segments sinuate or crenate, veins anastomosing freely; sori marginal, either small and confined to the bases of sinuses or elongate and extending around the sinuses and lobes; indusia marginal, membranous.

A genus of 16 spp., all in Africa, Madagascar and the Mascarene Is. except for 1 American sp. A taxonomically difficult genus because of the variation in dissection exhibited by fronds of different sizes in the same species and of inadequate collections of complete fronds.

Pinnae nearly uniformly incised from apex to base; rhizome creeping:
  Pinna lobes very close together, with narrow sinuses between them; frond chartaceous
                                               1. *crenata*
  Pinna lobes spaced with deep broad sinuses between them, and with pinna costae winged for much of their length; frond thinly herbaceous -   -   - 2. *glabra*
Pinnae progressively more deeply incised from apex to base; rhizome erect:

Rhizome with pale-brown hairs; stipe c. 1 m. long; sori subcircular (sometimes with a few larger lunulate ones in the larger sinuses) - - - 3. *natalensis*
Rhizome with ferrugineous hairs; stipe up to 60 cm. long; sori linear - 4. *currorii*

1. **Blotiella crenata** (Alston) Schelpe in Bol. Soc. Brot., Sér. 2, **41**: 211 (1967). TAB. **23**. Type from the Congo.
   Lonchitis crenata Alston in Bol. Soc. Brot., Sér. 2, **30**: 18 (1956). Type as above.

Rhizome c. 1·5 cm. in diam., creeping, with closely spaced fronds and with reddish-brown hairs c. 5 mm. long. Stipe stramineous, densely set at first with small pale hairs c. 0·8 mm. long, becoming glabrous with age. Lamina lanceolate in outline, pinnate in smaller fronds to deeply 2-pinnatifid or rarely 2-pinnate, chartaceous, apex with a prominent deeply crenate cultrate segment grading into petiolulate pinnae; pinnae mostly petiolulate, narrowly triangular often with the lowest segments the largest, acute, sinuate to deeply lobed into sinuate-oblong, obtuse to bluntly acute lobes or rarely cut into subsessile, very narrowly oblong sinuate acute pinnules; ventral surface dark-green almost shining but with curved white hairs set densely along the costa and thinly along the veins; dorsal surface paler dull-green with white hairs up to 2 mm. long on the costa, costules and veins and infrequently in the areoles; rhachis and costa stramineous and pilose with squarrose white scales 2–4 mm. long. Sori up to 2 mm. in diam., mostly semi-circular, set in the smaller sinuses but with larger lunulate sori in the larger sinuses; indusia pale, membranous.

Zambia. N: Abercorn Distr., Abercorn-Mporokoso road, 28.ix.1956, *Richards* 6320 (K). W: c. 48 km. W. of Mwinilunga, 26.x.1966, *Leach & Williamson* 13480 (BOL; SRGH).
Also in Angola, Congo and Uganda. Shaded streambanks in fringing forest and in swamp forest, 1200–1650 m.

2. **Blotiella glabra** (Bory) Tryon in Contr. Gray Herb. **191**: 99 (1962). Type from Réunion.
   Lonchitis glabra Bory, Voy. Quatre Princ. Iles, **1**: 321 (1804). Type as above.
   Lonchitis stenochlamys Fée, Mém. Fam. Foug. **5**: 142 (1852). Type from S. Africa.
   Pteris glabra (Bory) Mett., Fil. Hort. Bot. Lips.: 59, t. 25 fig. 29 (1856). Type as for *Blotiella glabra*.
   Lonchitis pubescens sensu Sim, Ferns S. Afr. ed. 2: 261, pro parte quoad t. 132 (1915) non Willd. ex Kaulf. (1824).
   Lonchitis gracilis Alston in Exell, Cat. Vasc. Pl. S. Tomé, Suppl.: 7 (1956). Type from Fernando Po.

Rhizome creeping, massive, with closely spaced fronds and clothed with a felt of reddish-brown hairs up to 6 mm. long. Stipe pale-brown, up to 80 cm. long, thinly pubescent with pale hairs 0·5–2 mm. long at first, later becoming subglabrous. Lamina up to 1·3 ×0·8 m., elliptic in outline with the lowest pinnae reduced, 2-pinnatifid to 3-pinnatifid, incised to nearly the same depth from apex to base, and with the pinna costa narrowly winged for at least ¾ its length; pinna lobes oblong, adnate to the rhachis, acute to acuminate, crenate, sinuate or pinnatifid into sinuate lobes, the pinna lobes separated by broad sinuses, pubescent with pale soft hairs up to 1·5 mm. long on the costae and costules and less densely on veins and occasionally in the areoles, on both surfaces. Sori up to 2 mm. in diam., mostly semicircular in the small sinuses of the frond segments, longer and lunulate in the larger sinuses; indusia membranous.

Zambia. N: Kawambwa Distr., Ntimbacushi Falls, 22.vi.1957, *Robinson* 2402 (SRGH). **Rhodesia.** E: Umtali Distr., Vumba Mts., 28.iii.1956, *Chase* 6048 (B; BM; BOL; K; P; SRGH). **Mozambique.** MS: Gorongosa Mt., Gogogo Peak, 5.vii.1955, *Schelpe* 5505 (BM; BOL).
Also in S. Africa, the tropical African mountains and Réunion. Moist and shaded forest floors, 1600–1770 m.

3. **Blotiella natalensis** (Hook.) Tryon in Contr. Gray Herb. **191**: 99 (1962). Type from S. Africa.
   Lonchitis natalensis Hook., Sp. Fil. **2**: 57, t. 89 fig. B (1851). Type as above.
   Lonchitis pubescens sensu Sim, Ferns S. Afr. ed. 2: 261, pro parte quoad t. 131 (1915) non Willd. ex Kaulf. (1824).

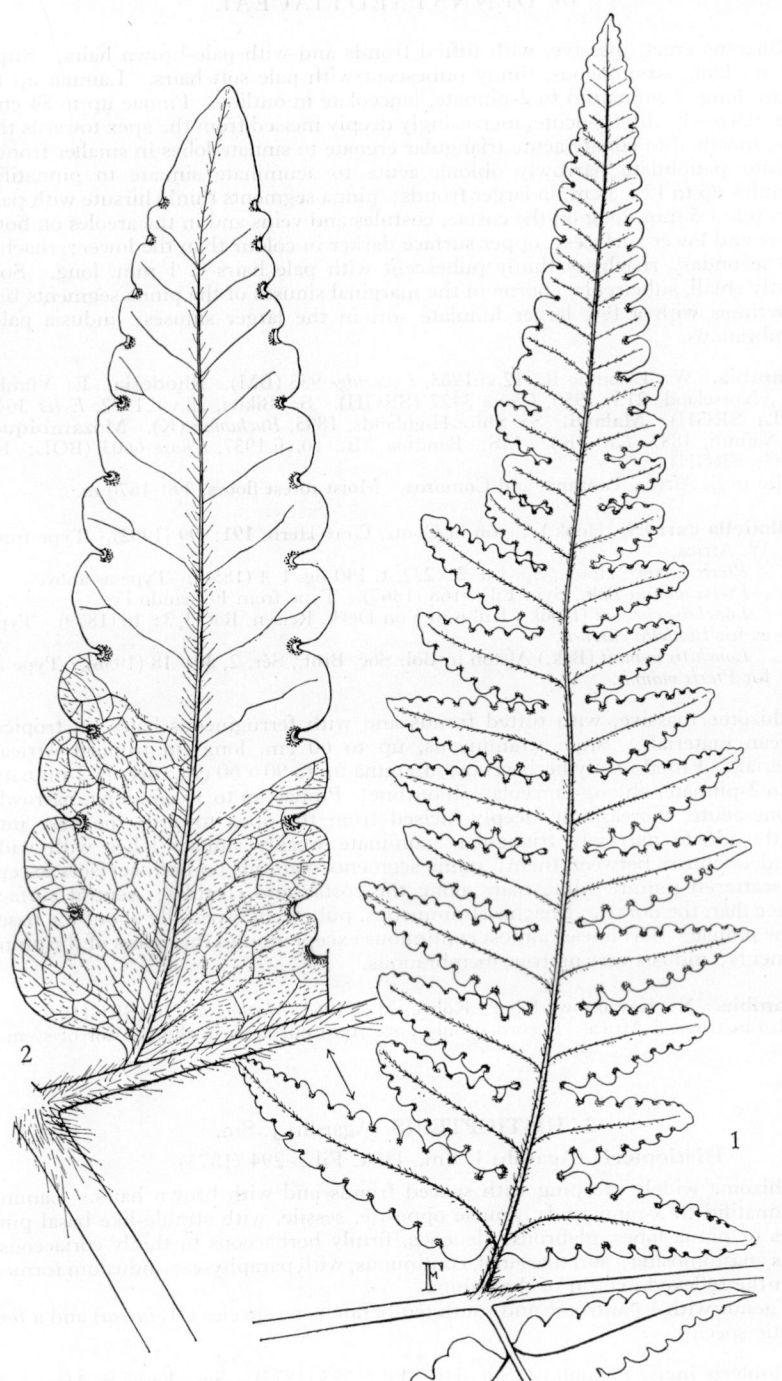

Tab. 23. BLOTIELLA CRENATA. 1, lower pinna (× ⅔); 2, enlargement of pinna lobe (×2), both from *White* 3174.

Rhizome erect, massive, with tufted fronds and with pale-brown hairs. Stipe c. 1 m. long, stramineous, thinly pubescent with pale soft hairs. Lamina up to 1·5 m. long, 2-pinnatifid to 2-pinnate, lanceolate in outline. Pinnae up to 54 cm. long, narrowly oblong-acute, increasingly deeply incised from the apex towards the base, mostly into adnate acute triangular crenate to sinuate lobes in smaller fronds or into petiolulate narrowly oblong acute to acuminate sinuate to pinnatifid pinnules up to 12 × 3 cm. in larger fronds; pinna segments thinly hirsute with pale hairs 0·5–1·5 mm. long on the costae, costules and veins and in the areoles on both upper and lower surfaces; upper surface darker in colour than the lower; rhachis and secondary rhachises thinly pubescent with pale hairs c. 1 mm. long. Sori mostly small, subcircular, borne in the marginal sinuses of the pinna segments but sometimes with a few larger lunulate sori in the larger sinuses; indusia pale, membranous.

Zambia. W: Lisombo R., 12.vi.1963, *Loveridge* 935 (BM). **Rhodesia.** E: Vumba Mts., Norseland, 21.iv.1950, *Chase* 3427 (SRGH). S: Bikita, 28.viii.1922, *Eyles* 3648 (BOL; SRGH). **Malawi.** S: Shire Highlands, 1885, *Buchanan* (K). **Mozambique.** Z: Namúli, 1887, *Last* (K). MS: Bandula Mt., 10.vii.1957, *Chase* 6603 (BOL; K; LISC; SRGH).

Also in S. Africa, Tanzania and Comoros. Moist forest floors, 390–1670 m.

4. **Blotiella currorii** (Hook.) Tryon in Contr. Gray Herb. **191**: 99 (1962). Type from W. Africa.

> *Pteris currorii* Hook., Sp. Fil. **2**: 232, t. 140 fig. 1–4 (1858). Type as above.
> *Pteris mannii* Bak., Syn. Fil.: 168 (1867). Type from Fernando Po.
> *Lonchitis currorii* (Hook.) Kuhn in Von Deck. Reisen, Bot. **3**, 3: 10 (1879). Type as for *Blotiella currorii*.
> *Lonchitis mannii* (Bak.) Alston in Bol. Soc. Brot., Sér. 2, **30**: 18 (1956). Type as for *Pteris mannii*.

Rhizome massive, with tufted fronds and with ferrugineous hairs (in tropical African material). Stipe stramineous, up to 60 cm. long (in tropical African material). Frond firmly herbaceous. Lamina up to 90 × 60 cm., deeply 2-pinnatifid to 2-pinnate, oblong-lanceolate in outline. Pinnae up to 35 cm. long, narrowly oblong-acute, increasingly deeply incised from the apex towards the base into mostly adnate narrowly triangular acuminate weakly sinuate lobes with wide rounded sinuses between them; pinna segments glabrous on both surfaces except for scattered minute white hairs along the costae and costules; ventral surface darker than the dorsal. Rhachis stramineous, pubescent especially about the bases of the pinnae. Sori linear, almost continuous except around the apices of the pinna segments; indusia very narrow, membranous.

Zambia. N: Kawambwa Distr., Kolwe, 21.ix.1963, *Mutimushi* 430 (SRGH).

Also in tropical Africa. Recorded only once from our area, in the interior of swamp forest.

2. HISTIOPTERIS (Agardh) J. Sm.

Histiopteris (Agardh) J. Sm., Hist. Fil.: 294 (1875).

Rhizome widely creeping with spaced fronds and with brown hairs. Lamina 2-pinnatifid to 3-pinnatifid; pinnae opposite, sessile, with stipule-like basal pinnules or pinna lobes, glabrous, glaucous, firmly herbaceous to thinly coriaceous; veins anastomosing; sori marginal, continuous, with paraphyses; indusium formed from the reflexed margin of the lamina.

A genus with 1 pantropic and south temperate zone species (*H. incisa*) and a few Asiatic species.

Histiopteris incisa (Thunb.) J. Sm., Hist. Fil.: 295 (1875).—Sim, Ferns S. Afr. ed. 2: 263, t. 133 (1915). TAB. **24.** Type from S. Africa.

> *Pteris incisa* Thunb., Prodr. Pl. Cap.: 171 (1800). Type as above.
> *Pteris vespertilionis* Labill., Nov. Holl. Pl. Sp. **2**: 96, t. 245 (1806). Type from Australia.
> *Pteris brunoniana* Endl., Prodr. Fl. Norfolk.: 12 (1833). Type from Norfolk I. (New Zealand).

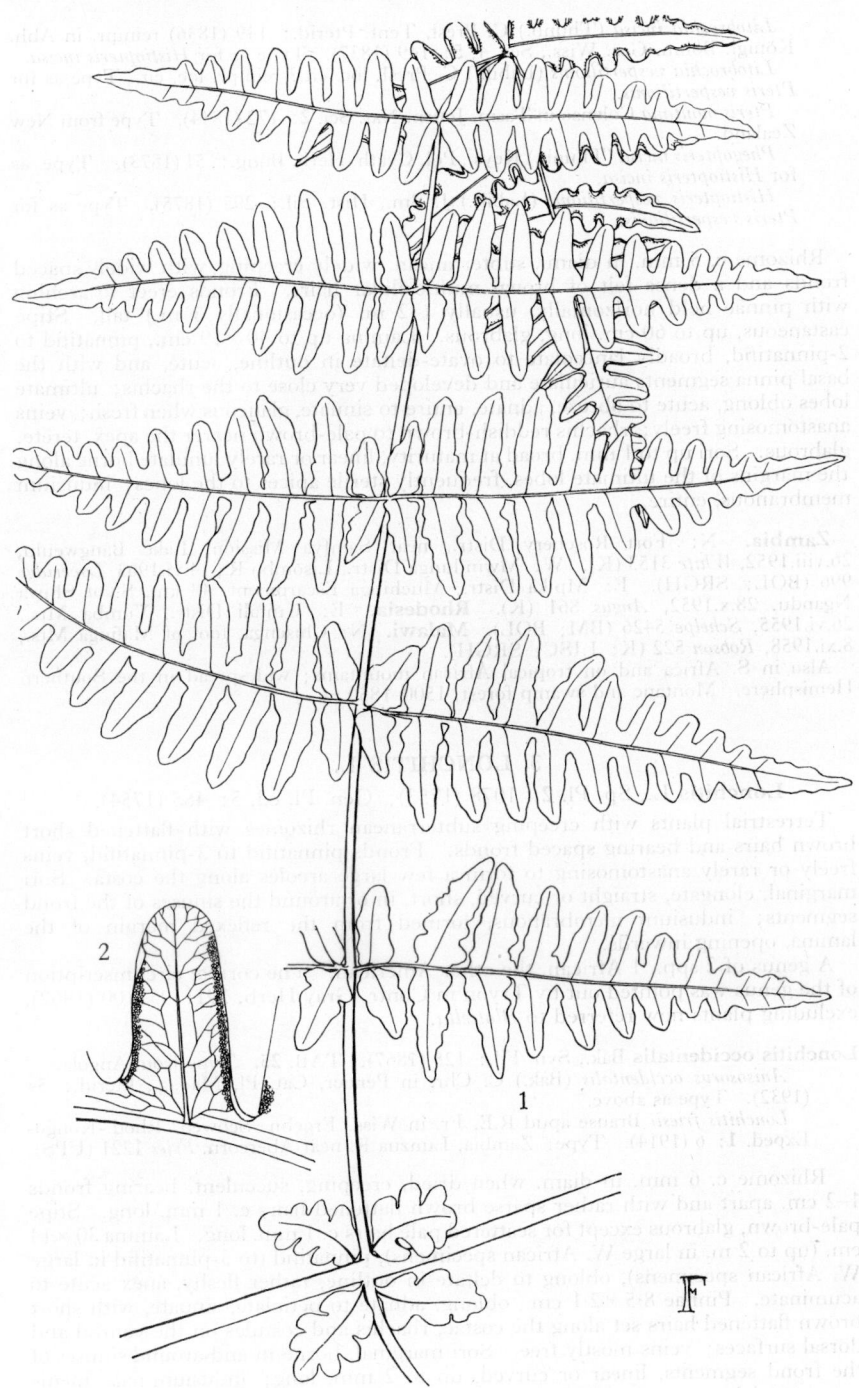

Tab. 24. HISTIOPTERIS INCISA. 1, pinna (×⅔); 2, fertile pinna segment (×2), both from
Fisher & Schweickerdt 336.

Litobrochia incisa (Thunb.) C. Presl, Tent. Pterid.: 149 (1836) reimpr. in Abh.
Königl. Böhm. Ges. Wiss., Ser. 4, **5**: 149 (1837). Type as for *Histiopteris incisa.*
Litobrochia vespertilionis (Labill.) C. Presl, loc. cit.? reimpr. loc. cit. Type as for
Pteris vespertilionis.
Pteris montana Colenso in Tasm. Journ. Nat. Sci. **2**: 172 (1844). Type from New
Zealand.
Phegopteris incisa (Thunb.) Keys., Pol. Cyath. Herb. Bung.: 51 (1873). Type as
for *Histiopteris incisa.*
Histiopteris vespertilionis (Labill.) J. Sm., Hist. Fil.: 295 (1875). Type as for
Pteris vespertilionis.

Rhizome c. 5 mm. in diam., subterranean, widely creeping, with widely spaced
fronds and a dense felt of brown multicellular hairs. Fronds erect to arching
with pinnae held horizontally, usually 1–2 m. (occasionally 3 m.) tall. Stipe
castaneous, up to 60 cm. long, glabrous. Lamina up to 40 × 19 cm., pinnatifid to
2-pinnatifid, broadly lanceolate to ovate-deltate in outline, acute, and with the
basal pinna segments auriculate and developed very close to the rhachis; ultimate
lobes oblong, acute to obtuse, adnate, entire to sinuate, glaucous when fresh; veins
anastomosing freely; rhachis reddish-brown to pale-brown nearer the apex, terete,
glabrous. Sori up to 1 mm. broad at maturity, linear or rarely lunulate borne along
the margins of the ultimate lobes, frequently sterile apices to the lobes; indusium
membranous, entire.

Zambia. N: Fort Rosebery Distr., near Samfya Mission, Lake Bangweulu,
26.viii.1952, *White* 3153 (K). W: Mwinilunga Distr., Lisombo R., 14.vi.1963, *Loveridge*
996 (BOL; SRGH). E: Mpika Distr., Muchinga Escarpment, 48 km. S. of Shiwa
Ngandu, 28.x.1952, *Angus* 861 (K). **Rhodesia.** E: Umtali Distr., Vumba Mts.,
26.vi.1955, *Schelpe* 5426 (BM; BOL). **Malawi.** N: Chisenga, foot of Mafinga Mts.,
8.xi.1958, *Robson* 522 (K; LISC; SRGH).
Also in S. Africa and on tropical African mountains; widespread in the Southern
Hemisphere. Montane and swamp forest, 1500–1850 m.

3. LONCHITIS L.

Lonchitis L., Sp. Pl. **2**: 1078 (1753); Gen. Pl. ed. 5: 485 (1754).

Terrestrial plants with creeping subterranean rhizomes with flattened short
brown hairs and bearing spaced fronds. Fronds pinnatifid to 3-pinnatifid, veins
freely or rarely anastomosing to form a few large areoles along the costa. Sori
marginal, elongate, straight or curved, short, in or around the sinuses of the frond
segments; indusium membranous, formed from the reflexed margin of the
lamina, opening inwards.

A genus of 2 spp., 1 African, the other American. The correct circumscription
of the genus was pointed out by Tryon in Contr. Gray Herb. **191**: 93–100 (1962),
excluding plants now referred to *Blotiella.*

Lonchitis occidentalis Bak., Syn. Fil.: 128 (1867). TAB. **25**. Type from Angola.
Anisosorus occidentalis (Bak.) C. Chr. in Perrier, Cat. Pl. Madag., Ptérid.: 54
(1932). Type as above.
Lonchitis friesii Brause apud R.E. Fr. in Wiss. Ergebn. Schwed. Rhod.-Kongo-
Exped. **1**: 6 (1914). Type: Zambia, Lunzua R. near Abercorn, *Fries* 1221 (UPS).

Rhizome c. 6 mm. in diam. when dried, creeping, succulent, bearing fronds
1–2 cm. apart and with rather sparse brown flattened hairs c. 1 mm. long. Stipe
pale-brown, glabrous except for scattered pale hairs c. 1 mm. long. Lamina 30 × 14
cm. (up to 2 m. in large W. African specimens), pinnatifid (to 3-pinnatifid in large
W. African specimens), oblong to deltate in outline, rather fleshy, apex acute to
acuminate. Pinnae 8·5 × 2·1 cm., oblong, adnate to petiolate, sinuate, with short
brown flattened hairs set along the costae, rhachis and costules on the ventral and
dorsal surfaces; veins mostly free. Sori marginal, borne in and around sinuses of
the frond segments, linear or curved, up to 2 mm. long; indusium pale, membranous, glabrous.

Zambia. N: Abercorn Distr., Lunzua Falls, 18.vii.1958, *Whellan* 1544 (K; SRGH).
Widespread in tropical Africa and in Madagascar. Recorded in our area from a river-
bank in the spray of a waterfall.

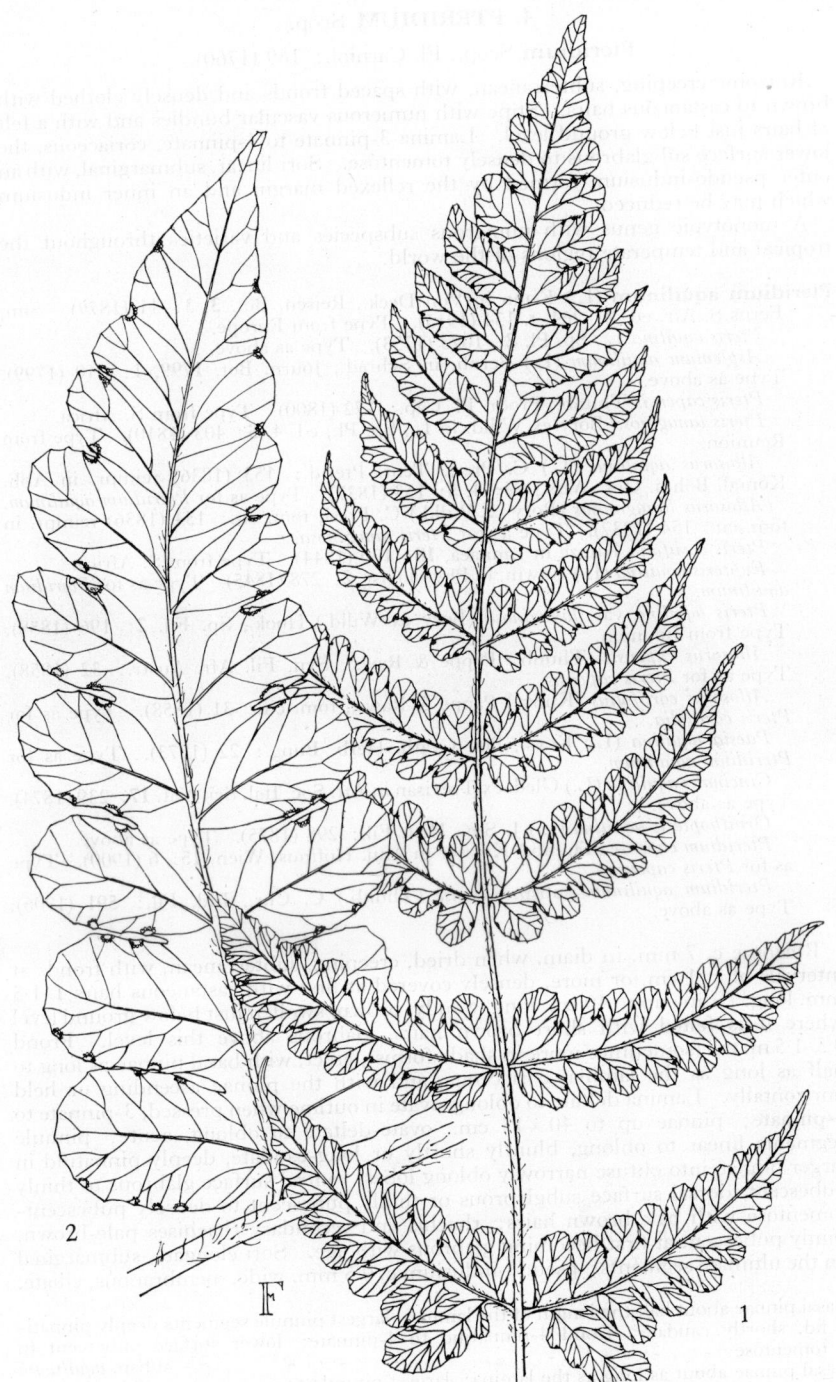

Tab. 25. LONCHITIS OCCIDENTALIS. 1, small frond (× ⅔); 2, enlargement of upper pinna of small frond (× 2), both from *Whellan* 1544.

4. PTERIDIUM Scop.

Pteridium Scop., Fl. Carniol.: 169 (1760).

Rhizome creeping, subterranean, with spaced fronds and densely clothed with brown to castaneous hairs. Stipe with numerous vascular bundles and with a felt of hairs just below ground level. Lamina 3-pinnate to 4-pinnate, coriaceous, the lower surface subglabrous to densely tomentose. Sori linear, submarginal, with an outer pseudo-indusium formed by the reflexed margin and an inner indusium which may be reduced.

A monotypic genus with numerous subspecies and varieties throughout the tropical and temperate regions of the world.

Pteridium aquilinum (L.) Kuhn in Von Deck., Reisen, Bot. **3**, 3: 11 (1879).—Sim, Ferns S. Afr. ed. 2: 264, t. 134 (1915). Type from Europe.
    Pteris aquilina L., Sp. Pl. **2**: 1073 (1753). Type as above.
    Asplenium aquilinum (L.) Bernh. in Schrad., Journ. Bot. **1799**, 1: 310 (1799). Type as above.
    Pteris capensis Thunb., Prodr. Pl. Cap.: 172 (1800). Type from S. Africa.
    Pteris lanuginosa Bory ex Willd. in L., Sp. Pl., ed. 4, **5**: 403 (1810). Type from Réunion.
    Allosorus aquilinus (L.) C. Presl, Tent. Pterid.: 153 (1836) reimpr. in Abh. Königl. Böhm. Ges. Wiss., Sér. 4, **5**: 153 (1837). Type as for *Pteridium aquilinum*.
    Allosorus lanuginosus (Bory ex Willd.) C. Presl, tom. cit.: 154 (1836) reimpr. in tom. cit.: 154 (1837). Type as for *Pteris lanuginosa*.
    Pteris coriifolia Kunze in Linnaea, **18**: 120 (1844). Type from S. Africa.
    Eupteris aquilina (L.) Newm. in Phytologist, **2**: 278 (1845). Type as for *Pteridium aquilinum*.
    Pteris aquilina var. *lanuginosa* (Bory ex Willd.) Hook., Sp. Fil. **2**: 196 (1858). Type from Réunion.
    Allosorus capensis (Thunb.) Pappe & Raws., Syn. Fil. Afr. Austr.: 32 (1858). Type as for *Pteris capensis*.
    Allosorus coriifolius (Kunze) Pappe & Raws., tom. cit.: 31 (1858). Type as for *Pteris coriifolia*.
    Paesia aquilina (L.) Keys., Pol. Cyath. Herb. Bung.: 22 (1873). Type as for *Pteridium aquilinum*.
    Cincinalis aquilina (L.) Gled. ex Trevisan in Att. Soc. Ital. Sci. Nat. **17**: 239 (1874). Type as above.
    Ornithopteris aquilina (L.) J. Sm., Hist. Fil.: 298 (1875). Type as above.
    Pteridium capense (Thunb.) Krasser in Ann. Hofmus. Wien, **15**: 6 (1900). Type as for *Pteris capensis*.
    Pteridium aquilinum subsp. *capense* (Thunb.) C. Chr., Ind. Fil.: 591 (1906). Type as above.

Rhizome c. 7 mm. in diam. when dried, creeping, subterranean, with fronds at intervals of c. 1 cm. or more, densely covered at first with castaneous hairs 1–1·5 mm. long. Stipe up to 40 cm. long, woody, brown, swollen just below ground level where it is felted with short brown hairs, glabrous above this level. Frond 0·5–1·5 m. tall (sometimes more in shade forms), erect, with basal pinnae as long to half as long as the lamina, coriaceous, and with the pinnae ascending or held horizontally. Lamina deltate to oblong-ovate in outline when pressed, 3-pinnate to 4-pinnate; pinnae up to 40 × 15 cm., ovate-deltate to oblong, acute; pinnule segments linear to oblong, bluntly shortly or long-caudate, deeply pinnatifid in larger pinnae into obtuse narrowly oblong lobes, ventral surface glabrous to thinly pubescent, dorsal surface subglabrous or thinly pubescent to densely pubescent-tomentose with pale-brown hairs; rhachis and secondary rhachises pale-brown, thinly pubescent at first becoming glabrous with age. Sori elongate, submarginal on the ultimate segments; pseudo-indusium c. 0·5 mm. wide, membranous, ciliate.

Basal pinnae about half the length of the lamina; largest pinnule segments deeply pinnatifid, shortly caudate; frond 4-pinnatifid to 4-pinnate; lower surface pubescent to tomentose - - - - - - - - - - - - - - - - subsp. *aquilinum*
Basal pinnae about as long as the lamina; largest pinnule segments pinnatifid and long-caudate or linear entire; frond 3-pinnate to 4-pinnatifid; lower surface subglabrous to pubescent - - - - - - - - - - - - - - - subsp. *centrali-africanum*

Subsp. **aquilinum**

Zambia. N: Kalombo Farm, Saisi Valley, 21.v.1952, *Richards* 1832 (K). W: Kitwe, 16.xi.1967, *Mutimushi* 2342 (NDO; SRGH). **Rhodesia.** N: Sinoia, Hamling's Farm, 2.xi.1931, *Rattray* 381 (BM; SRGH). W: Matopo Hills, v.1904, *Eyles* 79 (BM; SRGH). C: Domboshawa, *Mitchell* 80 (BOL). E: Melsetter Distr., Cashel, 9.vii.1953, *Schelpe* 3998 (BM; BOL). S: Fort Victoria Distr., Rungwa R., viii1932, *Cuthbertson* (BM; SRGH). **Malawi.** N: Vipya, Champoyo Forest, 1.ii.1962, *Chapman* 1573 (K; SRGH). C: Nchisi Mt., 3.ix.1929, *Burtt Davy* 1119 (K). S: Mt. Chiradzura, ix.1831, *Meller* (K). **Mozambique.** N: Source of R. Msinga, 31.viii.1931, *Gomes e Sousa* 738 (K; LISC). Z: Namúli Mt., 25.vii.1962, *Schelpe & Leach* 7003 (BOL). T: Marávia, Vila Vasco da Gama, 12.viii.1941, *Torre* 3268 (LISC). MS: Gorongosa Mt., Gogogo Peak, 5.vii.1955, *Schelpe* 5507 (BM; BOL). SS: Bilene, Praia de S. Martinho, *Torre* 6707 (LISC). LM: Rikatla, 18.v.1890, *Junod* 10 (G; LISC; PRE).

Also in S. Africa, Madagascar and the Mascarene and Comoro Is. and widespread at higher altitudes through tropical Africa northwards to Ethiopia and westwards to Sierra Leone. Also widespread in temperate Europe. Grasslands and forest margins, 750–2340 m.

Subsp. **centrali-africanum** Hieron. in Wiss. Ergebn. Schwed. Rhod.-Kongo-Exped. **1**: 7 (1914). TAB. **26**. Syntypes from Angola, Congo, Tanzania and Zambia.
    Pteridium aquilinum subsp. *caudatum* var. *africanum* Bonap., Not. Ptérid. **1**: 62 (1915). Type from SW. Tanzania.
    Pteridium aquilinum var. *africanum* (Bonap.) Tryon in Rhodora, **43**: 51 (1941). Type as above.
    Pteridium centrali-africanum (Hieron.) Alston in Bol. Soc. Brot., Sér. 2, **30**: 22 (1956). Syntypes as for *Pteridium aquilinum* subsp. *centrali-africanum*.

Zambia. N: Abercorn Distr., Chilongowelo, 8.ii.1952, *Richards* 955 (K). W: 6·4 km. SW. of Ndola, 16.iii.1961, *Drummond & Rutherford-Smith* 6935 (K; SRGH). S: E. of Mumbwa, 19.vi.1932, *Trapnell* 2063 (K). **Rhodesia.** N: Mtoroshanga Pass, E. entrance, 5.iii.1961, *Phipps* 2881 (K; SRGH). S: 6·4 km. E. of Zimbabwe, *Pole Evans* 2723 (PRE). **Malawi.** S: Manganja Highlands, iv.1859, *Kirk* (K). **Mozambique.** N: Vila Cabral, 17.v.1948; *Pedro & Pedrógão* 3661 (LMA).

Also in Angola, Tanzania, Congo and Kenya. Open glades in *Brachystegia* and other woodland, 1350–1800 m.

5. MICROLEPIA C. Presl

Microlepia C. Presl, Tent. Pterid.: 124 (1836) reimpr. in Abh. Königl. Böhm. Ges. Wiss., Ser. 4, **5**: 124 (1837).

Rhizome creeping, with closely spaced fronds and covered with brown hairs. Stipe with a single U-shaped vascular bundle and shortly pubescent. Lamina pinnate to 4-pinnatifid, often large, herbaceous, pubescent. Sori intramarginal, borne on a vein ending and with a small membranous cupped indusium opening outwards.

A genus of about 45 spp., mostly of the Old World tropics but with one variable pantropic species occurring in Africa. The Hawaiian *Microlepia setosa* (Sm.) Alston has apparently escaped from cultivation in the Umtali Distr. of Rhodesia (*Chase* 4996 (BM; SRGH)).

Microlepia speluncae (L.) Moore, Ind. Fil.: 93 (1857).—Sim, Ferns S. Afr. ed. 2: 129, t. 38 (1915).—Tardieu in Mém. Inst. Fr. Afr. Noire, **28**: 58, t. 7 fig. 3–5 (1953). TAB. **27**. Type from Ceylon.
    Polypodium speluncae (L.), Sp. Pl. **2**: 1093 (1753). Type as above.
    Aspidium speluncae (L.) Willd. in L., Sp. Pl., ed. 4, **5**: 269 (1810). Type as above.
    Davallia speluncae (L.) Bak., Syn. Fil.: 100 (1867). Type as above.

Rhizome up to 1 cm. in diam., widely creeping, with fronds spaced up to 6 cm. apart and covered with numerous pale multicellular hairs up to 4 mm. long. Stipe up to 1 m. long and 6 mm. in diam., pale-brown, often darker at the base, pubescent at first, soon becoming glabrous. Frond up to 3 m. long, erect to arching, herbaceous, fragile. Lamina 3–4-pinnatifid, pinnae up to 60×21 cm., narrowly oblong to oblong-lanceolate, acute, with oblong-obtuse, crenate to pinnatifid ultimate segments 4–12 mm. long, thinly pubescent on both surfaces along the veins and costules with pale multicellular hairs up to 0·4 mm. long; rhachis pale- to dark-brown, very thinly pubescent with minute hairs, eventually becoming glabrous. Sori c. 1 mm. in diam., small, round, clearly intramarginal, situated on a vein-ending which shows as a conspicuous hydathode on the upper surface; indusium small semi-transparent ovate-cupped, facing outwards.

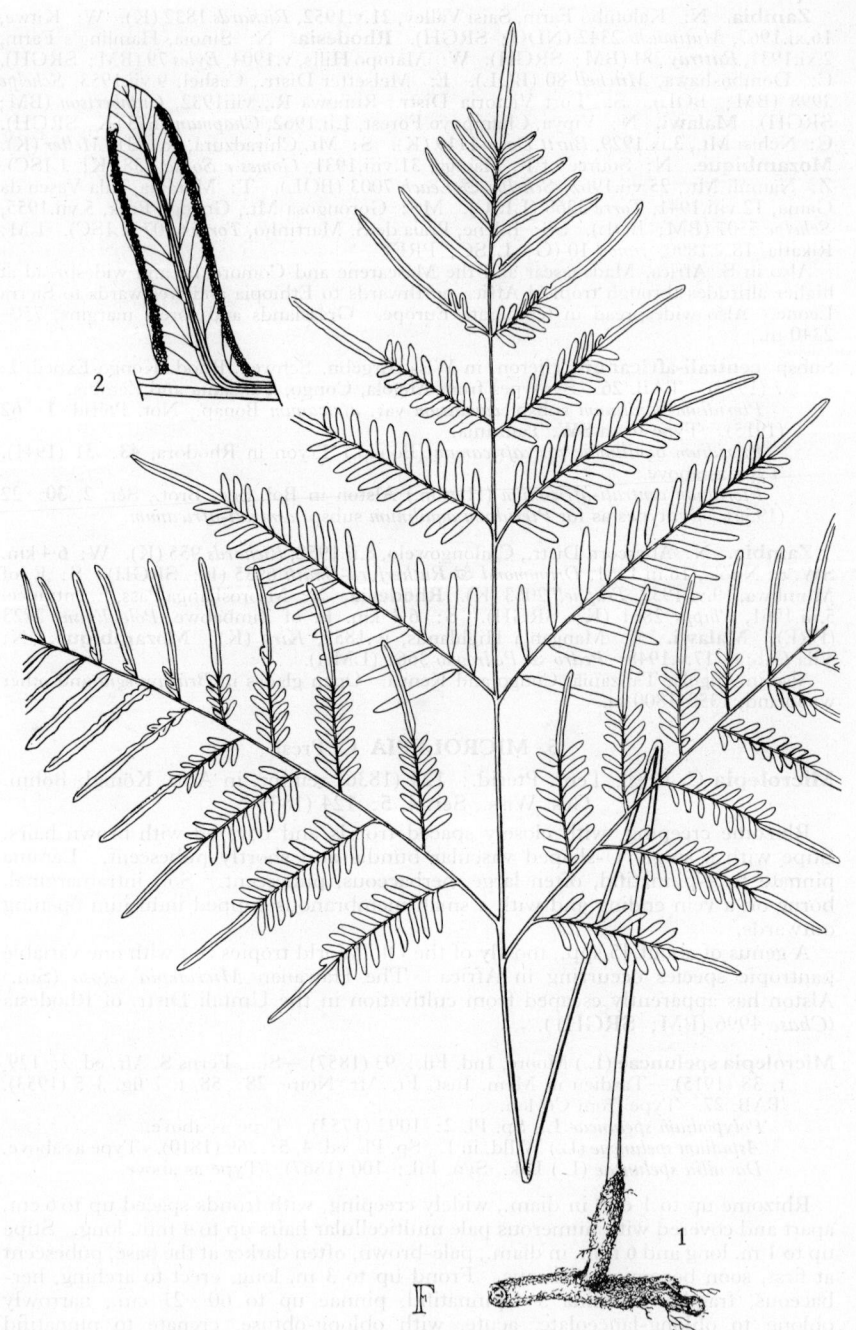

Tab. 26. PTERIDIUM AQUILINUM subsp. CENTRALI-AFRICANUM. 1, habit (×⅓); 2, fertile pinnule segment (×2), both from *Drummond & Rutherford-Smith 6935*.

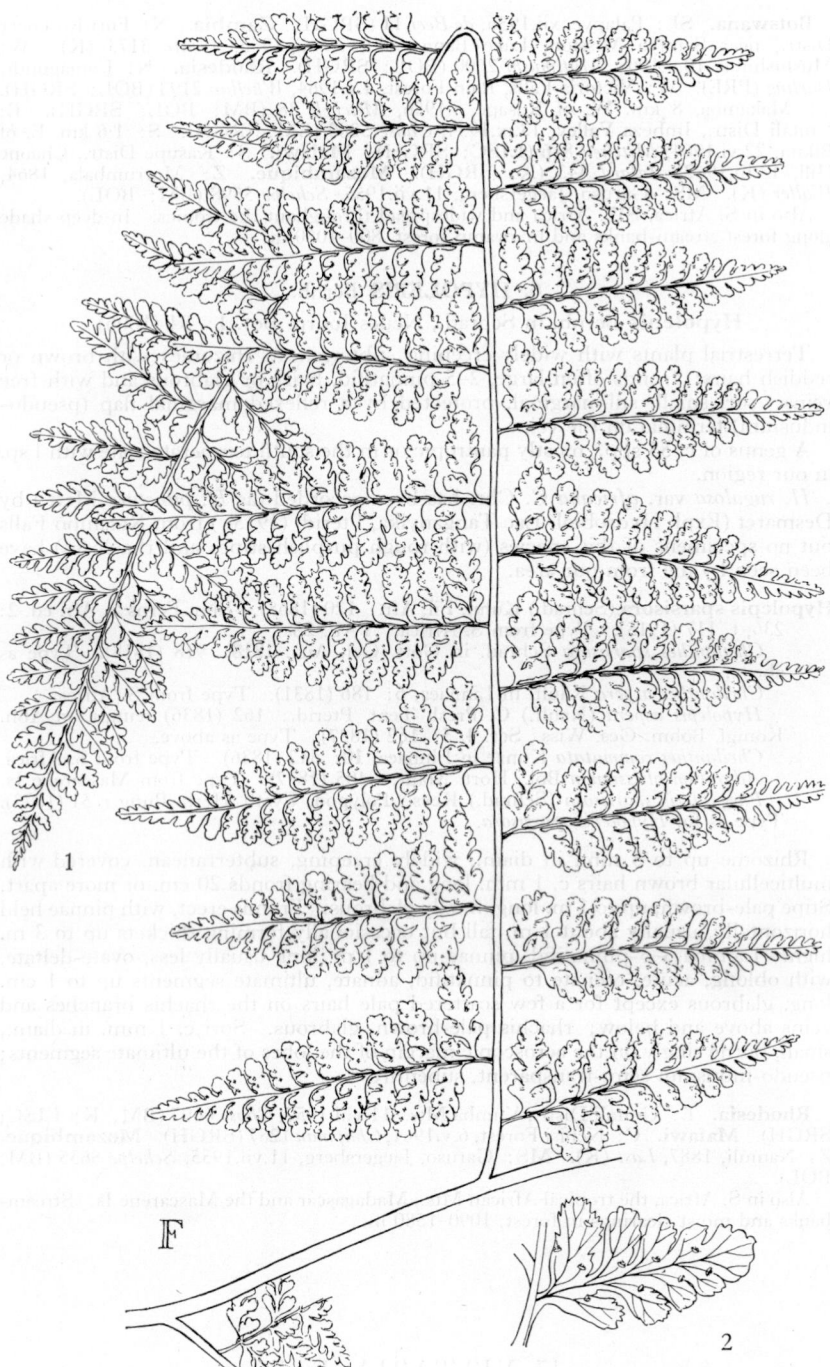

Tab. 27. MICROLEPIA SPELUNCAE. 1, pinna (× ⅔); 2, fertile pinnule segment (× 2), both from *Fisher & Schweickerdt* 499.

Botswana. SE: Palapye, xii.1956, *de Beer* 11 (SRGH). **Zambia.** N: Fort Rosebery Distr., near Samfya Mission, Lake Bangweulu, 30.viii.1952, *White* 3173 (K). W: Mushishima, 10.v.1968, *Mutimushi* 2600 (BOL; SRGH). **Rhodesia.** N: Lomagundi, *Darling* (PRE). W: Victoria Falls, Rain Forest, 4.x.1964, *Whellan* 2193 (BOL; SRGH). C: Makamba, 8 km. W. of Rusape, i.1956, *Mitchell* 38 (BM; BOL; SRGH). E: Umtali Distr., Imbeza Valley, 13.ix.1954, *Chase* 5280 (BM; SRGH). S: 1·6 km. E. of Bikita, 22.vi.1964, *Mitchell* 816 (LISC; SRGH). **Malawi.** S: Kasupe Distr., Chaone Hill, 13.iii.1962, *Adlard* 443 (K; SRGH). **Mozambique.** Z: Morrumbala, 1864, *Waller* (K). MS: Garuso, Jaegersberg, 11.vii.1955, *Schelpe* 5598 (BM; BOL).

Also in S. Africa, SW. Africa and widespread throughout the tropics. In deep shade along forest stream-banks and in swamp forest, 800–1050 m.

6. HYPOLEPIS Bernh.

Hypolepis Bernh. in Schrad., Neues Journ. Bot. **1**: 34 (1806).

Terrestrial plants with widely creeping subterranean rhizomes with brown or reddish hairs; fronds often large, 2–5-pinnatifid, hairy or glabrous and with free veins; sori small, submarginal, protected by a reflexed marginal flap (pseudo-indusium) terminal on the veins.

A genus of c. 55 spp., mostly pantropic or in the south temperate zone with 1 sp. in our region.

H. rugulosa var. *africana* C. Chr. has been recorded (as *H. punctata* Mett.) by Desmaret (Expl. Hydrobiol. Lac Tanganyika, Pterid. (1955)) from Kalambo Falls but no specimens of this species (with rough purplish stipes and rhachises) have been seen by me from our area.

Hypolepis sparsisora (Schrad.) Kuhn, Fil. Afr.: 120 (1868).—Sim, Ferns S. Afr. ed. 2: 236, t. 117 (1915). Type from S. Africa. TAB. **28**.
  Cheilanthes sparsisora Schrad. in Gött. Gel. Anz. **1818**: 918 (1818). Type as above.
  Cheilanthes aspera Kaulf. in Linnaea, **6**: 186 (1831). Type from S. Africa.
  Hypolepis aspera (Kaulf.) C. Presl, Tent. Pterid.: 162 (1836) reimpr. in Abh. Königl. Böhm. Ges. Wiss., Ser. 4, **5**: 162 (1837). Type as above.
  Cheilanthes commutata Kunze in Linnaea, **10**: 542 (1836). Type from S. Africa.
  Adiantum altissimum Boj., Hort. Maur.: 405 (1837). Type from Mascarene Is.
  Phegopteris sparsisora (Schrad.) Keys., Polypod. Cyath. Herb. Bung.: 51 (1873). Type as for *Hypolepis sparsisora*.

Rhizome up to 5 mm. in diam., widely creeping, subterranean, covered with multicellular brown hairs c. 1 mm. long and bearing fronds 20 cm. or more apart. Stipe pale-brown, up to 1 m. long, finally glabrous. Fronds erect, with pinnae held horizontally, usually about 1 m. tall but occasionally forming thickets up to 3 m. high. Lamina 3–5-pinnatifid, pinnae up to 1 m. long usually less, ovate-deltate, with oblong, acute, crenate to pinnatifid, adnate, ultimate segments up to 1 cm. long, glabrous except for a few scattered pale hairs on the rhachis branches and veins above and below; rhachis pale-brown, glabrous. Sori c. 1 mm. in diam., small, borne singly on the acroscopic margin of the lobes of the ultimate segments; pseudo-indusium semi-transparent, subentire.

Rhodesia. E: Umtali Distr., Vumba Mts., 11.viii.1957, *Chase* 6697 (BM; K; LISC; SRGH). **Malawi.** C: Nchisi Forest, 6.v.1961, *Chapman* 1287 (SRGH). **Mozambique.** Z: Namúli, 1887, *Last* (K). MS: Garuso, Jaegersberg, 11.vii.1955, *Schelpe* 5635 (BM; BOL).

Also in S. Africa, the tropical African Mts., Madagascar and the Mascarene Is. Stream-banks and moist clearings in forest, 1090–1580 m.

17. VITTARIACEAE

Epiphytic or lithophytic plants with creeping or suberect rhizomes often embedded in masses of roots with copious hairs and with brown to greyish-brown rhizome-scales. Fronds simple, linear, narrowly elliptic to ovate, often pendulous,

LMR

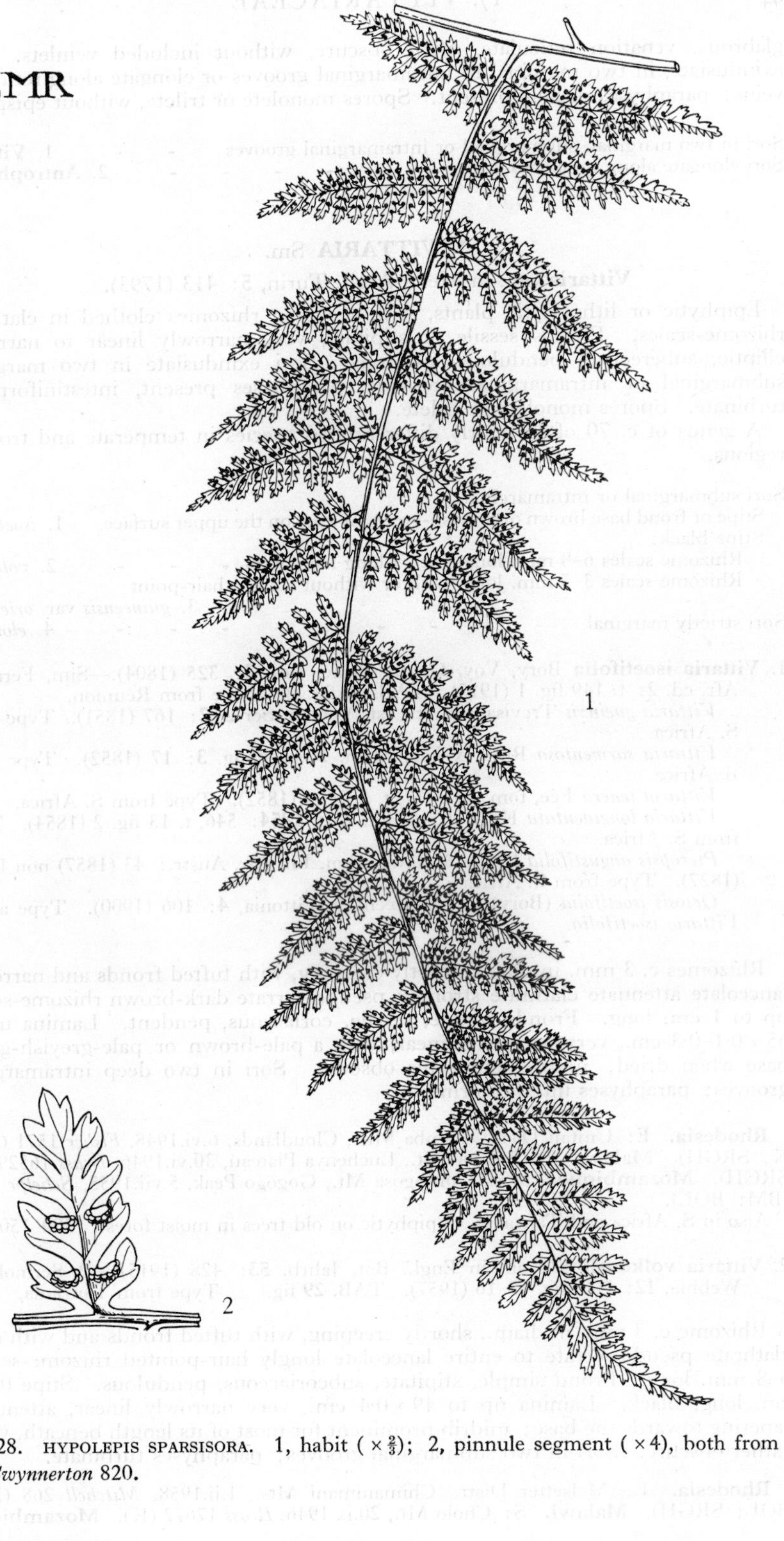

1

2

Tab. 28. HYPOLEPIS SPARSISORA. 1, habit (×⅔); 2, pinnule segment (×4), both from
Swynnerton 820.

glabrous, venation reticulate, often obscure, without included veinlets. Sori exindusiate, in two marginal to intramarginal grooves or elongate along reticulate veins; paraphyses usually present. Spores monolete or trilete, without epispore.

Sori in two marginal, submarginal or intramarginal grooves   - -   **1. Vittaria**
Sori elongate along reticulate veins - - - - - -   **2. Antrophyum**

1. VITTARIA Sm.

Vittaria Sm. in Mem. Acad. Turin, **5**: 413 (1793).

Epiphytic or lithophytic plants, with creeping rhizomes clothed in clathrate rhizome-scales. Fronds sessile or stipitate, very narrowly linear to narrowly elliptic, suberect to pendulous, glabrous. Sori exindusiate in two marginal, submarginal or intramarginal grooves; paraphyses present, intestiniform to turbinate. Spores monolete or trilete.

A genus of c. 70 often poorly differentiated species in temperate and tropical regions.

Sori submarginal or intramarginal:
  Stipe or frond base brown or greyish-green, at least on the upper surface   1. *isoetifolia*
  Stipe black:
    Rhizome scales 6–8 mm. long, with a long hair-point - - -   2. *volkensii*
    Rhizome scales 3–5 mm. long, with or without a short hair-point
                                            3. *guineensis* var. *orientalis*
Sori strictly marginal   - - - - - - - - - -   4. *elongata*

1. **Vittaria isoetifolia** Bory, Voy. Quatre Princ. Iles, **2**: 325 (1804).—Sim, Ferns S. Afr. ed. 2: t. 149 fig. 1 (1915). TAB. **29** fig. B. Type from Réunion.
  Vittaria gueinzii Trevisan in Att. Ist. Veneto, Ser 2, **2**: 167 (1851). Type from S. Africa.
  Vittaria sarmentosa Ruiz ex Fée, Mém. Fam. Foug. **3**: 17 (1852). Type from S. Africa.
  Vittaria tenera Fée, tom. cit.: 17, t. 2 fig. 1 (1852). Type from S. Africa.
  Vittaria longidentata K. Müll. in Bot. Zeit. **1854**: 546, t. 13 fig. 2 (1854). Type from S. Africa.
  Pteropsis angustifolia Pappe & Raws., Syn. Fil. Afr. Austr.: 43 (1857) non Desv. (1827). Type from S. Africa.
  Oetosis isoetifolia (Bory) E. L. Greene in Pittonia, **4**: 106 (1900). Type as for *Vittaria isoetifolia*.

Rhizomes c. 3 mm. in diam., shortly creeping, with tufted fronds and narrowly lanceolate attenuate clathrate strongly pseudo-serrate dark-brown rhizome-scales up to 1 cm. long. Fronds simple, sessile, coriaceous, pendent. Lamina up to 65 × 0·1–0·3 cm., very narrowly linear, with a pale-brown or pale-greyish-green base when dried; midrib and veins obscure. Sori in two deep intramarginal grooves; paraphyses intestiniform.

Rhodesia. E: Umtali Distr., Vumba Mts., Cloudlands, 6.vi.1948, *Fisher* 1571 (BM; K; SRGH). **Malawi.** S: Mlanje Mt., Luchenya Plateau, 30.vi.1946, *Brass* 16529 (K; SRGH). **Mozambique.** MS: Gorongosa Mt., Gogogo Peak, 5.vii.1955, *Schelpe* 5525 (BM; BOL).
Also in S. Africa and Tanzania. Epiphytic on old trees in moist forest, 1300–1500 m.

2. **Vittaria volkensii** Hieron. in Engl., Bot. Jahrb. **53**: 428 (1915).—P.-Sermolli in Webbia, **12**: 700, fig. 15–16 (1957). TAB. **29** fig. Type from Tanzania.

Rhizome c. 3 mm. in diam., shortly creeping, with tufted fronds and with dark clathrate pseudo-serrate to entire lanceolate longly hair-pointed rhizome-scales, 6–8 mm. long. Frond simple, stipitate, subcoriaceous, pendulous. Stipe 0·5–2 cm. long, black. Lamina up to 49 × 0·4 cm., very narrowly linear, attenuate, tapering towards the base; midrib prominent for most of its length beneath, veins rather obscure. Sori in two submarginal grooves; paraphyses turbinate.

Rhodesia. E: Melsetter Distr., Chimanimani Mts., 1.ii.1958, *Mitchell* 268 (BM; BOL; SRGH). **Malawi.** S: Cholo Mt., 20.ix.1946, *Brass* 17672 (K). **Mozambique.**

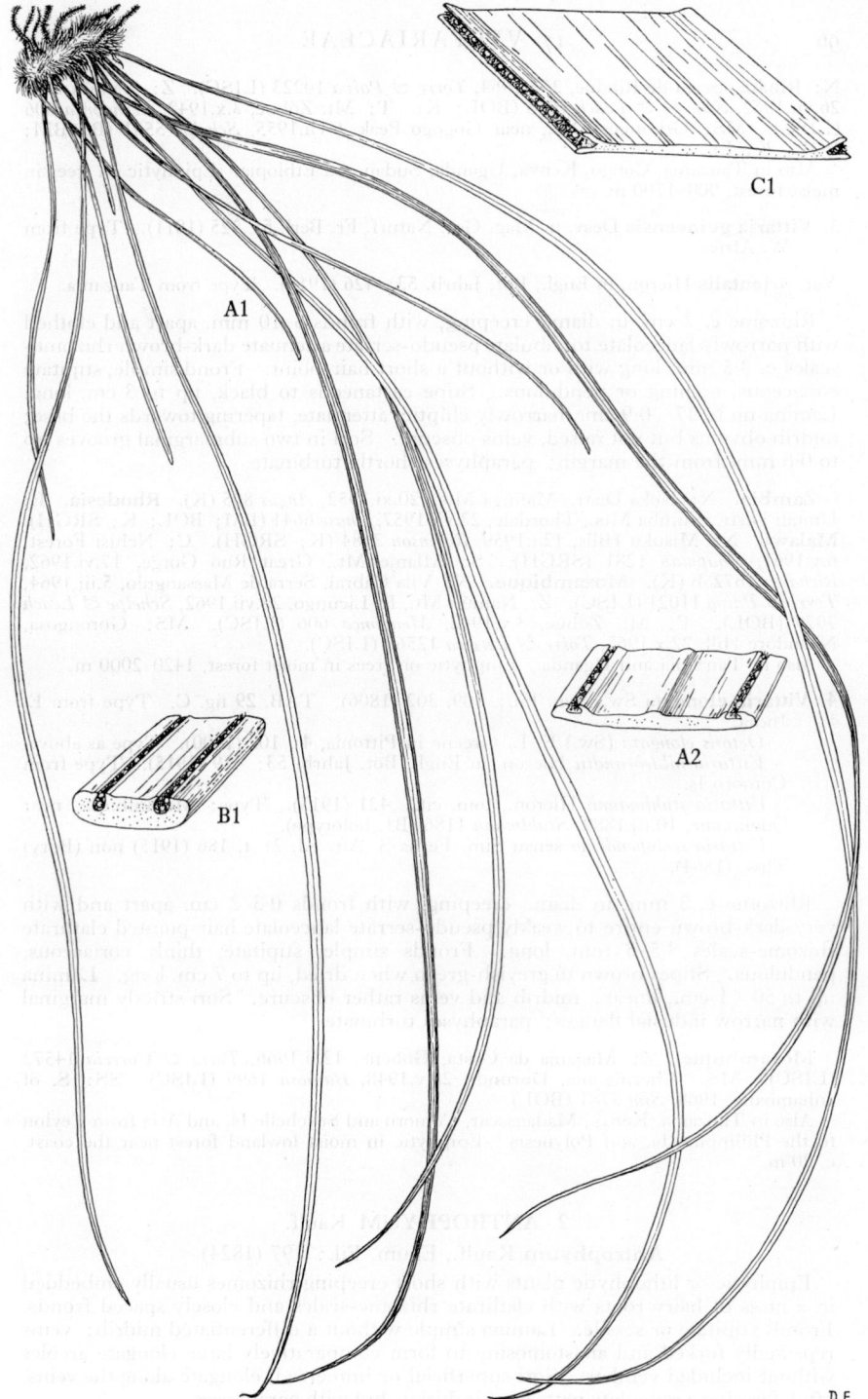

Tab. 29. A.—VITTARIA VOLKENSII. A1, habit (× 2/5); A2, cross-section of frond (× 5), both from *Richards* 14289. B.—VITTARIA ISOETIFOLIA. Cross-section of frond (× 5) *Richards* 16646. C.—VITTARIA ELONGATA. Cross-section of frond (× 5) *Gomes e Sousa* 4376.

N: Ribáuè, Serra de Ribáuè, 25.i.1964, *Torre & Paiva* 10223 (LISC). Z: Namúli Mt., 26.vii.1962, *Schelpe & Leach* 7045 (BOL; K). T: Mt. Zóbuè, 3.x.1942, *Mendonça* 606 (LISC). MS: Gorongosa Mt., near Gogogo Peak, 6.vii.1955, *Schelpe* 5531 (B; BM; BOL; K; P).

Also in Tanzania, Congo, Kenya, Uganda, Sudan and Ethiopia. Epiphytic on trees in moist forest, 900–1700 m.

3. **Vittaria guineensis** Desv. in Mag. Ges. Naturf. Fr. Berl. **5**: 325 (1811). Type from W. Africa.

Var. **orientalis** Hieron. in Engl., Bot. Jahrb. **53**: 426 (1915). Type from Tanzania.

Rhizome c. 2 cm. in diam., creeping, with fronds 3–10 mm. apart and clothed with narrowly lanceolate to subulate pseudo-serrate attenuate dark-brown rhizome-scales c. 3·5 mm. long with or without a short hair-point. Frond simple, stipitate coriaceous, arching or pendulous. Stipe castaneous to black, up to 3 cm. long. Lamina up to 37 × 0·9 cm., narrowly elliptic, attenuate, tapering towards the base; midrib obvious but not raised, veins obscure. Sori in two submarginal grooves up to 0·6 mm. from the margin; paraphyses shortly turbinate.

Zambia. N: Isoka Distr., Mafinga Mts., 20.xi.1952, *Angus* 805 (K). **Rhodesia.** E: Umtali Distr., Vumba Mts., Thordale, 22.vii.1957, *Chase* 6641 (BM; BOL; K; SRGH). **Malawi.** N: Misuku Hills, 12.i.1959, *Robinson* 3184 (K; SRGH). C: Nchisi Forest, 6.v.1961, *Chapman* 1281 (SRGH). S: Mlanje Mt., Great Ruo Gorge, 17.vi.1962, *Richards* 16725b (K). **Mozambique.** N: Vila Cabral, Serra de Massangulo, 5.iii.1964, *Torre & Paiva* 11021 (LISC). Z: Namúli Mt., R. Licungo, 29.vii.1962, *Schelpe & Leach* 7079 (BOL). T: Mt. Zóbuè, 3.x.1942, *Mendonça* 606 (LISC). MS: Gorongosa, Nhandore Hill, 22.x.1965, *Torre & Pereira* 12569 (LISC).

Also in Tanzania and Uganda. Epiphytic on trees in moist forest, 1420–2000 m.

4. **Vittaria elongata** Sw., Syn. Fil.: 109, 302 (1806). TAB. **29** fig. C. Type from E. India.

    Oetosis elongata (Sw.) E. L. Greene in Pittonia, **4**: 103 (1900). Type as above.
    Vittaria hildebrandtii Hieron. in Engl., Bot. Jahrb. **53**: 419 (1915). Type from Comoro Is.
    Vittaria stuhlmannii Hieron., tom. cit.: 421 (1915). Type: Mozambique, near Quelimane, 10.iii.1889, *Stuhlmann* 1186 (B†, holotype).
    Vittaria scolopendrina sensu Sim, Ferns S. Afr. ed. 2: t. 186 (1915) non (Bory) Thw. (1864).

Rhizome c. 3 mm. in diam., creeping, with fronds 0·3–2 cm. apart and with very-dark-brown entire to weakly pseudo-serrate lanceolate hair-pointed clathrate rhizome-scales 3·5–6 mm. long. Fronds simple, stipitate, thinly coriaceous, pendulous. Stipes brown to greyish-green when dried, up to 7 cm. long. Lamina up to 50 × 1 cm., linear; midrib and veins rather obscure. Sori strictly marginal with narrow indusial flanges; paraphyses turbinate.

Mozambique. Z: Maganja da Costa, Gobene, 12.ii.1966, *Torre & Correia* 14572 (LISC). MS: Cheringoma, Durúndi, 28.v.1948, *Barbosa* 1699 (LISC). SS: S. of Inhambane, 1908, *Sim* 5781 (BOL).

Also in Tanzania, Kenya, Madagascar, Comoro and Seychelle Is. and Asia from Ceylon to the Philippine Is. and Polynesia. Epiphytic in moist lowland forest near the coast, c. 20 m.

2. ANTROPHYUM Kaulf.

Antrophyum Kaulf., Enum. Fil.: 197 (1824).

Epiphytic or lithophytic plants with short creeping rhizomes usually embedded in a mass of hairy roots with clathrate rhizome-scales and closely spaced fronds. Fronds stipitate or sessile. Lamina simple without a differentiated midrib; veins repeatedly forked and anastomosing to form comparatively large elongate areoles without included veinlets. Sori superficial or immersed, elongate along the veins, often forming a reticulate pattern, exindusiate but with paraphyses.

A genus of c. 50 spp. distributed throughout the tropics, represented by only 1 sp. in our area.

Antrophyum mannianum Hook., Sec. Cent. Ferns: t. 73 (1861). TAB. **30**. Type from Fernando Po.

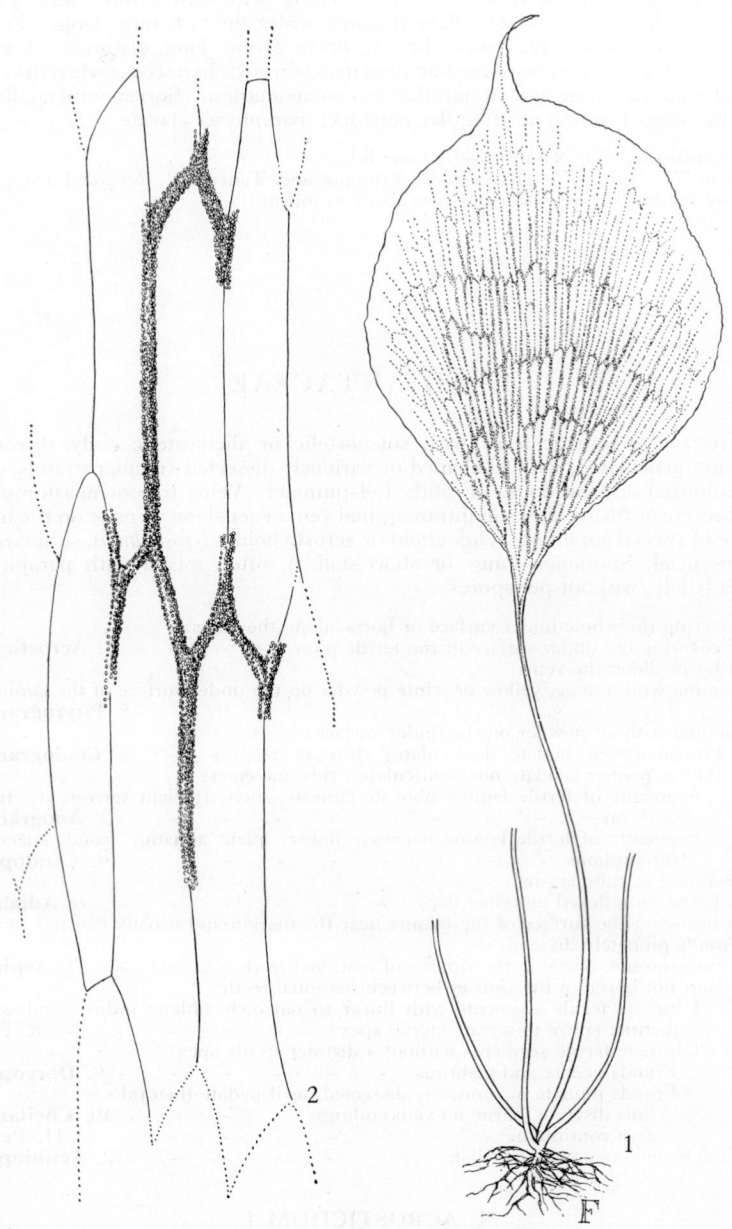

Tab. 30. ANTROPHYUM MANNIANUM.   1, habit (×⅓) *Last* s.n.; 2, enlargement of frond
(×4) *Ussher* s.n.

Rhizome up to 8 mm. in diam., short, creeping, with tufted fronds and clothed with black clathrate subulate ciliate rhizome scales up to 6 mm. long. Fronds pendent, simple, stipitate. Stipes brown, up to 21 cm. long, glabrous. Lamina up to 22 × 14·5 cm., simple, rotund-acuminate, glabrous, chartaceous when dry, with a broadly cuneate base and an undulate to crenate margin. Sori superficial, linear, along the veins forming an irregular network; paraphyses clavate.

Mozambique. Z: Namúli, 1887, *Last* (K).
Also in W. Africa, Congo, Uganda, Ethiopia and Tanzania. Recorded from East Africa as growing on wet rocks in dense shade in forest.

18. ADIANTACEAE

Terrestrial or aquatic ferns with solenostelic or dictyostelic scaly rhizomes; stipes not articulated, with U-shaped or variously dissected vascular strands, often dark-coloured and nitidous. Fronds 1–4-pinnate. Veins free or anastomosing. Sori discrete or fusing along an intramarginal vein or set along veins or on the under surface of special soral flaps (*Adiantum*) or acrostichoid (*Acrostichum*), submarginal or superficial. Sporangia long- or short-stalked, often mixed with paraphyses. Spores trilete, without perispores.

Sori covering the whole under surface or borne along the veins:
  Sori covering the under surface of the fertile pinnae - -   1. **Acrostichum**
  Sori borne along the veins:
    Lamina with orange, yellow or white powder on the under surface of the lamina
                    5. **Pityrogramma**
    Lamina without powder on the under surface:
      Pinnae oblong-caudate, denticulate; rhizome creeping - 3. **Coniogramme**
      Pinnae neither caudate nor denticulate; rhizome erect:
        Segments of fertile lamina obovate-cuneate, incised; plant terrestrial; fronds
          uniform - - - - - - - - - 2. **Anogramma**
        Segments of fertile lamina narrowly linear; plant aquatic; fronds succulent,
          dimorphous - - - - - - - - 4. **Ceratopteris**
Sori marginal or submarginal:
  Sori borne on reflexed marginal flaps - - - - - 6. **Adiantum**
  Sori borne on the surface of the lamina near the margin and usually covered by it:
    Fronds pinnately divided:
      Sori minute, borne in the sinuses of marginal teeth - - - 7. **Aspidotis**
      Sori not borne in the sinuses between marginal teeth:
        Ultimate fertile segments with linear to narrowly oblong indusia and with a
          distinct entire to serrate sterile apex - - - - - - 8. **Pteris**
        Ultimate fertile segments without a distinct sterile apex:
          Fronds pedate and glabrous - - - - - 9. **Doryopteris**
          Fronds pinnate or pinnately dissected, or if pedate then pilose:
            Sori discrete, borne on vein endings - - - - 10. **Cheilanthes**
            Sori continuous - - - - - - - 11. **Pellaea**
    Fronds dichotomously flabellate - - - - - 12. **Actiniopteris**

1. ACROSTICHUM L.

Acrostichum L., Sp. Pl. **2**: 1067 (1753); Gen. Pl. ed. 5: 484 (1754).
(" Acrosticum ").

Rhizome erect to procumbent, massive, with tufted large fronds with large tough rhizome scales and thick fleshy roots. Lamina large, simply pinnate with large entire coriaceous petiolate pinnae with freely anastomosing veins but without included veinlets. Sori acrostichoid, borne on the under surface of undifferentiated apical pinnae with paraphyses. Spores large, trilete.
A genus of c. 5 spp. along tropical and subtropical coasts, usually in mangrove swamps. 1 sp. occurs in our area.

Acrostichum aureum L., Sp. Pl. **2**: 1069 (1753).—Sim, Ferns S. Afr. ed. 2: 292, t. 153 (1915). TAB. **31**. Syntypes from Jamaica and S. Domingo.

    Acrostichum inaequale Willd. in L., Sp. Pl., ed. 4, **5**: 117 (1810). Type from E. India.

    Chrysodium inaequale (Willd.) Fée, Mém. Fam. Foug. **2**: 100 (1845). Type as above.

    Chrysodium vulgare Fée, tom. cit.: 97 (1845) *nom. illegit.* Type as for *Acrostichum aureum.*

    Chrysodium aureum (L.) Mett., Fil. Hort. Bot. Lips.: 21 (1856). Type as for *Acrostichum aureum.*

    Acrostichum guineense Gandog. in Bull. Soc. Bot. Fr. **66**: 305 (1919) non Carr. (1901). Type from Príncipe.

Rhizome erect to procumbent, massive, with tufted fronds and with hard subulate rhizome-scales c. 1 cm. long, with a thick black median area and narrow pale clathrate borders. Fronds erect, with fertile pinnae borne towards the apex of the frond. Stipe brown, up to 50 cm. long, shallowly channelled above. Lamina up to 1·5 × 0·4 m., simply pinnate; sterile pinnae petiolate, 8–36 × 1–5 cm., coriaceous, glabrous, linear to cultrate, entire or irregularly undulate, acuminate to truncate, base unequally cuneate, with the costa raised and prominent below, reticulate venation apparent; fertile pinnae similar to the sterile pinnae but with the under surface (except the costa) covered with sporangia; paraphyses present among the sporangia.

Mozambique. N: Cabo Delgado, Tungue, Palma, 21.x.1942, *Mendonça* 1026 (BM; LISC). Z: Luabo mouth of R. Zambese, xi.1860, *Kirk* (K). T: between Muatize and Tete, 23.vi.1949, *Barbosa & Carvalho* 3257 (BOL; LISC; LM; SRGH). MS: R. Idunda, Lower Buzi, 20.xii.1906, *Swynnerton* 838 (K). SS: Inhambane Distr., R. Inhanombe, 14.i.1954, *Schelpe* 4433 (B; BM; BOL; K; P). LM: Inhaca I., i–ii. 1962, *Mogg* 29794 (K; SRGH).

Also subtropical S. Africa and tropical African, American and Asiatic coasts. About the inland limit of mangrove swamps, 2–3 m.

2. ANOGRAMMA Link

Anogramma Link, Fil. Sp.: 137 (1841).

Rhizome very small, possibly produced annually from a persistent relatively large gametophyte, and with small brown scales. Stipe dark-brown, glabrous. Frond 2–3-pinnate with incised decurrent pinnules, glabrous or thinly pubescent, membranous; veins free. Sori borne along the forked veins, exindusiate.

A genus of c. 5 spp., only 1 occurring in Africa.

Anogramma leptophylla (L.) Link, Fil. Sp.: 137 (1841).—Sim, Ferns S. Afr. ed. 2: 193 t. 109 (1915).—Tardieu, Fl. Madag., Polypod. **1**: 120, t. 17 fig. 6–9 (1958). TAB. **32**. Type from S. Europe.

    Polypodium leptophyllum L., Sp. Pl. **2**: 1092 (1753). Type as above.

    Asplenium leptophyllum (L.) Sw., Obs. Bot.: 403 (1791). Type as above.

    Osmunda leptophylla (L.) Savigny in Lam., Encycl. Méth. Bot. **4**: 657 (1797). Type as above.

    Acrostichum leptophyllum (L.) Lam. & DC., Fl. Fr. ed. 3, **2**: 565 (1805). Type as above.

    Grammitis leptophylla (L.) Sw., Syn. Fil.: 23, 218, t. 1 fig. 6 (1806). Type as above.

    Gymnogramma leptophylla (L.) Desv. in Ges. Naturf. Berl. Mag. **5**: 305 (1811). Type as above.

    Hemionitis leptophylla (L.) Lagasca, Gen. & Sp. Pl.: 33 (1816). Type as above.

    Gymnogramma novae-zeylandiae Colenso in Tasm. Journ. Nat. Sci. **2**: 165 (1846). Type from New Zealand.

    Dicranodium leptophyllum (L.) Newm., Hist. Brit. Ferns, ed. 3: 13 (1854). Type as for *Anogramma leptophylla.*

    Gymnogramma schwackeana Christ in Schwacke, Pl. Nov. Mineir. **2**: 18 (1900); in Bull. Herb. Boiss., Sér. 2, **2**: 365 (1902). Type from Brazil.

Rhizome minute, annual, with a few tufted small fronds and pale minute linear entire rhizome scales c. 1 mm. long. Stipe castaneous, up to 8 cm. long, glabrous except for a few small pale scales near the base. Frond erect, herbaceous. Lamina up to 7 × 2·8 cm., oblong-ovate in outline to narrowly triangular, 2-pinnatifid to

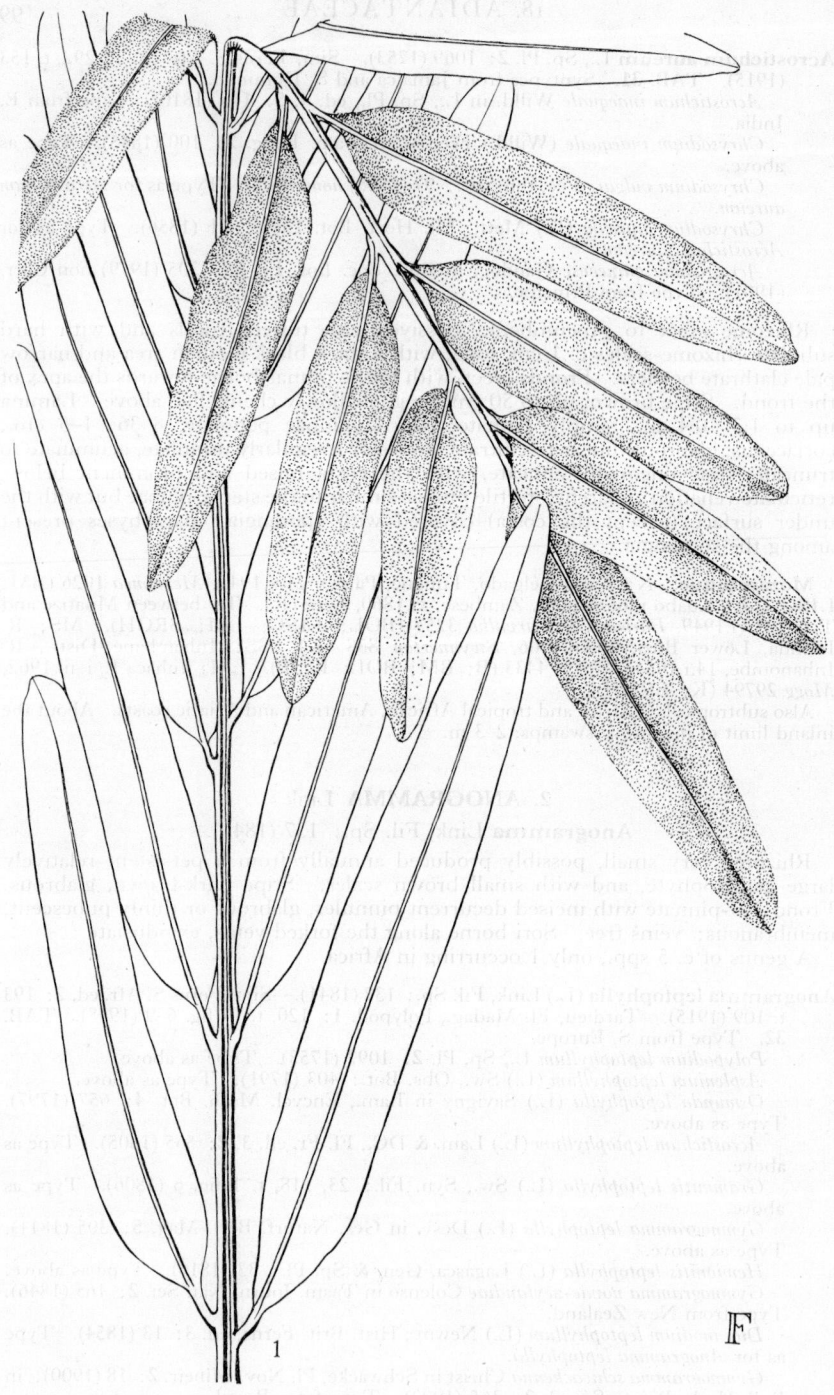

F

Tab. 31. ACROSTICHUM AUREUM. 1, fertile frond (× ⅔) *Schelpe* 4433.

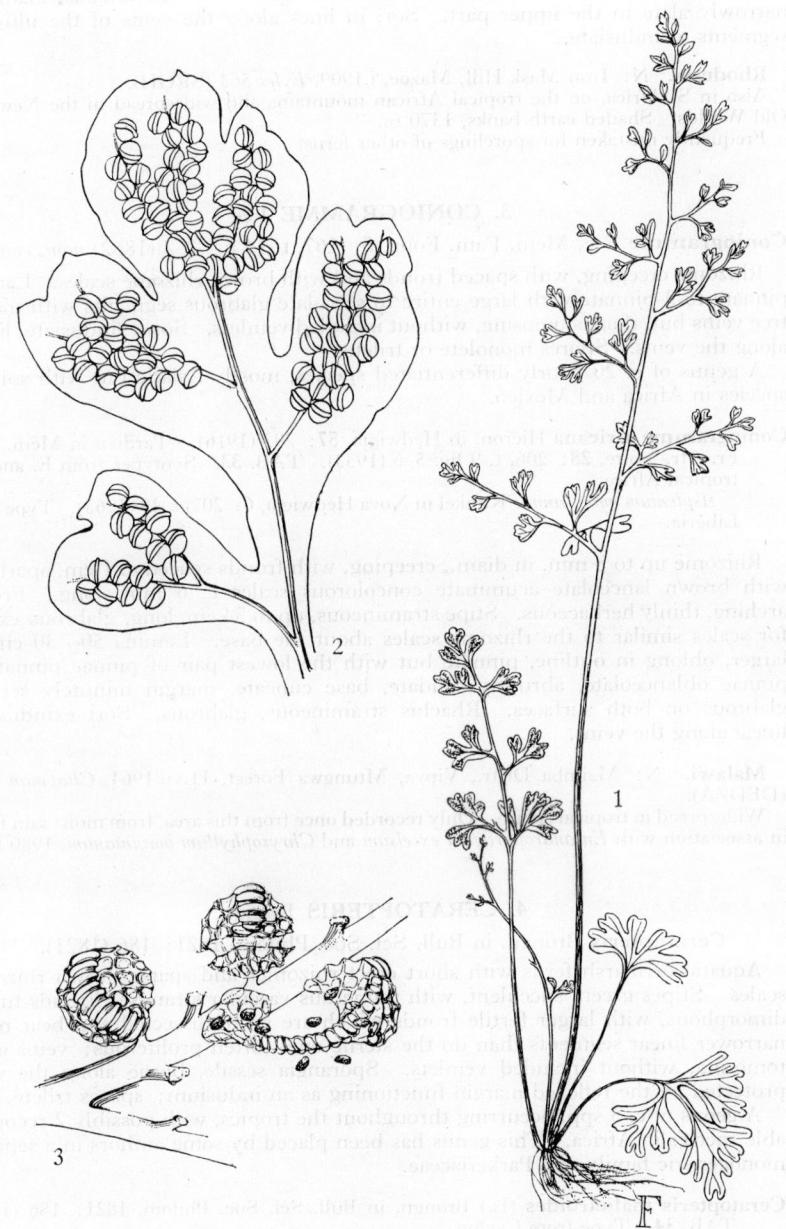

Tab. 32. ANOGRAMMA LEPTOPHYLLA. 1, Habit (× ⅔) *Taylor* 1205; 2, pinna of small frond
(×7); 3, sporangia (×30), both from *Schelpe* 4535.

3-pinnatifid (rarely pinnate in very small plants); pinnae ovate-triangular in outline; ultimate segments broadly to narrowly cuneate, emarginate or shallowly lobed, glabrous on both surfaces; rhachis castaneous to stramineous, glabrous, narrowly alate in the upper part. Sori in lines along the veins of the ultimate segments, exindusiate.

Rhodesia. N: Iron Mask Hill, Mazoe, i.1909, *Eyles* 563 (SRGH).
Also in S. Africa, on the tropical African mountains and widespread in the New and Old Worlds. Shaded earth banks, 1370 m.
Frequently mistaken for sporelings of other ferns.

3. CONIOGRAMME Fée

Coniogramme Fée, Mém. Fam. Foug. **5**: 167, t. 14 fig. 1–2 (1852) *nom. conserv.*

Rhizome creeping, with spaced fronds and with brown rhizome scales. Lamina pinnate to 3-pinnate with large entire to serrulate glabrous segments with mostly free veins but, if anastomosing, without included veinlets. Sori exindusiate, linear along the veins. Spores monolete or trilete.
A genus of c. 20 poorly differentiated species, mostly Asiatic but with solitary species in Africa and Mexico.

Coniogramme africana Hieron. in Hedwigia, **57**: 293 (1916).—Tardieu in Mém. Inst. Fr. Afr. Noire, **28**: 206, t. 9 fig. 5–6 (1953). TAB. **33**. Syntypes from E. and W. tropical Africa.
   Asplenium mary-annae Kunkel in Nova Hedwigia, **6**: 207, t. 49 (1963). Type from Liberia.

Rhizome up to 8 mm. in diam., creeping, with fronds spaced c. 1 cm. apart and with brown lanceolate acuminate concolorous scales c. 6 mm. long. Fronds arching, thinly herbaceous. Stipe stramineous, up to 34 cm. long, glabrous except for scales similar to the rhizome scales about the base. Lamina 50 × 30 cm. or larger, oblong in outline, pinnate but with the lowest pair of pinnae pinnatifid; pinnae oblanceolate, abruptly caudate, base cuneate, margin minutely serrate, glabrous on both surfaces. Rhachis stramineous, glabrous. Sori exindusiate, linear along the veins.

Malawi. N: Mzimba Distr., Vipya, Mtungwa Forest, 11.vii.1964, *Chapman* 2256 (DEDZA).
Widespread in tropical Africa. Only recorded once from this area, from moist rain forest in association with *Entandrophragma excelsum* and *Chrysophyllum boivinianum*, 1980 m.

4. CERATOPTERIS Brongn.

Ceratopteris Brongn. in Bull. Sci. Soc. Philom. **1821**: 186 (1821).

Aquatic or marsh ferns with short erect rhizomes and sparse brown rhizome-scales. Stipes green, succulent, with numerous vascular strands. Fronds tufted, dimorphous, with larger fertile fronds which are more dissected and bear much narrower linear segments than do the sterile ones, often proliferous; veins anastomosing, without included veinlets. Sporangia sessile, borne along the veins protected by the reflexed margin functioning as an indusium; spores trilete.
A genus of c. 5 spp. occurring throughout the tropics, with possibly 2 recognisable species in Africa. This genus has been placed by some authors in a separate monogeneric family, the Parkeriaceae.

Ceratopteris thalictroides (L.) Brongn. in Bull. Sci. Soc. Philom. **1821**: 186 (1821). TAB. **34**. Type from Ceylon.
   Acrostichum thalictroides L., Sp. Pl. **2**: 1070 (1753). Type as above.
   Acrostichum siliquosum L., loc. cit. Type from Ceylon.
   Pteris thalictroides (L.) Sw. in Schrad., Journ. Bot. **1800**, 2: 65 (1801). Type as for *Ceratopteris thalictroides*.
   Pteris siliquosa (L.) Beauv., Fl. Owar. & Ben. **1**: 63 (1806). Type as for *Acrostichum siliquosum*.
   Ellobocarpus oleraceus Kaulf., Enum. Fil.: 148 (1824) *nom. illegit.* Type from Ceylon.

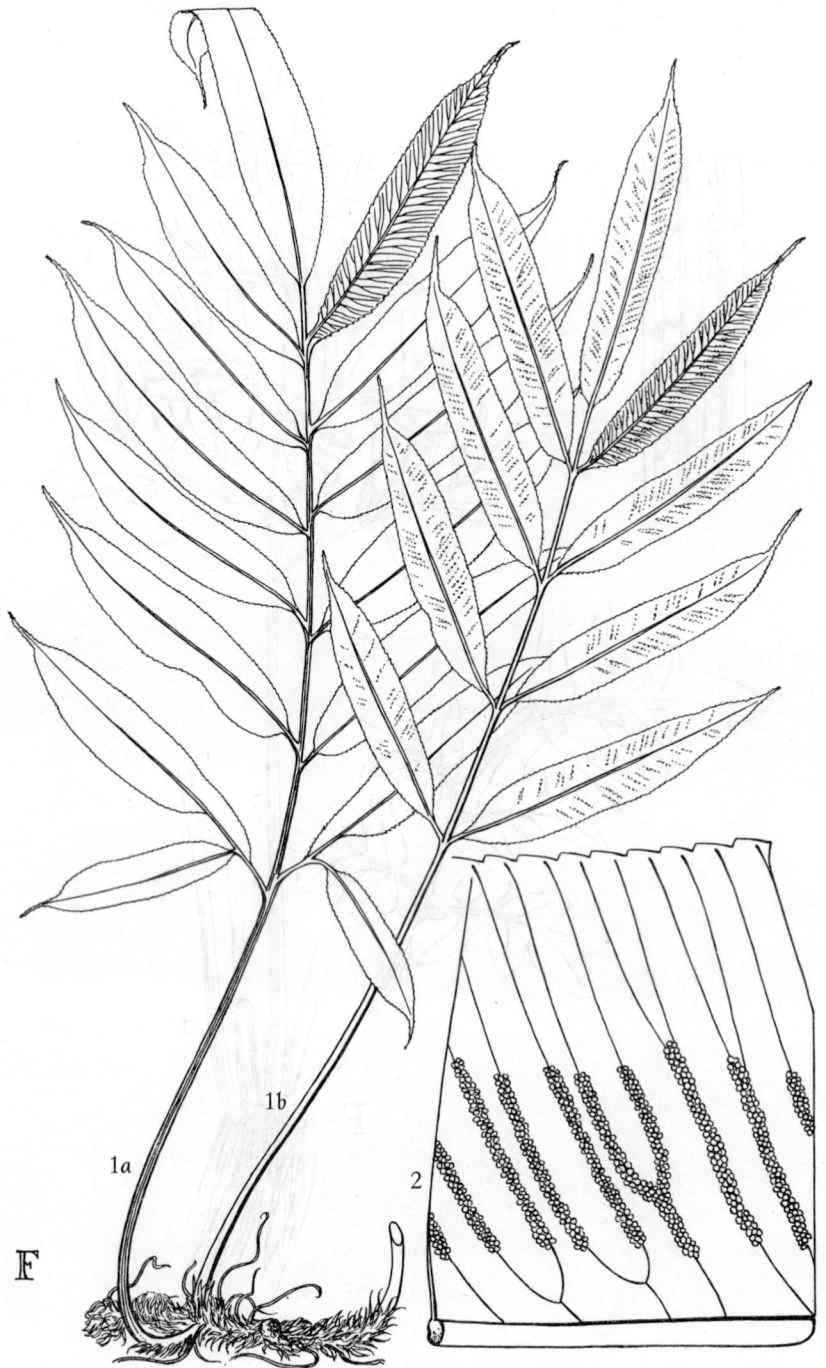

F

Tab. 33.   CONIOGRAMME AFRICANA.   1, habit (a) sterile frond (b) fertile frond (×⅕);
2, part of fertile pinna showing arrangement of sori (×4), both from *Adams* 1044.

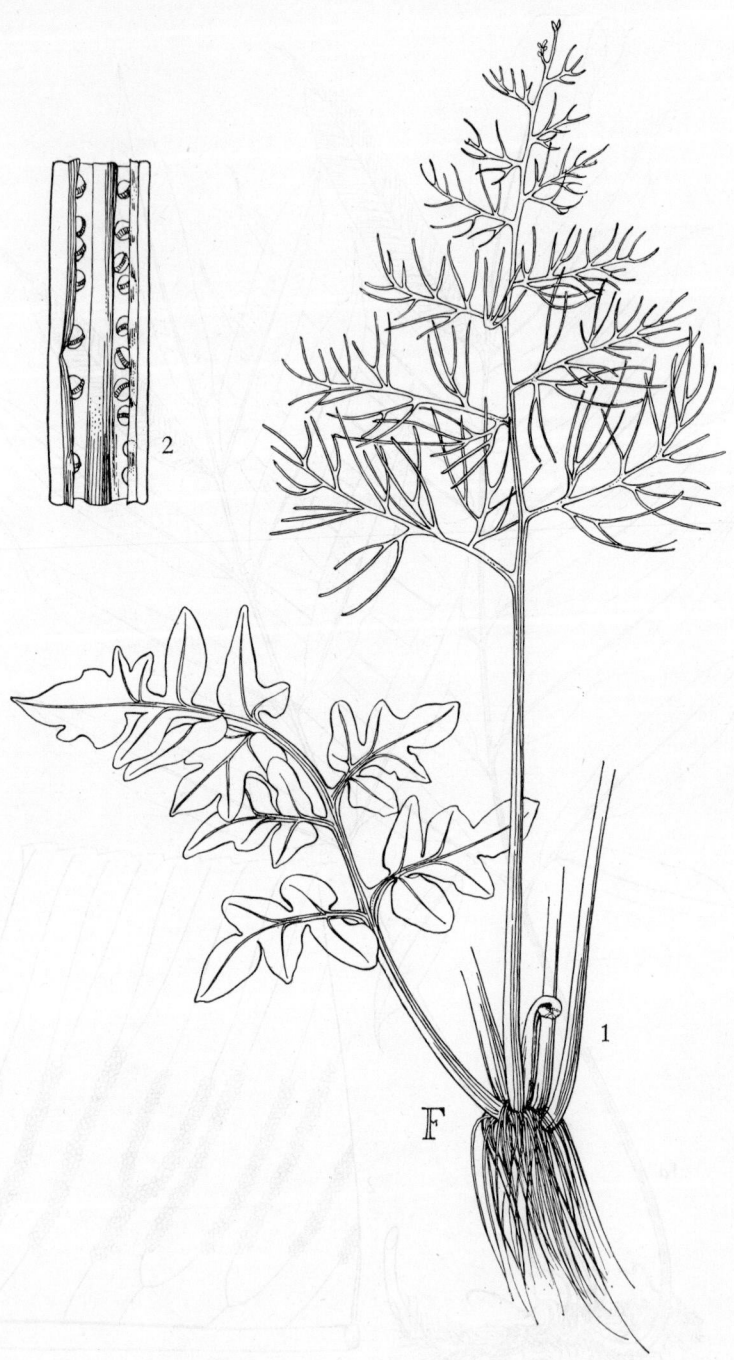

Tab. 34. CERATOPTERIS THALICTROIDES. 1, habit ($\times \frac{2}{3}$); 2, enlargement of fertile segment ($\times 10$), both from *Faulkner* 90.

Furcaria tharictroides (L.) Desv. in Mém. Soc. Linn. Par. **6**, 2: 292 (1827).
Type as for *Ceratopteris thalictroides*.
Pteris succulenta Roxb. in Calc. Journ. Nat. Hist. **4**: 508 (1844). Type from India.
Ceratopteris siliquosa (L.) Copel. in Philipp. Journ. Sci. **56**: 107 (1936). Type as
for *Acrostichum siliquosum*.

Rhizome short, erect, with tufted fronds and sparse brown scales. Stipe up to
16 cm. long, succulent, brittle when fresh, eventually glabrous. Frond erect to
arching, succulent, brittle, dimorphous. Sterile lamina up to 20 × 7 cm., oblong,
ovate or narrowly triangular in outline, usually 2-pinnatifid into acute to obtuse
triangular or lanceolate lobes, glabrous on both surfaces. Fertile lamina up to
24 × 12 cm. narrowly to broadly ovate in outline usually 3-pinnatifid into narrowly
linear acute lobes up to 4 × 0·1–0·15 cm., glabrous on both surfaces. Sporangia
borne sparsely along the veins, indusium marginal, entire, membranous.

Zambia. N: Abercorn Distr., Nimkole, Lake Tanganyika, 11.vi.1961, *Richards* 15232
(K; SRGH). S: Lake Kariba, 29.vii.1960, *Soulsby* 4 (SRGH). **Rhodesia.** W:
Victoria Falls, 3.vi.1962, *Schweickerdt* 2612 (BOL). **Mozambique.** Z: Mocuba,
Lugela Distr., *Faulkner* (K). MS: Manga, ii.1923, *Honey* 813 (BOL; K). LM:
Maputo, *Junod* (PRE).
Widespread through tropical Africa and Asia. Aquatic or rooted in mud in sluggish
streams and marshes, near sea-level to 780 m.

5. PITYROGRAMMA Link

Pityrogramma Link, Handb. Gewächs. **3**: 19 (1833).

Rhizome erect or creeping, with tufted or spaced fronds and with linear attenuate
brown scales. Stipe castaneous, glabrous except for a few scales towards the base.
Frond 2–4-pinnatifid, membranous to thinly coriaceous with white, pink, yellow or
orange powder on the dorsal surface, glabrous on the ventral surface; veins free.
Sori exindusiate, borne along the length of the veins.
A genus of c. 40 spp. most of which occur in tropical America but with a few
species in Africa.

Rhizome short; powder on dorsal surface of lamina white, pink or yellow:
  Frond broadly triangular, brittle; ultimate segments obtuse - - - 1. *argentea*
  Frond oblong-lanceolate, robust; ultimate segments acute-acuminate   3. *calomelanos*
Rhizome widely creeping; powder on dorsal surface orange; frond narrowly triangular to
narrowly oblong - - - - - - - - - 2. *aurantiaca*

1. **Pityrogramma argentea** (Willd.) Domin in Publ. Fac. Sci. Univ. Charlest. **88**: 6
(1928).—Tardieu, Fl. Madag., Polypod. **1**: 118 (1958). TAB. **35**. Type from
Réunion.
  Hemionitis argentea Willd. in L., Sp. Pl., ed. 4, **5**: 132 (1810). Type as above.
  Gymnogramma rosea Desv. in Mag. Ges. Naturf. Fr. Berl. **5**: 306 (1811). Type
from Réunion.
  Gymnogramma thiebautii Desv. in Mém. Soc. Linn. Par. **6**, 2: 215 (1827) *nom.
illegit.* Type from Réunion.
  Gymnogramma conspersa Kunze in Linnaea, **18**: 116 (1844). Type from S. Africa.
  Anogramma rosea (Desv.) Fée, Mém. Fam. Foug. **5**: 184 (1852). Type as for
Gymnogramma rosea.
  Anogramma conspersa (Kunze) Fée, loc. cit. Type as for *Gymnogramma conspersa*.
  Gymnogramma argentea (Willd.) Mett. ex Kuhn, Fil. Afr.: 59 (1868).—Sim,
Ferns S. Afr. ed. 2: 194, t. 87 (1915). Type as for *Pityrogramma argentea*.
  Ceropteris argentea (Willd.) Kuhn, Von Deck. Reisen, Bot. **3**, 3: 18 (1879). Type
as for *Pityrogramma argentea*.

Rhizome c. 2 mm. in diam., short, suberect to procumbent, with tufted fronds
and with linear attenuate entire brown rhizome-scales up to 2 mm. long. Stipe
castaneous, up to 15 cm. long, glabrous, nitidous, with a few scales near the base.
Frond arching, herbaceous, fragile. Lamina ovate-triangular in outline, 3–4-
pinnatifid with the lowest pinnae ¼–½ the length of the lamina, up to 21 × 16 cm.;
pinnae up to 12 cm. long, narrowly ovate-triangular; pinnules of upper pinnae and
pinnule segments of lower pinnae cuneate to broadly oblong-ovate, deeply pinnatifid
into obtuse or emarginate entire to crenate lobes c. 1 mm. broad, with white, pink
or yellow powder on the dorsal surface; rhachis and secondary rhachises nitidous,

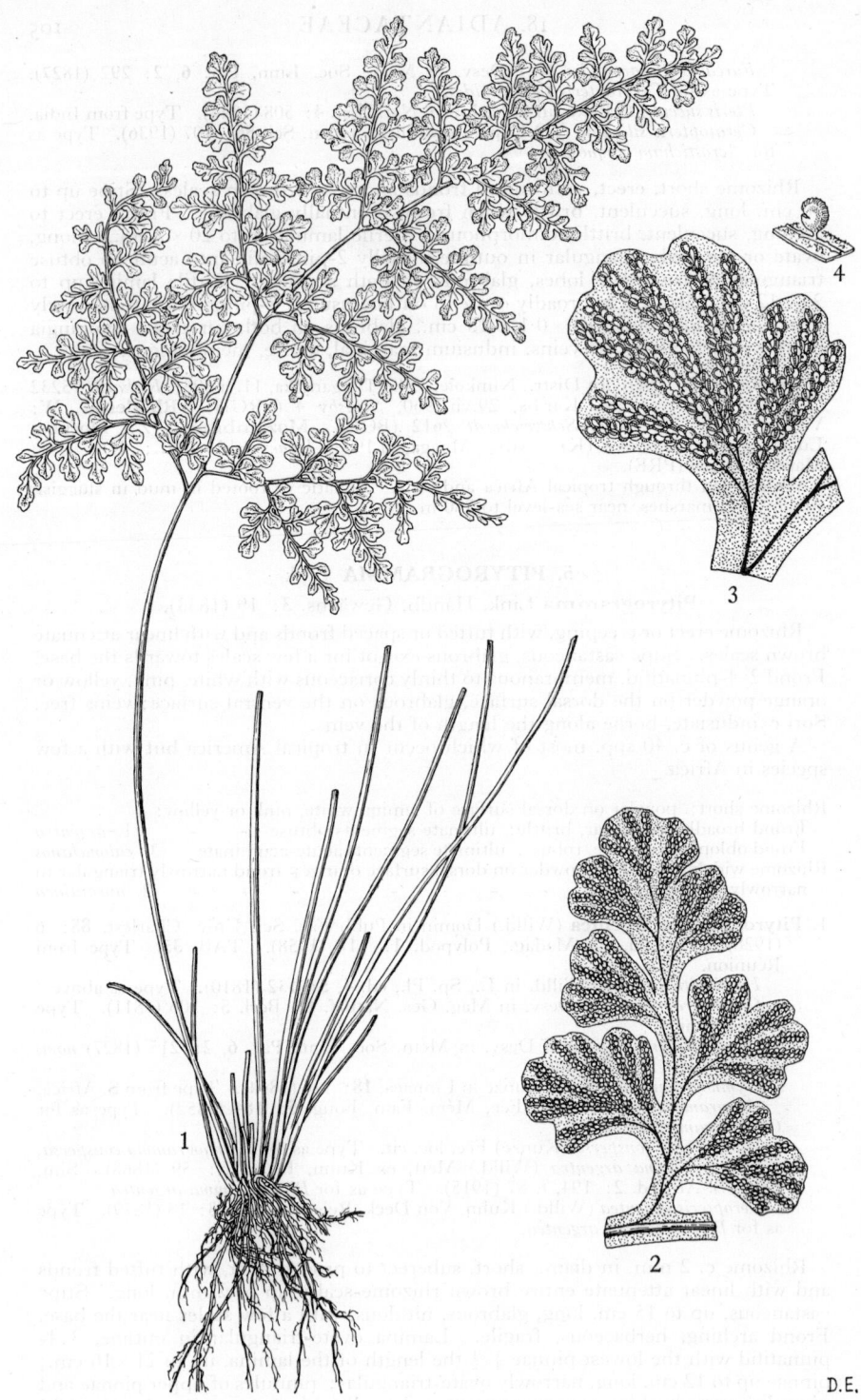

Tab. 35. PITYROGRAMMA ARGENTEA. 1, habit (× ½), rhizome from *Newman & Whitmore* 505, frond from *Chapman* 421; 2, pinnule (×3); 3, lobe of pinnule showing arrangement of sori (×6); 4, sorus (×12), 2–4 from *Chapman* 421.

D.E.

the latter often narrowly winged for some distance from the apex. Sori up to 2 mm. long, linear, along veins, exindusiate.

Rhodesia. N: Mazoe, Iron Mask Hill, v.1907, *Eyles* 547 (BM; SRGH). E: Inyanga Distr., 8 km. N. from Chipungu Falls, 14.vii.1955, *Schelpe* 5662 (BM; BOL; K; P). **Malawi.** S: Mlanje Mt., 21.viii.1956, *Newman & Whitmore* 505 (SRGH). **Mozambique.** MS: Chimanimani Mts., near Martin Falls, 9. ii.1958, *Mitchell* 333 (BM; COI; SRGH).

Sporadic in S. Africa, Angola, E. Congo, Madagascar and Réunion. In rock crevices and around boulder bases, 1700–2100 m.

Both the white-powdered and yellow-powdered forms are known from the Iron Mask Hills and the Chimanimani Mts.

2. **Pityrogramma aurantiaca** (Hieron.) C. Chr., Ind. Fil. Suppl. **3**: 138 (1934). Type from Tanzania.

Gymnogramma aurantiaca Hieron. in Engl., Bot. Jahrb. **46**: 383 (1911). Type as above.

Rhizome c. 2·5 mm. in diam., long, creeping, with closely spaced fronds and with lanceolate entire dark-brown scales c. 3 mm. long. Stipe castaneous, glabrous, nitidous except for a few scales similar to the rhizome-scales near the base. Frond erect, firmly membranous to subcoriaceous. Lamina up to 12 × 7 cm. (larger in other tropical African specimens), oblong to ovate-oblong in outline, 3-pinnatifid; pinnae up to 4 cm. long, triangular-ovate; pinna segments of upper pinnae and pinnule segments of lower pinnae up to 6 mm. broad, oblong to broadly cuneate, obtuse, very weakly crenate and with the margins sometimes revolute, with bright-orange powder on the dorsal surface, glabrous on the ventral surface; rhachis castaneous, nitidous; secondary rhachises narrowly winged for most of their length. Sori 1–2 mm. long, linear, along the veins, exindusiate.

Malawi. N: Nyika Plateau, 19.viii.1946, *Brass* 17331 (BM; K; SRGH). S: Mt. Mlanje, 1897, *Adamson* (K).

Also on Kilimanjaro and Ruwenzori. Rare, forming tussocks in high-altitude marshes, 2340 m.

3. **Pityrogramma calomelanos** (L.) Link, Handb. Gewächs. **3**: 20 (1833).—Tardieu in Aubréville, Fl. Camér., **3**, Ptérid.: 134, t. 17 figs. 1–2 (1964). Type from tropical America.

Acrostichum calomelanos L., Sp. Pl. **2**: 1072 (1753). Type as above.

Acrostichum ebeneum L., tom. cit.: 1071 (1753). Type from Jamaica.

Gymnogramma calomelanos (L.) Kaulf., Enum. Fil.: 76 (1824). Type as for *Pityrogramma calomelanos*.

Gymnogramma distans Link, Hort. Berol. **2**: 53 (1833). Type from Brazil.

Ceropteris calomelanos (L.) Underw. in Bull. Torrey Bot. Club, **29**: 632 (1902).— Sim, Ferns S. Afr. ed. 2: t. 88 (1915). Type as for *Pityrogramma calomelanos*.

Pityrogramma chamaesorbus Domin in Publ. Fac. Sci. Univ. Charlest. **88**: 6 (1928). Type from Guiana.

Pityrogramma insularis Domin, loc. cit. Syntypes from Príncipe and Fernando Po.

Rhizome short, procumbent, c. 8 mm. in diam., with tufted fronds and with concolorous light-brown entire linear rhizome-scales up to 4 mm. long. Stipe up to 33 cm. long, atrocastaneous, nitidous at maturity with a few scales at the base. Frond erect to arching, firmly herbaceous to thinly coriaceous. Lamina up to 37 × 14 cm., oblong-lanceolate in outline, 2-pinnate to 3-pinnatifid with the lowest pinnae not reduced; pinnae up to 9 × 2·2 cm., lanceolate, acute-acuminate, with the pinna segments oblong-trapeziform, the larger up to 1·7–0·5 cm., slightly auriculate, serrate (apparently entire if margin involuted), acute, acuminate, set at an acute angle to the costa and with white or yellow powder on the dorsal surface; rhachis atrocastaneous, nitidous at maturity. Sori up to 3 mm. long, set along the veins, exindusiate.

Powder on dorsal surface whitish   -   -   -   -   -   -   - var. *calomelanos*
Powder on dorsal surface yellow   -   -   -   -   -   -   - var. *aureoflava*

Var. calomelanos

Mozambique. Z: Namúli Mt., Chá Moçambique, 29.vii 1962, *Schelpe & Leach* 7074 (BOL).

Common weed in the moist tropics, introduced from tropical America. Weed in tea plantation, c. 730 m.

Var. **aureoflava** (Hook.) Weath. ex Bailey, Man. Cult. Pl.: 64 (1926). Type from S. America.
   Gymnogramma calomelanos var. *aureoflava* Hook., Garden Ferns: t. 50 (1862). Type as above.
   Pityrogramma austroamericana Domin in Publ. Fac. Sci. Univ. Charlest. **88**: 7 (1928). Lectotype from Bolivia.
   Pityrogramma calomelanos var. *austroamericana* (Domin) Farw. in Amer. Midl. Nat. **12**: 280 (1931). Type as above.

Zambia. W: Kitwe, 6.v.1967, *Fanshawe* 10052 (BOL; SRGH).
A variety introduced from higher elevations than the typical variety in S. America, which has become naturalised in Natal as well as around Kitwe.

6. ADIANTUM L.

Adiantum L., Sp. Pl. **2**: 1094 (1753); Gen. Pl. ed. 5: 485 (1754).

Rhizome shortly or widely creeping with tufted or spaced fronds and with brown rhizome-scales. Stipe dark-brown or black, polished. Fronds simply pinnate to 4-pinnate, with flabellate, dimidiate or cuneate segments, glabrous or pilose, or occasionally with yellow powder on the under surface, veins free. Sori borne on the inner surface of marginal reflexed lobes which also serve as indusia.

A cosmopolitan genus of over 200 spp. with a large proportion in S. America.

Frond 1-pinnate:
  Pinnae pilose - - - - - - - - - - - 1. *incisum*
  Pinnae glabrous:
    Stipe and rachis with a narrow brown wing - - - - 2. *mettenii*
    Stipe and rachis not winged:
      Pinna margin entire - - - - - - - - 3. *philippense*
      Pinna margin denticulate - - - - - - 4. *mendoncae*
Frond 2-pinnate to 4-pinnate:
  Frond pedate or repeatedly and unequally dichotomously divided:
    Frond glabrous - - - - - - - - - 5. *patens* subsp. *oatesii*
    Frond hispid - - - - - - - - - - 6. *hispidulum*
  Frond pinnately divided, not pedate:
    Veins of sterile pinnules ending in the marginal teeth - - 7. *capillus-veneris*
    Veins of sterile pinnules ending in the sinuses between marginal crenations:
      Bases of ultimate segments broadly obtuse to reniform, articulated to the petiolules; indusial flaps oblong to lunate - - - - - - 8. *poiretii*
      Bases of ultimate segments cuneate, not articulated to the petiolules; indusial flaps subcircular to reniform - - - - - - - 9. *raddianum*

1. **Adiantum incisum** Forsk., Fl. Aegypt.-Arab.: CXXV, 187 (1775). TAB. 36 fig. C. Type from Yemen.
   Adiantum capillus-gorgonis Webb, Spic. Gorg. in Hook., Niger Fl.: 192 (1849). Type from Cape Verde Is.
   Adiantum caudatum sensu Sim, Ferns S. Afr. ed. 2: 241, t. 118 fig. 2 (1915).

Rhizome short, erect, with tufted fronds and with brown subulate rhizome-scales c. 4 mm. long. Stipe castaneous to black, up to 9 cm. long, with numerous brown hairs. Fronds up to 11 cm. long, arching, membranous, often proliferous at the apex of a naked extension of the rhachis. Lamina up to 26 × 4 cm., linear to cultrate in outline, pinnate, attenuate; pinnae up to 2 × 1 cm., mostly oblong, but reduced and obcuneate towards the apex of the frond, shortly petiolate, incised irregularly on the acroscopic margin into mostly emarginate lobes, thinly set on both surfaces with pale-brown multicellular hairs c. 1·3 mm. long. Sori borne at the apices of the pinna lobes; indusial flaps lunate to oblong, glabrous to thinly pilose.

Zambia. N: Mpongwe Distr., 96 km. S. of Luanshya, Lake Kashiba, 3.ii.1960, *Robinson* 3326 (K; SRGH). C: 9·6 km. E. of Lusaka, 20.ii.1958, *King* 422 (K). E: Chadiza Hill, 1.xii.1958, *Robson* 791 (BM; K; LISC; SRGH). S: Victoria Falls, Knife Edge, 6.vii.1953, *Schelpe* 3943 (BM; BOL; K). **Rhodesia.** N: near Sinoia, 21.iv.1948, *Rodin* 4365 (SRGH). W: Victoria Falls, 1904, *Allen* 17 (K). C: Hartley Distr., Umsweswe, *Borle* 171 (PRE). E: Inyanga Distr., Mpanga R., v.1948, *Chase* 1005 (BM:

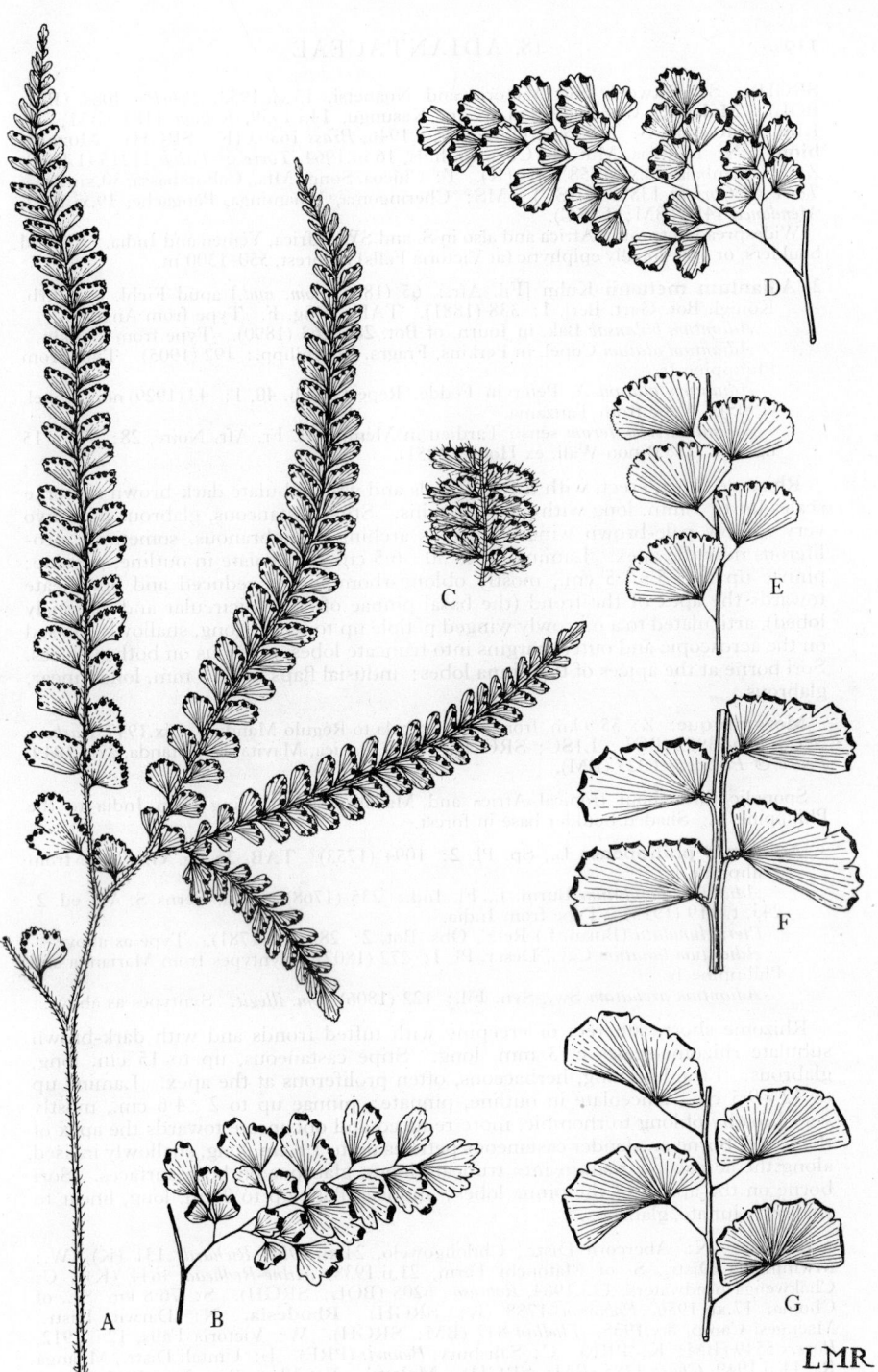

LMR

Tab. 36. A.—ADIANTUM HISPIDULUM. Part of fertile frond (× ⅔) *Mendonça* 294.
B.—ADIANTUM CAPILLUS-VENERIS. Pinna (× ⅔) from *Eyles* 68. C.—ADIANTUM
INCISUM. Part of lamina (× ⅔) *Eyles* 478. D.—ADIANTUM POIRETII var.
POIRETII. Pinna (× ⅔) *Nash* 182. E.—ADIANTUM MENDONCAE. Part of lamina
(× ⅔) *Pedro & Pedrógão* 7487. F.—ADIANTUM METTENII. Part of lamina (× ⅔)
Pedro & Pedrógão 7528. G.—ADIANTUM PHILIPPENSE. Part of lamina (× ⅔) *Chase*
3812.

SRGH). S: Between Fort Victoria and Nuanetsi, 13.vii.1953, *Schelpe* 4084 (BM; BOL). **Malawi.** C: c. 6·4 km. N. of Kasungu, 14.i.1959, *Robson* 1183 (BM; K; LISC; SRGH). S: Likubula Gorge, 21.vi.1946, *Brass* 16393 (K; SRGH). **Mozambique.** N: Malema, Mutuali, Cucuteia hills, 16.iii.1964, *Torre & Paiva* 11215 (LISC). Z: Morrumbala, 30.xii.1858, *Kirk* (K). T: Chicoa, Songa Mts., Caborabassa, 30.xii.1965, *Torre & Correia* 13899 (LISC). MS: Cheringoma, Inhaminga, Pangache, 19.xi.1942, *Mendonça* 1457 (BM; LISC).

Widespread in tropical Africa and also in S. and SW. Africa, Yemen and India. Around boulders, or occasionally epiphytic (at Victoria Falls) in forest, 550–1300 m.

2. **Adiantum mettenii** Kuhn [Fil. Afr.: 65 (1868) *nom. nud.*] apud Eichl. in Jahrb. Königl. Bot. Gart. Berl. **1**: 338 (1881). TAB. **36** fig. F. Type from Angola.

    Adiantum balansae Bak. in Journ. of Bot. **28**: 263 (1890). Type from Tonkin.

    Adiantum alatum Copel. in Perkins, Fragm. Fl. Philipp.: 192 (1905). Type from Philippine Is.

    Adiantum alatum A. Peter in Fedde, Repert. Beih. **40**, 1: 43 (1929) non. Copel. (1905). Type from Tanzania.

    Adiantum soboliferum sensu Tardieu in Mém. Inst. Fr. Afr. Noire, **28**: 94, t. 15 fig. 1–2 (1953) non Wall. ex Hook. (1851).

Rhizome short, erect, with tufted fronds and with subulate dark-brown rhizome scales up to 3 mm. long with paler margins. Stipe castaneous, glabrous with two very narrow pale-brown wings. Fronds arching, membranous, sometimes proliferous near the apex. Lamina up to 30 × 6·5 cm., lanceolate in outline, pinnate; pinnae up to 3·5 × 1·5 cm., mostly oblong-rhombic but reduced and obcuneate towards the apex of the frond (the basal pinnae often semicircular and variously lobed), articulated to a narrowly winged petiole up to 6 mm. long, shallowly incised on the acroscopic and outer margins into truncate lobes, glabrous on both surfaces. Sori borne at the apices of the pinna lobes; indusial flaps up to 5 mm. long, linear, glabrous.

Mozambique. Z: 55·9 km. from Morrumbala to Régulo Mandiua, 1.ix.1949, *Barbosa & Carvalho* 3952 (BOL; LISC; SRGH). MS: Manica, Mavita, R. Rotanda, 27.v.1949, *Pedro & Pedrógão* 6025 (BM).

Sporadic throughout tropical Africa and Madagascar extending from India to the Philippine Is. Shaded boulder base in forest.

3. **Adiantum philippense** L., Sp. Pl. **2**: 1094 (1753). TAB. **36** fig. G. Type from Philippine Is.

    Adiantum lunulatum Burm. f., Fl. Ind.: 235 (1768).—Sim, Ferns S. Afr. ed. 2: 243, t. 119 (1915). Type from India.

    Pteris lunulata (Burm. f.) Retz., Obs. Bot. **2**: 28, t. 4 (1781). Type as above.

    Adiantum lunatum Cav., Descr. Pl. **1**: 272 (1802). Syntypes from Marianna and Philippine Is.

    Adiantum arcuatum Sw., Syn. Fil.: 122 (1806) *nom. illegit.* Syntypes as above.

Rhizome short, suberect or creeping with tufted fronds and with dark-brown subulate rhizome-scales c. 3 mm. long. Stipe castaneous, up to 15 cm. long, glabrous. Frond arching, herbaceous, often proliferous at the apex. Lamina up to 42 × 9·5 cm., lanceolate in outline, pinnate; pinnae up to 2 × 4·6 cm., mostly very broadly oblong to rhombic, more reduced and obcuneate towards the apex of the frond, borne on slender castaneous petioles up to 1·8 cm. long, shallowly incised along the acroscopic margin into truncate lobes, glabrous on both surfaces. Sori borne on the apices of the pinna lobes; indusial flaps up to 2 cm. long, linear to shallowly lunate, glabrous.

Zambia. N: Abercorn Distr., Chilongowelo, 24.xii.1951, *Richards* 131 (K) W.: Mwinilunga Distr., S. of Matonchi Farm, 21.ii.1938, *Milne-Redhead* 4644 (K). C: Chakwenga headwaters, 19.i.1964, *Robinson* 6208 (BOL; SRGH). S: 76·8 km. SE. of Choma, 17.xii.1956, *Robinson* 1788 (K; SRGH). **Rhodesia.** N: Darwin Distr., Msengesi Camp, 8.v.1955, *Whellan* 847 (BM; SRGH). W: Victoria Falls, 12.ii.1912, *Rogers* 5549 (BM; K; PRE). C: Salisbury, *Bennett* (PRE). E: Umtali Distr., Mpanga R., 11.ii.1949, *Chase* 3285 (BM; SRGH). **Malawi.** N: Nkata Bay, 20.vi.1967, *Pawek* 1213 (SRGH). C: Dedza Distr., Mua Livulezi Forest, 22.iii.1962, *Adlard* 459 (SRGH). S: Malawe Hill, W. of Port Herald, 23.iii.1960, *Phipps* 2648 (K; SRGH). **Mozambique.** N: Ribáuè Mt., 19.vii.1962, *Schelpe & Leach* 6953 (BOL). Z: Ile, Errego, 3.iii.1966, *Torre & Correia* 14971 (LISC). MS: Serra do Garuso, 4.iv.1948, *Garcia* 877 (BM; LISC).

Pantropical. Terrestrial in moist forest, 610–1370 m.

4. **Adiantum mendoncae** Alston in Contr. Conhec. Fl. Moçamb. **2**: 19, t. 6–7 (1954). TAB. **36** fig. E. Type: Mozambique, Chimoio, Catarata do Revuè, *Mendonça* 2558 (BM, isotype; COI, isotype; LISC, holotype).

Rhizome short, erect to procumbent with tufted fronds and with dark-brown subulate rhizome-scales up to 3 mm. long. Fronds erect or arching, membranous, often proliferous at the end of a naked extension of the rhachis up to 8 cm. long. Stipe castaneous, glabrous for most of its length. Lamina up to 30 × 4 cm., linear to very narrowly oblong in outline, pinnate; pinnae up to 1·4–2 cm., mostly very broadly obovate, reduced towards the apex of the frond, articulate on filiform castaneous petioles up to 7 mm. long, finely denticulate on the outer margin, glabrous on both surfaces. Sori borne on the outer pinna margins; indusial flaps up to 1·4 mm. long, oblong, glabrous.

Rhodesia. E: Melsetter Distr., Haroni Gorge, 16.ii.1958, *Mitchell* 404 (BM; BOL; K; SRGH). S: 1·6 km. N. of Lundi R. bridge, 17.i.1958, *Mitchell* 456 (SRGH). **Mozambique.** N: Monte da Mesa between Nacala and Mossuril, 28.x.1952, *Barbosa & Balsinhas* 5217 (BM). MS: Chimoio, Catarata do Revuè, 31.x.1944 *Mendonça* 2558 (BM; COI; LISC).
Known only from Mozambique and the eastern border of Rhodesia. Shaded rock faces in forest, 500–600 m.

5. **Adiantum patens** Willd. in L., Sp. Pl., ed. 4, **5**: 439 (1810). Type from America.

Subsp. **oatesii** (Bak.) Schelpe in Bol. Soc. Brot., Sér. 2, **41**: 203 (1967). Type: Rhodesia, Victoria Falls, *Oates* (K).
   Adiantum oatesii Bak. in Oates, Matabele Land, App.: 369 (1881).—Sim, Ferns S. Afr. ed. 2: 244, t. 120 (1915). Type as above.
   Adiantum pedatum A. Peter in Fedde, Repert. Beih. **40**, 1: 45, descr. 4, t. 5 fig. 1 (1929) non L. (1753). Type from Tanzania.
   Adiantum patens var. *oatesii* (Bak.) Ballard in Kew Bull. **1937**: 31 (1937). Type as for *Adiantum patens* subsp. *oatesii*.

Rhizome up to 4 mm. in diam., creeping, with tufted fronds and with brown subulate weakly ciliate rhizome-scales c. 2 mm. long. Stipe castaneous, up to 39 cm. long, glabrous. Frond membranous, erect and then arching. Lamina up to 32 × 40 cm., approximately deltate in outline, pedately divided into up to 11 pinnate cultrate pinnae up to 29 × 3·5 cm., decreasing in length outwards; pinnules mostly falcately rhombic with a curved entire basiscopic margin and an irregularly crenate acroscopic and outward margin up to 2·3 × 0·9 cm., glabrous on both surfaces, borne on filiform castaneous petioles c. 1 mm. long. Sori borne on the acroscopic and outer margins of the pinnules; indusial flaps up to 2 mm. in diam., subcircular to reniform, glabrous.

Zambia. N: Abercorn Distr., above Chilongowelo, 24.xii.1951, *Richards* 362 (K). W: Mwinilunga Distr., Matonchi R., 15.ii.1938, *Milne-Redhead* 4581 (K). C: Kafulafuta R., 29.xii.1907, *Kassner* 2244 (K). **Rhodesia.** N: Lomagundi, *Darling* (PRE). W: Victoria Falls, 6.vii.1953, *Schelpe* 3942 (BOL).
Sporadic throughout tropical Africa. Terrestrial around boulder bases and termitaria in forest and on streamsides in riverine forest, c. 900 m.

6. **Adiantum hispidulum** Sw. in Schrad., Journ. Bot. **1800**, 2: 82 (1801). TAB. **36** fig. A. Type probably from Australasia.
   Adiantum lindsaea Cav., Descr. Pl. **1**: 271 (1802). Type probably from Australia.
   Adiantum lobulatum Kunze ex Ettingsh., Farnkr.: 82, fig. 32, t. 47 fig. 7 (1865). Type from Comoro Is.

Rhizome short, erect to procumbent, with tufted fronds and with castaneous lanceolate entire scales up to 1·5 mm. long. Stipe up to 29 cm. long, castaneous, hispid. Frond erect, pedate or repeatedly unequally dichotomously divided. Lamina up to 24 × 20 cm., approximately deltate in outline, pedately divided into up to 8 linear pinnae up to 20 cm. long, the longest in the centre, the others decreasing in length outwards; pinnules up to 10 × 6 mm., mostly rhombic, becoming reduced and obcuneate towards the pinna apex, firmly herbaceous, dark-green, the acroscopic and outer margins irregularly crenate-dentate, the basiscopic margin entire and ascending, thinly hispid on the lower surface only. rhachis castaneous-hispid with pale stiff hairs up to 0·5 mm. long. Sori borne on the acroscopic and outer margins of the pinnules; indusial flaps up to 1 mm. in diam., pilose, round to oblong.

Malawi. S: Shire Highlands, 1897, *Adamson* (K). **Mozambique.** MS: Chimoio, Gondola, R. Nhamissanguere, 19.ii.1948, *Garcia* 294 (BM; LISC).

Sporadic throughout east tropical Africa, Asia and Australasia. Terrestrial in deep shade in moist forest.

7. **Adiantum capillus-veneris** L., Sp. Pl. **2**: 1096 (1753).—Sim, Ferns S. Afr. ed. 2: 245, 246, t. 121–122 (1915). TAB. **36** fig. B. Type from southern Europe.
　　Adiantum coriandrifolium Lam., Fl. Fr. **1**: 29 (1778) *nom. illegit.* Type from Europe.
　　Adiantum fontanum Salisb., Prodr.: 404 (1796) *nom. illegit.* Type from Europe.
　　Adiantum capillus Sw. in Schrad., Journ. Bot. **1800**, 2: 83 (1801). Type from Europe.
　　Adiantum marginatum Schrad. in Gött. Gel. Anz. **1818**: 918 (1818). Type from S. Africa.
　　Adiantum pseudocapillus Fée, Mém. Fam. Foug. **5**: 118 (1852). Type from S. Africa.
　　Adiantum paradiseae Bak. in Gard. Chron., Ser. 3, **6**: 558 (1889). Type from S. Africa.

Rhizome up to 4 mm. in diam., creeping, with fronds borne at intervals of up to 1 cm., and with subulate reddish-brown entire rhizome-scales c. 3 mm. long. Stipe castaneous or ebeneous, glabrous. Fronds arching, herbaceous. Lamina up to 26 × 19 cm., usually narrowly ovate-deltate in outline, 3-pinnate; pinnules up to 2·7 × 2 cm., cuneate, entire or irregularly shallowly to deeply lobed, petiolulate, minutely crenate-dentate on the outer margin of the sterile pinnules, with the veins ending in the teeth, glabrous on both surfaces, often glaucous-green, thinly to firmly herbaceous; rhachis ebeneous, glabrous. Sori borne along the outer margin of the pinnules; soral flaps up to 5 × 1–1·5 mm., lunate to oblong, glabrous.

Zambia. W: Lower Solwezi Falls, 7.vi.1930, *Milne-Redhead* 447 (K). C: 11·2 km. SE. of Lusaka, 22.v.1955, *King* 8 (K). S: near Mumbwa, 1911, *Macaulay* 8 (K). **Rhodesia.** N: E. side of Umvukwe Mts., near Dawson, 28.iv.1948, *Rodin* 4464 (K; SRGH). W: Victoria Falls, 30.viii.1947, *Greenway & Brenan* 8027 (BM; BOL; K; SRGH). C: Selukwe Distr., Dunraven Falls, 17.iii.1964, *Wild* 6420 (SRGH). E: Inyanga, Nyawutari R., 4.viii.1957, *Chase* 6694 (LISC; SRGH). S: Belingwe Distr., Ngobi Dip, 18.viii.1948, *West* 2769 (SRGH). **Malawi.** S: Cholo, 30.ix.1946, *Brass* 17878 (BM; K; SRGH). **Mozambique.** Z: Massingire (Morrumbala), 3.viii.1942, *Torre* 4488 (LISC). MS: Cheringoma, Inhaminga, 25.v.1948, *Mendonça* 4382 (BM; LISC). SS: Rio das Pedras, xi.1935, *Gomes e Sousa* 1682 (COI; K).

Cosmopolitan. Shaded moist rock faces and crevices, often in semi-arid areas, 820–1220 m.

8. **Adiantum poiretii** Wikstr. in Kongel. Vet. Akad. Handl. **1825**: 443 (1826). Type from Tristan da Cunha.
　　Adiantum thalictroides Willd. ex Schlechtend., Adumbr. Pl. **5**: 53 (1832). Type from Mauritius.
　　Adiantum cycloides Zenker, Pl. Ind.: 11, t. 11 (1835). Type from India.
　　Adiantum pellucidum Mart. & Galeotti in Mém. Acad. Roy. Brux. **15**: 72, t. 19 (1842). Type from Mexico.
　　Adiantum aethiopicum sensu Sim, Ferns S. Afr. ed. 2: 248, t. 124 (1915) non L. (1759).

Rhizome slender, wide creeping, producing at intervals short thicker creeping branch-rhizomes with closely spaced fronds and with appressed castaneous lanceolate acuminate slightly ciliate rhizome-scales c. 8 × 0·8 mm. Stipe castaneous, up to 25 cm. long, glabrous. Frond arching, thinly herbaceous. Lamina 3–4-pinnate, up to 45 × 33 cm., broadly ovate-deltate in outline; pinnules up to 1·5 × 2 cm., very broadly obcuneate to semicircular or shallowly reniform, articulated at the apex of filiform petiolules, membranous, minutely crenate on the outer margins of the sterile pinnules with the veins mostly ending in the sinuses, glabrous on both surfaces or with yellow powder sparingly to densely produced on the lower surface or only about the sori; rhachis and secondary rhachises castaneous, glabrous; pinnules deciduous with age leaving the bare petiolules attached to the secondary rhachises of the old fronds. Sori borne along the acroscopic margins of the pinnules; indusial flaps up to 2·5 mm. long, lunate.

Under surface of pinnules without yellow powder - - - - - var. *poiretii*
Under surface of pinnules, especially about the sori, with yellow powder   var. *sulphureum*

Var. **poiretii**.   TAB. **36** fig. D.

Zambia. N: Abercorn Distr., above Chilongowelo, 24.xii.1951, *Richards* 363 (K). E: Kapatamoyo, near Fort Jameson, 5.i.1959, *Robson* 1032 (BM; K; LISC; SRGH). **Rhodesia.** N: Mazoe Distr., Iron Mask Hill, vi.1906, *Eyles* 328 (BM; BOL; PRE; SRGH). W: Essexvale, 25.i.1921, *Borle* 107 (NBG; SRGH). E: Umtali Distr., Tsetsera Range, 7.viii.1957, *Chase* 6692 (BM; SRGH). **Malawi.** N: Vipya, Chikana-gawa Forest, 7.ii.1962, *Chapman* 1580 (K; SRGH). C: Nchisi Forest, 5.v.1961, *Chapman* 1273 (SRGH). S: Shire Highlands, 1906, *Adamson* 257 (K). **Mozambique.** N: Maniamba, 29.v.1948, *Pedro & Pedrógáo* 4068 (LMA). T: Tete, Angónia, Domuè, 9.iii.1964, *Torre & Paiva* 11114 (LISC).
Throughout tropical and temperate Africa, Madagascar, the Mascarene Is., India, Tristan da Cunha, Central and S. America from Mexico to Uruguay. Shaded floors of montane forest, 1220–1980 m.

Var. **sulphureum** (Kaulf.) Tryon in Amer. Fern Journ. **47**: 139 (1957). Type from Chile.
 Adiantum sulphureum Kaulf., Enum. Fil.: 207 (1824). Type as above.
 Adiantum williamsii Moore in Gard. Chron. **10**: 45, fig. 4 (1878). Type from Chile.

Rhodesia. W: Victoria Falls, *Sim* (PRE). C: Makoni Distr., Silverbow, 17.vi.1957, *Chase* 6536 (BM; K; SRGH). E: Melsetter Distr., between Cashel and Melsetter, vii.1953, *Schelpe* 4012 (BM; BOL).
Sporadic in S. and E. Africa and also in S. America. Undergrowth of montane forest, 1520–1700 m.

9. **Adiantum raddianum** C. Presl, Tent. Pterid.: 158 (1836), reimpr. in Abh. Königl. Böhm. Ges. Wiss., Ser. 4, **5**: 158 (1837). Type from Brazil.
 Adiantum cuneatum Langsd. & Fisch., Ic. Fil.: 23, t. 26 (1810) non Forst. (1786). Type from Brazil.

Rhizome slender, short, creeping, with tufted fronds and with castaneous broadly lanceolate entire scales up to 1·5 mm. long. Stipe up to 30 cm. long, castaneous, glabrous. Frond arching, thinly herbaceous. Lamina up to 25 × 22 cm., 3–4-pinnate, broadly ovate-deltate in outline; ultimate segments up to 1·1 × 0·8 cm., obcuneate to trapeziform, with filiform petiolules but not articulated, minutely crenate-serrate on the outer margins with the veins ending in the sinuses, glabrous on both surfaces; rhachis and secondary rhachises castaneous, glabrous. Sori borne on the acroscopic margins of the ultimate segments; indusial flaps c. 1·5 mm. in diam., subcircular to reniform.

Rhodesia. E: Umtali, Imbeza Valley, 2.iii.1957, *Chase* 6343 (BOL; LISC; SRGH). **Malawi.** S: Mlanje Distr., Likabula Valley near Likabula House, 11.ii.1967, *Berrie* 89 (BOL). **Mozambique.** Z: Milange, Tumbine Mts., near Vila Masseti, 18.i.1966, *Correia* 462 (LISC). MS: Manica Distr., Border Farm, 25.viii.1963, *Chase* 8042 (BOL; SRGH).
A widely cultivated S. American species which has escaped in many moist subtropical localities in south, south-east and east tropical Africa. In our area it has become natural-ised on stream-banks in forest between 910 and 1310 m.

7. ASPIDOTIS (Nutt. ex Hook.) Copel.

Aspidotis (Nutt. ex Hook.) Copel., Gen. Fil.: 68 (1947).

Rhizome creeping, with tufted fronds and brown to almost black subulate hair pointed rhizome scales. Stipe castaneous, polished except at the base. Frond 4-pinnate, broadly deltate to ovate-deltate, glabrous, with small narrow dentate serrate segments. Sori minute, borne in the sinuses of the teeth of the ultimate segments; without paraphyses and with few sporangia; indusium formed from the reflexed margin.
A small genus of 3 spp., 1 in Africa, the other 2 in California and Mexico respectively.

Aspidotis schimperi (Kunze) P.-Sermolli in Webbia, **7**: 326 (1950). TAB. **37** fig. A. Type from Ethiopia.
 Cheilanthes schimperi Kunze, Farrnkr. **1**: 52, t. 26 (1840). Type as above.
 Hypolepis schimperi (Kunze) Hook., Sp. Fil. **2**: 70 (1852).—Sim, Ferns S. Afr. ed. 2: 239, t. 117 (1915). Type as above.

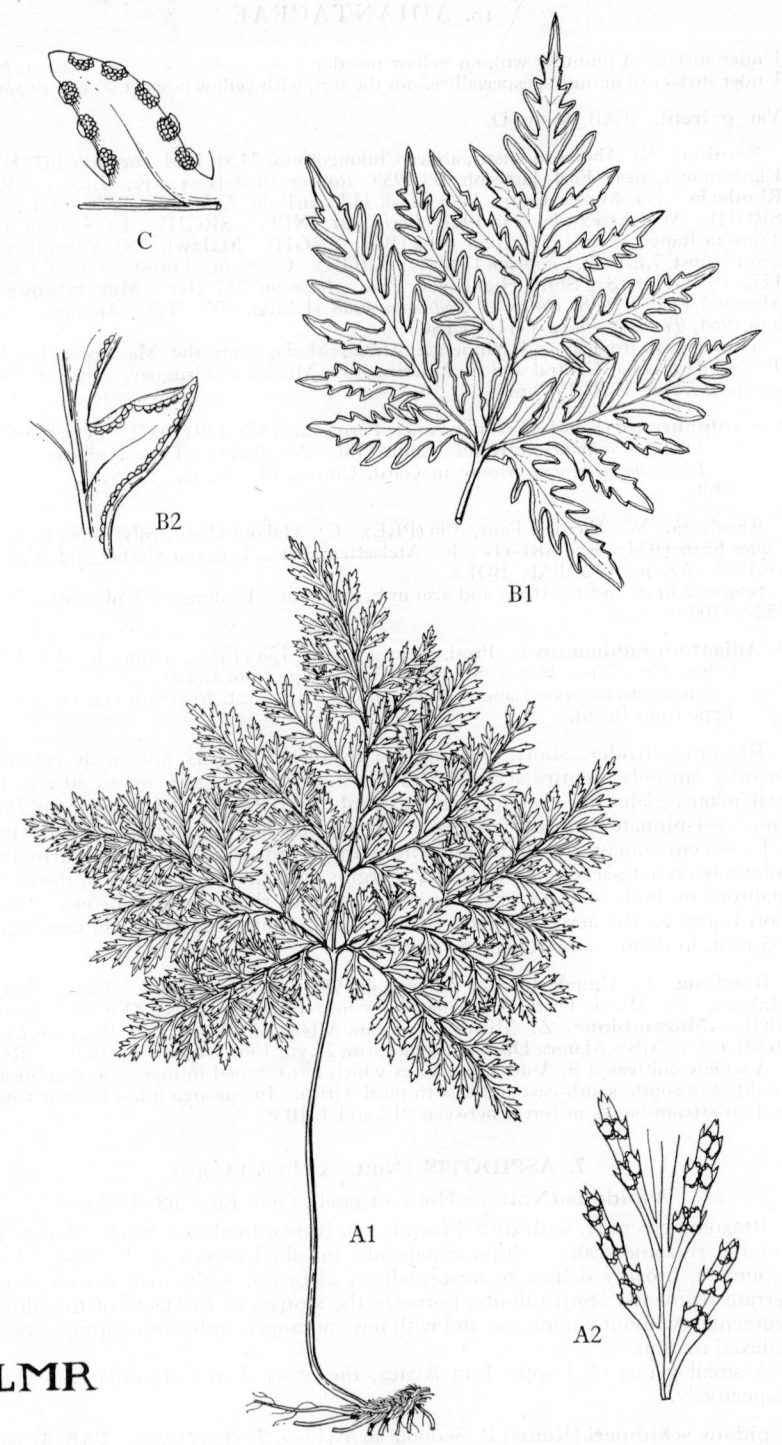

LMR

Tab. 37. A.—ASPIDOTIS SCHIMPERI. A1, habit (×⅔); A2, pinnule segment (×4), both from *Fisher & Schweickerdt* 435. B.—DORYOPTERIS CONCOLOR var. NICKLESII. B1, habit (×⅔); B2, pinnule segment (×4), both from *Chase* 31602. C.—DORYOPTERIS CONCOLOR var. KIRKII. Pinnule segment (×4) *E.M. & W.* 1063.

Rhizome c. 5 mm. in diam., subterranean, creeping, usually surrounded by a dense mat of roots, with closely spaced fronds and with subulate entire brown rhizome scales which are darker towards their hair-point apices. Stipe castaneous. up to 28 cm. long, glabrous except at the base. Frond erect, thinly coriaceous, Lamina broadly deltate in outline, 4-pinnatifid with the lowest pair of pinnae about as long as the rest of the lamina and much developed basiscopically; pinnules set at an acute angle to the castaneous secondary rhachises; ultimate segments up to 1 cm. long, cultrate, glabrous, strongly dentate-serrate, adnate to the rhachis and decurrent at the base. Sori c. 0·5 mm. in diam., minute, solitary in the sinuses of the marginal teeth of the ultimate segments; indusium minute, broadly ovate, irregularly crenate, semi-transparent.

Zambia. N: Abercorn Distr., Mukoma, 7.iv.1962, *Richards* 16282 (K; SRGH). W: Kalenda Ridge, W. of Matonchi Farm, 22.i.1938, *Milne-Redhead* 4266 (K). C: 9·6 km. E. of Lusaka, 16.i.1956, *King* 269 (K). S: 60·8 km. NE. of Choma, 17.xii.1956, *Robinson* 1791 (BM; K; SRGH). **Rhodesia.** N: Sipolilo, 29.i.1948, *Whellan* 293 (BM, SRGH). C: Ngomokurira, 23.i.1952, *Wild* 3753 (BM; SRGH). E: Umtali Distr., Laurenceville, 21.xii.1952, *Chase* 4747 (BM; BOL; SRGH). **Malawi.** N: Livingstonia, 9.i.1959, *Robinson* 3131 (K; SRGH). S: near Blantyre, 1887, *Last* (K). **Mozambique.** N: Moçambique, Malema, Mutuali, Cucuteia hill, c. 750 m., 16.iii.1964, *Torre & Paiva* 11214 (LISC). T: Macanga, 17.iii.1966, *Sarmento & Marques* 1799 (LMU).
Also in Tanzania, Uganda, Ethiopia, Sudan and N. Nigeria. On streambanks, among boulders and on termite mounds in open forest and *Brachystegia* woodland, 850–1750 m.

8. PTERIS L.

Pteris L., Sp. Pl. **2**: 1073 (1753); Gen. Pl. ed. 5: 484 (1754).

Rhizome erect or shortly to widely creeping, dictyostelic, with tufted to widely spaced fronds and linear to ovate rhizome-scales sometimes with a dark central stripe. Frond simply pinnate to 4-pinnatifid, thinly herbaceous to thinly coriaceous, glabrous, sometimes with spines on the rhachis costae and costules dorsally, often with costal spines ventrally at the junctions of costules; veins free or anastomosing. Sori almost marginal, confluent into soral lines but not extending to the apex of the ultimate segments, borne on an almost marginal vein, covered by a continuous indusium formed from the reflexed margin; paraphyses usually present. A genus of over 250 spp. mostly tropical but some in temperate regions.

Upper pinnae simple not pinnatifid, lower pinnae simple or with 1–3 basiscopic lobes:
  Lower pinnae all simple, gradually reduced - - - - - - 1. *vittata*
  Lower pinnae with 1–2 basiscopic lobes - - - - - - 2. *cretica*
Upper pinnae pinnatifid or 2-pinnatifid:
  Veins anastomosing at least in the more broadly winged parts of the costae:
    Fronds broadly triangular, about as broad as long; rhizomes widely creeping; costa
      not spinose - - - - - - - - - - 8. *buchananii*
    Fronds narrowly triangular; rhizome short; costae spinose - - 9. *hamulosa*
  Veins free:
    Rhachis castaneous, with stout spines - - - - - 4. *intricata*
    Rhachis without spines or if with slender spines then not castaneous:
      Fronds tripartite, sori c. ⅓ the length of the ultimate lobe margins
                  3. *pteridioides*
      Fronds pinnately divided, sori usually more than ½ the length of the ultimate lobe
      margins:
        Sterile apices of fertile and sterile lobes crenate-dentate - - 5. *dentata*
        Sterile apices of fertile and sterile lobes entire or subentire:
          Costules as well as costae with spines on the ventral surface 6. *catoptera*
          Only costae with spines on the ventral surface - - - - 7. *friesii*

1. **Pteris vittata** L., Sp. Pl. **2**: 1074 (1753).—Tardieu in Mém. Inst. Fr. Afr. Noire, **28**: 69, t. 10 fig. 1–2 (1953); Fl. Madag., Polypod. **1**: 85 (1958). Type from China.
  Polypodium trapezoides Burm. f., Fl. Ind.: t. 66 fig. 2 (1768). Type from Java.
  Pteris obliqua Forsk., Fl. Aegypt.-Arab.: CXXIV, 185 (1775). Type from Yemen.
  Pteris lanceolata Desf., Fl. Atlant. **2**: 401 (1800). Type from Algeria.
  Pteris ensifolia Poir., Encycl. Méth. Bot. **5**: 711 (1804). Type from Spain.
  Pteris diversifolia Sw., Syn. Fil.: 96, 288 (1806) *nom. illegit.* Type from Java.
  Pteris costata Bory ex Willd. in L., Sp. Pl., ed. 4, **5**: 367 (1810). Type from Mauritius.

Pteris inaequilateralis Poir., Encycl. Méth. Bot., Suppl. **4**: 601 (1816). Type from Réunion.

Pteris aequalis C. Presl, Reliq. Haenk.: 54 (1827). Type probably from Philippine Is.

Pteris alpinii Desv. in Mém. Soc. Linn. Par. **6**, 2: 295 (1827). Type from the Orient.

Pteris acuminatissima Bl., Enum. Pl. Jav. **2**: 208 (1828). Type from Java.

Pteris microdonta Gaudich. in Freyc., Voy., Bot.: 387 (1829). Type from Timor.

Pteris guichenotiana Gaudich. loc. cit. Type from the Moluccas.

Pteris tenuifolia Brack., U.S. Explor. Exped. **16**: 112 (1854). Type from Tonga Is.

Pteris vulcanica Bertol., Misc. Bot. **18**: 21 (1858). Type from Italy.

Pycnodoria vittata (L.) Small, Ferns S. E. States: 102, 468 (1938). Type as for *Pteris vittata*.

Rhizome up to 8 mm. in diam., creeping, with fronds spaced up to 1 cm. apart and with linear-lanceolate attenuate pale-brown rhizome scales. Stipe up to 12 cm. long, pale-brown, terete, glabrous for most of its length but with numerous scales similar to those on the rhizome about the base. Frond erect to arching, firmly membranous. Lamina up to 115 × 40 cm., elliptic-oblong in outline, simply pinnate, tapering towards the base; pinnae up to 16 × 1·4 cm., linear-attenuate, glabrous, sterile margins minutely crenate, lower pinnae petiolate, the upper sessile; veins free; rhachis pale-brown, glabrous, sulcate ventrally. Sori in submarginal lines with paraphyses extending for most of the length of the fertile pinnae; indusium membranous, subentire.

Zambia. W: near Solwezi Boma, 12.ix.1952, *Angus* 420 (BM; BOL; K). C: Mt. Makulu, 15.iv.1956, *Robinson* 1489 (K; SRGH). S: near Mumbwa, 1911, *Macaulay* 9 (K). **Rhodesia.** N: Mazoe Dam, 24.v.1934, *Gilliland* 203 (BM; SRGH). W: near Bulawayo, v.1904, *Eyles* 66 (BM; SRGH). C: Salisbury Distr., 28·8 km. NE. of Salisbury, vii.1955, *Mitchell* 5 (BOL). E: Umtali Commonage, 12.iv.1950, *Chase* 3148 (BM; SRGH). **Malawi.** S: near Lake Chilwa, iv.1859, *Kirk* (K). **Mozambique.** T: Massanga, 27.x.1947, *Barbosa* 580 (LMA). MS: Cheringoma, Inhaminga, 19.x.1942, *Mendonça* 1451 (BM; LISC). SS: Rio das Pedras, vii.1936, *Gomes e Sousa* 1787 (COI; K). LM: Marracuene, Bobole, 15.xii.1944, *Mendonça* 3424 (BM; LISC).

Widespread in the warmer regions of the Old World. Rock crevices in shade, 150–750 m.

2. **Pteris cretica** L., Syst. Nat. **2**: 688 (1767); Mant. Pl.: 130 (1767).—Sim, Ferns S. Afr. ed. 2: 253, t. 126 (1915).—Tardieu, Fl. Madag., Polypod. **1**: 86, t. 14 fig. 1–4 (1958). Type from S. Europe.

Pteris semiserrata Forsk., Fl. Aegypt.-Arab.: CXXIV, 186 (1775). Type from Yemen.

Pteris nervosa Thunb., Fl. Jap.: 332 (1784). Type from Japan.

Pteris serraria Sw. in Schrad., Journ. Bot. **1800**, 2: (1801). Type from S. Africa.

Pteris pentaphylla Willd. in L., Sp. Pl., ed. 4, **5**: 362 (1810). Type from Réunion.

Pycnodoria cretica (L.) Small, Ferns Florida: 91 (1932). Type as for *Pteris cretica*.

Rhizome up to 1 cm. in diam., creeping, with tufted fronds and with lanceolate acuminate entire concolorous dark-brown rhizome-scales c. 3 mm. long. Stipe up to 65 cm. long, stramineous to light-brown, shallowly sulcate, glabrous. Fronds often dimorphous with the fertile fronds taller and with narrower pinnae than in the sterile fronds, erect, firmly membranous to chartaceous. Lamina ovate to triangular in outline, mostly pinnate but with the basal pinnae unequally 2-fid, the shorter lobe produced basiscopically; pinnae and lobes of the basal pinnae linear-attenuate, the lower sessile the upper adnate, decurrent, the sterile margins of the sterile and fertile pinnae bluntly to sharply serrate-dentate, sometimes undulate; sterile pinnae up to 11 × 2·4 cm., fertile pinnae up to 20 × 0·9 cm.; veins free; rhachis stramineous, shallowly sulcate, glabrous. Sori in marginal lines extending for most of the length of the fertile pinnae, with paraphyses; indusium linear, subentire, membranous.

Zambia. E: Nyika Plateau, 3.i.1959, *Robinson* 3020 (K; SRGH). **Rhodesia.** N: Mazoe, xii.1906, *Eyles* 498 (BM; BOL; SRGH). W: Victoria Falls, Rain Forest, 4.x.1964, *Whellan* 2195 (BOL; SRGH). C: Salisbury, *Sim* (PRE). E: Vumba Mts., Thordale, 4.ii.1950, *Chase* 3525 (BM; SRGH). **Malawi.** N: Nyika Plateau, 4 km. SW. of Rest House, 24.x.1958, *Robson* 312 (BM; K; LISC; SRGH). C: Dedza Distr., Domwe Hill, iv.1961, *Chapman* 1211 (SRGH). S: Blantyre, viii.1870, *Buchanan* 144 (K).

Widespread in eastern S. Africa and the mountains of E. tropical Africa, Ascension I.,

Madagascar, Réunion and in Asia eastwards to Japan, and in southern Europe. In shade in forest undergrowth, usually on streambanks, 1200–2150 m.

3. **Pteris pteridioides** (Hook.) Ballard in Kew Bull. **1937**: 348 (1937).—Tardieu in Mém. Inst. Fr. Afr. Noire, **28**: 80, t. 11 fig. 3–4 (1953). Type from Fernando Po.
Hypolepis pteridioides Hook., Sec. Cent. Ferns: t. 59 (1861). Type as above.
Pteris brevisora Bak., Syn. Fil.: 162 (1867) *nom illegit.*—Sim, Ferns S. Afr. ed. 2: 258, t. 128 (1915). Type from Fernando Po.

Rhizome up to 1 cm. in diam., erect, with tufted fronds and with lanceolate to narrowly ovate attenuate entire brown rhizome-scales up to 5 mm. long. Stipe stramineous, up to 58 cm. long, glabrous except for a few scales similar to those on the rhizome near the base. Frond herbaceous, arching. Lamina up to 43 × 36 cm., 3-partite, 2–3-pinnatifid, with the lowest basiscopic pinnules narrowly oblong-acuminate, up to 10 × 2 cm., glabrous on both surfaces but with spines up to 2·5 mm. long on the costa ventrally at the junction of main veins and with smaller spines on the costules ventrally; ultimate lobes up to 1 × 0·4 cm., narrowly oblong, obtuse, crenate in the upper half; veins free; rhachises stramineous, glabrous, smooth, sulcate ventrally. Sori marginal, very narrowly oblong, borne between the base and apex of the ultimate lobes; indusium subentire, membranous.

Rhodesia. E: Chirinda, 24.x.1947, *Wild* 2199 (K; SRGH). **Mozambique.** MS: Garuso, Jaegersberg, Bandula Mt., 11.vii.1955, *Schelpe* 5612 (B; BM; BOL; K; P).
Widespread in tropical Africa. In shade in undergrowth of forest, 1060–1160 m.

4. **Pteris intricata** C. H. Wright in Kew Bull. **1906**: 252 (1906). Type from Uganda.
Pteris adamii Tardieu in Mém. Inst. Fr. Afr. Noire, **28**: 76, t. 12 fig. 1 (1953) Type from Guinée.

Rhizome up to 1·2 cm. in diam., erect, with tufted fronds and with linear-lanceolate entire dark-brown rhizome-scales up to 1 cm. long. Stipe 63 × 0·8 cm., castaneous to atrocastaneous, with sparse stout erect or recurved spines up to 3 mm. long, glabrous except for numerous scales similar to those on the rhizome about the base. Frond arching, herbaceous. Lamina up to 1 × 0·9 m., ovate to broadly ovate in outline, 3–4-pinnatifid, with the lowest pinnae up to 56 cm. long and developed basiscopically; pinnae narrowly triangular, petiolate; the costae and costules of larger pinnae sparsely set with stout spines dorsally and with a spine at the junction of main veins ventrally; ultimate lobes narrowly oblong to linear, acute to obtuse, c. 4 mm. broad, with the sterile margins of the fertile and sterile segments serrate, glabrous on both surfaces; veins free; rhachis and secondary rhachises castaneous, with sparse stout erect spines up to 2 mm. long. Sori in a marginal line, usually confined to the basal half of the lobe margins; indusium entire, membranous.

Zambia. N: Kawambwa Distr., near Kawambwa Boma, 7.xi.1952, *Angus* 742 (BM; K). **Mozambique.** MS: Chimanimani Mts., below Martin Falls, 8.ii.1958, *Mitchell* 326 (BM; BOL; K; SRGH).
Sporadic throughout tropical Africa. Rare, in shade of wet forest or swamp forest, about 1400 m.

5. **Pteris dentata** Forsk., Fl. Aegypt.-Arab.: CXXIV, 186 (1775).—Sim, Ferns S. Afr. ed. 2: 255, t. 129 (1915). Type from Yemen.

Subsp. **flabellata** (Thunb.) Runemark in Bot. Notis. **115**: 190 (1962). Type from S. Africa.
Lonchitis adscensionis Forst. in Comm. Soc. Reg. Gott. **9**: 72 (1789). Type from Ascension I.
Pteris flabellata Thunb., Prodr. Pl. Cap.: 172 (1800). Type as for *Pteris dentata* subsp. *flabellata*.
Pteris adscensionis (Forst.) Sw. in Schrad., Journ. Bot. **1800,** 2: 67 (1801). Type as for *Lonchitis adscensionis*.
Asplenium adscensionis (Forst.) Bernh. in Schr. Akad. Erfurt, **1802**: 18, fig. 8 (1802). Type as above.
Pteris semiserrata Roxb. in Beatson, St. Helena: 319 (1816) non Forsk. (1775). Type from St. Helena.
Pteris arguta var. *flabellata* (Thunb.) Mett. in Kuhn, Fil. Afr.: 76 (1868). Type as for *Pteris dentata* subsp. *flabellata*.

Rhizome up to 1 cm. in diam., erect to procumbent, with tufted fronds and with linear-lanceolate atrocastaneous shining rhizome scales up to 5 mm. long with sparse pale marginal cilia. Stipe stramineous becoming atrocastaneous at the base, up to 50 cm. long, glabrous. Frond herbaceous, arching. Lamina up to 1×0.4 m., ovate in outline, 3-pinnatifid, rarely 4-pinnatifid, with the lowest pinnae markedly developed basiscopically; middle pinnae narrowly oblong, attenuate, deeply pinnatifid into broadly linear often falcate decurrent lobes up to 3.2×0.5 cm. serrate towards the apices, glabrous on both surfaces and with spines on the costa ventrally at the junctions of main veins; veins free; rhachis stramineous or tending towards castaneous, glabrous, smooth, sulcate ventrally. Sori in a marginal line extending for most of the length of the ultimate lobes; indusia erose or subentire, membranous.

Zambia. N: Mpongwe, 11.x.1962, *Fanshawe* 7088 (SRGH). W: Kitwe, Ichimpi, 13.vi.1962, *Mutimushi* 173 (SRGH). **Rhodesia.** E: Umtali Distr., Imbesa Forest Estate, 10.xi.1956, *Chase* 6240 (BM; K; SRGH). **Malawi.** S: Blantyre, 1887, *Last* (K). **Mozambique.** MS: Penhalonga Waterfall, 24.xi.1950, *Chase* 3248 (SRGH).

Also in S. Africa and on E. tropical African mountains, Fernando Po, Ascension I., St. Helena and the Mascarene Is. Occasional in forest, 1160–1580 m.

6. **Pteris catoptera** Kunze in Linnaea, **18**: 119 (1844). Type from S. Africa.
   Pteris biaurita sensu Sim, Ferns S. Afr. ed. 2: 257, t. 127 (1915).

Rhizome c. 1 cm. in diam., erect to procumbent, with tufted fronds and with linear-attenuate, atrocastaneous rhizome scales c. 3·5 mm. long with pale ciliate margins. Stipe stramineous tending to castaneous at the base, up to 90 cm. long, smooth or set with slender spines, glabrous except for brown scales up to 1·5 mm. long and similar to those on the rhizome at the base. Frond herbaceous, arching. Lamina up to 90×60 cm., oblong-ovate in outline 3-pinnatifid with the lower pinnae much developed basiscopically; upper pinnae very narrowly oblong, acute, glabrous on both surfaces but with spines on the costae and costules ventrally at the junction of main veins and with or without slender spines on the costae dorsally; ultimate lobes linear to broadly linear, obtuse, decurrent; sterile margins of sterile and fertile lobes subentire; veins free; rhachis stramineous, glabrous, smooth or spiny, sulcate ventrally. Sori in marginal lines extending for most of the length of the fertile lobes; indusia erose, membranous.

Rhachis and dorsal surface of costae without spines - - - - - var. *catoptera*
Rhachis and dorsal surface of costae spiny - - - - - var. *horridula*

Var. **catoptera**

Zambia. E: Nyika Plateau, Kangampande Mt., 6.v.1952, *White* 2713 (BM; K). **Rhodesia.** E: Chirinda, 24.x.1947, *Wild* 2204 (K; SRGH). **Malawi.** N: Misuku Hills, 12.i.1959, *Robinson* 3188 (K; SRGH). C: Nchisi Mt., 30.vii.1946, *Brass* 17039 (K; SRGH). S: Cholo Mt., 24.ix.1946, *Brass* 17794 (K; SRGH). **Mozambique.** N: Ribáuè Mt., 19.vii.1962, *Schelpe & Leach* 6938 (BOL). Z: Morrumbala, 30.xii.1858, *Kirk* (K). T: Zobué Mts., 3.x.1942, *Mendonça* 607 (LISC). MS: Garuso, Jaegersberg, 11.vii.1955, *Schelpe* 5637 (BM; BOL).

Also in S. Africa and Angola. Specimens of the *Pteris quadriaurita* Retz. complex with ventral costular spines are known to occur throughout tropical Africa. Forest undergrowth, 800–1770 m.

Var. **horridula** Schelpe in Bol. Soc. Brot., Sér. 2, **41**: 216 (1967). TAB. **38**. Type: Rhodesia, Umtali Distr., Inyamatshira Range, *Chase* 4890 (BM, isotype; BOL, holotype; SRGH, isotype).

Rhodesia. E: Umtali Distr., Vumba Mts., Excelsior, 24.ix.1959, *Chase* 7171 (BOL; K; SRGH). **Malawi.** S: Kasupe Distr., Chaone Hill, 13.iii.1962, *Adlard* 446 (BM; BOL; SRGH). **Mozambique.** N: Maniamba, Mt. Geci, 29.v.1948, *Pedro & Pedrógão* 4085 (LMA).

Known only from Rhodesia, Malawi, and Mozambique. In forest, 1220–1600 m.

7. **Pteris friesii** Hieron. in Wiss. Ergebn. Schwed. Rhod.-Kongo-Exped. **1**: 5 (1914). Type: Zambia, near Abercorn, *Fries* 1220 (UPS).
   Pteris abrahamii Hieron. in Engl., Bot. Jahrb. **53**: 409 (1915). Type from S. Africa.

Rhizome c. 1·5 cm. in diam., erect to procumbent with tufted fronds and with

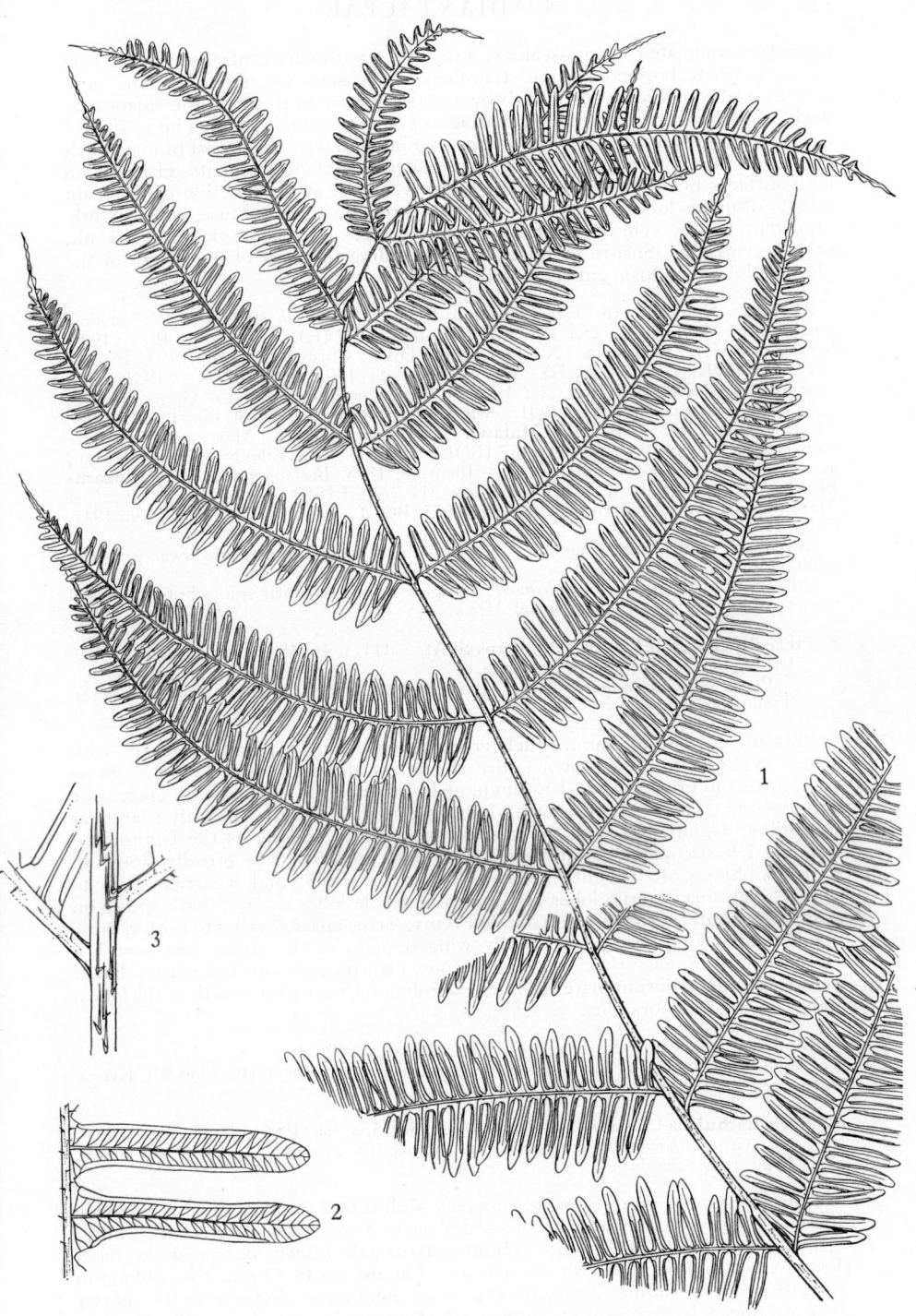

Tab. 38. PTERIS CATOPTERA var. HORRIDULA. 1, frond ($\times \frac{2}{3}$); 2, enlargement of pinnule ($\times 2\frac{1}{2}$); 3, enlargement of rhachis ($\times 2\frac{1}{2}$), all from *Chase* 4890.

lanceolate-attenuate rhizome-scales c. 3 mm. long with dark central stripe and pale ciliate-fimbriate borders. Stipe stramineous to castaneous, up to 60 cm. long, smooth, glabrous except for dark-brown scales similar to those on the rhizome at the base. Frond erect to arching, chartaceous to coriaceous. Lamina up to 70 × 48 cm., narrowly ovate-triangular in outline, 3-pinnatifid with the lowest pinnae much developed basiscopically; upper pinnae narrowly oblong-attenuate, glabrous on both surfaces but with spines on the costae ventrally at the junction of the main veins; ultimate lobes up to 2·8 × 0·5 cm., broadly linear, obtuse, only slightly decurrent, entire; veins free; rhachis stramineous to castaneous, glabrous, smooth, sulcate ventrally. Sori in marginal lines extending for most of the length of the ultimate lobes; indusia entire, membranous.

Zambia. N: Abercorn Distr., Mwengo, 10.vi.1951, *Bullock* 3956 (K). W: Mwinilunga Distr., Matonchi R., 25.x.1937, *Milne-Redhead* 2941 (K). C: Kashitu R., vii.1909, *Rogers* 8305 (K). **Rhodesia.** N: Goromonzi Distr., Chishawasha, 20.i.1960, *Mitchell* 563 (BM; BOL; LISC; SRGH). W: Victoria Falls, 1905, *Allen* 181 (K; SRGH). C: near Salisbury, *Holland* (NBG). E: Umtali Distr., Umtali Commonage, Upper Park R., 7.viii.1948, *Chase* 2036 (BM; BOL; SRGH). S: 1·6 km. E. of Bikita, 22.ii.1964, *Mitchell* 817 (LISC; SRGH). **Malawi.** N: Mzimba Distr., Marymount, towards Lunyangwa R., 6.x.1967, *Pawek* 1467 (BOL; SRGH). C: Lilongwe Distr., Kasitu, 4.xi.1962, *Chapman* 1734 (SRGH). S: Blantyre, 1878, *Buchanan* 142 (K). **Mozambique.** N: Vila Cabral, 30.x.1934, *Torre* 211 (K; LISC). Z: Namúli Mt., Chá Moçambique, 29.vii.1962, *Schelpe & Leach* 7075 (BOL). MS: Border Farm, 18.vii.1947, *Chase* 1105 (SRGH).

Also in S. Africa and Angola. Undergrowth of scrub, forest, and swamp forest, 1200–1750 m.

Other specimens of the *P. quadriaurita* complex without costular spines but with costal spines are known throughout tropical Africa.

8. **Pteris buchananii** Bak. ex Sim, Ferns S. Afr.: 111, t. 46 (1892); op. cit. ed. 2: 259, t. 130 (1915). Type from S. Africa.
    Pteris subquadripinnata Chiov., Lav. Ist. Bot. Univ. Modena, **6**: 147 (1935). Type from Kenya.

Rhizome c. 1 cm. in diam., widely creeping, with widely spaced fronds and with sparse ovate acute dark-brown entire rhizome-scales up to 4·5 × 3 mm. Stipe stramineous to brown, up to 1 m. high, glabrous, smooth. Frond with erect stipe and ascending lamina, herbaceous. Lamina up to 50 × 50 cm., broadly triangular in outline, 4-pinnatifid with the basal pinnae nearly as long as the lamina and developed basiscopically; ultimate lobes up to 2·6 × 0·5 cm., broadly linear to narrowly oblong, often slightly falcate, adnate to the rhachis and decurrent, acute to obtuse, glabrous; sterile lobes and apices of fertile lobes serrate; with spines on costules ventrally at the junction of main veins; veins anastomosing to form narrow areoles only along the more broadly winged parts of the costae and costules; rhachis and secondary rhachises stramineous, glabrous, smooth but with a darker ventral channel. Sori in marginal lines extending for nearly the length of the lobes; indusia erose, membranous.

Rhodesia. E: Chirinda, 26.x.1947, *Wild* 2216 (K; SRGH).
Also in S. Africa and in Tanzania on the Uluguru and Usambara Mts., and Mt. Kenya. Rare, deep shade in forest, 1160 m.

9. **Pteris hamulosa** Christ in Ann. Mus. Cong., Sér. 5, **3**: 30 (1909). Type from Congo.
    Pteris acanthoneura Alston in Contr. Conhec. Fl. Moçamb. **2**: 15, t. 4–5 (1954). Type from Uganda.

Rhizome up to 7 mm. in diam., suberect, with tufted fronds and with narrowly linear subentire to lacerate rhizome-scales up to 5 mm. long with a dark central stripe and pale margins. Stipe stramineous to pale brown, darker at the base. Frond erect to arching, thinly coriaceous. Lamina up to 43 × 25 cm., oblong to broadly ovate in outline, 2-pinnate except for 2-fid basal pinnae with the shorter lobe produced basiscopically; pinnae set at an angle of c. 45° to the rhachis, elliptic to narrowly oblong, shortly caudate, the upper pinnae narrowly decurrent; costae and sometimes costules spinose dorsally, not spinose ventrally at vein junctions; pinnae lobes narrowly oblong, obtuse to acute, slightly falcate, glabrous on both surfaces, sterile margins crenate-serrate; veins anastomosing only along the costa to form narrow areoles; rhachis stramineous, spinose. Sori marginal, extending for most of the length of the ultimate lobes; indusia entire, membranous.

Mozambique. MS: Cheringoma, Durúndi, 29.v.1948, *Mendonça* 4423 (BM; LISC). Also in Uganda, southern Congo, Sudan and Ivory Coast.

9. DORYOPTERIS J. Sm.

Doryopteris J. Sm. in Journ. of Bot. **3**: 404 (1841); op. cit. **4**: 162 (1841).

Rhizome creeping or procumbent, short or long, solenostelic, with tufted or spaced fronds and with linear rhizome-scales with a dark central stripe. Stipes atrocastaneous or ebeneous, glabrous, nitidous, with a few scales about the base. Frond firmly membranous to subcoriaceous, pedate in outline, 2–4-pinnatifid, glabrous; veins free or anastomosing. Sori marginal or submarginal, discrete or continuous, with discrete or continuous indusia.

A predominantly S. American genus of c. 35 spp. with 1 occurring in Africa.

Doryopteris concolor (Langsd. & Fisch.) Kuhn in Von Deck. Reisen, Bot., **3**, 3: 19 (1879). Type from Marquesas Archipelago.

    Pteris concolor Langsd. & Fisch., Ic. Fil.: 19, t. 21 (1810). Type as above.

Rhizome short, procumbent, with tufted fronds and with linear-lanceolate to subulate attenuate entire rhizome-scales up to 3 mm. long with a black central stripe and pale-brown margins. Stipe atrocastaneous or ebeneous, up to 22 cm. long, sulcate ventrally, glabrous except for occasional scales similar to those on the rhizome about the base. Frond suberect, thinly coriaceous. Lamina up to 20 × 18 cm., broadly triangular-hastate in outline, 2–3-pinnatifid, almost as broad as long, with the basiscopic lobes of the basal pinnae prominently developed; upper pinnae oblong, attenuate, decurrent at the base, deeply pinnatifid into entire to pinnatifid glabrous oblong acute to attenuate lobes; lowest pair of pinnae very unequally triangular; rhachis and dorsal surface of costae and larger costules atrocastaneous to ebeneous, glabrous. Sori marginal, discrete or forming continuous soral lines; indusia membranous, small, linear, discrete to continuous.

Indusium discontinuous; spores smooth - - - - - - - var. *kirkii*
Indusium continuous; spores rough to spinulose - - - - var. *nicklesii*

Var. **kirkii** (Hook.) R. E. Fr. in Wiss. Ergebn. Schwed. Rhod.-Kongo-Exped. **1**: 4 (1914). TAB. **37** fig. C. Type: Mozambique, *Kirk* (K).
    Cheilanthes kirkii Hook., Sec. Cent. Ferns: t. 81 (1861). Type as above.
    Doryopteris kirkii (Hook.) Alston in Bol. Soc. Brot., Sér. 2, **30**: 14 (1956). Type as above.

Botswana. SE: Bamangwato, 1869, *Todd* 10 (K). **Zambia.** N: Chilongowelo, 23.i.1955, *Richards* 4239 (K). E: Chipiri Hill near Chadiza, 29.xi.1958, *Robson* 782 (BM; K; LISC; SRGH). **Rhodesia.** N: Mazoe Distr., 9.ii.1947, *Wild* 1643 (BM; SRGH). W: Bulawayo, Khami Ruins, vii.1953, *Schelpe* 3914 (BOL). C: Makoni Distr., Silverbow, 13.vi.1957, *Chase* 6520 (BM; BOL; K; SRGH). E: Umtali Distr., Umtali Commonage, Inyamatshira Mt., 21.i.1958, *Chase* 6802 (BM; BOL; K; SRGH). S: Zimbabwe Ruins, 1.vii.1930, *Hutchinson & Gillett* 3330 (K). **Malawi.** S: Likabula Gorge, 21.vi.1946, *Brass* 16392 (K). **Mozambique.** N: Monte de Massangulo, xii.1932. *Gomes e Sousa* 1198 (COI; K). T: Macanga, Mt. Furancungo, 17.iii.1966, *Pereira, Sarmento & Marques* 1865 (LMU). Z: Morrumbala, 30.xii.1888, *Kirk* (K). MS: Garuso, iv.1935, *Gilliland* 1885 (BM).
    Also in S. Africa and widespread in tropical Africa. Woodland and sparse forest in shade, 700–1700 m.

Var. **nicklesii** (Tardieu) Schelpe in Bol. Soc. Brot., Sér. 2, **41**: 212 (1967). TAB. **37** fig. B. Type from Central African Republic.
    Doryopteris nicklesii Tardieu in Notul. Syst. **3**: 166 (1948). Type as above.

Zambia. W: Ndola, 23.iii.1967, *Fanshawe* 9988 (SRGH). S: Mazabuka Distr., Mapanza-Choma, 15.ii.1958, *Robinson* 2764 (K; SRGH). **Rhodesia.** N: Urungwe Distr., S. bank of Mauora R., 27.ii.1958, *Phipps* 921 (BM; BOL; SRGH). **Malawi.** N: Rumpi Distr., Southern Rukuru Gorge, 25.v.1967, *Pawek* 1109 (SRGH).
    Widespread in tropical Africa. In shade in woodland, 600–750 m.

10. CHEILANTHES Sw.

Cheilanthes Sw., Syn. Fil.: 5, 126 (1806) *nom. conserv.*

Rhizome erect or shortly creeping, with tufted fronds and linear scales, sometimes with a dark sclerotic central line. Stipe castaneous to black, glabrous and polished or pilose. Frond 2–5-pinnatifid, uniform or rarely dimorphous, glabrous, thinly pubescent, pilose or tomentose below, with or without lacerate scales on the costae and costules. Sori small, discrete, borne on vein endings near the margin, protected by a continuous or interrupted indusium or by the recurved lobes of the ultimate segments.

As construed here to include the genus *Notholaena* R. Br., *Cheilanthes* is a cosmopolitan genus of over 200 spp. Even among the African species previously referred to these two genera no characters appear to effect a generic separation between them.

Under surface of lamina with whitish or orange-yellow powder:
  Powder on lamina whitish or pale-yellow - - - - - - 1. *farinosa*
  Powder on lamina orange-yellow - - - - - - 2. *mossambicensis*
Under surface of lamina without powder:
  Lamina glabrous on both surfaces:
    Frond ovate, deltate, 3–4-pinnate - - - - - 3. *multifida*
    Frond narrowly elliptic, 2–3-pinnatifid - - - - 4. *angustifrondosa*
  Lamina pubescent, pilose or tomentose:
    Fronds strongly dimorphous, with the fertile fronds taller and more divided
                                     5. *similis*
    Fronds uniform:
      Lamina thinly pubescent on both surfaces; stipe pilose, not becoming glabrous
        with age:
        Rhizome erect; frond ovate-triangular, 4–5-pinnatifid - - - 6. *bergiana*
        Rhizome shortly creeping; frond narrowly elliptic, 2-pinnate - 7. *hirta*
      Lamina pilose or tomentose below:
        Pinnae tomentose and with lanceolate lacerate scales on the costae and costules
        below - - - - - - - - - - - 8. *eckloniana*
        Pinnae tomentose or pilose below but without lacerate scales:
          Fronds proliferous, pilose - - - - - - - 9. *leachii*
          Fronds not proliferous, pilose to tomentose below - - 10. *inaequalis*

1. **Cheilanthes farinosa** (Forsk.) Kaulf., Enum. Fil.: 212 (1824).—Sim, Ferns S. Afr. ed. 2: 235, t. 114 (1915). Type from Yemen.
  Pteris farinosa Forsk., Fl. Aegypt.-Arab.: CXXIV, 187 (1775). Type as above.
  Allosorus farinosus (Forsk.) C. Presl, Tent. Pterid.: 153 (1836) reimpr. in Abh. Königl. Böhm. Ges. Wiss., Ser. 4, 5: 153 (1837). Type as above.
  Cassebeera farinosa (Forsk.) J. Sm. in Hook., Lond. Journ. Bot. 4: 159 (1841). Type as above.
  Aleuritopteris farinosa (Forsk.) Fée, Mém. Fam. Foug. 5: 154, t. 12 B fig. 1 (1852). Type as above.

Rhizome up to 1 cm. in diam., erect, with tufted fronds and with dark-brown subulate entire rhizome-scales up to 7 mm. long with paler margins. Stipe castaneous, up to 28 cm. long, glabrous or with scales similar to the larger rhizome scales towards the base. Frond arching, thinly coriaceous or firmly herbaceous. Lamina up to 39 × 18 cm., oblong, acute in outline, 2–3-pinnatifid, dark-green above, the lower surface covered with pale-yellow to whitish powder; upper pinnae oblong, becoming unequally deltate towards the base of the lamina; pinna segments oblong and rounded or pinnatifid into oblong rounded minutely crenate lobes; rhachis and the under surface of the costae and costule castaneous, glabrous. Sori minute, c. 1 mm. in diam., closely set along the margins of the pinna segments; indusium small, semi-transparent, variously lacerate.

Zambia. W: Zambezi Rapids, 28.x.1966, *Leach & Williamson* 13517 (SRGH). S: Victoria Falls north bank, 28.viii.1947, *Greenway & Brenan* 7998 (BK; K). **Rhodesia.** W: Victoria Falls, Rain Forest, 6.vii.1953, *Schelpe* 3930 (B; BM; BOL; K; P). C: Salisbury Distr., Gilnockie Farm, 14.vii.1958, *Mitchell* 448 (SRGH). E: 9·6 km. N. of Inyanga Village, 28.iv.1967, *Rushworth* 863 (SRGH). **Malawi.** N: Mafinga Hills, 15.vii.1963, *Chapman* 1867 (BOL; SRGH).
  Widespread in tropical Africa, where both tetraploids and triploids are known. Mostly in undergrowth of moist forest, c. 900 m.

2. **Cheilanthes mossambicensis** Schelpe in Journ. S. Afr. Bot. **30**: 183, t. 1 fig. b (1964). Type: Mozambique, Ribáuè, 2·4 km. SW. of Poste Agrícola, *Schelpe & Leach* 11421 (BOL, holotype; K, isotype; SRGH, isotype).

Rhizome c. 2 mm. in diam., creeping, slender, bearing fronds at intervals of c. 5 mm., and with dark-brown subulate rhizome-scales c. 1·5 mm. long. Stipe castaneous, up to 22 cm. long, terete, with narrowly lanceolate brown scales when young, becoming glabrous with age. Frond erect, firmly herbaceous. Lamina up to 17 × 7 cm., oblong or narrowly oblong in outline, 2-pinnatifid, dark-green above, the lower surface covered with orange-yellow powder; pinnae ovate-deltate, pinnatifid with narrowly oblong crenate lobes up to 5 mm. broad; rhachis and the under surface of the costae and costules castaneous, glabrous. Sori minute, along the margins of the pinna segments; indusium very narrow, irregularly lacerate.

Rhodesia. C: Ngomokurira, 21.ii.1959, *Mitchell* 477 (BOL; K; SRGH). **Mozambique.** N: Eráti, Mt. Muchamapa, 14.xi.1963, *Torre & Paiva* 9578 (LISC). Also in S. Tanzania. Sporadic in rock crevices on outcrops, 550–1600 m.

3. **Cheilanthes multifida** (Sw.) Sw., Syn. Fil.: 129, 334 (1806).—Sim, Ferns S. Afr. ed. 2: 231, t. 113 (1915). Type from S. Africa.
> *Adiantum multifidum* Sw. in Schrad., Journ. Bot. **1800,** 2: 85 (1801). Type as above.
> *Adiantum globatum* Poir., Encycl. Méth., Bot., Suppl. **1**: 144 (1810). Type from S. Africa.
> *Cheilanthes bolusii* Bak. in Hook., Ic. Pl.: t. 1636 (1886). Type from S. Africa.

Rhizome short, creeping, with tufted fronds and with subulate entire rhizome-scales 2·5–4 mm. long with a dark central stripe and pale margins. Stipe brown-castaneous, up to 15 cm. long, shallowly sulcate, glabrous except for numerous pale-brown narrowly lanceolate scales at the base. Lamina up to 32 × 28 cm., ovate-deltate in outline, 3–4-pinnate with the lowest pinnae almost as long as the rest of the lamina and much developed basiscopically; pinnae up to 25 cm. long, unequally narrowly deltate, acute, acuminate; pinnule segments narrowly deltate, oblong to cultrate, obtuse, crenate to pinnatifid into rounded lobes, glabrous on both surfaces or with a few hair-like scales along the costae; veins apparent below; rhachis and secondary rhachises brown to castaneous, glabrous. Sori minute, less than 1 mm. in diam., on the margins of reflexed lamina lobes; indusium minute, ovate, membranous, entire to irregularly lacerate.

Zambia. N: Abercorn Distr., Chilongowelo escarpment, 21.i.1963, *Richards* 17530 (K). W: Mwinilunga Distr., 0·8 km. E. of Dobeka Bridge, 5.xi.1937, *Milne-Redhead* 3104 (K). E: Lundazi Distr., Nyika Plateau, Kangampande Mt., 8.v.1952, *White* 2802 (BM; K). **Rhodesia.** N: Umvukwes Mts., near Mtoroshanga Pass, iv.1948, *Rodins* 4432 (K). W: Matopo Hills, ix.1905, *Gibbs* 69 (K). C: Makoni Distr., Silverbow, 15.vi.1957, *Chase* 6527 (BM; K; SRGH). E: Melsetter Distr., Mt. Pene, 14.iv.1957, *Chase* 6404 (BM; K; SRGH). S: Bikita, Denga, 22.ii.1964, *Mitchell* 807 (SRGH). **Malawi.** N: Nyika Plateau, vii.1896, *Whyte* 115 (K). C: Dedza Mt., 10.ix.1924, *Burtt Davy* 1529/29 (K). S: Zomba Plateau, 31.v.1946, *Brass* 16123 (K; SRGH). **Mozambique.** Z: Namúli Mt., 26.vii.1962, *Schelpe & Leach* 7017 (BOL). T: Macanga, Mt. Furancungo, 15.iii.1966, *Pereira, Sarmento & Marques* 1694 (LMU). MS: Mozambique-Rhodesia border, Sheba, 25.vi.1957, *Chase* 6547 (BM; K; SRGH). LM: Libombos, near Namaacha, Mt. M'Ponduine, *E., M. & W.* 519 (LISC).
Also in S. Africa, E. tropical Africa and St. Helena. Rock crevices and around boulders on forest margins, 700–2000 m.

4. **Cheilanthes angustifrondosa** Alston in Bull. Brit. Mus. (Nat. Hist.) Bot. **1**: 48, t. 2 (1948). Type from Angola.

Rhizome c. 1 cm. in diam., creeping, with tufted fronds and with brown subulate rhizome-scales c. 5 mm. long with a dark central stripe and pale margins. Stipe castaneous to black, up to 14 cm. long, glabrous, very shallowly sulcate. Frond erect, firmly herbaceous. Lamina up to 49 × 8·5 cm., very narrowly elliptic in outline, 2-pinnatifid to 3-pinnatifid; pinnae up to 6 × 2·8 cm., broadly lanceolate, acute-acuminate; pinnae segments up to 1 × 0·35 cm., narrowly oblong, obtuse, crenate, sinuate to pinnatifid, glabrous, veins obscure; rhachis and costae in lower half of pinnae, castaneous, glabrous. Sori minute, less than 1 mm. in diam. on the margins of the pinna segments; indusium minute, membranous, entire.

Zambia. N: Luapula Distr., Mbereshi, 16.i.1960, *Richards* 12384 (BOL; K; SRGH).
Also in the Congo and Tanzania. Among rocks in woodland, 1000–1500 m.

5. **Cheilanthes similis** Ballard in Kew Bull. **12**: 47 (1957). Type: Zambia, *Milne-Redhead* 4351 (BM, isotype; K, holotype).
   Cheilanthes heterophylla sensu Tardieu in Mém. Inst. Fr. Afr. Noire, **28**: 89, t. 14 fig. 1–2 (1953).

Rhizome c. 8 mm. in diam., short, creeping, with tufted fronds and light-brown very narrowly linear entire concolorous rhizome-scales up to 10 × 1 mm. Fronds dimorphous, thinly coriaceous, the fertile more dissected and with longer stipes than in the sterile fronds. Stipe castaneous, thinly pubescent with pale hairs 1–2 mm. long, up to 1·7 cm. in sterile fronds and 12–22 cm. in fertile fronds. Sterile lamina up to 6·5 × 5·7 cm., triangular in outline, deeply pinnatifid above becoming pinnate at the base, the lowest pinnae with basiscopically developed lobes; pinna lobes oblong, obtuse to acute, entire to crenate, thinly pilose on both surfaces with pale flexuous hairs; venation obscure. Fertile lamina up to 10·7 × 7·6 cm., triangular in outline, pinnate above to pinnatifid below, the lowest pinnae with basiscopically developed entire to sinuate lobes; middle pinnae pinnatifid into oblong to triangular obtuse entire lobes thinly pilose on both surfaces with pale flexuous hairs. Sori small, submarginal, discrete, developing into a soral line later; indusium membranous, hyaline, mostly continuous, longly ciliate.

Zambia. N: Luapula Distr., Mbereshi, 16.i.1960, *Richards* 12385 (BOL; K; SRGH). W: Mwinilunga Distr., Luakera Falls, 25.i.1938, *Milne-Redhead* 4351 (BM; K).
Also in the Congo. Around rock in open *Brachystegia* woodlands, 1000–1500 m.

6. **Cheilanthes bergiana** Schlechtend., Adumbr. Pl.: 51 (1832). Type from S. Africa.
   Cheilanthes elata Kunze in Linnaea, **10**: 542 (1836). Type from S. Africa.
   Hypolepis bergiana (Schlechtend.) Hook., Sp. Fil. **2**: 67 (1852).—Sim, Ferns S. Afr. ed. 2: 238, t. 115 (1915). Type as for *Cheilanthes bergiana*.
   Cheilanthes streetiae Bak. in Journ. of Bot. **18**: 327 (1880). Type from Madagascar.

Rhizome up to 1 cm. in diam., short, erect, with tufted fronds and with dark-brown subulate rhizome-scales up to 9 mm. long with pale-brown margins. Stipe dark-brown, up to 64 cm. long, densely pubescent with short brown hairs less than 0·7 mm. long. Frond arching, herbaceous. Lamina up to 30 × 28 cm., ovate-triangular in outline, 3-pinnate to 5-pinnatifid with the lowest pinnae almost as long as the rest of the lamina and much developed basiscopically; upper pinnae oblong, acute; ultimate segments up to 1·5 × 0·7 cm., oblong, obtuse, pinnatifid into oblong rounded lobes, thinly pubescent on both surfaces along the costa and veins with very pale multicellular hairs up to 1 mm. long; costae castaneous, thinly pubescent. Sori minute, less than 1 mm. in diam., borne on the margins of the lobes of the ultimate segments; indusia minute, subentire, almost transparent.

Rhodesia. E: Umtali Distr., Vumba Mts., Elephant Forest, 1.iii.1956, *Chase* 5990 (BM; BOL; SRGH). **Malawi.** S: near Blantyre, Shire Highlands, 1887, *Last* (K). **Mozambique.** N: Ribáuè, Mepaluè mt., 25.i.1964, *Torre & Paiva* 10227 (LISC). Z: Morrumbala, 1864, *Waller* (K). MS: Garuso, Jaegersberg, 11.vii.1955, *Schelpe* 5615 (BM; BOL).
Also in S. Africa, Tanzania, Uganda and Congo. Forest floors and margins in deep shade 900–1500 m.

7. **Cheilanthes hirta** Sw., Syn. Fil.: 128, 329 (1806).—Sim, Ferns S. Afr. ed. 2: 227, t. 110 (1915). Type from S. Africa.
   Adiantum caffrorum Sw. in Schrad., Journ. Bot. **1800**, 2: 85 (1801) non L.f. (1781). Type from S. Africa.
   Adiantum hirtum (Sw.) Poir., Encycl. Méth. Bot., Suppl. **1**: 142 (1810). Type as for *Cheilanthes hirta*.
   Notholaena capensis Spreng., Tent. Suppl. Syst. Veg.: 32 (1828). Type from S. Africa.

Cheilanthes hirta var. *intermedia* Kunze in Linnaea, **10**: 539 (1836). Type from S. Africa.
Cheilanthes hirta var. *laxa* Kunze, tom. cit.: 540 (1836). Type from S. Africa.
Notholaena hirta (Sw.) J. Sm. in Hook., Lond. Journ. Bot. **4**: 50 (1841). Type as for *Cheilanthes hirta.*
Myriopteris intermedia (Kunze) Fée, Mém. Fam. Foug. **5**: 149 (1852). Type from S. Africa.
Cheilanthes glandulosa Pappe & Raws., Syn. Fil. Afr. Austr.: 35 (1858) non Sw. (1817). Type from S. Africa.
Myriopteris hirta (Sw.) J. Sm., Ferns Brit. & Foreign: 174 (1866). Type as for *Cheilanthes hirta.*

Rhizome c. 5 mm. in diam., short, creeping, with tufted fronds and with entire subulate rhizome-scales c. 3·5 mm. long with a dark sclerotic central stripe and pale transparent margins. Stipe castaneous, up to 7·5 cm. long, pilose with patent brown hairs up to 1 mm. long and with large pale-brown narrowly lanceolate scales up to 6 mm. long about the base. Frond erect, herbaceous. Lamina up to 34 × 5·5 cm., linear to narrowly elliptic in outline, 2-pinnate to 3-pinnatifid; pinnae narrowly oblong to oblong, acute or obtuse, sessile (adnate) to petiolulate, crenate to pinnatifid to pinnate into obtuse crenate lobes, sparsely pubescent with glandular hairs on both surfaces; veins obscure; rhachis and costae castaneous, pilose with patent brown hairs. Sori minute, less than 1 mm. in diam., set on the margin of incurved pinnule lobes; indusium minute or absent.

Botswana. N: near Shashi R., 9.v.1963, *Drummond* 8234 (SRGH). SE: Dikoma Di Kai, 26.ii.1960, *Wild* 5177 (SRGH). **Rhodesia.** N: near Darwendale, Umvukwes Mts., 20.iv.1948, *Rodin* 4351 (K; SRGH). W: Matopo Hills, ix.1905, *Gibbs* 90 (K). C: 3·2 km. E. of Selukwe, 17.ii.1964, *Mitchell* 715 (SRGH). E: above Konda, Sabi Valley, 16.ii.1960, *Goodier* 912 (K; SRGH). S: E. of Zimbabwe, 2.vii.1930, *Hutchinson & Gillett* 3355 (K). **Mozambique.** LM: Maputo, Goba, 23.xi.1944, *Mendonça* 3072 (LISC).
Widespread in S. Africa and also in Angola, Kenya and Madagascar. Around boulders in woodland, 1220–1520 m.

8. **Cheilanthes eckloniana** (Kunze) Mett. in Abh. Senckenb. Nat. Ges. **3**: 66 (1859) reimpr. in Mett., Farngatt., Cheil.: 22 (1859). Type from S. Africa.
Notholaena eckloniana Kunze in Linnaea, **10**: 501 (1836).—Sim, Ferns S. Afr. ed. 2: 222, t. 107 (1915). Type as above.
Notholaena krebsiana C. Presl, Tent. Pterid.: 224 (1836) reimpr. in Abh. Königl. Böhm. Ges. Wiss., Ser. 4, **5**: 224 (1837) *nom. nud.*

Rhizome c. 4 mm. in diam., creeping, with tufted fronds and with narrowly lanceolate subentire attenuate rhizome-scales up to 3 mm. long, some concolorous, mixed among others with a darker central stripe and pale borders. Stipe up to 10·5 cm. long, terete black, tomentose at first with pale linear hair-like scales gradually becoming subglabrous with age, with a mass of concolorous reddish scales at the base. Frond erect, thinly coriaceous. Lamina up to 11 × 3 cm., narrowly oblong in outline, acute, not decrescent below, 2-pinnate to 3-pinnatifid; pinnae deltate, usually not much developed basiscopically; pinna segments deeply pinnatifid into obtuse lobes; dorsal surface densely tomentose with matted hairs mixed with lacerate lanceolate attenuate concolorous pale-brown scales, the latter covering the costae and costules; rhachis terete, black, tomentose with pale-brown hairs and hair-like scales at first and with much of the tomentum persistent. Sori marginal, forming an interrupted soral line; indusium very small, membranous, ciliate or absent.

Rhodesia. C: near Salisbury, 1896, *Bryce* (K). E: 1·6 km. W. of Inyanga Township, 26.ii.1964, *Mitchell* 831 (SRGH).
Also in S. Africa. Moist granite outcrops, c. 1700 m.

9. **Cheilanthes leachii** (Schelpe) Schelpe in Bol. Soc. Brot., Sér. 2, **41**: 212 (1967). Type: Mozambique, Ribáuè, *Schelpe & Leach* 6973 (BOL, holotype).
Notholaena leachii Schelpe in Journ. S. Afr. Bot. **30**: 185, t. 1 fig. a (1964). Type as above.

Rhizome c. 4 mm. in diam., shortly creeping, with tufted fronds and with narrowly lanceolate brown rhizome-scales often with a dark central stripe and paler margins. Fronds proliferous below the apex and occasionally on the costae of basal pinnae, fertile fronds usually suberect, sterile fronds arching over to ground level, firmly herbaceous. Stipe castaneous, up to 19 cm. long, terete, ± pilose, later glabrous. Lamina up to 9 × 5·5 cm., narrowly ovate or lanceolate in outline, 2-pinnate or 2-pinnatifid, pilose on both surfaces; pinnae ovate to unequally deltate with entire crenate or pinnatifid lobes, somewhat produced basiscopically; costae and costules of the lower pinnae dark-castaneous, pilose below; rhachis terete, atrocastaneous-pilose. Sori marginal, forming an interrupted soral line at first; indusia very narrow, membranous.

Zambia. C: Mkushi Distr., Fiwila, 7.i.1958, *Robinson* 2670 (K; SRGH). **Rhodesia.** C: Makoni Distr., 9 km. E. of Rusape, 30.xi.1930, *F.N. & W.* 3354 (LD). **Malawi.** S: Mlanje Mt., Tuchila Plateau, v.1901, *Purves* 44 (K). **Mozambique.** N: Monte de Massangulo, xii.1952, *Gomes e Sousa* 1193 (COI; K). Z: Namúli Mt., 26.vii.1952, *Schelpe & Leach* 7018 (BOL). T: Macanga, Mt. Furancungo, 15.iii.1966, *Pereira, Sarmento & Marques* 1723 (LMU). MS: Garuso, 28.i.1949, *Fisher & Schweickerdt* 525 (K; SRGH).

Also in Tanzania. Forming mats around sheltered boulder bases, frequent on large granite outcrops in N. Mozambique, 900–1850 m.

10. **Cheilanthes inaequalis** (Kunze) Mett. in Abh. Senckenb. Naturf. Ges. **3**: 68, t. 3 fig. 4 (1859) reimpr. in Mett., Farngatt., Cheil.: 24, t. 3 fig. 4 (1859).—Tardieu in Mém. Inst. Fr. Afr. Noire, **28**: 89, t. 14 fig. 3 (1953). Type from S. Africa.
    Notholaena inaequalis Kunze, Farnkr. **1**: 146, t. 64 fig. 1 (1844).–Sim, Ferns S. Afr. ed. 2: 221, t. 108 (1915).—Alston, Ferns W. Trop. Afr.: 43 (1959). Type as above.
    Notholaena tricholepis Bak. in Journ. of Bot. **21**: 245 (1883). Type from Tanzania.
    Notholaena bipinnata sensu Sim, Ferns S. Afr. ed. 2: 224 (1915) pro parte excl. t. 109 fig. 2, non Liebm. (1849).

Rhizome c. 5 mm. in diam., short, creeping, with tufted fronds and with narrowly linear attenuate entire concolorous reddish-brown rhizome-scales up to 1·4 cm. long. Stipe terete, atrocastaneous, thinly pubescent when young with short white hairs or ferrugineous scales similar to those on the rhizome at the base. Frond erect, herbaceous to coriaceous. Lamina up to 22 × 11 cm., narrowly oblong to ovate-deltate in outline (deltate in juveniles) 2-pinnate towards the apex, 3-pinnatifid at the base, not decrescent below; pinnae narrowly to broadly, unequally to almost equally deltate, lower pinnae usually much developed basiscopically; pinna segments oblong to very narrowly oblong, subentire, crenate or pinnatifid, obtuse, pilose to densely tomentose on the dorsal surface with long soft hairs, white at first, becoming ferrugineous with age; ventral surface thinly pubescent or pilose, becoming subglabrous with age or not; rhachis terete, atrocastaneous, thinly pubescent at first, later becoming glabrous, secondary rhachises atrocastaneous on the dorsal surface. Sori marginal discrete but forming a continuous soral line at maturity; indusium continuous or irregularly discontinuous, narrow, ciliate, membranous.

Dorsal surface of lamina densely tomentose - - - - - - - var. *inaequalis*
Dorsal surface of lamina pilose - - - - - - - - - var. *buchananii*

Var. **inaequalis.** TAB. **39** fig. A.

Zambia. N: Abercorn Distr., Kalambo R. Gorge, 21.v.1952, *Richards* 1772 (K). W: Mwinilunga Distr., 11·2 km. N.W. of Kalene Mission, 11.xi.1962, *Richards* 17169 (K). C: Mkushi Distr., Fiwila, 9.i.1958, *Robinson* 2714 (K; SRGH). E: Lundazi Distr., Nyika Plateau, Kangampande Mt., 8.v.1952, *White* 2801 (BM; K). S: Mumbwa Distr., Miukuiukuil, 10.x.1965, *Mitchell* 3010 (SRGH). **Rhodesia.** N: Umvukwes Mts., 8 km. N. of Banket, 23.iv.1948, *Rodin* 4404 (K; SRGH). W: Matobo, Besna Kobila, iv.1955, *Miller* 2787 (BM; SRGH). C: Rusape road, 22.i.1949, *Fisher & Schweickerdt* 476 (BM; K; SRGH). E: Vumba Mts., Norseland, 13.ii.1950, *Chase* 3501 (BM; SRGH). **Malawi.** C: Nchisi Mt., 3.ix.1929, *Burtt Davy* 1111 (K).

Widespread in tropical Africa. Among rocks in open woodland, 1440–2140 m.

Var. **buchananii** (Bak.) Schelpe in Bol. Soc. Brot., Sér. 2, **41**: 211 (1967). TAB. **39** fig. B. Type from S. Africa.
    Notholaena buchananii Bak., Syn. Fil.: 373 (1868).—Sim, Ferns S. Afr. ed. 2: 222, t. 108 (1915). Type as above.

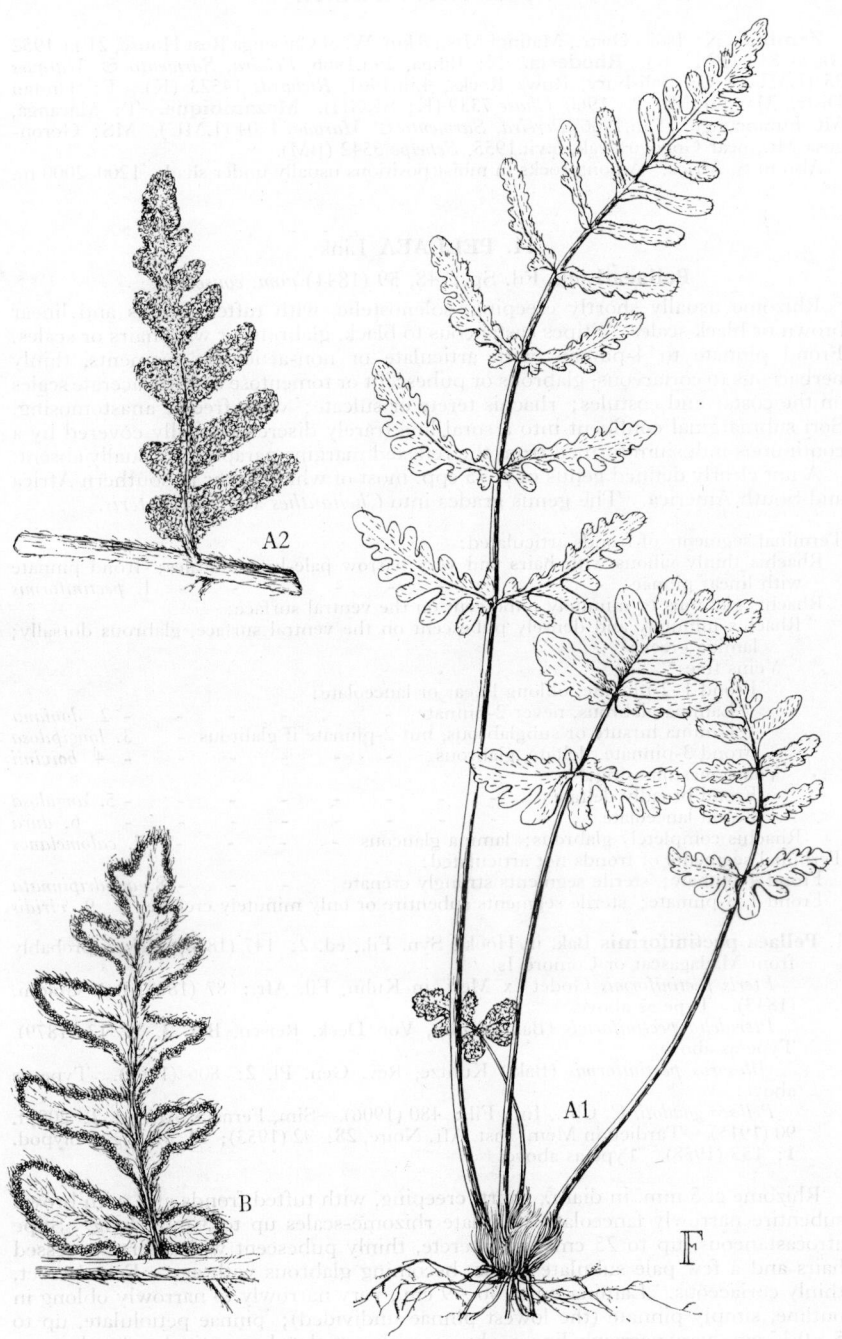

Tab. 39. A.—CHEILANTHES INAEQUALIS var. INAEQUALIS. A1, habit (× ⅔) *Fisher & Schweickerdt* 476; A2, dorsal surface of pinna (× 2) *Chase* 3501. B.—CHEILANTHES INAEQUALIS var. BUCHANANII. Dorsal surface of pinna (× 2) *Angus* 813.

Zambia. N: Isoka Distr., Mafingi Mts., 8 km. W. of Chisenga Rest House, 21.xi.1952 *Angus* 813 (BM; K). **Rhodesia.** N: Binga, 29.i.1966, *Pereira, Sarmento & Marques* 35 (LMU). C: Salisbury, Ruwa Rocks, 4.iii.1961, *Richards* 14523 (K). E: Umtali Distr., Mandini Mt., 8.v.1960, *Chase* 7339 (K; SRGH). **Mozambique.** T: Macanga, Mt. Furancungo, 15.iii.1966, *Pereira, Sarmento & Marques* 1704 (LMU). MS: Gorongosa Mt., near Gogogo Peak, 6.vii.1955, *Schelpe* 5542 (BM).

Also in S. Africa. Among rocks in moist positions usually under shade, 1200–2000 m.

11. PELLAEA Link

Pellaea Link, Fil. Sp.: 48, 59 (1841) *nom. conserv.*

Rhizome usually shortly creeping, solenostelic, with tufted fronds and linear brown or black scales. Stipes castaneous to black, glabrous or with hairs or scales. Frond pinnate to 4-pinnate with articulate or non-articulate segments, thinly herbaceous to coriaceous, glabrous or pubescent or tomentose or with lacerate scales on the costae and costules; rhachis terete or sulcate; veins free or anastomosing. Sori submarginal confluent into a soral line (rarely discrete) usually covered by a continuous indusium formed from the reflexed margin; paraphyses usually absent.

A not clearly defined genus of c. 75 spp. most of which occur in southern Africa and South America. The genus grades into *Cheilanthes* and *Doryopteris*.

Terminal segments of fronds articulated:
　Rhachis thinly villous with hairs and very narrow pale-brown scales; frond pinnate
　　with linear pinnae　-　-　-　-　-　-　-　-　-　1. *pectiniformis*
　Rhachis glabrous or minutely pubescent on the ventral surface:
　　Rhachis minutely and densely pubescent on the ventral surface, glabrous dorsally;
　　　lamina green:
　　　Veins free:*
　　　Frond 1–2-pinnate, oblong linear or lanceolate:
　　　　Lamina glabrous, never 2-pinnate　-　-　-　-　-　-　2. *doniana*
　　　　Lamina hirsute or subglabrous, but 2-pinnate if glabrous　-　　3. *longipilosa*
　　　Frond 3-pinnate, deltate, glabrous　-　-　-　-　-　-　4. *boivinii*
　　　Veins anastomosing:*
　　　Frond broadly deltate　-　-　-　-　-　-　-　-　5. *angulosa*
　　　Frond lanceolate　-　-　-　-　-　-　-　-　-　6. *dura*
　　Rhachis completely glabrous; lamina glaucous　-　-　-　-　7. *calomelanos*
Terminal segments of fronds not articulated:
　Frond 4-pinnate; sterile segments strongly crenate　-　-　-　8. *quadripinnata*
　Frond 1–3-pinnate; sterile segments subentire or only minutely crenate　-　9. *viridis*

1. **Pellaea pectiniformis** Bak. in Hook., Syn. Fil., ed. 2: 147 (1874). Type probably
　　from Madagascar or Comoro Is.
　　　Pteris pectiniformis Godet ex Mett. in Kuhn, Fil. Afr.: 87 (1868) non Goldm.
　　(1843). Type as above.
　　　Pteridella pectiniformis (Bak.) Kuhn, Von Deck. Reisen, Bot. **3**, 3: 13 (1879).
　　Type as above.
　　　Allosorus pectiniformis (Bak.) Kuntze, Rev. Gen. Pl. **2**: 806 (1891). Type as
　　above.
　　　Pellaea goudotii C. Chr., Ind. Fil.: 480 (1906).—Sim, Ferns S. Afr. ed. 2: 200, t.
　　90 (1915).—Tardieu in Mém. Inst. Afr. Noire, **28**: 92 (1953); Fl. Madag., Polypod.
　　1: 155 (1958). Type as above.

Rhizome c. 5 mm. in diam., short, creeping, with tufted fronds and with brown subentire narrowly lanceolate attenuate rhizome-scales up to 3 mm. long. Stipe atrocastaneous, up to 25 cm. long, terete, thinly pubescent with short appressed hairs and a few pale subulate scales, becoming glabrous with age. Frond erect, thinly coriaceous. Lamina up to 30 × 9 cm., very narrowly to narrowly oblong in outline, simply pinnate (the lowest pinnae undivided); pinnae petiolulate, up to 5 × 0·15 cm., very narrowly linear, obtuse, with a cordate base articulated at the apex of a short atrocastaneous petiolule; ventral surface glabrous, dorsal surface glabrous except for pale flexuous hairs along the costa; rhachis terete, atrocastaneous, persistently thinly villous with hairs and very narrow pale-brown scales. Sori in a continuous line at maturity; indusium continuous, erose, membranous.

* Only visible after clearing in aqueous chlorine solution or strong aqueous boiling potassium hydroxide solution.

Zambia. N: Luapula Distr., Mbereshi, 16.i.1960, *Richards* 12383 (K). W: Luan-shya, 12.v.1962, *Mutimushi* 153 (SRGH). C: Mkushi Distr., Fiwila, 7.i.1958, *Robinson* 2672 (K; SRGH). S: Choma, c. 1909, *Rogers* 8086a (K). **Rhodesia.** N. Mazoe, Iron Mask Hills, iii.1906, *Eyles* 250 (BM; SRGH). C: Ngomokurira, 21.ii.1959, *Mitchell* 474 (BOL; SRGH). E: Umtali Distr., Norseland, Vumba Mts., 26.vi.1959, *Schelpe* 5397 (BM; BOL). S: Belingwe Mt., 20.x.1959, *Wild* 4845 (SRGH). **Malawi.** N: Nyika Plateau, Mwanda Mt., 24.xii.1963, *Lemon* 1034 (SRGH). C: Dedza, Nchinji Hill, 22.i.1959, *Robson* 1296 (BM; K; LISC). **Mozambique.** N: Vila Cabral, road to Litunde, 20.i.1935, *Torre* 726 (LISC). T: Macanga, Furancungo, 7.iii.1966, *Pereira, Sarmento & Marques* 1806 (LMU). MS: Musapa Gap, 6.x.1950, *Chase* 3028 (SRGH).

South and tropical Africa and Madagascar. Around rocks in woodland, 900–1530 m.

2. **Pellaea doniana** J. Sm. ex Hook., Sp. Fil. **2**: 137, t. 125 fig. A (1858).—Sim, Ferns S. Afr. ed. 2: 211, t. 102 (1915).—Alston, Ferns W. Trop. Afr.: 43 (1959). TAB. **40** fig. C. Type from S. Tomé.

   Pteris doniana (J. Sm. ex Hook.) Kuhn, Fil. Afr.: 80 (1868). Type as above.
   Pteridella doniana (J. Sm. ex Hook.) Kuhn, Von Deck. Reisen, Bot. **3**, 3: 13 (1879). Type as above.
   Allosorus doniana (J. Sm. ex Hook.) Kuntze, Rev. Gen. Pl. **2**: 806 (1891). Type as above.

Rhizome up to 1 cm. in diam., creeping, with tufted fronds and with linear attenuate subentire brown rhizome-scales up to 7 mm. long with paler margins. Stipe atrocastaneous, up to 43 × 0·4 cm., terete with brown scales similar to those on the rhizome in the lower half when young, becoming glabrous with age. Frond erect, thinly coriaceous. Lamina up to 55 × 21 cm., narrowly oblong to oblong in outline, imparipinnate with the lowest pinnae the same size as those above; pinnae up to 14 × 2·5 cm., lanceolate, acuminate, with cordate bases, minutely crenate, articulated to the apex of the petiolule, glabrous on both surfaces, dark-green above, paler green below; veins free; rhachis and petiolules atrocastaneous-pubescent with short stiff hairs on both surfaces. Sori forming a marginal soral line c. 1 mm. broad at maturity; indusium c. 0·5 mm. broad, continuous, entire, membranous.

Zambia. N: Abercorn Distr., Dhulumiti Kloof, 28.i.1952, *Richards* 781 (K). W: Mwinilunga Distr., Matonchi R., 11.xi.1938, *Milne-Redhead* 3184 (BM; K). C: 9·6 km. E. of Lusaka, 17.i.1956, *King* 279 (K). **Rhodesia.** E: Umtali Commonage, 13.v.1961, *Chase* 7491 (BM; K; LISC; SRGH). **Malawi.** N: Livingstonia, Kazeweziwe R., 9.i.1959, *Robinson* 3123 (K; SRGH). S: Cholo Distr., Nswadzi R., 29.ix.1946, *Brass* 17865 (K). **Mozambique.** N: Ribáuè Mt., 19.vii.1962, *Schelpe & Leach* 6951 (BOL). Z: Namúli Mt., 25.vii.1962, *Schelpe & Leach* 7004 (BOL). T: Macanga, Furancungo, 17.iii.1966, *Pereira, Sarmento & Marques* 1840 (LMU). MS: Vila Gouveia, 28.vi.1941, *Torre* 2948 (LISC).

Widespread in tropical Africa. Undergrowth of forest, 600–1500 m.

3. **Pellaea longipilosa** Bonap., Not. Ptérid. **15**: 33 (1924). TAB. **40** fig. D. Type from the Sudan.

Rhizome up to 8 mm. in diam., short, creeping, with tufted fronds and with narrowly linear subentire dark-brown rhizome-scales 4–5 mm. long with paler margins. Stipe castaneous to ebeneous, up to 25 cm. long, terete, glabrous or pilose with erect hairs and with numerous pale-brown scales about the base similar to those on the rhizome. Frond erect, coriaceous. Lamina up to 39 × 12 cm., narrowly oblong to lanceolate in outline, simply pinnate or 2-pinnate; pinnae and pinnules of lower pinnae in 2-pinnate fronds, linear to lanceolate or oblong, entire obtuse with cordate bases, up to 5·5 × 1·0 cm., articulated at the apex of the petiole or petiolule, both surfaces glabrous or pilose, the ventral surface a darker green than the dorsal; veins free, obscure; rhachis and secondary rhachises castaneous to ebeneous, pubescent with short stiff hairs ventrally, glabrous or pilose dorsally. Sori forming a marginal soral line; indusium continuous, membranous, entire.

Zambia. N: Chilongowelo, 28.i.1955, *Richards* 4271 (K). W: Kitwe, 10.iii.1963, *Fanshawe* 7738 (SRGH). C: 9·6 km. E. of Lusaka, 16.i.1956, *King* 270 (K). E: Chipiri Hill near Chadiza, 29.xi.1958, *Robson* 783 (BM; K; LISC; SRGH). S: 4·8 km. NE. of Mapanza, 10.iv.1955, *Robinson* 1229 (K; SRGH). **Rhodesia.** N: Umvukwes Mts., 24–27. iv.1948, *Rodin* 4427 (BOL; K; SRGH). C: Salisbury, Rumani Park, 23.iii.1950, *Chase* 3449 (BM; SRGH). **Malawi.** S: near Blantyre, 1887, *Last* (K). **Mozambique.**

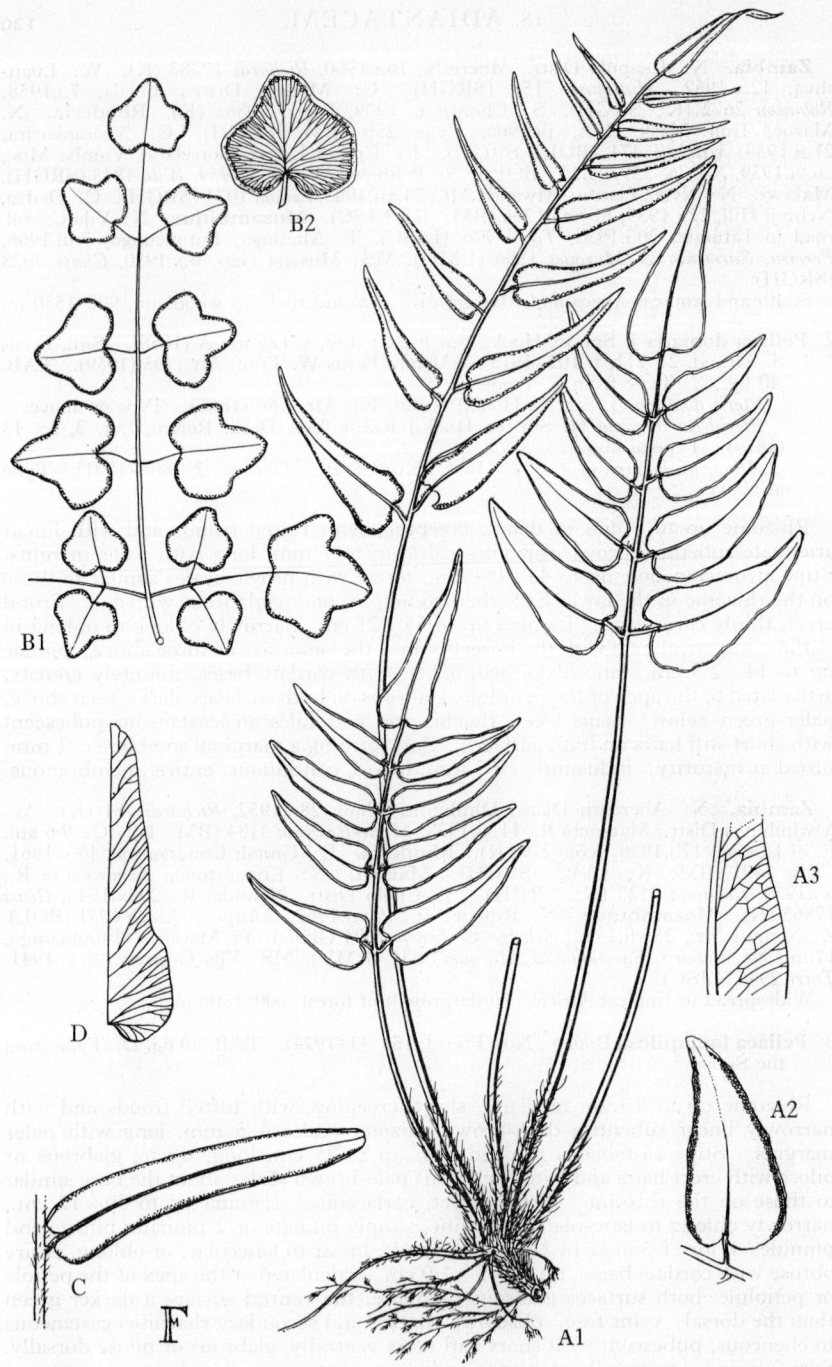

Tab. 40. A.—PELLAEA ANGULOSA. A1, habit (×⅓); A2, pinna (×⅔); A3, part of pinna showing venation (×3), all from *Bruce 132*. B.—PELLAEA CALOMELANOS var. SWYNNERTONIANA. B1, frond (×⅓); B2, pinnule showing venation (×⅔). both from *Schelpe 5393*. C.—PELLAEA DONIANA. Pinna (×⅔) *Chase 4013*. D.—PELLAEA LONGIPILOSA. Part of pinna showing venation (×2) *Chase 3449*.

N: Moçambique, Ribáuè, Mepaluè, 27.i.1964, *Torre & Paiva* 10255 (LISC). T: between Zobuè and Vila Mouzinho, 20.vii.1949, *Barbosa & Carvalho* 3716 (LISC). Widespread in E. tropical Africa to the Sudan. Among rocks in woodland, 600–1430 m.

4. **Pellaea boivinii** Hook., Sp. Fil. **2**: 147, t. 118 fig. A (1858).—Sim, Ferns S. Afr. ed. 2: 204, t. 93 (1915).—Tardieu, Fl. Madag., Polypod. **1**: t. 22 fig. 6–8 (1958). Type from Madagascar.
    Pteris boivinii (Hook.) Bedd., Ferns S. India: t. 36 (1865). Type as above.
    Pteridella adiantoides Desv. ex Kuhn, Von Deck. Reisen, Bot., **3**, 3: 14 (1879). Syntypes from Africa and India.
    Pellaea adiantoides (Desv. ex Kuhn) Prantl in Engl., Bot. Jahrb. **3**: 418 (1882) non (Bory ex Willd.) J. Sm. (1856). Syntypes as above.
    Allosorus boivinii (Hook.) Kuntze, Rev. Gen. Pl. **2**: 806 (1891). Type as for *Pellaea boivinii*.

Rhizome short, creeping, with tufted fronds and with shining ferrugineous minutely serrate very narrowly linear attenuate rhizome scales up to 1 cm. long with paler margins. Stipe atrocastaneous, terete, glabrous dorsally at maturity, pubescent ventrally in upper parts and with some pale-brown scales similar to those on the rhizome about the base. Frond erect, coriaceous. Lamina up to 18 × 14 cm., triangular in outline, 3-pinnate with the basal pinnae 2-pinnate, somewhat developed basiscopically; pinnules of central pinnae, c. 1 × 0·3 cm., very narrowly oblong, entire, obtuse, with a cordate base, glabrous on both surfaces, articulated to the apices of the petiolules; veins free, obscure; rhachis and secondary rhachises atrocastaneous, terete, shortly pubescent ventrally, glabrous dorsally. Sori forming a marginal line; indusium continuous, erose, membranous.

Zambia. W: Kitwe, 24.v.1967, *Fanshawe* 10084 (BOL; SRGH). S: Choma, Chimwani Hill, 23.xi.1962, *Astle* 1683 (SRGH). **Rhodesia.** N: Mrewa, Shawanoya, 16.vi.1957, *Whellan* 1297 (SRGH). W: Matobo, Besna Kobila, iv.1955, *Miller* 2802 (SRGH). C: Domboshawa, 16.i.1958, *Mitchell* 216 (BOL; SRGH). E: Umtali Distr., Zimunya's Reserve, 5.viii.1956, *Chase* 6163 (BM; BOL; K; SRGH). S: 30·4 km. S. of Belingwe, 19.ii.1964, *Mitchell* 735 (SRGH).
Also in S. Africa, Madagascar, Mauritius, Ceylon and India. Among rocks, woodland. 1000–1500 m.

5. **Pellaea angulosa** (Bory ex. Willd.) Bak., Syn. Fil., ed. 2: 153 (1874).—Tardieu, Fl. Madag., Polypod. **1**: 160, t. 22 fig. 9 (1958). TAB. **40** fig. A. Type from Réunion.
    Pteris angulosa Bory ex Willd. in L., Sp. Pl., ed. 4, **5**: 377 (1810). Type as above.
    Pteris articulata Kaulf. ex Spreng. in L., Syst. Veg., ed. 16, **4**: 76 (1827). Type from Mascarene Is.
    Litobrochia articulata (Kaulf. ex Spreng.) C. Presl, Tent. Pterid.: 149 (1836) reimpr. in Abh. Königl. Böhm. Ges. Wiss., Ser. 4, **5**: 149 (1837). Type as above.
    Doryopteris articulata (Kaulf. ex Spreng.) Fée, Mém. Fam. Foug. **5**: 133 (1852). Type as above.
    Pellaea articulata (Kaulf. ex Spreng.) Bak., Syn. Fil.: 153 (1867). Type as above.
    Pellaeopsis articulata (Kaulf. ex Spreng.) J. Sm., Hist. Fil.: 290 (1875). Type as above.
    Pteridella angulosa (Bory ex Willd.) Mett. ex Kuhn, Von Deck. Reisen, Bot., **3**, 3: 15 (1879). Type as for *Pellaea angulosa*.
    Allosorus angulosus (Bory ex Willd.) Kuntze, Rev. Gen. Pl. **2**: 806 (1891). Type as above.

Rhizome up to 7 mm. in diam., short, creeping, with tufted fronds and with linear-attenuate subentire dark-brown shining rhizome-scales up to 4·5 mm. long. Stipe eneous, up to 20 cm. long, terete, glabrous except for a short pubescence on the ventral surface in the upper half. Frond erect and arching, coriaceous. Lamina up to 36 × 34 cm., triangular in outline, 2-pinnate with the basal pinnae developed basiscopically; upper pinnae and pinnules of lower pinnae up to 6·5 × 1·5 cm., lanceolate, sometimes falcate, entire, attenuate, with cordate bases articulated to the apices of their petioles or petiolules; veins anastomosing, obscure; rhachis and secondary rhachises ebeneous, shortly pubescent ventrally, glabrous dorsally. Sori forming a marginal soral line; indusium continuous, entire, membranous.

Mozambique. MS: Chimanimani Mts., below Martin Falls. 8.ii.1958. *Mitchell* 330 (BM; BOL; SRGH).
Also in Tanzania, Madagascar and the Mascarene Is. Undergrowth of moist forest, 1370 m.

6. **Pellaea dura** (Willd.) Hook., Sp. Fil. 2: 139 (1858).—Sim, Ferns S. Afr. ed. 2: 198, t. 90 (1915).—Tardieu, Fl. Madag., Polypod. 1: 159 (1958). Type from Mauritius.
   Pteris dura Willd. in L., Sp. Pl., ed. 4, 5: 376 (1810). Type as above.
   Allosorus durus (Willd.) C. Presl, Tent. Pterid.: 153 (1836) reimpr. in Abh. Königl. Böhm. Ges. Wiss., Ser. 4, 5: 153 (1837). Type as above.
   Litobrochia dura (Willd.) Moore, Ind. Fil.: 44 (1857). Type as above.
   Pteris burkeana Hook., Sp. Fil. 2: 213, t. 126 fig. B (1858). Type from S. Africa.
   Pellaea burkeana (Hook.) Bak., Syn. Fil.: 153 (1867). Type as above.
   Pellaeopsis burkeana (Hook.) J. Sm., Hist. Fil.: 290 (1875). Type as above.
   Pteridella dura (Willd.) Kuhn, Von Deck. Reisen, Bot. 3, 3: 14 (1879). Type as for *Pellaea dura*.

Rhizome 5–7 mm. in diam., short, creeping, with tufted fronds and with linear attenuate shining dark-brown concolorous subentire rhizome-scales c. 5 mm. long. Stipe ebeneous, terete, subglabrous, usually as long or longer than the lamina, with a few narrowly lanceolate entire scales c. 5 mm. long with paler margins towards the base. Frond erect, coriaceous. Lamina up to 25 × 11 cm., lanceolate to ovate in outline, 2-pinnate (rarely simply pinnate); upper pinnae and pinnules of lower pinnae up to 6 × 1·1 cm., very narrowly to broadly linear, entire, obtuse, with cordate bases articulated to the apices of their petioles and petiolules; veins anastomosing, obscure; rhachis and secondary rhachises terete, ebeneous, minutely pubescent ventrally, glabrous dorsally. Sori forming a marginal line; indusium continuous, entire, membranous.

Zambia. S: near Mumbwa, 1911, *Macaulay* 6 (K). **Rhodesia.** N: Mazoe, iv.1906, *Eyles* 263 (BM; PRE; SRGH). W: Matobo, Besna Kobila, iv.1954, *Miller* 2323 (PRE; SRGH). C: Domboshawa, i.1957, *Mitchell* 83 (BOL; SRGH). E: Umtali Distr., 17·6 km. S. on Melsetter road, 4.xi.1956, *Chase* 6230 (BM; BOL; K; SRGH). S: Victoria Distr., Kyle Dam, 21.ii.1964, *Mitchell* 793 (SRGH). **Malawi.** N: Mzimba, Vipya, 24.ii.1961, *Vesey-FitzGerald* 3041 (SRGH). C: Nchisi Mt., 26.vii.1946, *Brass* 16964 (K; SRGH). S: Zomba Plateau, 2.vi.1946, *Brass* 16151 (K; SRGH). **Mozambique.** N: Moatize, Zóbuè, 12.i.1966, *Correia* 419 (LISC). Z: Namúli Mt., 27.vii.1962, *Schelpe & Leach* 7072 (BOL). MS: Garuso, Jaegersberg, 11.vii.1955, *Schelpe* 5600 (BM; BOL). T: Macanga, Mt. Furancungo, 15.iii.1966, *Pereira, Sarmento & Marques* 1739 (LMU).
Also in S. Africa, Tanzania, Angola, Madagascar and Mauritius. Among rocks in woodland, 900–1500 m.

7. **Pellaea calomelanos** (Sw.) Link, Fil. Sp. Hort. Berol.: 51 (1841).—Tardieu, Fl. Madag., Polypod. 1: 163, t. 23 fig. 5–6 (1958). Type from S. Africa.
   Pteris calomelanos Sw. in Schrad., Journ. Bot. **1800**, 2: 70 (1801). Type as above.
   Allosorus calomelanos (Sw.) C. Presl, Tent. Pterid.: 153 (1836) reimpr. in Abh. Königl. Böhm. Ges. Wiss., Ser. 4, 5: 153 (1837). Type as above.
   Platyloma calomelanos (Sw.) J. Sm. in Curt. Bot. Mag. 72 Comp.: 21 (1846). Type as above.
   Notholaena calomelanos (Sw.) Keys., Pol. Cyath. Herb. Bung.: 29 (1873). Type as above.
   Pellaea hastata sensu Sim, Ferns S. Afr. ed. 2: 211, t. 100 (1915) non (L.f.) Link (1841).

Rhizome c. 6 mm. in diam., short, creeping, with tufted fronds and with linear attenuate entire black rhizome-scales up to 4 mm. long with pale-brown margins. Stipe ebeneous, up to 30 cm. long, terete, glabrous except for a few hairs similar to the rhizome-scales at the base. Frond erect to arching, thinly to thickly coriaceous. Lamina up to 43 × 25 cm., narrowly ovate to ovate or triangular in outline, 2-pinnate; pinnules 0·6–5·4 × 0·4–5·2 cm., rotund to broadly hastate with 3–5 acute or obtuse points, entire, glaucous, glabrous on both surfaces, articulate at the apex of the ebeneous petiolules 0·5–26 mm. long; veins free, obscure; rhachis and secondary rhachises ebeneous, terete, glabrous. Sori forming a marginal soral line up to 1·2 mm. broad at maturity; indusium continuous, entire, membranous.

Pinnules less than 3 cm. broad, frond ovate in outline; basal pinnae with more than 5 pinnules - - - - - - - - - - - - - var. *calomelanos*
Pinnules more than 3 cm. broad; frond triangular in outline; basal pinnae with 3–5 pinnules - - - - - - - - - - - var. *swynnertoniana*

Var. **calomelanos**

Botswana. N: Kwebe Hills, 15.iv.1898, *Lugard* 225 (K). SE: Lobatsi, Molopo Ranch, 24.ii.1965, *McConnell* (BOL). **Rhodesia.** N: Umvukwes Mts., near Darwendale, 20.iv.1948, *Rodin* 4352 (K; SRGH). W: Matobo, Besna Kobila, iv.1956, *Miller* 2786 (BM; BOL; SRGH). C: Epworth Mission, Ruwa Rocks, 4.iii.1961, *Richards* 14530 (K). E: Umtali Distr., Mt. Nhuri, 31.vii.1955, *Chase* 5696 (BM; BOL; SRGH). S: 16 km. NE. from Lundi R. bridge, 13.vii.1953, *Schelpe* 4082 (BM; BOL). **Malawi.** C: Dedza, Nchinji Hill, 22.i.1959, *Robson* 1297 (BM; K; LISC; SRGH). S: Shire Highlands, 1885, *Buchanan* 4 (K). **Mozambique.** N: Cabo Delgado Distr., Mt. Mkota, 1907, *Stocks* 149 (K). Z: Massingire, Metalola, 24.v.1943, *Torre* 5385 (LISC). T: Zóbuè Mts., 12.vii.1942, *Torre* 4397 (LISC). MS: Chimoio, Belas Hills, 1.iii.1948, *Garcia* 458 (LISC). LM: Maputo, Goba Hills, 15.xi.1940, *Torre* 2028 (LISC).

Also in S. Africa, Angola, Tanzania, Kenya, Madagascar, the Mascarene Is., NE Spain and N. India. Among rocks in woodland and grassland, 900–1500 m.

Var. **swynnertoniana** (Sim) Schelpe in Journ. S. Afr. Bot. **30**: 187 (1964). TAB. **40** fig. B. Type: Rhodesia, Chirinda Forest, *Swynnerton* 850 (K).

Pellaea swynnertoniana Sim, Ferns S. Afr. ed. 2: 213, t. 101 (1915) Type as above.

Rhodesia. E: Vumba Mts., Norseland, 26.vi.1955, *Schelpe* 5393 (B; BM; BOL; K; P). **Malawi.** S: Zomba, 1901, *Manning* 116 (K). **Mozambique.** MS: Bandula Peak, Jaegersberg, 23.x.1957, *Chase* 6724 (COI; K; SRGH).

The variety is known only from our area. Among granite rocks in woodland, 1100–1400 m.

8. **Pellaea quadripinnata** (Forsk.) Prantl in Engl., Bot. Jahrb. **3**: 420 (1882).—Sim, Ferns S. Afr. ed. 2: 202, t. 92 (1915).—Tardieu in Mém. Inst. Fr. Afr. Noire, **28**: 92, t. 14 fig. 5 (1953); Fl. Madag., Polypod. **1**: 166, t. 23 fig. 7 (1958).—Alston, Ferns W. Trop. Afr.: 44 (1959). TAB. **41**. Type from Yemen.

Pteris quadripinnata Forsk., Fl. Aegypt.-Arab.: CXXIV, 186 (1775). Type as above.

Allosorus quadripinnatus (Forsk.) C. Presl, Tent. Pterid.: 154 (1836) reimpr. in Abh. Königl. Böhm. Ges. Wiss., Ser. 4, **5**: 154 (1837). Type as above.

Pteris consobrina Kunze in Linnaea, **10**: 526 (1836). Type from S. Africa.

Cheilanthes triangula Kunze, tom. cit.: 536 (1836). Type from S. Africa.

Cheilanthes atherstonei Hook., Sp. Fil. **2**: 107 (1852). Type from S. Africa.

Cheilanthes firma Moore in Journ. of Bot. **5**: 225 (1853). Type from S. Africa.

Cheilanthes linearis Moore, tom. cit.: 226 (1853). Type from S. Africa.

Pellaea consobrina (Kunze) Hook., Sp. Fil. **2**: 145 (1858). Type as for *Pteris consobrina*.

Allosorus consobrinus (Kunze) Pappe & Raws., Syn. Fil. Afr. Austr.: 31 (1858). Type as above.

Cheilanthes quadripinnata (Forsk.) Kuhn, Fil. Afr.: 74 (1868). Type as for *Pellaea quadripinnata*.

Pteridella quadripinnata (Forsk.) Mett. ex Kuhn, Von Deck. Reisen, Bot., **3**, 3: 16 (1879). Type as above.

Rhizome up to 6 mm. in diam., short, creeping, with tufted fronds and with narrowly attenuate entire rhizome-scales up to 5 mm. long, some concolorous pale-brown mixed with others with a black central stripe and pale margins. Stipe castaneous, up to 40 cm. long, glabrous, shallowly channelled on the ventral surface. Frond erect to arching, thinly coriaceous. Lamina up to 58 × 40 cm., triangular in outline, mostly 4-pinnatifid (3-pinnate in juvenile fronds) to 4-pinnate with the lowest pinnae well developed basiscopically; ultimate segments 1–2 cm. long, oblong, subacute to acute, pinnatifid with oblong to triangular lobes, subentire when fertile, evidently crenate when sterile, both surfaces glabrous; venation free, evident; rhachis and secondary rhachises castaneous, glabrous, channelled on the ventral surface. Sori forming a marginal soral line; indusium continuous, subentire, membranous.

Rhodesia. E: Inyanga, 19.x.1946, *Wild* 1397 (K; SRGH). **Malawi.** N: Karonga Distr., Nganda Hill, Nyika Plateau, 6.ix.1962, *Tyrer* 836 (SRGH). S: Shire Highlands, 1885, *Buchanan* 31 (K). **Mozambique.** MS: Gorongosa Mt., Gogogo Peak, 5.vii.1953, *Schelpe* 5520 (BM; BOL).

Also in S. Africa, Tanzania, Kenya, Ethiopia, Cameroon, Arabia, Madagascar and Comoro Is. Around boulders on forest margins and in grassland, 1700–2050 m.

9. **Pellaea viridis** (Forsk.) Prantl in Engl., Bot. Jahrb. **3**: 420 (1882).—Sim, Ferns S. Afr. ed. 2: 207, t. 96 (1915).—Tardieu, Fl. Madag., Polypod. **1**: 162, t. 23 fig. 1 (1958). Type from Yemen.

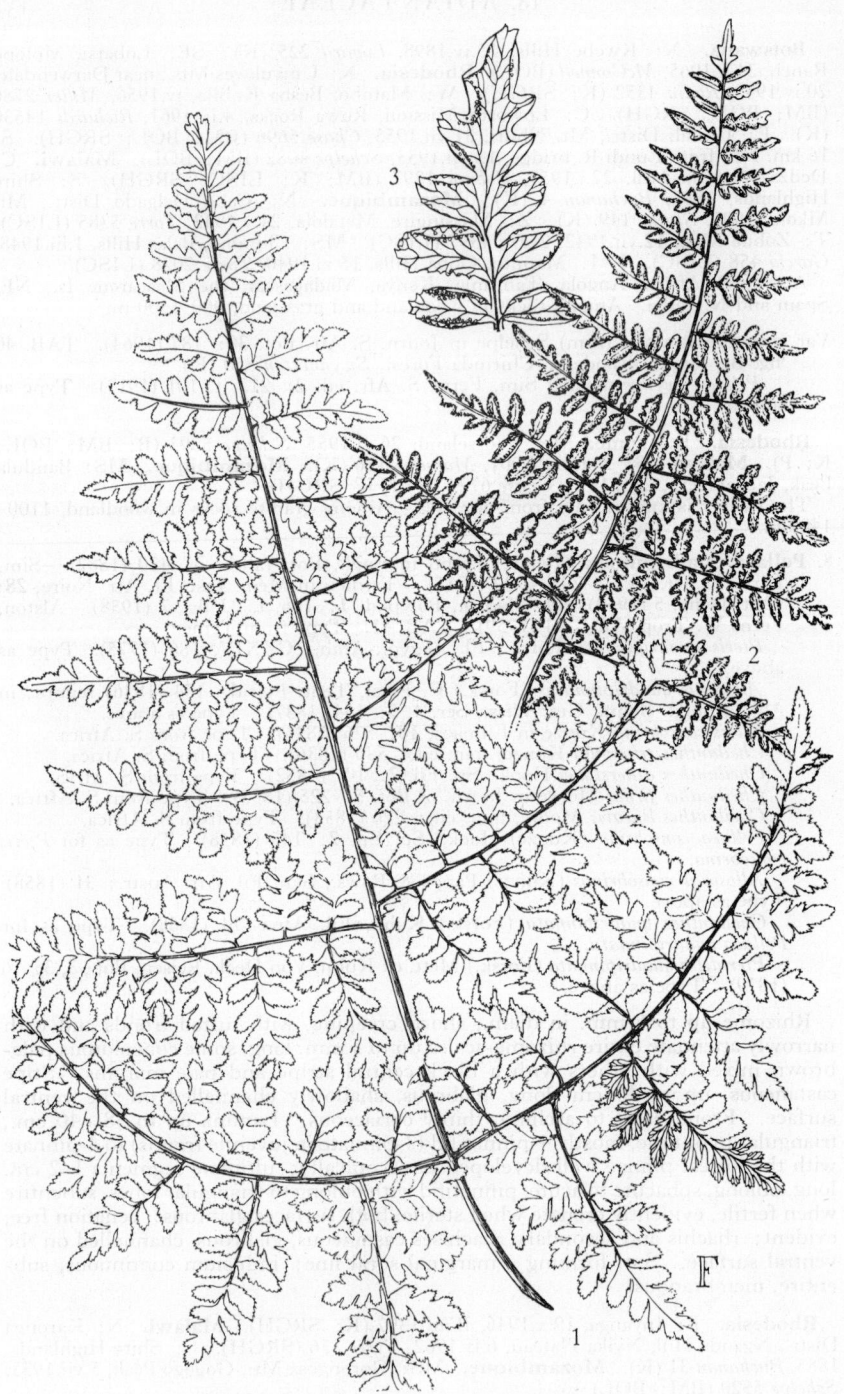

Tab. 41. PELLAEA QUADRIPINNATA. 1, sterile frond (×⅔); 2, fertile frond (×⅔); 3, enlargement of fertile pinnule (×2), 1 from *Schelpe* 5730, 2–3 from *Schelpe* 5661.

Pteris viridis Forsk., Fl. Aegypt.-Arab.: CXXIV, 186 (1775). Type as above.
Adiantum viride (Forsk.) Vahl, Symb. Bot. **3**: 104 (1794). Type as above.
Pteris polymorpha Poir., Encycl. Méth., Bot. **5**: 719 (1804). Type from Mauritius.
Cheilanthes viridis (Forsk.) Sw., Syn. Fil.: 127 (1806). Type as for *Pellaea viridis*.
Pteris hastifolia Schrad. in Gött. Gel. Anz. **1818**: 917 (1818). Type from S. Africa.
Cheilanthes hastata var. *stenophylla* Kunze in Linnaea, **10**: 533 (1836). Type from S. Africa.
Allosorus viridis (Forsk.) Kuntze, Rev. Gen. Pl. **2**: 806 (1891). Type as for *Pellaea viridis*.

Rhizome c. 5 mm. in diam., short, creeping, with tufted fronds and with linear-attenuate entire ciliate or not ciliate black rhizome-scales c. 3 mm. long with pale margins often mixed together with pale-brown concolorous scales. Stipe castaneous to ebeneous, 2·5–40 cm. long, terete below, shallowly channelled above, glabrous or with patent hair-like scales. Frond erect to arching, membranous to coriaceous. Lamina up to 50 × 24 cm., broadly linear to lanceolate or broadly triangular-pentagonal in outline, 2-pinnatifid to 3-pinnate, sometimes with the basal pinnae the largest and much developed basiscopically; ultimate segments narrowly oblong, lanceolate or ovate, attenuate to obtuse, minutely crenate, petiolulate on a short glabrous or pubescent petiolule except for the terminal adnate segments, glabrous on both surfaces or sometimes pilose on the dorsal surface along the costa and veins; veins free, evident or obscure; rhachis and secondary rhachises castaneous to ebeneous, glabrous or shortly pubescent ventrally or densely set with patent hair-like scales, channelled ventrally with concolorous or paler lateral ridges. Sori in a marginal soral line; indusium continuous, subentire, membranous.

Rhachis set with patent or squarrose hair-like scales - - - - var. *involuta*
Rhachis glabrous or pubescent with short unicellular hairs towards the apices of the rhachis and secondary rhachises:
Frond 2-pinnate to 3-pinnatifid - - - - - - - var. *viridis*
Frond 3-pinnate - - - - - - - - - - var. *glauca*

Var. **viridis**

Rhodesia. W: Embakwe, 15.ii.1942, *Feiertag* (SRGH). E: Chirinda, 25.x.1947, *Wild* 2187 (K; SRGH). S: Zimbabwe, 9.viii.1929, *Priestley & Scott* (BM). **Malawi.** S: Mlanje Mt., Lukulezi Valley, 2.iii.1958, *Chapman* 522 (BM; SRGH). **Mozambique.** Z: Milange, Tumbine Mts., 18.i.1966, *Correia* 464 (LISC). T: Mt. Zóbuè, 3.x.1942, *Mendonça* 591 (BM; LISC). MS: Cheringoma, Serração de Durúndi, 22.v.1942, *Torre* 4175 (BM; LISC). SS: Inhambane Distr., Môngoè, 1953, *Schelpe* 4476 (BM; BOL). LM: Namaacha, 20.i.1958, *Barbosa & Lemos* 8235 (COI; K; LISC; LMA).
Also in S. Africa, E. tropical Africa, Madagascar, Arabia and India. Undergrowth and margins of forest and among rocks in woodland, 820–1580 m.

Var. **glauca** (Sim) Sim, Ferns S. Afr. ed. 2: 209, t. 97–98 (1915). Type from S. Africa.
Pteris adiantoides Bory ex Willd. in L., Sp. Pl., ed. 4, **5**: 391 (1810). Type from Réunion.
Allosorus adiantoides (Bory ex Willd.) C. Presl, Tent. Pterid.: 153 (1836) reimpr. in Abh. Königl. Böhm. Ges. Wiss., Ser. 4, **5**: 153 (1837). Type as for *Pteris adiantoides*.
Pellaea adiantoides (Bory ex Willd.) J. Sm., Cat. Kew Ferns: 4 (1856). Type as for *Pteris adiantoides*.
Platyloma adiantoides (Bory ex Willd.) Lowe, Ferns, **3**: t. 33 (1857). Type as above.
Pellaea hastata var. *glauca* Sim, Karrarian Ferns: 30, t. 19 (1891). Type as for *Pelloea voides* var. *glauca*.

Botswana. N: Zhilo Hill, Shashi R., 5.v.1963, *Drummond* 8036 (SRGH). **Rhodesia.** N: Umvukwes Mts., N. of Banket, 23.iv.1948, *Rodin* 4406 (SRGH). W: Matopos, World's View, 1953, *Schelpe* 3918 (BM; BOL). C: Makoni Distr., 19 km. NE. of Badderley, 14.ii.1960, *Chase* 7268 (K; SRGH). E: Umtali Distr., E. of Rhodes View, 1.ii.1957, *Chase* 6313 (BM; BOL; SRGH). S: 83 km. E. of Fort Victoria, 9.vii.1953, *Schelpe* 3992 (BM; BOI). **Malawi.** S: Zomba, 25.iii.1950, *Sturgeon* F.I (BM). **Mozambique.** N: Moçambique, Nampula, Murrumpula, 15.vii.1936, *Torre* 785 (LISC). Z: Milange, Tumbine Mts., 19.i.1966, *Correia* 482 (LISC). T: Moatize, Zóbuè, 2.iii.1964, *Torre & Paiva* 11129 (LISC). MS: Chidoco, Madana, Mossurize, 9.ii.1907, *Johnson* 110 (K). SS: between Vila João Belo and the mouth of Limpopo R., 10.ii.1942, *Torre* 3915 (LISC). LM: Namaacha, 5.vii.1967, *Marques* 2057 (LMU).
Widespread in S. Africa and E. tropical Africa and in Madagascar and the Mascarene Is. Among rocks in woodland and in grassland and in open outcrops, 100–1600 m.

Var. **involuta** (Sw.) Schelpe in Bol. Soc. Brot., Sér. 2, **41**: 214 (1967). Type from S. Africa.
　　Pteris involuta Sw. in Schrad., Journ. Bot. **1800**, 2: 69 (1801). Type as above.
　　Cheilanthes cornuta Kunze in Linnaea, **10**: 534 (1836). Type from S. Africa.
　　Pellaea involuta (Sw.) Bak., Syn. Fil., ed. 2: 148 (1874). Type as for *Pellaea viridis* var. *involuta*.
　　Pteridella involuta (Sw.) Mett. ex Kuhn, Von Deck. Reisen, Bot., **3**, 3: 15 (1879). Type as above.

Zambia. S: Mapanza, 2.ii.1957, *Robinson* 2129 (K; SRGH). **Rhodesia.** N: Umvukwes Range, Mtoroshanga Pass, ii.1957, *Mitchell* 114 (BOL). W: Bulawayo Distr., Hillside, xii.1956, *Miller* 4002 (BM; BOL; SRGH). C: Gilnockie, 32 km. E. of Salisbury, i.1956, *Mitchell* 17 (BOL). E: Vumba Mts., Norseland, 14.ii.1950, *Chase* 3469 (BM; SRGH). S: 59 km. E. of Fort Victoria, 1953, *Schelpe* 3994 (BM; BOL). **Malawi.** N: Rumpi Distr., S. Rukuru Gorge, 25.v.1967, *Pawek* 1120 (SRGH). S: Shire Highlands, ii.1888, *Scott* (K). **Mozambique.** T: between Zóbuè and Vila Mouzinho, 20.ix.1949, *Barbosa & Carvalho* 3717 (BOL). MS: Chimoio, Chicama Hill, 24.iv.1948, *Andrada* 1176 (LISC). LM: Namaacha, Mt. M'Ponduine, 20.xi.1966, *Moura* 121a (LMU).
　　Also in S. Africa and E. tropical Africa and Madagascar. Among rocks in woodland and grassland, c. 1000 m.

12. ACTINIOPTERIS Link
Actiniopteris Link, Fil. Sp.: 79 (1841).

Rhizome creeping, with tufted fronds and with linear attenuate entire rhizome-scales with or without a dark central stripe. Stipe usually stramineous. Fronds flabellate to obcuneate, dichotomously divided into linear segments, green or glaucous with or without scales dorsally, fertile fronds usually taller than sterile fronds and sometimes differently dissected; veins free. Sori borne in a submarginal line; indusia continuous, membranous, entire.
A predominantly African genus of 5 spp. with 2 extending to India.

Fronds dimorphic with the fertile fronds much taller than the sterile fronds; rhizome-scales with or without a thin central black stripe　-　　-　　-　　-　1. *dimorpha*
Fronds uniform or weakly dimorphic; rhizome-scales black with or without a narrow pale border:
　　Segments of fertile frond closely set together and with 2–6 teeth near the apex
　　　　　　　　　　　　　　　　　　　　　　　　　　　2. *radiata*
　　Segments of fertile frond widely set apart with usually a single tooth at the apex
　　　　　　　　　　　　　　　　　　　　　　　　　　　3. *pauciloba*

1. **Actiniopteris dimorpha** P.-Sermolli in Webbia, **17**: 18, t. 2 fig. a–c (1962). TAB. 42 fig. A. Type from Tanzania.

Rhizome c. 5 mm. in diam., creeping, with densely tufted fronds and with linear-attenuate entire rhizome-scales up to 4·5 mm. long, some concolorous pale-brown, others with a black central stripe and pale-brown borders. Stipe glaucous becoming castaneous at the base, up to 19 cm. long, mostly glabrous except for a few linear concolorous pale-brown scales towards the base. Frond erect, coriaceous, dimorphous, with the sterile fronds half the length of the fertile fronds, with the lamina declinate when desiccated. Lamina flabellate, repeatedly dichotomously divided into up to 16 linear glaucous segments, glabrous ventrally but with a few persistent brown linear hair-pointed scales dorsally and with usually reflexed margins; fertile frond segments up to 6·5 cm. long, very narrowly linear, entire; sterile frond segments narrowly linear, entire below but serrate with up to 7 teeth near the apex. Sori in marginal lines; indusium continuous, entire, membranous.

Zambia. C: Mkushi Distr., Fiwila, 3.i.1958, *Robinson* 2581 (K; SRGH). **Rhodesia.** N: Umvukwes Mts., 8 km. N. of Banket, 23.iv.1948, *Rodin* 4407 (BOL; K; SRGH). W: Matobo Distr., Besna Kobila, iv.1954, *Miller* 2361 (SRGH). C: Domboshawa, 1.ii.1965, *Loveridge* 1339 (BOL; SRGH). E: Dora R. bridge, near Umtali, 15.i.1949, *Fisher & Schweickerdt* 426 (BM; K). S: SE. of Zimbabwe, *Holland* (NBG). **Malawi.** C: Dedza Distr., Changoni Forest, 19.xii.1966, *Salubeni* 474 (BOL; SRGH). S: Mt. Mlanje, Likubula, 18.ii.1957, *Chapman* 436 (BM). **Mozambique.** N: Ribáuè, 2·2 km. SW. of Posto Agrícola, 21.vii.1962, *Schelpe & Leach* 6972 (BOL). Z: Montes de Ile

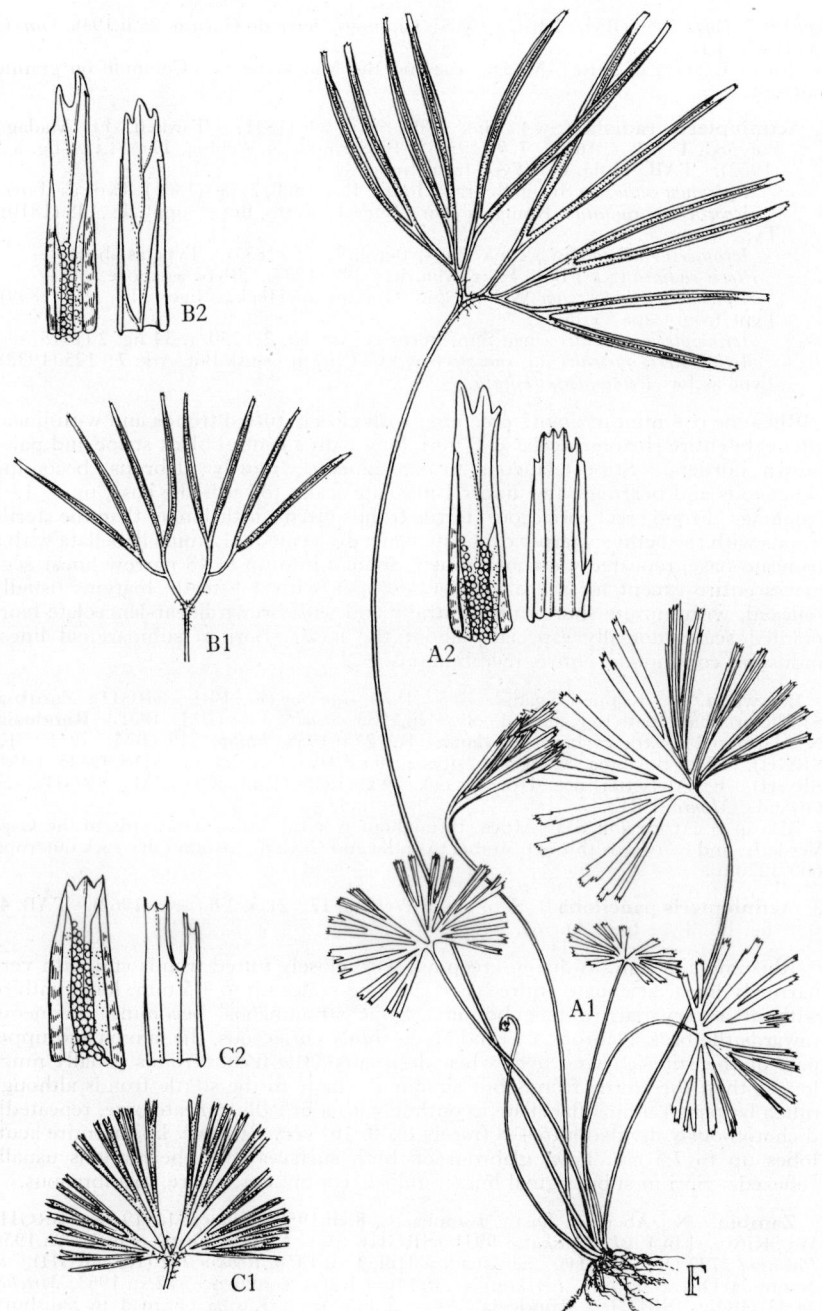

Tab. 42. A.—ACTINIOPTERIS DIMORPHA. A1, habit (×1); A2, enlargement of fertile frond (×10) both from *Chapman* 436. B.—ACTINIOPTERIS PAUCILOBA. B1, fertile frond (×⅔); B2, enlargement of fertile frond (×10) both from *Phipps* 989. C.—ACTINIOPTERIS RADIATA. C1, fertile frond (×⅔); C2, enlargement of fertile frond (×10) both from *Chase* 2009.

2.iv.1943, *Torre* 5056 (BM; LISC). MS: Chimoio, Serra do Garuso, 25.ii.1948, *Garcia* 356 (BM; LISC).

Also in E. tropical Africa, Madagascar and the Mascarene Is. Common on granite outcrops.

2. **Actiniopteris radiata** (Sw.) Link, Fil. Sp.: 80 (1841).—Tardieu, Fl. Madag., Polypod. **1**: 81, t. 16 fig. 7–9 (1958).—P.-Sermolli in Webbia, **17**: 8, t. 1 fig. a–f (1962). TAB. **42** fig. C. Type from India.

  Asplenium radiatum Sw. in Schrad., Journ. Bot. **1800**, 2: 50 (1801). Type as above.

  Acrostichum radiatum Konig ex Poir., Encycl. Méth., Bot., Suppl. **1**: 128 (1810). Type as above.

  Acropteris radiata (Sw.) Link, Hort. Berol. **2**: 56 (1833). Type as above.

  Pteris radiata (Sw.) Boj., Hort. Maurit.: 399 (1837). Type as above.

  Asplenium polydactylon Webb, Spic. Gorgon. in Hook., Niger Fl.: 193 (1849). Type from Cape Verde Is.

  Actiniopteris australis sensu Sim, Ferns S. Afr. ed. 2: 250, t. 34 fig. 2 (1915).

  Actiniopteris australis var. *radiata* (Sw.) C. Chr. in Dansk Bot. Ark. **7**: 125 (1932). Type as for *Actiniopteris radiata*.

Rhizome c. 4 mm. in diam., creeping, with closely tufted fronds and with linear attenuate entire rhizome-scales c. 3 mm. long with a central black stripe and pale-brown borders. Stipe glaucous to stramineous, mostly glabrous, becoming castaneous and bearing a few linear lanceolate scales towards the base, up to 13·5 cm. long. Frond erect coriaceous; fertile fronds often a little longer than the sterile fronds with the lamina sharply declinate when desiccated. Lamina flabellate with a truncate base, repeatedly dichotomously divided into up to 48 narrow linear segments entire except for a sharply dentate apex with 2–6 teeth, margins usually reflexed, with minute short hairs ventrally and with brown linear-lanceolate hair-pointed scales dorsally especially about the base. Sori in submarginal lines; indusium continuous, entire, membranous.

Botswana. N: Kopjies, Ngamiland, 5.v.1930, *Van Son* (K; PRE; SRGH). **Zambia.** S: Victoria Falls, path to Katonta Pools, 6.vii.1953, *Schelpe* 3945 (BM; BOL). **Rhodesia.** N: Urungwe Distr., S. bank of Mauora R., 27.ii.1958, *Phipps* 919 (BM; BOL; K; SRGH). W: Bulalima Mangwe, Isibabe, 6.v.1942, *Feiertag* in GHS 45445 (BM; SRGH). E: Melsetter, near Hot Springs, 29.xii.1948, *Chase* 2004 (BM; SRGH). S: Gwanda, *Monro* (PRE).

Also in S. Africa and SW. Africa, throughout tropical Africa westwards to the Cape Verde Is. and eastwards through Arabia to India and Ceylon. Around dry rock outcrops, 600–1220 m.

3. **Actiniopteris pauciloba** P.-Sermolli in Webbia, **17**: 21, t. 3 fig. a–c (1962). TAB. **42** fig. B. Type from Angola.

Rhizome c. 3 mm. in diam., creeping, with closely tufted fronds and with very narrowly linear attenuate entire black rhizome-scales up to 4·5 mm. long, with or without a very narrow pale border. Stipe stramineous, becoming castaneous towards the base, glabrous. Frond erect, thinly coriaceous, the lamina and upper part of the stipe curving over when desiccated, the fertile fronds usually much longer than the sterile fronds but similar in shape to the sterile fronds although much larger. Lamina flabellate in outline with a broadly cuneate base, repeatedly dichotomously divided into 4–8 (rarely up to 16) very narrowly linear entire acute lobes up to 7·5 cm. long, glabrous on both surfaces with the margins usually reflexed. Sori in submarginal lines; indusia continuous, entire, membranous.

Zambia. N: Abercorn Distr., Lupuba R., 8.xii.1959, *Richards* 11919 (K; SRGH). W: Kitwe, 1.iii.1967, *Fanshawe* 9931 (SRGH). C: Mkushi Distr., Fiwila, 9.i.1958, *Robinson* 2727 (K; SRGH). E: Chadiza Hill, 1.xii.1958, *Robson* 792 (K; SRGH). S: Namwala Distr., Bayewa, 6·4 km. N. of Musa-Kafue confluence, 22.xii.1963, *Mitchell* 24/33 (BOL; SRGH). **Rhodesia.** N: 32 km. from Kariba on road to Salisbury, 11.i.1960, *Mitchell* 560 (BOL; LISC; SRGH).

Also in Angola and Tanzania. In shade in *Brachystegia* woodland, 600–1220 m.

Some of the specimens from northern and eastern Zambia appear to be intermediates between *A. pauciloba* and *A. dimorpha*.

19. LINDSAEACEAE

Terrestrial or epiphytic plants. Rhizome creeping, protostelic or solenostelic, with non-peltate rhizome-scales grading to hairs. Stipe not articulated, with 2 C-shaped vascular strands back to back. Lamina pinnate to 2-pinnate, glabrous; veins free, or anastomosing without included veinlets. Sori marginal or submarginal, linear along both margins to oblong only along the acroscopic margins of the pinnae; indusium opening outwards; paraphyses absent. Spores trilete, without peri-spores.

Only one genus is recognised as occurring in continental Africa, as the genus *Lindsaea* is construed here in the wide sense. *L. ensifolia* with linear marginal sori has previously been referred to the genera *Schizoloma* Fée and *Schizolegnia* Alston.

LINDSAEA Dryand. apud Sm.

Lindsaea Dryand. apud Sm. in Mem. Acad. Turin, **5**: 413 (1793).

Rhizome creeping, with tufted or spaced fronds and with non-peltate rhizome-scales grading into hairs. Stipes not articulated. Lamina pinnate (rarely simple or 2-pinnatifid); pinnae symmetrical to dimidiate, glabrous; veins free to infre-quently anastomosing. Sori marginal or submarginal, linear or oblong, indusiate. A pantropic genus of about 200 spp. with only 2 spp. in continental Africa.

Pinnae entire to shallowly crenate; sori linear on both acroscopic and basiscopic margins
1. *ensifolia*
Pinnae with the acroscopic margin incised into short oblong lobes; sori oblong on the acroscopic pinna lobes - - - - - - - - - - - 2. *odorata*

1. **Lindsaea ensifolia** Sw. in Schrad., Journ. Bot. **1800**, 2: 77 (1801). TAB. **43**. Type from Mauritius.
   Schizoloma ensifolia (Sw.) J. Sm. in Hook., Lond. Journ. Bot. **3**: 414 (1841). Type as above.
   Lindsaea membranacea Kunze in Linnaea, **18**: 121 (1844). Type from S. Africa.
   Schizolegnia ensifolia (Sw.) Alston in Bol. Soc. Brot., Sér. 2, **30**: 24 (1956). Type as for *Lindsoea ensifolia*.

Rhizome c. 1·5 mm. in diam., slender, creeping, with fronds spaced up to 3 cm. apart and with dark-brown lanceolate attenuate entire scales up to 2·5 mm. long. Frond erect, membranous. Stipe brown, nitidous, up to 36 cm. long, glabrous except for scales similar to the rhizome scales at the extreme base. Lamina up to 45 × 22 cm., simply pinnate; pinnae up to 12·5 × 1·5 cm., very narrowly oblong-obtuse, base unequally cuneate, entire to shallowly crenate, glabrous on both surfaces, veins anastomosing. Sori up to 0·5 mm. broad, linear, marginal. Indusia linear, semi-transparent, erose.

Mozambique. MS: Cheringoma, 2·4 km. E. of R. Tambani, 22.x.1949, *Pedro & Pedrógão* 8872 (BM).
Also from Mascarene Is., Pemba, Seychelles, Nigeria and S. Africa. Wet situations in shade in forest.

2. **Lindsaea odorata** Roxb. in Calc. Journ. **4**: 511, t. 34 (1844). Type from India.
   Lindsaea loherana Christ in Bull. Herb. Boiss. **6**: 144, t. 4 fig. 6 (1898). Type from Philippine Is.

Rhizome c. 1 mm. in diam., creeping, with fronds spaced up to 1 cm. apart and with squarrose lanceolate attenuate brown shining scales up to 2 mm. long. Frond erect, herbaceous. Stipe pale- to dark-brown, up to 5 cm. long, shallowly sulcate, glabrous except for a few scales about the base similar to the rhizome scales. Lamina up to 20 × 3 cm., pinnate, narrowly elliptic in outline, acute-acuminate with the lowest 4 pairs of pinnae reduced. Pinnae petiolate, rhombic, incised on the acroscopic margin into 4–6 shallow truncate lobes; basiscopic margin entire, ascending, glabrous on both surfaces. Rhachis channelled ventrally, glabrous, stramineous to castaneous. Sori oblong to narrowly oblong, sub-marginal,

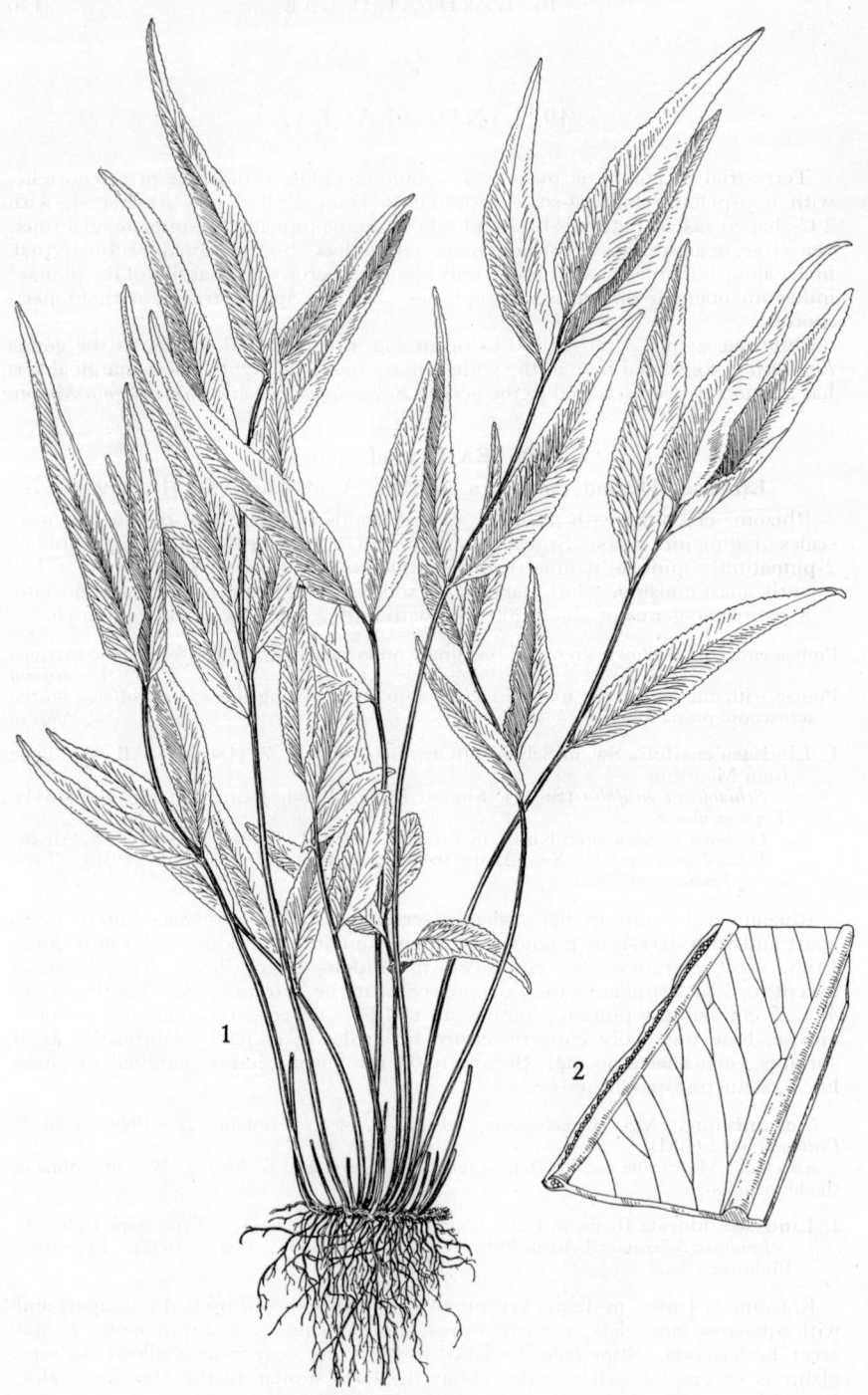

Tab. 43. LINDSAEA ENSIFOLIA. 1, habit (×½) *Gerrard* 1940; 2, enlargement of fertile frond (×3) *Buchanan* 160.

usually 1 per pinna lobe, up to 2 mm. long. Indusium oblong to narrowly oblong, erose, extending from the base of the sorus to the apex of the truncate pinna lobe.

Rhodesia. E: Melsetter, Chimanimani Mts., Long Gully, 25.v.1957, *Ball* 695 (BOL; K; SRGH).
Also in Madagascar and Réunion, and widely distributed in tropical Asia from Ceylon to New Guinea.
Wet streamside rocks in forest, c. 390 m.

20. GRAMMITIDACEAE

Small epiphytic or lithophytic plants with short erect to widely creeping rhizomes set with narrow brown to black rhizome-scales. Stipes not articulated to the rhizome, often with spreading multicellular hairs. Fronds simple, pinnatifid, pinnate to 2-pinnatifid, with entire lobes, glabrous or set with hairs or, rarely, with chalky white vesicles; veins simple or forked not anastomosing. Sori round to elliptic, exindusiate, superficial. Spores trilete.

Fronds simple, entire to subentire - - - - - - - 1. **Grammitis**
Fronds pinnatifid more than halfway to the midrib - - - - 2. **Xiphopteris**

1. GRAMMITIS Sw.

Grammitis Sw. in Schrad., Journ. Bot. **1800**, 2: 17 (1801).

Small epiphytic or lithophytic plants with short creeping or suberect rhizomes with brown rhizome-scales. Fronds linear to narrowly spathulate, entire to shallowly crenate, glabrous or set with multicellular hairs, membranous to coriaceous; veins forked, free. Sori round to elliptic, in a row on either side of the midrib.
A genus of about 150 spp. distributed throughout the tropics and southern hemisphere.

Grammitis nanodes (A. Peter) Ching in Bull. Fan Mem. Inst. Biol. Bot. **10**: 241 (1941).
TAB. **44** fig. A. Type from Tanzania.
Polypodium nanodes A. Peter in Fedde, Repert. Beih. **40**, 1: 27, descr. 3, t. 1 fig. 7–9 (1929). Type as above.

Rhizome short, erect, with tufted fronds and clothed with subulate reddish-brown non-clathrate rhizome-scales up to 1·6 mm. long. Fronds simple, subsessile. Lamina simple, narrowly spathulate to linear, apex subacute, base long-tapering, margin and under surface of midrib with forked or stellate reddish-brown hairs which pale or fall off with age, glabrous on remaining lamina surfaces; midrib concolorous with the lamina, veins obscure. Sori up to 2 mm. long at maturity, oval, without hairs, borne close to the midrib.

Rhodesia. E: Chimanimani Mts. 11.ii.1958, *Mitchell* 364 (BOL; SRGH). **Mozambique.** MS: Gorongosa Mt., Gogogo Peak, 6.vii.1955, *Schelpe* 5536 (BM; BOL).
Also in Tanzania. Occasional mid-level epiphyte in mist forest, 1770–2060 m.

2. XIPHOPTERIS Kaulf.

Xiphopteris Kaulf. in Berl. Jahrb. Pharm. **21**; 35 (1820).

Small epiphytic or lithophytic plants with short erect to widely creeping rhizomes with brown to grey-brown rhizome-scales. Fronds linear, deeply pinnatifid, glabrous or villous. Sori 1–8 per lobe, round to oval, with or without paraphyses.
A genus of over 200 spp. distributed throughout the tropics.

Lamina villous:
Frond clearly stipitate; lamina coriaceous, thinly villous with castaneous hairs, midrib concolorous with lamina - - - - - - - - - 1. *villosissima*

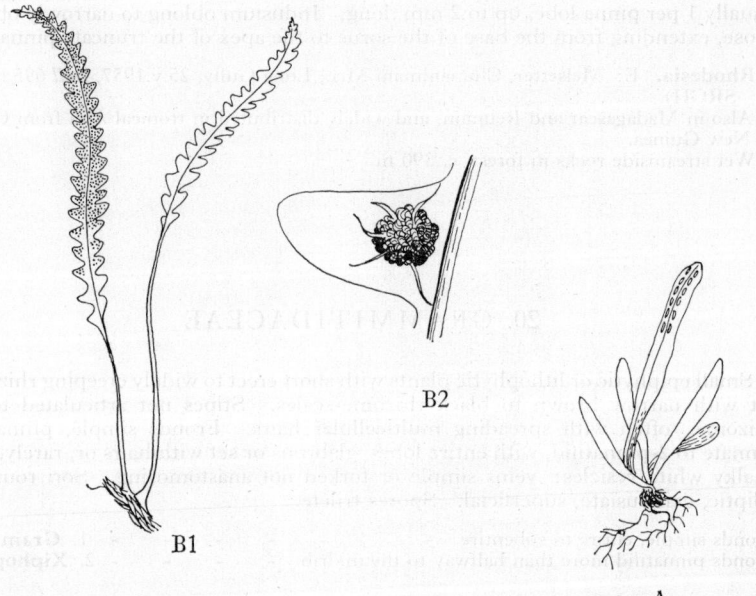

Tab. 44. A.—GRAMMITIS NANODES. Habit (× ⅔) *Schelpe* 5536. B.—XIPHOPTERIS FLABELLIFORMIS. B1, habit (× ¼) *Whellan* 991. B2, fertile lobe showing sorus (× 5) *Schelpe* 2847.

Frond with very short stipe; lamina membranous, villous with pale-brown hairs
   midrib black - - - - - - - - - - - - 2. *cultrata*
Lamina glabrous:
   Sori confluent in the upper crenate ½ of the frond; lower ½ pinnatifid to ½-way to the
      midrib with a conspicuous vein in each acute lobe - - - 5. *serrulata*
   Sori discrete on the lobes of the frond; fronds pinnatifid for their whole length:
      Fronds with stipes ½–⅓ the length of the lamina; rhizome creeping; rhizome scales
         metallic grey-brown; young sori with castaneous paraphyses - 3. *flabelliformis*
      Fronds subsessile; rhizome erect or shortly creeping; rhizome scales brown; young
        sori without paraphyses - - - - - - - - - 4. *oosora*

1. **Xiphopteris villosissima** (Hook.) Alston in Bol. Soc. Brot., Sér. 2, **30**; 27 (1956).
   Type from W. Africa.
     Polypodium villosissimum Hook., Sp. Fil. **4**: 197 (1862). Type as above.
     Polypodium villosissimum var. *majus* Reim. in Notizbl. Bot. Gart. Berl. **11**: 937
   (1933). Type from Tanzania.
     Ctenopteris villosissima (Hook.) W. J. Harley in Contr. Gray Herb. **177**: 92 (1955).
   Type as for *Xiphopteris villosissima*.

Rhizome c. 1·5 mm. in diam., erect to shortly creeping, with tufted fronds and
with brown subulate rhizome-scales. Fronds deeply pinnatifid, stipitate. Stipes
brown, up to 3·5 cm. long, villous with stiff patent reddish-brown hairs up to 3
mm. long. Lamina oblong to oblong-lanceolate in outline, abruptly decrescent
below, pinnatifid almost to the midrib into oblong entire lobes up to 7 mm. long,
villous on both surfaces with stiff reddish-brown hairs up to 3 mm. long; veins and
midrib obscure. Sori submarginal, up to 8 per lobe.

Malawi. S: Mlanje Distr., Mlanje Mt., Luchenya Plateau, 30.vi.1946, *Brass* 16549
(K).
   Also in Sierra Leone, Liberia, Guinea, Fernando Po, S. Tomé, Congo and Tanzania.
Recorded as an epiphyte on mossy upper branches of a *Widdringtonia* at 1890 m. altitude
on Mlanje Mt.
   In W. tropical Africa the fronds are usually much larger with stipes up to 7 cm. long
and laminae up to 11·5 × 2·8 cm.

2. **Xiphopteris cultrata** (Bory ex Willd.) Schelpe in Bol. Soc. Brot., Sér. 2, **41**: 217 (1967). Type from Jamaica.
   Polypodium cultratum Bory ex Willd. in L., Sp. Pl., ed. 4, **5**: 187 (1810). Type as above.
   Polypodium elasticum Bory ex Willd. in L., tom. cit.: 183 (1810). Type from Réunion.
   Polypodium cultratum var. *elasticum* (Bory ex Willd.) Bak., Syn. Fil.: 327 (1874). Type from Réunion.
   Ctenopteris elastica (Bory ex Willd.) Copel. in Philipp. Journ. Sci. **84**: 426 (1936). Type as for *Polypodium elasticum.*
   Ctenopteris cultrata (Bory ex Willd.) Copel., Gen. Fil.: 219 (1947). Type as for *Xiphopteris cultrata.*
   Xiphopteris elastica (Bory ex Willd.) Alston in Bol. Soc. Brot., Sér. 2, **30**: 26 (1956). Type as for *Polypodium elasticum.*

Rhizome c. 1 mm. in diam., erect or very shortly creeping, with tufted fronds and with numerous pale-brown hair-like scales up to 1 mm. long. Fronds membranous, pinnatifid almost to the midrib, stipitate. Stipes c. 1 mm. long, dark-brown, set with pale-brown patent hairs up to 1 mm. long. Lamina up to 8 × 1·2 cm., linear-oblong in outline, long-tapering below, pinnatifid almost to the midrib into oblong entire rounded to acute lobes up to 7 mm. long, thinly villous with soft pale hairs up to 1 mm. long; midrib black, veins obscure. Sori up to 7 per lobe; sporangia sometimes setose.

Malawi. S: Shire, *Kirk* (K).
Also in Central America, Cameroon, Seychelles, Madagascar and Mascarene Is.

3. **Xiphopteris flabelliformis** (Poir.) Schelpe in Bol. Soc. Brot., Sér. 2, **41**: 217 (1967). TAB. **44** fig. B. Lectotype from Réunion.
   Polypodium flabelliforme Poir. in Lam., Encycl. Méth. Bot., **5**: 519 (1804). Type as above.
   Polypodium rigescens Bory ex Willd. in L., Sp. Pl., ed. 4, **5**: 184 (1810). Type from Réunion.
   Ctenopteris rigescens (Bory ex Willd.) J. Sm., Hist. Fil.: 184 (1875).—Tardieu, Fl. Camér., **3** Ptérid.: 327, t. 52 fig. 1–3 (1964). Type as above.
   Xiphopteris rigescens (Bory ex Willd.) Alston in Bol. Soc. Brot., Sér. 2, **30**: 26 (1956). Type as above.
   Grammitis flabelliformis (Poir.) Morton in Contr. U.S. Nat. Herb. **38**: 57 (1967). Type as for *Xiphopteris flabelliformis.*

Rhizome creeping, with spaced fronds and with metallic grey-brown lanceolate acuminate rhizome-scales up to 4 mm. long. Frond pinnatifid, stipitate, subcoriaceous to coriaceous. Stipe light-brown to black, up to 6 cm. long, very narrowly winged. Lamina up to 21 × 1 cm., linear in outline, glabrous on both sides, decrescent below, pinnatifid to the midrib into rounded oblong to quadrate lobes up to 6 mm. long; midrib black, veins usually obscure. Sori 1–6 per lobe, intramarginal, set with castaneous hair-like paraphyses.

Rhodesia. E: Chimanimani Mts., 11.ii.1958, *Mitchell* 366 (BM; BOL; SRGH).
Malawi. N: Nyika Plateau, 6.viii.1946, *Brass* 17242 (K). S: Mlanje Mt., Luchenya Plateau, 3.vii.1946, *Brass* 16655 (K). **Mozambique.** MS: Chimanimani Mts., Nhamadura Summit, 18.iv.1960, *Goodier & Phipps* 367 (SRGH).
Also in Cameroon, Fernando Po, Congo, Uganda, Kenya, Tanzania, S. Africa (Natal), Réunion and tropical America. Epiphytic in montane forest or lithophytic on shaded rock faces at altitudes between 2200 m. and 2350 m.

4. **Xiphopteris oosora** (Bak.) Alston in Bol. Soc. Brot., Sér. 2, **30**: 26 (1956). Type from S. Tomé.
   Polypodium oosorum Bak. apud Henriq. in Bol. Soc. Brot. **4**: 154, t. 2 fig. A (1887). Type as above.
   Polypodium newtonii Bak. in Kew Bull. **1896**: 41 (1896). Type from Fernando Po.

Rhizome c. 1 mm. in diam., shortly creeping, with closely set fronds and with brown lanceolate acuminate clathrate rhizome-scales c. 1·5 mm. long. Fronds linear, pinnatifid, subcoriaceous, sessile or subsessile. Stipe less than ⅐ the length of the lamina. Lamina up to 5·5 × 0·5 cm., linear in outline, deeply pinnatifid into decurrent cuneate to oblong lobes up to 2 mm. long, glabrous on both surfaces; midrib often darker than the lamina below, veins simple and ending in a conspicuous pale hydathode on the upper surface. Sori round to oval, one to each lobe.

Malawi. S: Mlanje Mt., Luchenya Plateau, 5.vii.1946, *Brass* 16670 (K; SRGH).
Also in Sierra Leone, Guinée, Cameroon, Fernando Po, S. Tomé and Tanzania.
Locally frequent on shaded mossy rocks in streambeds between altitudes of 1680 m. and
1890 m. on Mlanje Mt.

5. **Xiphopteris serrulata** (Sw.) Kaulf., Enum. Fil.: 85 (1824). Type from Jamaica.
Acrostichum serrulatum Sw., Prodr. Veg. Ind. Occ.: 128 (1788). Type as above.
Grammitis serrulata (Sw.) Sw. in Schrad., Journ. Bot. **1800**, 2: 18 (1801). Type
as above.
Asplenium serrulatum (Sw.) Bernh. in Schrad., Neues Journ. **2**, 2: 48 (1806).
Type as above.
Micropteris serrulata (Sw.) Desv. in Mém. Soc. Linn. Par. **6**, 2: 217 (1827).
Type as above.
Micropteris orientalis Desv., loc. cit. Type from the Mascarene Is.
Polypodium serrulatum (Sw.) Mett., Fil. Hort. Lips.: 30 (1856) non Sw. (1801).
Type as for *Xiphopteris serrulata*.
Xiphopteris orientalis (Desv.) Fourn. in Compt. Rend. Par. **81**: 1140 (1875).
Type as for *Micropteris orientalis*.
Xiphopteris extensa Fée, Mém. Fam. Foug. **11**: 14, t. 19 fig. 3 (1886). Type from
Mexico (Guadeloupe).
Polypodium duale Maxon in Contr. U.S. Nat. Herb. **16**: 61 (1912). Type from
Jamaica.

Rhizome c. 0·5 mm. in diam., erect or ascending, with tufted fronds and with
brown lanceolate rhizome scales up to 1 mm. long. Fronds firmly membranous,
subsessile. Stipe c. 1 mm. long, brown, glabrous. Lamina up to $17·5 \times 2$ mm.,
linear to very narrowly oblanceolate in outline, base decrescent, pinnatifid about $\frac{1}{2}$
way to the midrib in the lower $\frac{1}{2}$ into acute triangular lobes each with a single
conspicuous vein, crenate obtuse in the upper $\frac{1}{2}$; midrib immersed. Sori up to 8,
set at an acute right-angle to the midrib, confluent at maturity and borne in the
widened upper part of the frond.

Rhodesia. E: Chimanimani Mts., 14.i.1969, *Williams* 11 (BOL; SRGH).
Also in Réunion, Madagascar, Congo (Kivu), W. tropical Africa and tropical America.
On rock in river-bed below flood level at 422 m.
The fronds in the Rhodesian specimen are much shorter than in most west African
specimens which have laminae up to 40×2 mm.

21. POLYPODIACEAE

Epiphytic or less frequently terrestrial plants, with creeping or climbing rhizomes
with peltate often clathrate rhizome-scales. Fronds simple or pinnatifid to pinnate,
often articulated to the rhizome; veins anastomosing to form areoles with included
veinlets. Sori round, elongate or acrostichoid, without, or more rarely with, paraphyses, exindusiate. Sporangia with long stalks; spores monolete, without
perispores. Gametophytes thalloid, cordate or elongate.

Fronds with stellate hairs:
  Fronds dimorphous with sterile " nest leaves " and arching or pendulous fertile fronds;
    sori acrostichoid - - - - - - - - - - 1. **Platycerium**
  Fronds uniform, simple; sori discrete round, not acrostichoid - 2. **Pyrrosia**
Fronds without stellate hairs:
  Fronds dimorphous with sterile sinuate " nest leaves " and deeply pinnatifid fertile
    fronds; frond margins distantly serrulate - - - - - 3. **Drynaria**
  Fronds uniform; frond margins entire or shallowly crenate:
    Sori elongate or round, superficial, not acrostichoid:
      Sori elongate, linear - - - - - - - 4. **Loxogramme**
      Sori round:
        Soral paraphyses present:
          Soral paraphyses peltate, conspicuous in young sori; fronds simple
                                              5. **Pleopeltis**
          Soral paraphyses not peltate; fronds pinnatifid - - 6. **Phymatodes**
          Soral paraphyses absent:

Fronds simple; veins anastomosing:
  Only costal areoles with included veinlets; epiphyte with long climbing
  rhizomes - - - - - - - - 7. **Microgramma**
All areoles with included veinlets; lithophytic or terrestrial
                             8. **Microsorium**
Fronds deeply pinnatifid; veins free   -   -   - - 9. **Polypodium**
Sori acrostichoid on a narrowly linear apical segment -   -   - 10. **Belvisia**

1. PLATYCERIUM Desv.

Platycerium Desv. in Mém. Linn. Soc. Par. **6**, 2: 213 (1827).

Large epiphytes with short rhizomes. Fronds dimorphous. Sterile fronds
(" nest leaves ") sessile, persistent, circular to oblong, entire to shallowly lobed,
soon becoming brown and enclosing a mass of roots, old sterile fronds and plant
detritus. Fertile fronds simple and broadly cuneate to dichotomously divided,
arching or pendulous, densely set on the under surface with stellate hairs. Sori
acrostichoid covering a large area.

A genus of about 12 spp. in the tropics of Australasia, Asia, Africa and S.
America.

Fertile frond broadly cuneate, entire - - - - - - - 1. *elephantotis*
Fertile frond dichotomously divided - - - - - - - 2. *alcicorne*

1. **Platycerium elephantotis** Schweinf. in Bot. Zeit. **29**: 361, cum fig. (1878).
FRONTISP. Type from Sudan.
    Platycerium angolense Welw. ex Bak. in Hook. & Bak., Syn. Fil.: 425 (1868) *in
synon.*—Sim, Ferns S. Afr. ed. 2: 294, t. 155 (1915). Type from Angola.
    Alcicornium angolense Welw. ex Underw. in Bull. Torrey Bot. Club, **32**: 593
(1905). Type as above.
    Platycerium velutinum C. Chr. in Dansk Bot. Ark. **9**, 3: 69 (1937). Type from
Congo.

Rhizome c. 1 cm. in diam., with subulate ciliate scales up to 1·5 cm. long, brown
with a central black stripe. Sterile fronds c. 45 × 30 cm., erect, oblong, prominently
veined in the upper ⅔, entire to irregularly undulate at the apex, covered at first
with grey stellate hairs, becoming glabrous with age. Fertile fronds 50 × 25 cm.,
broadly obcuneate, rounded, entire to subentire, coriaceous, subglabrous above,
densely set below with rufous stellate hairs c. 1 mm. in diam. Sporangia c. 25
cm. broad, borne in a single large oval soral area below the frond apex, densely set
with brown stellate hairs similar to those on the lamina.

Zambia. N: Kawambwa Distr., near Kafulwe Mission, Lake Mweru, 2.xi.1952,
Angus 691 (K). **Malawi.** S: Mlanje base, 1899, *Mahon* (K). **Mozambique.** Z:
between Alto Ligonha and Alto Molócuè, 13.x.1949, *Barbosa & Carvalho* 4412 (K). MS:
43·2 km. N. of Gondola, 1.vii.1955, *Schelpe* 5449 (B; BM; BOL; K; P).
    Also northwards to the Sudan and westwards to Sierra Leone. High-level epiphyte in
dry forest, about 350 m.

2. **Platycerium alcicorne** Desv. in Mém. Soc. Linn. Par. **6**, 2: 213 (1827).—de Joncheere
in Blumea, **15**, 2: 449 (1967). FRONTISP. Type from the Comoros.
    Acrostichum alcicorne Willemet apud Usteri in Ann. Bot. Leipz. **18**: 61 (1796) non
Sw. (1801). Type from Madagascar.
    Platycerium vassei Poisson in Rev. Hort. **82**: 530 (1910). Type: Mozambique,
from living material not extant.
    Platycerium bifurcatum sensu Sim, Ferns S. Afr. ed. 2: 293, t. 154 (1915).

Rhizome short, c. 1 cm. in diam., with brown subulate scales. Sterile fronds ovoid,
convex, forming an almost hemispherical mass, 32 cm. in diam., thinly set with
minute whitish stellate hairs at first, becoming glabrous with age. Fertile fronds
up to 60 cm. long, narrowly cuneate, 2–4 times dichotomously divided into rounded
to acute oblong to cultrate ultimate lobes up to 2·5 cm. broad, subglabrous above,
densely clothed below with minute white stellate hairs c. 0·2 mm. in diam.
Sporangia borne in soral areas on the ultimate and penultimate lobes, covered with
white stellate hairs similar to those on the lamina.

Rhodesia. E: Inyanga Distr., Lower Pungwe R., 11.ii.1957, *Coates Palgrave* (SRGH).
Mozambique: Z: Maganja da Costa, 12.ii.1966, *Torre & Correia* 14534 (LISC). MS:

43·2 km. N. of Gondola, 1.vii.1955, *Schelpe* 5448 (B; BM; BOL; K; P). SS: Rio das Pedras, vii.1936, *Gomes e Sousa* 1789 (COI; K).

Also in Madagascar, Seychelles and Comoro Is. High-level epiphyte in tall woodland and dry forest, 20–400 m.

2. PYRROSIA Mirb.

Pyrrosia Mirb. in Lam. & Mirb., Hist. Nat. Vég. **5**: 91 (1803).

Rhizome slender with non-clathrate scales. Fronds simple, entire, carnose-coriaceous, articulated and covered with stellate hairs, becoming glabrous on the upper surface with age. Sori round, borne in the upper half of the frond in numerous closely set rows on both sides of the midrib, covered at first with densely set stellate hairs.

A mainly tropical genus of about 80 spp. in the Old World.

Frond tomentum composed of stellate hairs of one kind only:
  Stellate hairs with short flat white arms:
    Rhizome scales ciliate; fronds widely spaced on a slender rhizome   1 *lanceolata*
    Rhizome scales not ciliate; fronds closely spaced   -   -   - 2. *schimperana*
  Stellate hairs with long thin brownish arms -   -   -   -   - 3. *rhodesiana*
Frond tomentum composed of stellate hairs of two kinds -   -   -   - 4. *stolzii*

1. **Pyrrosia lanceolata** (L.) Farw. in Amer. Midl. Nat. **12**: 245 (1931). Type from India.
    Acrostichum lanceolatum L., Sp. Pl. **2**: 1067 (1753). Type as above.
    Candollea lanceolata (L.) Mirb. in Lam. & Mirb., Hist. Nat. Vég. **5**: 89 (1803). Type as above.
    Polypodium adnascens Sw., Syn. Fil.: 25, 222, t. 2 fig. 2 (1806). Type from India.
    Polypodium spissum Bory ex Willd. in L., Sp. Pl., ed. 4, **5**: 146 (1810). Type from Réunion.
    Cyclophorus spissus (Bory ex Willd.) Desv. in Mag. Naturf. Berl. **5**: 301 (1811). Type as above.
    Niphobolus spissus (Bory ex Willd.) Kaulf., Enum. Fil.: 126 (1824). Type as above.
    Polypodium pertusum Hook., Exot. Fl. **2**: t. 162 (1825) pro parte. Type from China.
    Cyclophorus vittarioides C. Presl, Epim. Bot.: 129 (1851) reimpr. in Abh. Königl. Böhm. Ges. Wiss., Ser. 5, **6**: 489 (1851). Type from Malacca.
    Polypodium vittarioides Wall. ex Mett. in Abh. Senckenb. Nat. Ges. **2**: 126 (1856) reimpr. in Mett., Farngatt., Polypod.: 126 (1857). Type as above.
    Niphobolus fissus Bedd., Ferns S. India: t. 184 (1864) non Bl. (1828). Type from India.
    Niphobolus adnascens (Sw.) Bedd., Handb. Ferns Brit. Ind.: 325 (1883). Type as for *Polypodium adnascens*.
    Niphobolus lanceolatus (L.) Trimen in Journ. Linn. Soc., Bot. **24**: 152 (1886). Type as for *Pyrrosia lanceolata*.
    Niphobolus tener Giesenh., Niphobolus: 211 (1901). Type from Réunion.
    Cyclophorus tener (Giesenh.) C. Chr., Ind. Fil.: 201 (1905). Type as above.
    Niphobolus giesenhagenii Christ in Ann. Conserv. Jard. Bot. Genève, **7–8**: 330 (1905). Type from India (Garhwal).
    Cyclophorus giesenhagenii (Christ) C. Chr., Ind. Fil.: 199 (1905). Type as above.
    Cyclophorus spissus var. *continentalis* Hieron. in Engl., Bot. Jahrb. **46**: 399 (1911). Syntypes from Tanzania and Cameroon.
    Cyclophorus lanceolatus (L.) Alston in Journ. of Bot. **69**: 102 (1931). Type as for *Pyrrosia lanceolata*.
    Pyrrosia adnascens (Sw.) Ching in Bull. Chin. Bot. Soc. **1**: 45 (1935). Type as for *Polypodium adnascens*.

Rhizome c. 1·3 mm. in diam., widely creeping, with widely spaced fronds and with pale-brown to grey, linear-lanceolate, ciliate, rhizome-scales up to 4 mm. long. Frond stipitate, thinly coriaceous. Stipe up to 1·8 cm. long, tomentose, becoming glabrous with age. Lamina up to 16 × 1·5 cm., simple, linear to lanceolate to narrowly elliptic, narrowly acute to obtuse, entire but usually with a narrowly reflexed margin; upper surface more or less glabrous, lower surface appressed tomentose with pale-brown or grey stellate hairs with short flattened arms, becoming sub-glabrous with age. Sori emergent through the tomentum.

Mozambique. MS: Cheringoma, Serração de Durúndi, 25.v.1942, *Torre* 4189A (BM; COI; LISC).

Also in Tanzania, Uganda, Congo, Cameroon, Príncipe and the Mascarene Is. and in tropical Asia. Epiphytic in forest on the Mozambique Plain.

2. **Pyrrosia schimperana** (Mett. ex Kuhn) Alston in Journ. of Bot. **72,** Suppl. 2: 8 (1934). TAB. **45** fig. A. Type from Ethiopia.

Polypodium schimperanum Mett. ex Kuhn, Fil. Afr.: 152 (1868). Type as above.
Niphobolus schimperanus (Mett. ex Kuhn) Giesenh., Niphobolus: 112 (1901). Type as above.
Cyclophorus schimperanus (Mett. ex Kuhn) C. Chr., Ind. Fil.: 200 (1905). Type as above.
Cyclophorus mechowii Brause & Hieron. in Engl., Bot. Jahrb. **46:** 395 (1911). Type from Cameroon.
Pyrrosia schimperana var. *mechowii* (Brause & Hieron.) Schelpe in Journ. S. Afr. Bot. **18:** 129 (1952). Type as above
Pyrrosia mechowii (Brause & Hieron.) Alston in Contr. Conhec. Fl. Moçamb. **2:** 37 (1954). Type as for *Cyclophorus mechowii*.

Rhizome c. 2 mm. in diam., creeping, with fronds borne at intervals up to 1 cm. apart and with brown entire ovate-cucullate to lanceolate-acuminate rhizome-scales up to 6 mm. long. Frond stipitate, carnose-coriaceous. Stipe up to 2·8 cm. long. tomentose becoming glabrous with age. Lamina up to 28 × 1·4 cm., simple, linear-lanceolate, narrowly elliptic to oblanceolate, acute to acuminate, base narrowly cuneate-decurrent, entire; both surfaces tomentose with grey or greyish-brown stellate hairs with short flattened arms. Sori emergent through the tomentum.

Zambia. N: Mporokoso Distr., Lake Mweru, 5.xi.1952, *White* 3617 (K). W: Mwinilunga Distr., near Matonchi R., 19.xi.1962, *Richards* 17414 (K). C: Mkushi Distr., Fiwila, 3.i.1958, *Robinson* 2583 (K; SRGH). S: Mapanza W., 7.ii.1954, *Robinson* 512 (SRGH). **Rhodesia.** N: near Sinoia, Hunyani R., 22.iv.1948, *Rodin* 4382 (K; SRGH). C: Goromonzi Distr., Umwindsi R., 19.ix.1947, *Wild* 2007 (BM; SRGH). E: Umtali Distr., Dora R., 4.vii.1957, *Chase* 6577 (BM; BOL; SRGH). **Malawi.** C: Dedza, Mphunzi Hill, 15.i.1964, *Adlard* 616 (SRGH). S: Mlanje Mt., 7.iv.1958, *Chapman* 584 (K; SRGH). **Mozambique.** N: Ribáuè Mt., 19.vii.1962, *Schelpe &* *Leach* 6919 (BOL). Z: Gúruè, near Pico Namúli, 11.viii.1949, *Andrada* 1974 (COI; LISC). T: Zóbuè, 20.vii.1949, *Barbosa & Carvalho* 3727 (LISC). MS: 20·8 km. N. of Gondola, 1.vii.1955, *Schelpe* 5450 (BM; BOL).

Widespread throughout tropical Africa northwards to the Sudan and Ethiopia and westwards to Nigeria. Epiphytic and lithophytic in forest and tall woodland, 400–1100 m.

Specimens with longer acuminate rhizome-scales have been referred to *P. schimperana* and those with shorter blunt scales to *P. mechowii* but no clear distinction in this character appears to exist.

3. **Pyrrosia rhodesiana** (C. Chr.) Schelpe in Journ. S. Afr. Bot. **18:** 126 (1952). Type. Rhodesia, Umtali, *Eyles* 4472 (K, holotype).

Cyclophorus rhodesianus C. Chr. in Dansk Bot. Ark. **7:** 161 (1932). Type as above.

Rhizome c. 3 mm. in diam., creeping, with fronds up to 1·5 cm. apart and with shining light-brown narrowly lanceolate rhizome-scales up to 6 mm. long, denticulate below and subentire above. Frond stipitate, carnose-coriaceous. Stipe up to 8 cm. long, tomentose, becoming glabrous with age. Lamina up to 30 × 3 cm., simple, narrowly elliptic-lanceolate to oblanceolate, narrowly acute to obtuse, base narrowly cuneate, entire; upper surface subglabrous, lower surface tomentose with reddish-brown stellate hairs with long slender arms c. 1 mm. long. Sori sunk in the tomentum.

Rhodesia. E: Umtali Distr., SW. Vumba Mts., 26.v.1957, *Chase* 6507 (BM; BOL; LISC; SRGH). **Malawi.** C: Dedza Mt., 18.v.1963, *Chapman* 2052 (SRGH). S: Shire Highlands, Sotchi, xii.1894, *Scott Elliot* 8523 (K). **Mozambique.** N: Ribáuè Mt., 19.vii.1943, *Schelpe & Leach* 6922 (BOL). Z: Gúruè, R. Licungo, 7.iv.1943, *Torre* 5099 (LISC). MS: Bandula Peak, Jaegersberg, 23.x.1957, *Chase* 6722 (BM; BOL; K; SRGH).

Also in Uganda (Mt. Elgon). Epiphytic or lithophytic in forest, 800–1500 m.

4. **Pyrrosia stolzii** (Hieron.) Schelpe in Journ. S. Afr. Bot. **18:** 133 (1952). Type from Tanzania.

Niphobolus stolzii Hieron. in Engl., Pflanzenw. Afr. **2,** 1: 55 (1908) (" stoltzii "). Type as above.
Cyclophorus stolzii (Hieron.) Hieron. in Engl., Bot. Jahrb. **46:** 396 (1911). (" stoltzii "). Type as above.

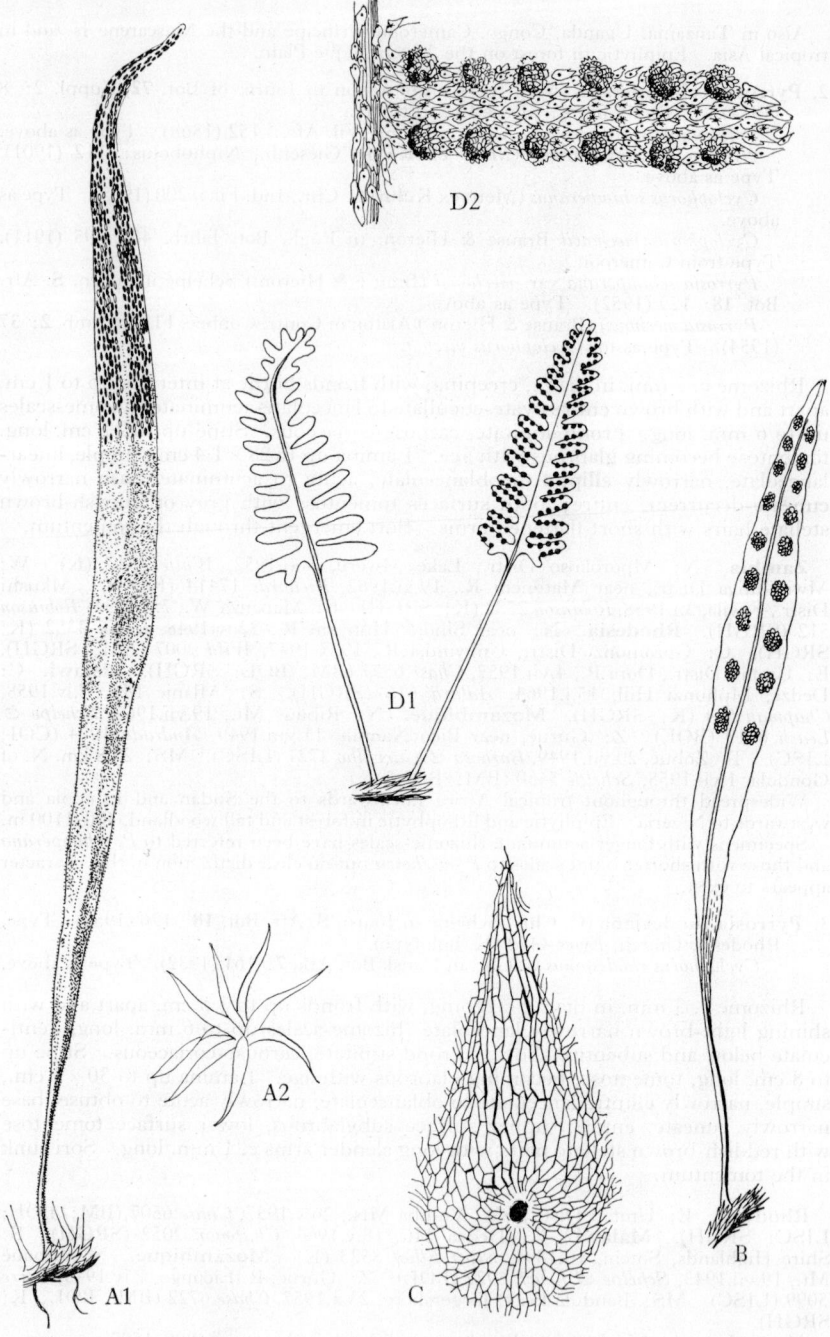

Tab. 45. A.—PYRROSIA SCHIMPERANA. A1, fertile frond (× ⅔); A2, stellate scale (× 20), both from *White* 3617. B.—PLEOPELTIS MACROCARPA. Fertile frond (× ⅓) *Chase* 1095. C.—PLEOPELTIS EXCAVATA. Rhizome-scale (× 14) *Schelpe* 5508. D.—POLYPODIUM POLYPODIOIDES subsp. ECKLONII. D1, habit (× ⅓); D2, dorsal surface of fertile lobe (× 6), both from *Chase* 6723.

Rhizome c. 1·5 mm. in diam., creeping, with fronds c. 1 cm. apart and with dull-brown lanceolate scales up to 2 mm. long subentire below and ciliate above. Frond stipitate, carnose-coriaceous. Stipe up to 9 cm. long, tomentose at first becoming glabrous with age. Lamina up to 25 × 3 cm., simple, lanceolate or elliptic to oblanceolate, acute, base narrowly cuneate-decurrent, entire; upper surface subglabrous, lower surface tomentose with two kinds of stellate hairs, the larger with straight arms c. 0·7 mm. long standing above more numerous smaller hairs with twisted matted arms. Sori emergent through the tomentum.

Zambia. E: Lundazi Distr., Nyika Plateau, upper slopes of Kangampande Mt., 6.v.1952, *White* 2709 (BM; K). **Malawi.** N: Misuku, Mugesse Forest, 20.iv.1963, *Chapman* 1919 (SRGH). C: Nchisi Forest, 5.v.1961, *Chapman* 1274 (SRGH). Also in Tanzania. Epiphyte in mist forest at 2100 m.

3. DRYNARIA (Bory) J. Sm.

Drynaria (Bory) J. Sm. in Hook., Journ. Bot. **4**: 60 (1841).

Large epiphytes with thick fleshy creeping rhizomes. Fronds dimorphous. Sterile fronds sessile, persistent, oblong, shallowly lobed, soon becoming brown and accumulating plant detritus. Fertile fronds deeply pinnatifid with adnate articulated pinna-like segments; venation obvious, finely reticulate with included veinlets. Sori round, superficial, without paraphyses.

A tropical genus of about 20 spp., occurring mainly in Asia and Australasia.

Drynaria volkensii Hieron. in Engl., Bot. Jahrb. **46**: 393 (1911).—Tardieu, Fl. Camér. 3, Ptérid.: 336, t. 53 fig. 4–6 (1964). **TAB 46.** Type from Tanzania (Kilimanjaro).

Rhizome up to 1 cm. in diam., creeping, often encircling the boles of large trees, with spaced fronds and with dark-castaneous spinose rhizome-scales up to 6 mm· long. Sterile " nest " fronds up to 24 × 18 cm., sessile, erect, ovate in outline, chartaceous, with prominent veins, pinnatifid into narrowly oblong acute to obtuse lobes, glabrous except for minute brown laciniate scales along the midrib. Fertile fronds erect to arching, firmly herbaceous, stipitate. Stipe pale brown, up to 15 × 0·3–0·4 cm., glabrous. Lamina up to 67 × 32 cm., ovate-oblong in outline, with a narrow wing decurrent on to the stipe, pinnatifid almost to the midrib into cultrate acute-acuminate lobes up to 23 × 4 cm. with abruptly broadened adnate bases, margin minutely and irregularly notched, thinly set with brown laciniate scales when young, glabrous at maturity. Sori 2–2·5 mm. in diam., round, borne close to the costa about 3 mm. apart.

Zambia. E: Nyika Plateau, c. 2·4 km. SW. of Rest House, 30.x.1958, *Robson* 474 (K; LISC; SRGH). **Malawi.** N: Rumpi Distr., Nyika Plateau, Rukuru R. Falls, 6.i.1959, *Richards* 10523 (K). **Mozambique.** N: Maniamba, 29.v.1948, *Pedro & Pedrógão* 4084 (EA).

Widespread in tropical Africa. Epiphytic in large masses in evergreen forest, 1800–2150 m.

4. LOXOGRAMME (Bl.) C. Presl

Loxogramme (Bl.) C. Presl, Tent. Pterid.: 214, t. 9 fig. 8 (1836) reimpr. in Abh. Königl. Böhm. Ges. Wiss., Ser. 4, **5**: 214, t. 9 fig. 8 (1837).

Rhizome slender, with masses of hairy roots and greyish clathrate scales. Fronds simple, usually entire, carnose-coriaceous, glabrous with immersed freely anastomosing veins without included veinlets. Sori elongate, set at an angle to the midrib, superficial, without paraphyses.

A genus of about 35 spp., mainly Asiatic.

Loxogramme lanceolata (Sw.) C. Presl, Tent. Pterid.: 215 (1836) reimpr. in Abh. Königl. Böhm. Ges. Wiss. Ser. 4, **5**: 215 (1837). **TAB. 48** fig. E. Type from Mascarene Is.

Grammitis lanceolata Sw. in Schrad., Journ. Bot. **1800**, 2: 18 (1801). Type as above.
Grammitis coriacea Kaulf. ex Spreng. in L., Syst. Veg., ed. 16, **4**: 41 (1827). Type from Mascarene Is.

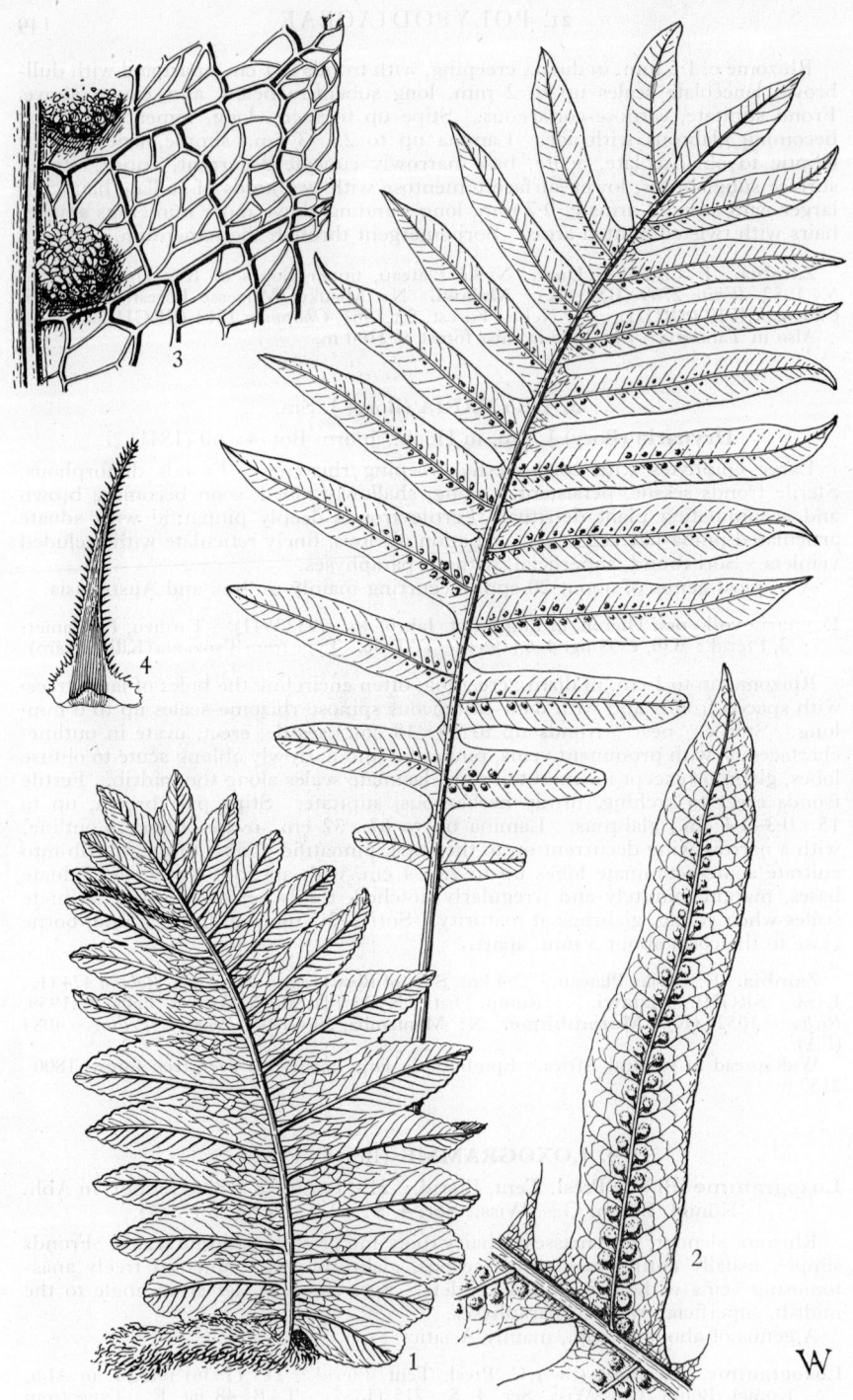

Tab. 46. DRYNARIA VOLKENSII. 1, habit (× ⅓) *White* 2726; 2, fertile pinna (× ⅔) *Meidner* 115; 3, part of fertile pinna (× 4) *Meidner* 115; 4, rhizome scale (× 10) *White* 2726.

Antrophyum lanceolatum (Sw.) Bl., Fl. Jav. Fil.: 84, t. 36 (1829). Type as for *Loxogramme lanceolata.*

Selliguea lanceolata (Sw.) Fée, Mém. Fam. Foug. 5: 177 (1852). Type as for *Loxogramme lanceolata.*

Selliguea coriacea (Kaulf. ex Spreng.) Fée, loc. cit. Type as for *Grammitis coriacea.*

Polypodium coriaceum (Kaulf. ex Spreng.) Mett. in Abh. Senckenb. Nat. Ges. 2: 112 (1856) reimpr. in Mett., Farngatt., Polypod.: 112 (1857). Type as above.

Polypodium loxogramme Mett., loc. cit., reimpr. loc. cit.—Sim, Ferns S. Afr. ed. 2: 281, t. 146 (1915). Type from Mascarene Is.

Gymnogramma lanceolata (Sw.) Hook., Sp. Fil. 5: 156 (1864). Type as for *Loxogramme lanceolata.*

Gymnogramma abyssinica Bak., Syn. Fil.: 517 (1874). Type from Ethiopia.

Loxogramme suberosa Christ in Ann. Mus. Cong., Bot., Sér. 5, 3: 37 (1909). Type from Congo.

Polypodium suberosum (Christ) C. Chr., Ind. Fil., Suppl.: 62 (1913). Type as above.

Loxogramme africana Copel. in Philipp. Journ. Sci. 11: 45 (1916). Type from Angola.

Rhizome widely creeping, with reddish-brown hairy roots and with fronds spaced up to 5 cm. apart, and with dark-grey-brown clathrate narrowly lanceolate acuminate pseudoserrate rhizome-scales. Stipe stramineous to purplish-brown, up to 3 cm. long. Frond stipitate, carnose-coriaceous. Lamina up to $33 \times 2 \cdot 8$ cm., simple, narrowly elliptic, entire to irregularly sinuate; midrib and veins immersed. Sori up to $2 \cdot 3 \times 0 \cdot 25$ cm. at maturity, linear, set at an angle of about 15° to the midrib, overlapping to less than $\frac{1}{4}$ of their length.

Zambia. N: Abercorn Distr., Lake Chila, 15.xi.1952, *White* 3665 (K). W: Mwinilunga Distr., Matonchi R., 19.ix.1937, *Milne-Redhead* 2859 (K). E: Nyika Plateau, 25.x.1958, *Robson* 346 (K). **Rhodesia.** E: Inyanga Distr., Pungwe Gorge, 25.vii.1957, *Chase* 6665 (BM; BOL; K; SRGH). **Malawi.** N: Nyika Plateau, 17.viii.1946, *Brass* 17287 (K). C: Mt. Dedza, 27.ii.1964, *Wild* 6336 (SRGH). S: Mlanje Mt., Luchenya Plateau, 12.vii.1946, *Brass* 16805 (K; SRGH). **Mozambique.** N: Ribáuè Mt., 19.vii.1962, *Schelpe & Leach* 6918 (BOL). MS: Gorongosa Mt., Gogogo Peak, 5.vii.1955 *Schelpe* 5513 (BM; BOL).

Also in S. Africa, Madagascar and the Mascarene Is. and widespread in tropical Africa. Epiphytic or lithophytic in deep shade in forest, 820–1830 m.

5. PLEOPELTIS Humb. & Bonpl. ex Willd.

Pleopeltis Humb. & Bonpl. ex Willd. in L., Sp. Pl., ed. 4, 5: 211 (1810).

Rhizome creeping, with closely or widely spaced fronds and clathrate scales. Fronds simple, entire, membranous to carnose-coriaceous, articulated to the rhizome, with or without peltate scales and with freely and irregularly anastomosing veins with included veinlets. Sori round (in African species) protected when young with prominent peltate paraphyses.

A temperate and tropical genus of about 40 spp.

Under surface of frond glabrous, or with a few scales along the midrib:
Frond membranous, deciduous; rhizome-scales concolorous - - - 1. *excavata*
Frond carnose-coriaceous, persistent; rhizome-scales with a prominent black central stripe - - - - - - - - - - 2. *schraderi*
Under surface of frond with numerous lacerate peltate scales - - 3. *macrocarpa*

1. **Pleopeltis excavata** (Bory ex Willd.) Sledge in Bull. Brit. Mus. (Nat. Hist.) Bot., 2, 5: 138 (1960). TAB. 45 fig. C. Type from Mascarene Is.

Polypodium simplex Sw. in Schrad., Journ. Bot. 1800, 2: 19 (1801) non Burm. f. (1768). Type from Mascarene Is.

Polypodium excavatum Bory ex Willd. in L., Sp. Pl., ed. 4, 5: 158 (1810). Type as for *Pleopeltis excavata.*

Phymatodes excavata (Bory ex Willd.) C. Presl, Tent. Pterid.: 196 (1836) reimpr. in Abh. Königl. Böhm. Ges. Wiss., Ser. 4, 5: 196 (1837). Type as above.

Phymatodes simplex C. Presl, loc. cit., reimpr., loc. cit. Type as for *Polypodium simplex.*

Drynaria excavata (Bory ex Willd.) Fée, Mém. Fam. Foug. 5: 270 (1852). Type as for *Pleopeltis excavata.*

Pleopeltis simplex (C. Presl) Bedd., Handb. Ferns Brit. Ind.: 347 (1883). Type as for *Polypodium simplex.*

Rhizome creeping, with a white waxy covering and with fronds borne 4–12 mm. apart, with brown clathrate lanceolate to narrowly ovate-acuminate rhizome-scales up to 4 mm. long with a paler subentire to weakly lacerate margin. Frond simple, membranous, deciduous, shortly to longly stipitate. Stipe stramineous, up to 7 cm. long, with a few scales similar to the rhizome-scales when young. Lamina up to 33 × 2·7 cm., narrowly lanceolate to linear, entire to weakly undulate, acuminate to obtuse, base widely to narrowly cuneate, completely glabrous or with a few scales similar to those on the stipe along the midrib when young; veins rather obscure. Sori up to 4 mm. in diam. at maturity, round, in a line on either side of the midrib in the upper $\frac{2}{3}$ of the lamina.

Zambia. N: Abercorn Distr., Ndundu, 19.i.1963, *Richards* 17527 (K). W: Mwinilunga Distr., Luao R., 28.xii.1937, *Milne-Redhead* 3855 (K). E: Nyika Plateau, 3.i.1959, *Robinson* 3019 (K; SRGH). **Rhodesia.** N: Lomagundi Distr., Dikwe, Lemon Forest, vii.1967, *Jacobson* 2473 (SRGH). E: Umtali Distr., SW. Vumba Mts., 24.ii.1957, *Chase* 6338 (BM; BOL; K; SRGH). **Malawi.** N: Mafinga Hills, 12.iii.1961, *Robinson* 4461 (BM; K; LISC; SRGH). C: Dedza, Chongoni, Ciwao Hill, 19.i.1959, *Robson* 1265 (BM; K; LISC; SRGH). S: Zomba Distr., Zomba Plateau, 28.v.1946, *Brass* 16060 (K). **Mozambique.** N: Serra de Ribáuè, 16.vi.1935, *Torre* 856 (COI; LISC). Z: Morrumbala (Massingire), Serra de Morrumbala, 6.viii.1942, *Torre* 4507 (LISC). T: Moatize, Zóbuè, *Correia* 409 (LISC). MS: Gorongosa Mt., Gogogo Peak, 5.vii.1955, *Schelpe* 5508 (B; BM; BOL; K; P).

Also in S. Africa, Madagascar, Mascarene Is. and widespread in tropical Africa. Usually epiphytic in montane forest, 1500–2300 m.

P. excavata is construed here in the wide sense as intergrading variation throughout the African material has been found with respect to rhizome-scale shape and margin and frond shape, and the presence or absence, and shape, of lamina-scales. West African plants have been referred to *P. preussii* Hieron., Ethiopian plants to *P. corradii* P.-Sermollined., Ruwenzori plants to *P. mildbraedii* Hieron., Tanzanian plants to *P. stolzii* Hieron., and *P. vesicularipalacea* Hieron. and Kenyan plants to *P. rotundum* Bonap.

2. **Pleopeltis schraderi** (Mett.) Tardieu, Fl. Madag., Polypod., **2**: 110 (1960). Type from S. Africa.

  Polypodium elongatum Schrad. in Gött. Gel. Anz. **1818**: 915 (1818) non Ait. (1789). Type from S. Africa.

  Polypodium schraderi Mett. in Abh. Senckenb. Naturf. Ges. **2**: 98, t. 2 fig. 11 (1856) reimpr. in Mett., Farngatt., Polypod.: 98, t. 2 fig. 11 (1857). Type as for *Pleopeltis schraderi*.

  Polypodium gueintzii Mett., tom. cit.: 91, t. 3 fig. 18–19 (1856) reimpr. in Mett., tom. cit.: 91, t. 3 fig. 18–19 (1857). Type from S. Africa.

  Phymatodes elongata (Schrad.) Pappe & Raws., Syn. Fil. Afr. Austr.: 41 (1858). Type as for *Polypodium elongatum*.

  Niphobolus schraderi (Mett.) Keys., Pol. Cyath. Herb. Bung.: 39 (1873). Type as for *Pleopeltis schraderi*.

  Polypodium lineare var. *schraderi* (Mett.) Sim, Ferns S. Afr. ed. 2: 276, t. 140 (1915). Type as above.

  Lepisorus schraderi (Mett.) Ching in Bull. Fan Mem. Inst. **4**: 51 (1933). Type as above.

  Lepisorus gueintzii (Mett.) Ching, loc. cit. Type as for *Polypodium gueintzii*.

Rhizome c. 2 mm. in diam., creeping, with fronds borne up to 1 cm. apart, with broadly ovate-acuminate black clathrate rhizome-scales with a black central stripe extending to the apex. Fronds simple, shortly stipitate, carnose-coriaceous. Stipe up to 2·3 cm. long, stramineous, glabrous. Lamina up to 33 × 2·3 cm., very narrowly elliptic, acuminate, base very narrowly cuneate, entire, glabrous on both surfaces at maturity. Sori up to 6 mm. in diam., round to oval, in a line on either side of the midrib in the upper $\frac{1}{2}$ of the lamina.

Rhodesia. E: Umtali Distr., SW. Vumba Mts., Chigadora, 30.vi.1951, *Chase* 6568 (BM; BOL; K; SRGH). S: Bikita Distr., Hunyani Mt., 22.ii.1964, *Mitchell* 810 (SRGH). **Malawi.** S: Mlanje Mt., 11.vi.1962, *Richards* 16636 (K). **Mozambique.** Z: Namúli Mt., 27.vii. 1962, *Schelpe & Leach* 7068 (BOL). MS: Gorongosa Mt., above Morambodzi Falls, 3.vii.1955, *Schelpe* 5472 (BOL).

Also in S. Africa and occasional in E. tropical Africa. Epiphytic in montane forest, 900–1200 m.

3. **Pleopeltis macrocarpa** (Bory ex Willd.) Kaulf. in Berl. Jahrb. Pharm. **21**: 41 (1820). TAB. **45** fig. B. Type from Mascarene Is.

  Polypodium lanceolatum L., Sp. Pl. **2**: 1082 (1753).—Sim, Ferns S. Afr. ed. 2: 278. t. 142 (1915) non *Pleopeltis lanceolata* Kaulf. Type from S. Domingo.

Polypodium macrocarpum Bory ex Willd. in L., Sp. Pl., ed. 4, **5**: 127 (1810). Type as for *Pleopeltis macrocarpa.*
Polypodium marginale Bory ex Willd. in L., tom. cit. 149 (1810). Type from Mascarene Is.
Polypodium adspersum Schrad. in Gött. Gel. Anz. **1818**: 915 (1818). Type from S. Africa.
Pleopeltis ensifolia Carm. ex Hook., Exot. Fl. **1**: t. 62 (1823). Type from S. Africa.
Pleopeltis marginalis (Bory ex Willd.) Kaulf. in Berl. Jahrb. Pharm. **21**: 41 (1820). Type as for *Polypodium marginale.*
Pleopeltis lanceolata Kaulf., Enum. Fil.: 245 (1824). Type from S. Africa.
Polypodium lepidotum Willd. ex Schlechtend., Adumbr.: 17, t. 8 (1825). Type from S. Africa.
Pleopeltis lepidota (Willd. ex Schlechtend.) C. Presl, Tent. Pterid.: 193 (1836) reimpr. in Abh. Königl. Böhm. Ges. Wiss., Ser. 4, **5**: 193 (1837). Type as above.
Pleopeltis kaulfussiana C. Presl, loc. cit. reimpr., loc. cit. Type from S. Africa.
Drynaria macrocarpa (Bory ex Willd.) Fée, Mém. Fam. Foug. **5**: 270 (1852). Type as for *Pleopeltis macrocarpa.*
Drynaria lepidota (Willd. ex Schlechtend.) Fée, loc. cit. Type as for *Polypodium lepidotum.*

Rhizome c. 2 mm. in diam., widely creeping, with fronds borne up to 2·5 cm. apart with lanceolate-acuminate brown rhizome-scales c. 3 mm. long with a dark central stripe and wide pale lacerate margins. Frond simple, stipitate, coriaceous. Stipe grey-brown at maturity, up to 8 cm. long, with a few small pale circular to lanceolate scales with a dark centre. Lamina up to 20 × 1·7 cm., narrowly elliptic, entire to weakly undulate (rarely sinuate to pinnatifid), upper surface glabrous, lower surface with numerous pale circular to lanceolate minutely lacerate scales under 1 mm. long with dark centres. Sori oval, up to 4 mm. in diam., in a line on either side of the midrib in the upper ½ of the lamina.

Rhodesia. N: Mt. Binga, 29.i.1966, *Pereira, Sarmento & Marques* 51 (LMU). C: Ngomokurira, 21.ii.1959, *Mitchell* 494 (SRGH). E: Umtali Distr., Nyachowa Falls, 20.vi.1948, *Chase* 1095 (BM; BOL; SRGH). S: Belingwe Distr., Mt. Buhwe, 10.xii.1953, *Wild* 4311 (BM; BOL; SRGH). **Malawi.** N: Nyika Plateau, 14.viii.1946, *Brass* 17227 (K; SRGH). C: Kota Kota Distr., Nchisi Mt., 3.vii.1946, *Brass* 17057 (K; SRGH). S: Zomba Distr., Zomba Plateau, 31.v.1946, *Brass* 16119 (K; SRGH). **Mozambique.** N: Vila Cabral, Serra de Massangulo, 25.ii.1964, *Torre & Paiva* 10789 (LISC). Z: Namúli Mt., R. Licungo, 29.vii.1962, *Schelpe & Leach* 7088 (BOL). T: Mt. Zóbuè, 3.x.1942, *Mendonça* 605 (LISC). MS: Gorongosa Mt., Gogogo Peak, 5.vii.1955, *Schelpe* 5526 (BM).
Also in S. Africa, Madagascar and Mascarene Is., and widespread in tropical Africa, America and India. Epiphytic and lithophytic in forest, 1000–2300 m.

6. PHYMATODES C. Presl

Phymatodes C. Presl, Tent. Pterid.: 195, t. 8 (1836) reimpr. in Abh. Königl. Böhm. Ges. Wiss., Ser. 4, **5**: 195, t. 8 (1837).

Rhizome creeping, epiphytic, lithophytic or terrestrial, with widely spaced fronds and clathrate scales. Fronds simple to deeply pinnatifid, entire (or rarely minutely notched), glabrous, coriaceous, with articulated stipes; venation reticulate with numerous included veinlets ending in hydathodes. Sori round, comparatively large, partially sunk in the lamina, with non-peltate paraphyses.

A tropical and subtropical genus of the Old World, included by Copeland in *Microsorium.*

Phymatodes scolopendria (Burm. f.) Ching in Contr. Inst. Bot. Nat. Acad. Peiping, **2**: 63 (1933).—Tardieu in Mém. Inst. Fr. Afr. Noire, **28**: 222, t. 43 fig. 4 (1953). TAB. **47**. Type from India.
Polypodium scolopendria Burm. f., Fl. Ind.: 232 (1768). Type as above.
Polypodium phymatodes L., Mant. Pl. Alt.: 306 (1771).—Sim, Ferns S. Afr. ed. 2: 273, t. 138 (1915). Type from India.
Polypodium grossum Langsd. & Fisch., Ic. Fil.: 9, t. 8 (1810). Type from Marquesas Is.
Phymatodes vulgaris C. Presl, Tent., Pterid.: 196 (1836) reimpr. in Abh. Königl. Böhm. Ges. Wiss., Ser. 4, **5**: 196 (1837) *nom. illegit.* Type from India.
Chrysopteris phymatodes (L.) Link, Fil. Sp.: 122 (1841). Type as for *Polypodium phymatodes.*

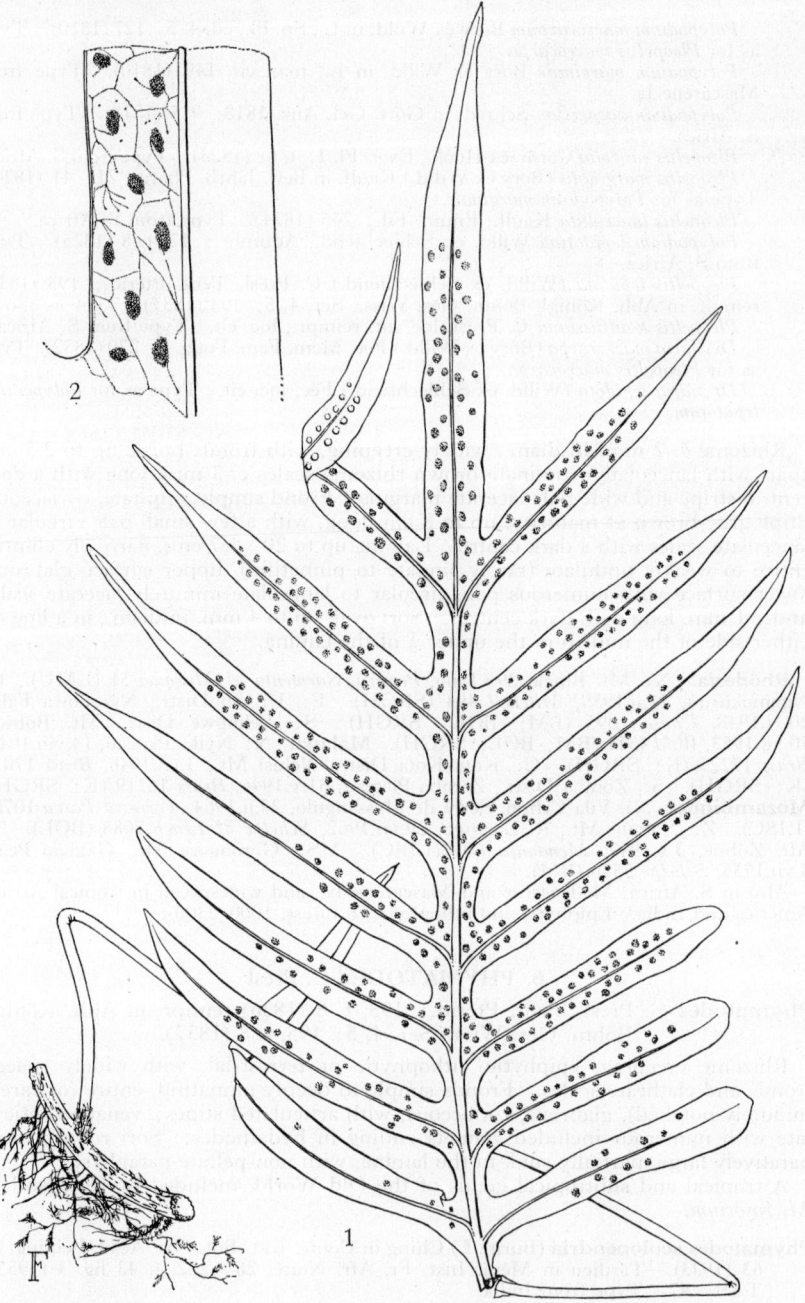

Tab. 47. PHYMATODES SCOLOPENDRIA. 1, frond ($\times\frac{1}{3}$); 2, part of frond ($\times 1\frac{1}{2}$), both from *Lemos & Balsinhas* 134.

Drynaria vulgaris (C. Presl) J. Sm. in Hook., Lond. Journ. Bot. **3**: 397 (1841).
Type as for *Phymatodes vulgaris.*
Drynaria phymatodes (L.) Fée, Mém. Fam. Foug. **5**: 270 (1852). Type as for
Polypodium phymatodes.
Phymatodes grossa (Langsd. & Fisch.) Fée, loc. cit. Type as for *Polypodium grossum.*
Pleopeltis phymatodes (L.) Moore, Ind. Fil.: 78 (1857). Type as for *Polypodium phymatodes.*
Phymatodes phymatodes (L.) Maxon in Contr. U.S. Nat. Herb. **9**: 352, t. 62 (1905)
nom. illegit. Type as above.
Polypodium astrosorum Christ in Journ. de Bot. **22**: 22 (1909). Type from S.
Tomé.

Rhizome up to 1 cm. in diam., widely creeping, with widely spaced fronds and with narrowly lanceolate acuminate pseudoserrate often squarrose rhizome-scales up to 4 mm. long, which are caducous leaving older rhizomes with a white coriaceous surface. Frond deeply pinnatifid, coriaceous, stipitate. Stipe pale-yellowish-green to grey, up to 40 cm. long. Lamina broadly oblong in outline, glabrous, deeply pinnatifid into narrowly oblong, acute to acuminate lobes up to 15 × 3 cm.; midrib prominent below. Sori 2–3 mm. in diam. at maturity, round or oval, in one or two rows on either side of the costae, somewhat sunken into the lamina.

Rhodesia. E: Umtali Distr., Vumba Mts., Valhalla, 22.vii.1957, *Chase* 6669 (BM; BOL; COI; K; SRGH). **Mozambique.** Z: near Luabo, 29.v.1858, *Kirk* (K). MS: Cheringoma, Inhaminga, Pangache, 19.xi.1942, *Mendonça* 1449 (LISC). SS: Inhambane Distr., near Môngoè N. of Maxixe, 20.i.1954, *Schelpe* 4473 (B; BM; BOL; K; P). LM: Inhaca I., xii.1956, *Mogg* 27042 (K; LMU; SRGH).
Also in S. Africa and widespread in tropical Africa and Asia. Epiphytic, lithophytic or terrestrial in forest, from sea-level to 760 m.

7. MICROGRAMMA C. Presl

Microgramma C. Presl, Tent. Pterid.: 213, t. 9 (1836) reimpr. in Abh. Königl.
Böhm. Ges. Wiss., Ser. 4, **5**: 213, t. 9 (1837).

Rhizome creeping, with widely spaced fronds and lanceolate to subulate scales. Fronds simple, sometimes somewhat dimorphous with the fertile fronds longer and narrower than the sterile fronds, entire, articulate. Venation reticulate, with included veinlets only in the costal areoles. Sori round, without paraphyses.
A genus of about 15 spp. mainly in the American tropics but with one variable species in Africa.

Microgramma lycopodioides (L.) Copel., Gen. Fil.: 185 (1947). TAB. **48** fig. C.
Type from the W. Indies.
Polypodium lycopodioides L., Sp. Pl. **2**: 1082 (1753).—Sim, Ferns S. Afr. ed. 2:
279, t. 144 fig. 1 (1915). Type as above.
Polypodium mauritianum Willd. in L., Sp. Pl., ed. 4, **5**: 150 (1810). Type from
Mauritius.
Polypodium owariense Desv. in Mag. Ges. Naturf. Fr. Berl. **5**: 314 (1811). Type
from W. Africa.
Polypodium myrtillifolium Kaulf., Enum. Fil.: 91 (1824). Type from Madagascar.
Polypodium ligustrifolium Desv. in Mém. Soc. Linn. Par. **6**, 2: 225 (1827). Type
from Madagascar.
Polypodium mackenii Bak., Syn. Fil.: 357 (1868). Type from S. Africa.
Polypodium lycopodioides var. *mackenii* (Bak.) Sim, Ferns S. Afr.: 203 (1892); op.
cit. ed. 2: 280, t. 144 fig. 2 (1915). Type from S. Africa.
Polypodium anguinum A. Peter in Fedde, Repert. Beih. **40**, 1: 29, descr. 8. t. 1 fig.
10 (1929). Type from Tanzania.
Polypodium lycopodioides var. *myrtillifolium* (Kaulf.) C. Chr. in Dansk Bot. Ark., **7**:
157 (1932). Type as for *P. myrtillifolium.*
Microgramma owariensis (Desv.) Alston in Bol. Soc. Brot., Sér. 2, **30**: 20 (1956).
Type as for *Polypodium owariense.*
Microgramma mauritiana (Willd.) Tardieu, Fl. Madag., Polypod. **2**: 108 (1960).
Type as for *Polypodium mauritianum.*

Rhizome up to 5 mm. in diam., widely creeping, with fronds spaced 1 cm. or more apart and with subulate entire pale-brown rhizome-scales up to 6 mm.

long turning grey with age. Frond simple, stipitate, sometimes somewhat dimorphous, with the fertile fronds longer and narrower than the sterile fronds.
Stipe up to 1·5 cm. long. Lamina up to 15 × 2·2 cm., narrowly oblong to elliptic,
entire, acute, obtuse or caudate, decurrent at the base; midrib pale, prominent
below, venation mostly obscure. Sori 2–2·5 mm. in diam., circular, in a line on
either side of the midrib about halfway between the midrib and the margin.

Rhodesia. E: Vumba Mts., Burma Valley, 5.xii.1961, *Wild & Chase* 5578 (K;
SRGH). **Mozambique.** Z: Maganja da Costa, 8.ii.1966, *Torre & Correia*, 14503
(LISC). MS: Amatongas Forest, 1.iv.1952, *Chase* 4483 (BM; SRGH). SS: Gaza,
Junod 352 (G). LM: Marracuene, Costa do Sol, 10.viii.1959, *Barbosa & Lemos* 8674
(COI; K; LISC).

Also in S. Africa and widespread in tropical Africa and America. Epiphytic in forest,
from near sea-level to 700 m.

8. MICROSORIUM Link

Microsorium Link, Hort. Berol. **2**: 110 (1833).

Rhizome creeping, with tufted or spaced fronds and with dark lanceolate
rhizome-scales. Frond simple (in African species) to pinnatifid, glabrous, entire,
subsessile to stipitate, articulated; venation reticulate with numerous areoles with
included veinlets. Sori circular, without paraphyses.

A genus of about 40 spp. mainly in the tropics of Asia but with 2 spp. in Africa.

M. scandens (Forst.) Tindale from New Zealand has apparently escaped from cultivation at Vumba Heights in the Umtali Distr. of Rhodesia (*Chase* 6570).

Frond subsessile; sori c. 1 mm. in diam., scattered over the surface of the lamina; lamina
often over 40 cm. long -   -   -   -   -   -   -   -   -   1. *punctatum*
Frond stipitate; sori 1·5–2·5 mm. in diam., occurring from the midrib to half-way to the
margin; lamina usually less than 30 cm. long   -   -   -   -   -   2. *pappei*

1. **Microsorium punctatum** (L.) Copel. in Univ. Calif. Publ. Bot. **16**: 111 (1929).
TAB. **48** fig. A. Type from China.

 Acrostichum punctatum L., Sp. Pl., ed. 2, **2**: 1524 (1763). Type as above.
 Polypodium punctatum (L.) Sw. in Schrad., Journ. Bot. **1800**, 2: 21 (1801). Type
as above.
 Polypodium polycarpon Cav. ex Sw., loc. cit.—Cav., Descr. Pl. **1**: 246 (1802).
Type from Luzon or Marianne Is.
 Polypodium irioides Poir., Encycl. Méth. Bot. **5**: 513 (1804). Type from Mauritius.
 Polypodium lingulatum Sw., Syn. Fil.: 30 (1806) *nom. illegit.* Type from China.
 Niphobolus polycarpus (Cav. ex Sw.) Spreng. in L., Syst. Veg., ed. 16, **4**: 45 (1827).
Type as for *Polypodium polycarpon.*
 Polypodium crassinerve Schumach. in Kongel. Dansk. Vid. Selsk. Afh. **4**: 227
(1829) non Bl. (1828). Type from West Africa.
 Microsorium irregulare Link, Hort. Berol. **2**: 110 (1833). Type of unknown
origin.
 Phymatodes polycarpa (Cav. ex Sw.) C. Presl, Tent. Pterid.: 198, t. 8 fig. 19 (1836)
reimpr. in Abh. Königl. Böhm. Ges. Wiss., Ser. 4, **5**: 198, t. 8 fig. 19 (1837). Type
as for *Polypodium polycarpon.*
 Phymatodes irioides (Poir.) C. Presl, tom. cit.: 198 (1836) reimpr., tom. cit.: 198
(1837). Type as for *Polypodium irioides.*
 Drynaria irioides (Poir.) J. Sm. in Hook., Journ. Bot. **3**: 398 (1841). Type as
above.
 Microsorium irioides (Poir.) Fée, Mém. Fam. Foug. **5**: 268 (1852). Type as for
Polypodium irioides.
 Pleopeltis irioides (Poir.) Moore, Ind. Fil.: 78 (1857). Type as above.
 Drynaria polycarpa (Cav. ex Sw.) Moore, Ind. Fil.: 78 (1857). Type as for
Polypodium polycarpon.
 Pleopeltis polycarpa (Cav. ex Sw.) Moore, loc. cit. Type as for *Polypodium
polycarpon.*
 Pleopeltis punctata (L.) Bedd., Ferns Brit. Ind.: 22 (1876). Type as for *Micro-
sorium punctatum.*
 Colysis irioides (Poir.) J. Sm., Hist. Fil.: 101 (1875). (" erioides "). Type as for
Polypodium irioides.
 Microsorium polycarpon (Cav. ex Sw.) Tardieu, Fl. Madag., Polypod. **2**: 114
(1960). Type as for *Polypodium polycarpon.*

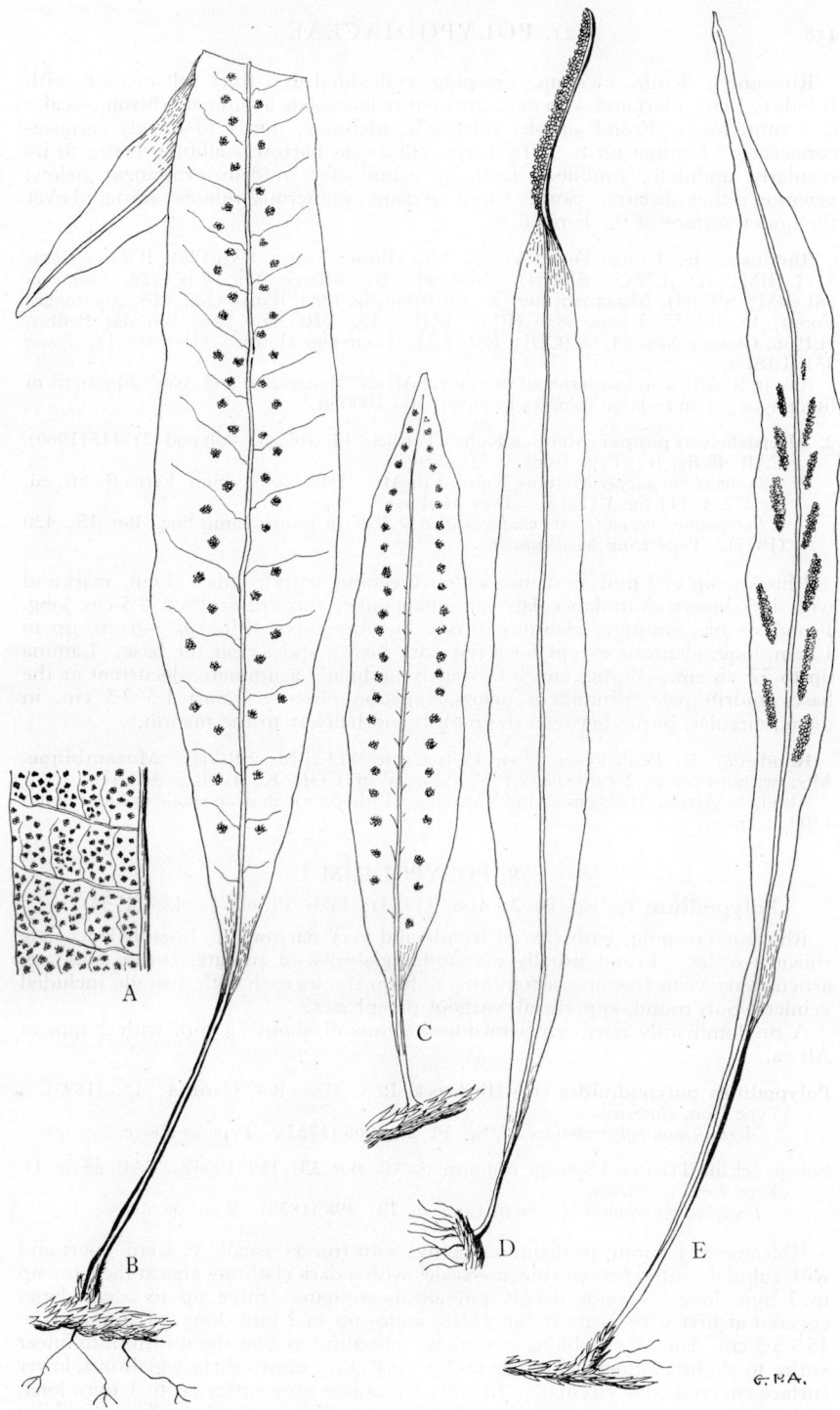

Tab. 48. A.—MICROSORIUM PUNCTATUM. Part of fertile lamina (× ⅓) *Chase 5571.* B.—MICROSORIUM PAPPEI. Fertile frond (× ⅔), *Chase 5613.* C.—MICROGRAMMA LYCOPODIOIDES. Fertile frond (× ⅔) *Chase 4483.* D.—BELVISIA SPICATA. Fertile frond (× ⅔) *Schelpe 5469.* E.—LOXOGRAMMA LANCEOLATA. Fertile frond (× ½) *Chase 6665.*

Rhizome c. 8 mm. in diam., creeping, embedded in a thick felt of roots with fronds c. 1 cm. apart and with dark-grey entire lanceolate acuminate rhizome-scales c. 3 mm. long. Frond simple, subsessile, glabrous, thinly to thickly carnose-coriaceous. Lamina up to 100 × 9 cm., elliptic to narrowly elliptic, entire to irregularly undulate, rounded, acute or acuminate; midrib prominent below, venation rather obscure. Sori c. 1 mm. in diam., numerous, minute, scattered over the under surface of the lamina.

Rhodesia. E: Umtali Distr., Vumba Mts., Burma Valley, 8.xii.1961, *Wild & Chase* 5571 (BM; K; LISC; SRGH). **Malawi.** S: Mlanje Mt., 7.iv.1958, *Chapman* 581 (BM; SRGH). **Mozambique.** Z: Morrumbala, 1864, *Waller* (K). MS: Amatongas Forest, 15.vii.1957, *Chase* 6630 (BM; BOL; K; SRGH). SS: Rio das Pedras, vi.1936, *Gomes e Sousa* 1760 (COI; K). LM: Lourenço Marques (?), 1917–18, *Junod* 351 (LISC).
Also in S. Africa and widespread in tropical Africa, Madagascar and Asia. Epiphytic or lithophytic, often in large colonies in forest, 100–1000 m.

2. **Microsorium pappei** (Mett. ex Kuhn) Tardieu, Fl. Madag., Polypod. **2**: 115 (1960). TAB. **48** fig. B. Type from S. Africa.
　　Polypodium pappei Mett. ex Kuhn, Fil. Afr.: 150 (1868).—Sim, Ferns S. Afr. ed. 2: 277, t. 141 fig. 1 (1915). Type as above.
　　Polypodium normale var. *madagascariense* Bak. in Journ. Linn. Soc., Bot. **15**: 420 (1877). Type from Madagascar.

Rhizome up to 3 mm. in diam., widely creeping, with fronds c. 3 cm. apart and with dark-brown ovate-lanceolate acuminate entire rhizome-scales c. 3·5 cm. long. Frond simple, stipitate, glabrous, firmly membranous. Stipe pale-green, up to 12 cm. long, glabrous except for a few dark-brown scales near the base. Lamina up to 37 × 8 cm., elliptic, entire to weakly undulate, acuminate, decurrent at the base, midrib pale, prominent below, venation obscure. Sori 1·5–2·5 cm. in diam., circular, borne between the midrib and halfway to the margin.

Rhodesia. E: Penhalonga, 27.viii.1950, *Chase* 3174 (BM; SRGH). **Mozambique.** MS: without precise locality, 8.ix.1957, *Chase* 6716 (COI; K; LISC; SRGH).
Also in S. Africa, Madagascar and Tanzania. Lithophytic in deep shade in forest about 1500 m., rare.

9. POLYPODIUM L.

Polypodium L., Sp. Pl. **2**: 1082 (1753); Gen. Pl. ed. 5: 485 (1754).

Rhizome creeping, with spaced fronds and very narrowly to broadly lanceolate rhizome-scales. Frond usually pinnatifid, glabrous or paleate, entire, stipitate, articulated; veins free or anastomosing to form areoles each with a single included veinlet. Sori round, superficial, without paraphyses.
A predominantly northern hemisphere genus of about 75 spp. with 2 spp. in Africa.

Polypodium polypodioides (L.) Hitchcock, Rep. Miss. Bot. Gard. **4**: 156 (1893). Type from America.
　　Acrostichum polypodioides L., Sp. Pl. **2**: 1068 (1753). Type as above.

Subsp. **ecklonii** (Kunze) Schelpe in Journ. S. Afr. Bot. **30**: 189 (1964). TAB. **45** fig. D. Type from S. Africa.
　　Polypodium ecklonii Kunze in Linnaea, **10**: 498 (1836). Type as above.

Rhizome c. 1·5 mm. in diam., creeping, with fronds usually c. 1 cm. apart and with subulate entire brown rhizome-scales with a dark clathrate area at the base, up to 3 mm. long. Fronds deeply pinnatifid, stipitate. Stipe up to 6 cm. long, covered at first with ovate to lanceolate scales up to 2 mm. long. Lamina up to 15 × 5·5 cm., lanceolate-oblong in outline, pinnatifid to near the midrib into linear entire to slightly sinuate lobes up to 2·5 × 0·4 cm.; upper surface glabrous, lower surface covered with circular to broadly lanceolate grey scales up to 1 mm. long, with a dark centre at maturity. Sori 1–1·5 mm. in diam., circular, submarginal, up to 12 per lobe, usually produced in the upper half of the lobe.

Rhodesia. C: Mt. Wedza, 22.v.1968, *Rushworth* 1077 (LMU). E: Chirinda Forest, 21.x.1947, *Wild* 2119a (K). S: Zimbabwe Ruins, 1.vii.1930, *Hutchinson & Gillett* 3342

(K). **Malawi.** C: Dedza, 12.vii.1961, *Chapman* 1424 (BM; BOL). S: Shire High-lands, iv.1906, *Adamson* 10 (K). **Mozambique.** Z: Namúli Mt., R. Licungo, 29.vii.1962, *Schelpe & Leach* 7085 (BOL). T: Zóbuè, 20.vii.1949, *Barbosa & Carvalho* 3729 (LISC). MS: Gorongosa Mt., Morambodzi Falls, 3.vii.1955, *Schelpe* 5470 (BM; BOL).

Also in S. Africa and Tanzania. Epiphytic in forest, 900–1500 m.

10. BELVISIA Mirb.

Belvisia Mirb. in Lam. & Mirb., Hist. Nat. Pl. **4**: 65 (1803).

Rhizome creeping, often embedded in a mass of roots, with closely spaced fronds and with clathrate scales. Fronds simple, usually narrowly elliptic, entire, glab-rous, stipitate, articulated; venation reticulate with included veinlets. Sporangia acrostichoid borne on a linear segment at the apex of the frond.

A genus of about 15 spp. in the Old World tropics with 1 sp. in Africa.

Belvisia spicata (L.f.) Mirb. in Lam. & Mirb., Hist. Nat. Vég. **5**: 111 (1803). TAB. **48** fig. D. Type from Mauritius.

Acrostichum spicatum L.f., Suppl. Pl.: 444 (1781). Type as above.
Hymenolepis spicata (L.f.) C. Presl, Epim. Bot.: 159 (1851) reimpr. in Abh. Königl. Böhm. Ges. Wiss. Ser. 5, **6**: 519 (1851). Type as above.
Hymenolepis spicata var. *occultivenia* C. Chr. in Dansk Bot. Ark. **6**: 57 (1929). Syntypes from Madagascar, Seychelles and S. Tomé.
Hymenolepis spicata var. *usambarensis* C. Chr., loc. cit. Type from Tanzania.

Rhizome c. 4 mm. in diam., shortly creeping, with tufted fronds and with dark-brown lanceolate acute-acuminate entire clathrate rhizome-scales c. 2·5 mm. long with a pale border. Frond simple, stipitate, with a linear fertile segment at the apex. Stipe up to 2 cm. long, glabrous. Lamina up to 18 × 2·3 cm., narrowly elliptic, entire, tapering towards the base; midrib pale and prominent below. Fertile segment up to 10 × 0·4 cm., linear; sori acrostichoid.

Rhodesia. E: Inyanga, Nyamingura R., Aberfoyle Tea Estate, 26.x.1959, *Chase* 7196 (BOL; K; SRGH). **Mozambique.** Z: Namúli Mt., R. Licungo, 29.vii.1962, *Schelpe & Leach* 7082 (BOL). MS: Gorongosa Mt., Morambodzi Falls, 2.vii.1955, *Schelpe* 5469 (B; BM; BOL; K; P).

Also in Tanzania, Cameroon, S. Tomé, Ivory Coast, Gabon, Madagascar, Seychelles, Comoro and Mascarene Is. A rare epiphyte or lithophyte in forest at about 900 m.

22. DAVALLIACEAE

Terrestrial, lithophytic or epiphytic plants. Rhizome erect or creeping, with tufted or spaced fronds, often producing perennating tubers, and with peltate rhizome-scales. Stipe with several vascular strands, articulated or not, in the latter case with articulated pinnae. Frond simple, pinnate or much dissected. Sori superficial or terminal on veins; indusium usually opening towards the margin. Spores monolete, without perispore.

Pinnae articulated to the rhachis; fronds pinnate to 2-pinnatifid:
Stipe not articulated; frond-bearing rhizome erect; plants often with tubers
  1. Nephrolepis
Stipe articulated; frond-bearing rhizome creeping; plants never with tubers
  2. Arthropteris

Fronds simple, or pinnae not articulated to the rhachis:
Frond simple; sori superficial on veins - - - - - **3. Oleandra**
Frond much dissected; sori terminal on veins - - - - **4. Davallia**

1. NEPHROLEPIS Schott

Nephrolepis Schott, Gen. Fil.: t. 3 (1834).

Rhizomes short, erect, sometimes stoloniferous and tuber-forming, with tufted fronds and with brown rhizome-scales. Stipe not articulated. Frond pinnate, pinnae articulated to the rhachis, hairy to subglabrous; veins free. Sori terminal on veins, circular and intramarginal to elongate-submarginal, with reniform to elongate indusia.

A genus of c. 35 spp., mostly pantropical.

The widely cultivated *Nephrolepis duffii* Moore, with small circular pinnae, has apparently escaped from cultivation around Rio das Pedras, Inhambane Distr. of Mozambique (*Gomes e Sousa* 1775 (COI)).

Sori linear, submarginal - - - - - - - - -   2. *acutifolia*
Sori circular or lunate, intramarginal:
  Pinnae narrowly oblong, markedly auriculate; sori opening towards the pinna apex;
    plant tuberous - - - - - - - - - - - 3. *undulata*
  Pinnae linear, not markedly auriculate; sori opening towards the pinna margin; plant
    not tuberous - - - - - - - - - - 1. *biserrata*

1. **Nephrolepis biserrata** (Sw.) Schott, Gen. Fil.: sub. t. 3 (1834).—Sim, Ferns S. Afr. ed. 2: 123, t. 35 (1915).—Tardieu, Fl. Camér.: 69, t. 7 fig. 11–12 (1964); Fl. Gabon: 53, t. 9 fig. 11–12 (1964). Type from Mauritius.
    Aspidium biserratum Sw. in Schrad., Journ. Bot. **1800**, 2: 32 (1801). Type as above.
    Aspidium acutum Schkuhr, Kr. Gewächse, **1**: 32, t. 31 (1806). Origin unknown.
    Aspidium splendens Willd. in L., Sp. Pl. ed. 4, **5**: 220 (1810). Type from Ceylon.
    Nephrodium biserratum (Sw.) C. Presl, Rel. Haenk. **1**: 31 (1825). Type as for *Nephrolepis biserrata*.
    Nephrodium acutum (Schkuhr) C. Presl, loc. cit. Origin unknown.
    Nephrodium splendens (Willd.) Desv. in Mém. Soc. Linn. Par. **6**, 2: 253 (1827). Type as for *Aspidium splendens*.
    Aspidium guineense Schumach. in Kongel. Dansk. Vid. Selsk. Afh. **4**: 229 (1829). Type from W. Africa.
    Hypopeltis biserrata (Sw.) Bory, Bélanger Voy. Bot. **2**: 65 (1833). Type as for *Nephrolepis biserrata*.
    Nephrolepis splendens (Willd.) C. Presl, Tent. Pterid.: 79 (1836) reimpr. in Abh. Königl. Böhm. Ges. Wiss., Ser. 4, **5**: 79 (1837). Type as for *Aspidium splendens*.
    Nephrolepis acuta (Schkuhr) C. Presl, loc. cit. reimpr. loc. cit. Origin unknown.
    Lepidoneuron biserratum (Sw.) Fée, Mém. Fam. Foug. **5**: 301 (1852). Type as for *Nephrolepis biserrata*.
    Nephrolepis punctulata var. *hirsuta* Mett. ex Kuhn, Fil. Afr.: 156 (1868). Type from W. Africa.
    Nephrolepis caudata Christ in Ann. Mus. Cong., Bot., Sér. 5, **3**: 27 (1909). Type from Congo.

Rhizome erect. Frond suberect to arching. Stipe pale brown, up to 22 cm. long, with narrowly lanceolate pale-brown scales up to 2 mm. long when young becoming subglabrous with age. Lamina 60 × 24 cm., pinnate, narrowly elliptic in outline, acute with the lower pinnae only slightly reduced. Pinnae up to 13 × 1·8 cm., shortly petiolate, up to 36 pairs, very narrowly oblong, attenuate, crenate with broadly cuneate base, thinly pilose with minute white hairs on both surfaces when young, becoming glabrous with age; submarginal hydathodes present but not conspicuous. Rhachis pale-brown, thinly pilose with minute white hairs and with scattered very narrowly lanceolate pale-brown scales. Sori c. 1 mm. in diam., circular, discrete, c. 4 mm. apart in a line $\frac{2}{3}$ the distance from the costa to the margin on either side of the costa, opening outwards at right angles to the veins. Indusium c. 0·6 mm. in diam., membranous, entire.

Zambia. N: Mporokosa Distr., L. Chisi, 24.ix.1956, *Richards* 6279 (K). **Rhodesia.** E: between Haroni and Makurupini rivers, 10.i.1969, *Biegel* 2779 (BOL; SRGH). **Mozambique.** SS: R. das Pedras, vi.1936, *Gomes e Sousa* 1776 (COI; K) (bifurcate. form). LM: Marracuene, Rikatla, vii.1917, *Junod* 1 (G; LISC). Pantropical and also in S. Africa. Wet shaded localities in forest, 10–1220 m.

2. **Nephrolepis acutifolia** (Desv.) Christ in Verh. Nat. Ges. Basel, **11**: 243 (1895). TAB. 49 fig. B. Type from Mascarene Is.
    Lindsaea acutifolia Desv. in Mém. Soc. Linn. Par. **6**, 2: 312 (1827). Type as above.
    Isoloma lanuginosa J. Sm. in Hook., Gen. Fil.: t. 102 (1842). Type from E. Indies.
    Lindsaea lanuginosa (J. Sm.) Hook., Sp. Fil. **1**: 210, t. 69 fig. A (1846). Type a above.

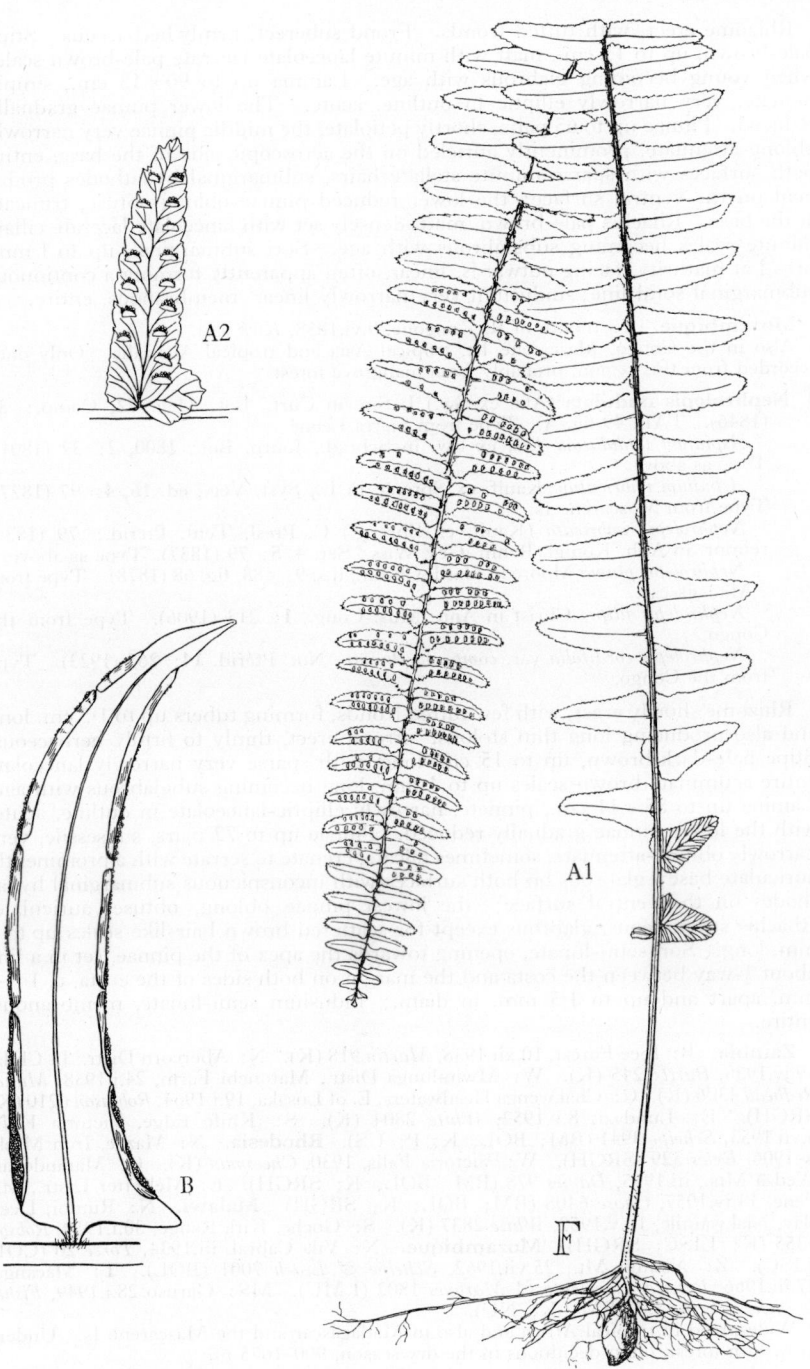

Tab. 49. A.—NEPHROLEPIS UNDULATA. A1, habit (× ⅔); A2, fertile pinna (× 2), both from *Fanshawe* 3005. B.—NEPHROLEPIS ACUTIFOLIA. Fertile pinna (× 2) *Lebrun* 6559.

Rhizome erect, with tufted fronds. Frond suberect, firmly herbaceous. Stipe pale-brown, up to 15 cm., matt with minute lanceolate lacerate pale-brown scales when young becoming glabrous with age. Lamina up to 90 × 13 cm., simply pinnate, very narrowly elliptic in outline, acute. The lower pinnae gradually reduced. Pinnae up to 65 pairs, shortly petiolate, the middle pinnae very narrowly oblong-attenuate, prominently auricled on the acroscopic side of the base, entire both surfaces with sparse minute stellate hairs, submarginal hydathodes prominent on the ventral surface; the lower reduced pinnae oblong-obtuse, truncate at the base. Rhachis pale-brown, matt, densely set with lanceolate lacerate ciliate minute scales becoming subglabrous with age. Sori submarginal, up to 1 mm. broad at maturity, facing outwards, linear, often apparently forming a continuous submarginal soral line; indusium very narrowly linear, membranous, entire.

Mozambique. Z: mouth of R. Kongone, 9.vi.1858, *Kirk* (K).

Also in the Congo, Mascarene Is., tropical Asia and tropical Australia. Only once recorded from this region, probably from mangrove forest.

3. **Nephrolepis undulata** (Afz. ex Sw.) J. Sm. in Curt., Bot. Mag. **72,** Comp.: 37 (1846). TAB. **49** fig. A. Type from Sierra Leone.

   Aspidium undulatum Afz. ex Sw. in Schrad., Journ. Bot., **1800,** 2: 32 (1801). Type as above.

   Aspidium imbricatum Kaulf. ex Spreng. in L., Syst. Veg., ed. 16, **4**: 97 (1827). Type from Mascarene Is.

   Nephrolepis imbricata (Kaulf. ex Spreng.) C. Presl, Tent. Pterid.: 79 (1836) reimpr. in Abh. Königl. Böhm. Ges. Wiss., Ser. 4, **5**: 79 (1837). Type as above.

   Nephrolepis pluma Moore in Gard. Chron., n.s. **9**: 588, fig. 68 (1878). Type from ? Madagascar.

   Nephrolepis filipes Christ in Ann. Mus. Cong., **1**: 213 (1906). Type from the Congo.

   Nephrolepis cordifolia var. *compacta* Bonap., Not. Ptérid. **14**: 265 (1923). Type from the Congo.

Rhizome shortly erect, with few tufted fronds, forming tubers up to 1·5 cm. long and also producing long thin stolons. Frond erect, thinly to firmly herbaceous. Stipe pale-dark-brown, up to 15 cm. long, with sparse very narrowly lanceolate entire acuminate brown scales up to 4 mm. long becoming subglabrous with age. Lamina up to 85 × 11 cm., pinnate, narrowly elliptic-lanceolate in outline, acute, with the lower pinnae gradually reduced. Pinnae up to 72 pairs, subsessile, very narrowly oblong-attenuate, sometimes falcate, crenate to serrate with a prominently auriculate base; glabrous on both surfaces with inconspicuous submarginal hydathodes on the ventral surface; the lowest pinnae oblong, obtuse, auriculate. Rhachis stramineous, glabrous except for scattered brown hair-like scales up to 2 mm. long. Sori semi-lunate, opening towards the apex of the pinnae, set in a line about ½-way between the costa and the margin on both sides of the costa, c. 1·5–3 mm. apart and up to 1·5 mm. in diam.; indusium semi-lunate, membranous, entire.

Zambia. B: Ipee Forest, 10.xii.1938, *Martin* 918 (K). N: Abercorn Distr., L. Chila, 19.iv.1936, *Burtt* 6245 (K). W: Mwinilunga Distr., Matonchi Farm, 24.i.1938, *Milne-Redhead* 4306 (K). C: Chakwenga Headwaters, E. of Lusaka, 19.i.1964, *Robinson* 6210 (K; SRGH). E: Lundazi, 8.v.1952, *White* 2804 (K). S: Knife Edge, Victoria Falls, 6.vii.1953, *Schelpe* 3941 (BM; BOL; K; P; US). **Rhodesia.** N: Mazoe, Iron Mask, iv.1906, *Eyles* 329 (SRGH). W: Victoria Falls, 1930, *Cheesman* (K). C: Marandellas, Wedza Mts., iii.1955, *Davies* 978 (BM; BOL; K; SRGH). E: Melsetter Distr., Mt. Pene, 14.iv.1957, *Chase* 6408 (BM; BOL; K; SRGH). **Malawi.** N: Rumpi, Deep Bay, Njakwamile, 12.v.1952, *White* 2837 (K). S: Goche, Kirk Range, 30.i.1959, *Robson* 1355 (K; LISC; SRGH). **Mozambique.** N: Vila Cabral, iii.1934, *Torre* 20 (COI; LISC). Z: Namúli Mt., 25.vii.1962, *Schelpe & Leach* 7001 (BOL). T: Macanga, 17.iii.1966, *Pereira, Sarmento & Marques* 1802 (LMU). MS: Garuso 28.i.1949, *Fisher & Schweickerdt* 535 (BM; K; NU).

Widespread in tropical Africa and also in Madagascar and the Mascarene Is. Undergrowth in forest, often deciduous in the dry season, 900–1675 m.

2. ARTHROPTERIS J. Sm.

Arthropteris J. Sm. in Hook. f., Fl. Nov. Zel. **2,** 2: 43, t. 82 (1854).

Rhizome creeping, solenostelic, with spaced fronds and with brown peltate rhizome-scales. Stipe articulated. Fronds pinnate to 2-pinnatifid; pinnae articulated to the rhachis; veins free. Sori circular, terminal on veins, intra-marginal; indusium reniform.

A widely distributed genus of c. 20 spp., of which 4 occur in continental Africa.

Stipe articulated in lower half; white dots absent on ventral pinna surface 1. *monocarpa*
Stipe articulated in the upper half; white dots usually present on the ventral pinna surface
2. *orientalis*

1. **Arthropteris monocarpa** (Cordem.) C. Chr. in Acad. Malgache, Cat. Pl. Madag., Ptérid.: 32 (1932). TAB. **50**. Type from Madagascar.
   Nephrodium monocarpum Cordem. in Bull. Soc. Sci. Art. Réunion, **1890-91**: 186 (1891). Type as above.

Rhizome c. 2 mm. in diam. when dry, widely creeping, with widely spaced fronds and with subcircular to broadly ovate peltate brown entire rhizome-scales up to 1·5 mm. long. Frond arching, thinly herbaceous. Stipe up to 18 cm. long, articulated near the base, pale-brown, glabrous at maturity. Lamina up to 33 × 12 cm., oblong-lanceolate, acute, the basal pinnae somewhat reduced, deeply 2-pinnatifid. Pinnae up to 7·5 × 1·5 cm., oblong-lanceolate, attenuate, very broadly oblong at the base, deeply pinnatifid into narrowly oblong obtuse undulate crenate lobes up to 1 × 0·3 cm., very thinly pubescent on the costa, costules and veins dorsally. Rhachis stramineous, pubescent with minute pale-brown hairs. Sori usually solitary in each lobe but if 2–3 then occurring along the acroscopic margin of each lobe, circular, up to 1·5 mm. in diam.; indusium c. 1 mm. in diam., membranous, entire.

Rhodesia. N: Mt. Binga, 29.i.1966, *Pereira, Sarmento & Marques* 47 (LMU). C: Salisbury, *Holland* (NBG). E: Umtali, Stapleford Forest Reserve, 29.iv.1952, *Chase* 4499 (BM; SRGH). **Malawi.** N: Nyika Plateau, 19.viii.1946, *Brass* 17336 (SRGH). C: Dedza Mt., 18.v.1963, *Chapman* 2051 (SRGH). S: Zomba Mt., 9.iii.1955, *E.M. & W.* 767 (LISC; SRGH). **Mozambique.** Z: Namúli Mt., 29.vii.1962, *Schelpe & Leach* 7091 (BOL). MS: Gorongosa Mt., Gogogo Peak, 5.vii.1955, *Schelpe* 5511 (BM; BOL).

Also in S. Africa and widespread in tropical Africa, Madagascar, Comoro Is. and Réunion. Lithophytic or epiphytic in shade in forest, 825–1770 m.

2. **Arthropteris orientalis** (J. F. Gmel.) Posthumus in Rec. Trav. Bot. Néerl. **21**: 218 (1924). Origin unknown.
   Polypodium pectinatum Forsk., Fl. Aegypt.–Arab.: CXXV, 185 (1775) non L. (1753). Origin unknown.
   Polypodium orientale J. F. Gmel. in L., Syst. Nat., ed. 13, **2**: 1312 (1791). Type as for *Arthropteris orientalis*.
   Aspidium albopunctatum Bory ex Willd. in L., Sp. Pl. ed. 4, **5**: 242 (1810). Type from Bourbon I. (Réunion).
   Nephrodium albopunctatum (Bory ex Willd.) Desv. in Mém. Soc. Linn. Par. **6, 2**: 255 (1827). Type as above.
   Aspidium thonningii Schumach. in Kongel. Dansk. Vid. Selsk., Afh. **4**: 229 (1829). Type from Guinée.
   Aspidium leucosticton Kunze in Linnaea, **23**: 227, 301 (1850). Type from Sierra Leone.
   Dryopteris orientalis (J. F. Gmel.) C. Chr., Ind. Fil.: 281 (1905). Type as for *Arthropteris orientalis*.

Rhizome up to 3 mm. in diam. when dried, widely creeping, frequently producing short lateral axes with fronds variously spaced and with brown peltate broadly ovate obtuse entire rhizome-scales up to 1·5 mm., darker around the point of attachment. Frond erect, seldom arching, firmly herbaceous to thinly coriaceous. Stipe stramineous, pale- to dark-brown, up to 26 cm. long, matt, glabrous. Lamina up to 40 × 17 cm., deeply 2-pinnatifid, lanceolate obtuse in outline. Basal pinnae somewhat or not reduced. Pinnae up to 9·5 × 1·8 cm., very narrowly oblong-attenuate, very broadly cuneate at the base, deeply pinnatifid into narrowly oblong obtuse entire lobes, very thinly pubescent on the lamina and densely pubescent on the costa, ventrally less so, on the costules, veins and lamina dorsally; lobes up to 9 × 3·5 mm. Rhachis stramineous to dark-brown, densely pubescent with minute white hairs. Sori c. 1 mm. in diam., up to 9 per lobe, circular, borne about ½-way between the costules and the margin; indusium c. 0·6 mm. in diam., glabrous, entire.

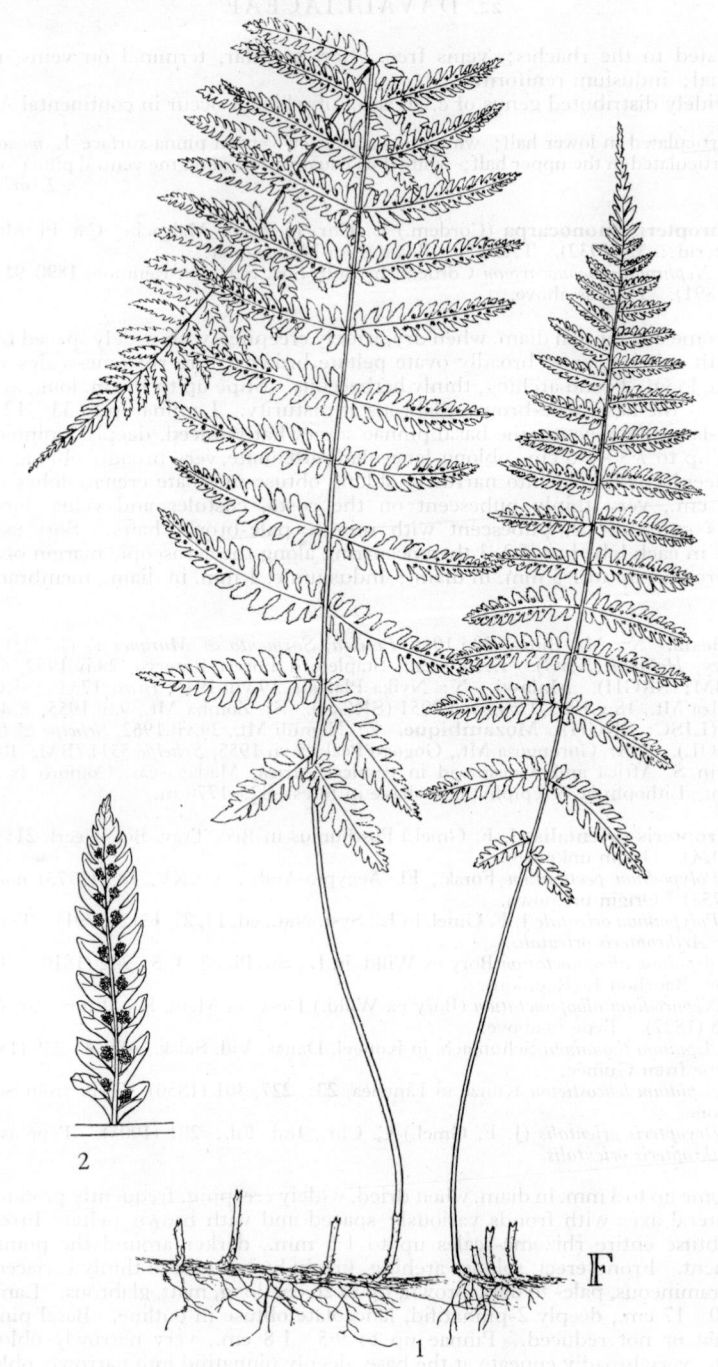

Tab. 50. ARTHROPTERIS MONOCARPA. 1, habit (× ⅔); 2, fertile pinna (×2), both from
Milne-Redhead 3854.

Zambia. N: Kasama Distr., Malike Rocks, Kasama, 2.iii.1960, *Richards* 12702 (K). W: Solwezi Distr., Chifubwa, 14.vi.1930, *Milne-Redhead* 497 (K). C: Mt. Makulu, 14.iv.1956, *Robinson* 1467 (K; SRGH). S: Choma, 5.i.1957, *Robinson* 2020 (K; SRGH). **Rhodesia.** N: Darwin, Umsengesi Camp, 9.v.1955, *Whellan* 882 (BM; SRGH). W: Between Shangani and Fort Rixon, 15.iv.1943, *Feiertag* in GHS 45435 (SRGH). C: Marandellas, 13.iv.1948, *Corby* 82 (BM; SRGH). E: Melsetter Distr., Mt. Pene, 14.iv.1957, *Chase* 6406 (BM; K; SRGH). S: Kyle Dam, 32 km. SE. of Fort Victoria, 21.ii.1964, *Mitchell* 792 (LISC; SRGH). **Malawi.** N: Vipya, 61 km. S. of Mzuzu, 16.iv.1967, *Pawek* 980 (BOL; SRGH). S: Ncheu Distr., Chirobwe, 18.iii.1955, *E.M. & W.* 1009 (LISC; SRGH). **Mozambique.** N: Ribáuè Mt., 19.vii.1962, *Schelpe & Leach* 6942 (BOL). Z: Namúli Mt., 25.vii.1962, *Schelpe & Leach* 6997 (BOL). T: Moatize, Zóbuè Mt., 12.i.1966, *Correia* 411 (LISC). MS: Gorongosa Mt., Morambodzi Falls, 2.vii.1955, *Schelpe* 5466 (BM; BOL).

Also widespread in tropical Africa, Madagascar and the Mascarene Is. Terrestrial in shade of woodland or scrub, 825–1740 m.

3. OLEANDRA Cav.

Oleandra Cav., Ann. Hist. Nat. **1**: 115 (1799).

Rhizome long, creeping, dictyostelic, with closely or widely spaced fronds and with brown attenuate peltate rhizome-scales; roots sparingly produced at wide intervals. Stipe articulated. Fronds simple, entire, uniform (rarely dimorphic); veins free. Sori superficial on the veins, in a row on each side of the midrib; indusium reniform.

A genus of c. 40 tropical spp., of which 4 occur in continental Africa.

Oleandra distenta Kunze in Bot. Zeit. **1851**: 347 (1851). TAB. **51.** Type from S. Africa.

> *Aspidium articulatum* Willd. in L., Sp. Pl., ed. 4, **5**: 212 (1810) non Sw. (1801). Type from Mauritius.
> *Oleandra densifrons* Kunze, loc. cit. Type from S. Africa.
> *Oleandra articulata* var. *welwitschii* Bak., Syn. Fil.: 303 (1867). Type from Angola.
> *Oleandra africana* Bonap., Not. Ptérid. **14**: 257 (1923). Type from Uganda (Ruwenzori).
> *Oleandra distenta* var. *hirsuta* Tardieu in Notul. Syst. **14**: 333 (1953). Type from Ivory Coast.
> *Oleandra distenta* var. *villosa* Tardieu in Mém. Inst. Fr. Afr. Noire, **28**: 157 (1953). Syntypes from W. Africa.

Rhizome up to 4 mm. in diam., very widely creeping, sometimes producing short side branches with fronds spaced or clustered and with appressed (rarely squarrose) narrowly lanceolate attenuate variously ciliate light to medium brown rhizome-scales c. 5 mm. long with a darker area around the point of attachment. Frond erect or arching, thinly membranous to thinly coriaceous, articulated, deciduous. Stipe up to 4·5 cm. long, stramineous, with or without lanceolate ciliate scales articulated near the base. Lamina up to 33 × 6·5 cm., simple, usually with a caudate apex and a broadly cuneate base, subentire to undulate with the ventral surface glabrous at maturity and the dorsal surface glabrous or thinly pubescent with minute white hairs and with or without broadly lanceolate ciliate pale-brown hairs along the costa. Costa prominent below, stramineous to light castaneous, veins free. Sori up to 2 mm. in diam., circular, set in an irregular line on either side of the costa, nearer the costa than the margin; indusia up to 1·5 mm. in diam., glabrous to glandular, brown, entire.

Zambia. N: Abercorn Distr., Vomo Gorge near Kawimbe, 2.ix.1960, *Richards* 13183 (K). W: Mwinilunga Distr., Luakera Falls, 25.i.1938, *Milne-Redhead* 4335 (K). **Rhodesia.** N: Mt. Binga, 29.i.1966, *Pereira, Sarmento & Marques* 49 (LMU). E: Chimanimani Mts., Stonehenge, 9.v.1958, *Chase* 6912 (BM; BOL; K; SRGH). **Malawi.** N: Rumpi Distr., Nyika Plateau, Rukuru R. Falls, 6.i.1959, *Richards* 10532 (K; SRGH). S: Mlanje Distr., Simpson's Peak, 30.iii.1960, *Phipps* 2788 (BOL; K; SRGH). **Mozambique.** Z: Namúli Mt., 26.vii.1962, *Schelpe & Leach* 7044 (BOL). MS: Gorongosa Mt., 4.vii.1955, *Schelpe* 5491 (BM; BOL).

Widespread in tropical Africa and also in S. Africa. Lithophytic or epiphytic in forest, usually deciduous during the dry season, 1370–1850 m.

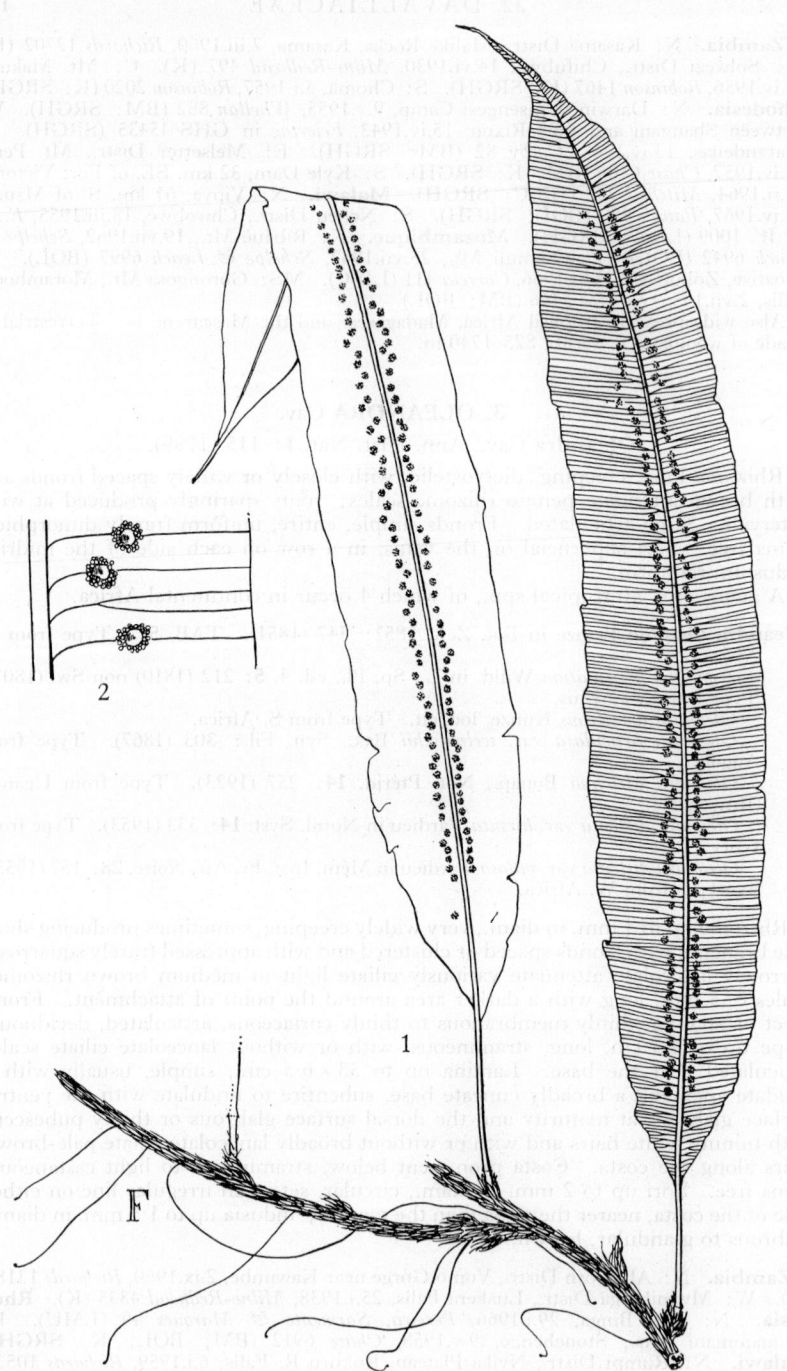

Tab. 51. OLEANDRA DISTENTA. 1, habit (× ⅔); 2, enlargement of fertile frond (× 2), both from *Milne-Redhead* 4335.

4. DAVALLIA Sm.

Davallia Sm. in Mem. Acad. Turin, **5**: 414 (1793).

Rhizome creeping, dictyostelic, with spaced fronds and with attenuate peltate ciliate rhizome-scales. Stipe articulated. Fronds much dissected, lamina triangular to ovate in outline, coriaceous, glabrous; veins free. Sori terminal on veins; indusium elongate, fused to the lamina at the base and sides.

A predominantly Asiatic and Polynesian genus of c. 40 spp., of which only 1 occurs in tropical and S. Africa.

Davallia chaerophylloides (Poir.) Steud., Nomencl. Bot. Pl. Crypt.: 146 (1824) TAB. **52.** Type from Madagascar.

    Trichomanes chaerophylloides Poir., Encycl. Bot. **8**: 80 (1808). Type as above.
    Humata chaerophylloides (Poir.) Desv. in Mém. Soc. Linn. Par. **6**, 2: 325 (1827). Type as above.
    Davallia nitidula Kunze in Linnaea, **10**: 545 (1836). Type from S. Africa.
    Davallia vogelii Hook., Sp. Fil. **1**: 169, t. 59 (1868). Type from Fernando Po.
    Davallia denticulata var. *intermedia* Mett. ex Kuhn, Fil. Afr.: 158 (1868). Syntypes from Madagascar and S. Africa.
    Davallia schnellii Tardieu in Notul. Syst. **13**: 372 (1948). Type from Guinée.

Rhizome up to 1·5 cm. in diam., widely creeping, with widely spaced fronds and with brown fimbriate narrowly lanceolate attenuate hair-pointed scales up to 8 mm. long with pale margins. Frond arching, thinly coriaceous. Stipe up to 44 cm., glabrous at maturity, with tufted scales similar to those on the rhizome about the extreme base. Lamina up to 70 × 50 cm., deeply 4–5-pinnatifid, ovate-triangular in outline, acute, with the basal pinnae the largest and developed basiscopically. Pinnae triangular to oblong in outline, acute-acuminate; ultimate segments narrowly triangular to trapeziform, incised into spathulate to cuneate lobes up to 1–1·5 mm. broad, glabrous on both surfaces. Rhachis and secondary rhachises stramineous, glabrous. Sori up to 1 mm. in diam., solitary in the lobes on vein endings, subtended by blunt or sharp teeth; indusium very broadly oblong, membranous, entire.

Rhodesia. E: Umtali, Vumba Mts., Monkfield, 9.ii.1950, *Chase* 2190 (BM; SRGH).
Mozambique. Z: Ile, Errego, 3.iii.1966, *Torre & Correia* 14973 (LISC). MS: Cheringoma, Serração de Durúndi, 19.vii.1941, *Torre* 3047 (BM; COI; LISC; LMA). LM: Maputo, Salamango, Forest of Magala, 27.xi.1947, *Mendonça* 3542 (LISC).
Also in S. Africa, widespread in tropical Africa, in Madagascar, Comoro Is. and the Seychelles. Epiphytic on fallen tree trunks or lithophytic on boulders in shade in forest, from near sea-level to 1400 m.

23. ASPLENIACEAE

Epiphytic or terrestrial plants with creeping or erect rhizomes and with clathrate usually dark-coloured rhizome-scales. Fronds simple or variously pinnately dissected; stipes not articulated to the rhizome and with 2 vascular strands at the base which unite upwards to form a 4-armed strand; veins free or anastomosing marginally. Sori usually linear, borne on the costal side of a vein and with a narrow or obsolete indusium. Sporangia with long stalks; spores monolete, with a perispore.

Under surface of pinnae glabrous or with scattered small scales -   -   1. **Asplenium**
Under surface of pinnae with dense imbricate scales   -   -   -   2. **Ceterach**

1. ASPLENIUM L.

Asplenium L., Sp. Pl. **2**: 1078 (1753); Gen. Pl. ed. **5**: 485 (1754).

Rhizome erect or creeping, dictyostelic, with clathrate scales. Stipe matt or nitidous, black, castaneous or greyish-green, glabrous, or with hairs, or with clathrate scales. Frond simple to 4-pinnatifid, glabrous, pubescent or with

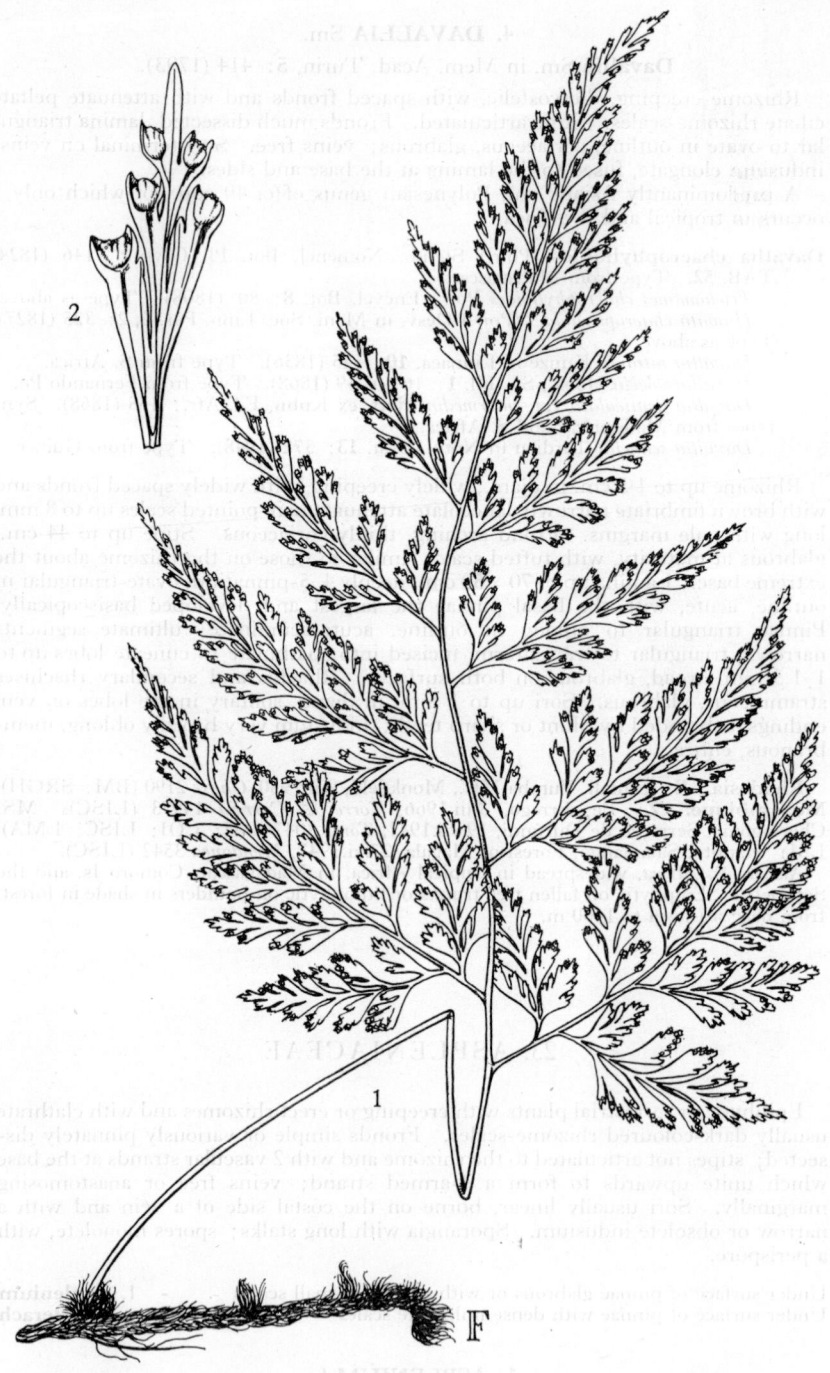

Tab. 52. DAVALLIA CHAEROPHYLLOIDES. 1, habit (× ⅔); 2, fertile ultimate segment (× 4), both from *Chase* 2190.

scattered scales, veins pinnate or flabellate, free.   Sori usually elongate, but about
as long as broad in some spp., borne on the costal side of a vein; indusium narrow.
 A cosmopolitan genus of over 600 spp.

Frond simple, entire -   -   -   -   -   -   -   -   -   -   1. *holstii*
Frond pinnate to 3-pinnate:
  Plant stoloniferous with naked stolons -   -   -   -   -   -   -   37. *mannii*
  Plant not stoloniferous:
    Frond pinnate to 2-pinnatifid:
      Frond gemmiferous near the apex:
        Proliferating bud at the end of an extension of the rhachis   -   33. *sandersonii*
        Proliferating bud on the rhachis on or below the terminal segment:
          Rhachis pubescent below   -   -   -   -   -   -   20. *protensum*
          Rhachis not pubescent below:
            Pinnae incised more than ½-way to the costa:
              Apical segment of the frond above the proliferating bud with less than
                5 lobes   -   -   -   -   -   -   -   -   34. *dregeanum*
              Apical segment of the frond above the proliferating bud with more than
                5 lobes   -   -   -   -   -   -   35. *pseudoauriculatum*
            Pinnae entire, crenate or dentate:
              Stipe and rhachis black   -   -   -   -   -   -   24. *torrei*
              Stipe and rhachis greyish-green or brown:
                Fronds dimorphous, the fertile longer than the sterile; pinnae apices
                rounded to acute:
                  Pinnae of fertile fronds gradually becoming smaller and more distant
                  towards the apex   -   -   -   -   -   -   -   3. *christii*
                  Pinnae of fertile fronds not gradually becoming smaller or more
                  distant towards the apex   -   -   -   -   4. *mossambicense*
                Fronds not dimorphous; pinna apices acuminate, attenuate or
                caudate:
                  Pinna margin shallowly crenate to subentire; terminal pinna similar
                  in size and shape to the subterminal pinna   -   6. *gemmiferum*
                  Pinna margin incised with each lobe 2-fid at least in the lower part
                  of the pinnae; terminal pinna much larger than the subterminal
                  pinnae -   -   -   -   -   -   -   -   -   5. *boltonii*
      Frond not gemmiferous near the apex:
        Pinnae dimidiate (with the basiscopic side of the pinna bordered by the costa for
        most of its length):
          Stipe and rhachis castaneous   -   -   -   -   -   9. *unilaterale*
          Stipe and rhachis matt, greyish-green   -   -   -   -   8. *obscurum*
        Pinnae not dimidiate:
          Rhachis and stipe castaneous; fronds tufted:
             Pinnae narrowly trapeziform in outline:
              Pinnae longly caudate   -   -   -   -   -   25. *megalura*
              Pinnae acute, not caudate -   -   -   -   -   -   31. *chaseanum*
            Pinnae not narrowly trapeziform in outline or long-caudate:
              Pinnae deeply incised acroscopically into simple or 2-fid lobes; rhachis
              with narrow pale-brown wings   -   -   -   -   22. *formosum*
             Pinnae crenate or dentate:
              Pinnae subcircular   -   -   -   -   -   -   10. *trichomanes*
             Pinnae oblong:
                Pinnae with only 1 or 2 sori on the basiscopic side of the costa
                  11. *monanthes*
                Pinnae with 3 or more sori on the basiscopic side of the costa
                  12. *normale*
         Rhachis matt-green or greyish-green, or if almost castaneous the rhizome
          creeping:
           Rhizome creeping; fronds widely spaced; sori borne very close to the costa
             19. *friesiorum*
           Rhizome erect; fronds tufted; sori borne at an angle to the costa:
             Frond apex with a terminal pinna resembling the middle pinnae:
              Pinnae attenuate; pinna margin regularly crenate -   2. *anisophyllum*
              Pinna caudate to abruptly acute-acuminate; pinna margin irregu-
              larly undulate   -   -   -   -   -   -   7. *atroviride*
            Frond apex with gradually decrescent pinnae:
              Venation of pinnae flabellate (with no distinct costa):   23. *ramlowii*
              Venation of pinnae pinnate (with a distinct costa):
                  Stipe and rhachis set with numerous black hair-like scales
                  21. *pellucidum*

Stipe bases set with scales similar to those on the rhizome but
  rhachis glabrous:
  Lamina not or hardly decrescent at the base:
    Stipe black on the under surface; pinnae deeply incised
      towards the base -   -   -   -   -   17. *auritum*
    Stipe matt-grey or green; pinnae serrate to crenate
                                      14. *inaequilaterale*
  Lamina gradually tapering towards the base -   - 13. *erectum*
Frond 2-pinnate to 4-pinnatifid:
  Sori submarginal, solitary on the ultimate segments of the frond:
    Sori almost terminal on the ultimate segments   -   -   39. *theciferum*
    Sori lateral on the ultimate segments:
      Sori more than twice as long as broad   -   -   -   36. *rutifolium*
      Sori about as long as broad   -   -   -   -   38. *hypomelas*
  Sori intramarginal, 2 or more on the ultimate segments:
    Lamina, rhachis and stipe thinly pubescent with white hairs -   18. *pumilum*
    Lamina glabrous or set with small clathrate scales:
      Frond gemmiferous near the apex -   -   -   -   - 30. *blastophorum*
      Frond not gemmiferous, or proliferating only at the base of the lamina:
        Lamina 4-pinnatifid:
          Rhizome erect; ultimate lobes acute -   -   -   -   15. *lobatum*
          Rhizome creeping; ultimate lobes sharply serrate:
            Ultimate segments linear   -   -   -   -   - 32. *linckii*
            Ultimate segments cuneate -   -   -   -   - 28. *aethiopicum*
        Lamina 2-pinnate to 3-pinnate:
          Veins ending in a linear pale hydathode ventrally   -   17. *auritum*
          Veins not ending in a conspicuous linear pale hydathode ventrally:
            Rhizome erect; fronds herbaceous; venation pinnate:
              Pinnule lobes crenate; lamina elliptic -   -   -   15. *lobatum*
              Pinnule lobes dentate; lamina narrowly ovate   -   - 16. *varians*
            Rhizome creeping; fronds thinly coriaceous; venation flabellate:
              Rhizome scales black: -   -   -   -   -   - 26. *simii*
              Rhizome scales brown:
                Rhizome scales less than 3 mm. long, pale-brown, concolorous
                                      27. *lividum*
                Rhizome scales more than 4 mm. long, clathrate:
                  Rhachis with numerous hair-like scales at least on the young
                    fronds   -   -   -   -   -   -   28. *aethiopicum*
                Rhachis with shortly hair-pointed scales on the rhachis
                                      29. *buettneri*

1. **Asplenium holstii** Hieron. in Engl., Bot. Jahrb. **46**: 348 (1911). TAB. **53** fig. H.
Type from Tanzania.
    Asplenium africanum var. *holstii* (Hieron.) Tardieu in Mém. Inst. Fr. Afr. Noire,
28: 172 (1953). Type as above.

Rhizome up to 2 cm. in diam. and 7 cm. long, erect, with tufted fronds and with
brown, subulate, irregularly and sparsely fringed rhizome-scales up to 10×0.5
mm. at the base. Frond erect. Stipe matt-pale-grey-brown, up to 9 cm.
long and 4 mm. in diam., at first clothed with scales, smaller but similar to the
rhizome-scales, later becoming subglabrous. Lamina up to 65×3-6·5 cm.,
simple, elliptic to linear-oblong, subcoriaceous, glabrous, apex usually acute-
acuminate or occasionally rounded or rarely caudate, base narrowly cuneate. Sori
up to 3·3 cm. long at maturity and c. 1·5 mm. broad, along the veins, usually extending from near the costa to$\frac{1}{2}$-$\frac{3}{4}$ of the distance to the margin. Indusium c. 0·9 mm.
broad, membranous, entire.

Rhodesia. E: Chimanimani Mt., Timbiri R., 13.ii.1958, *Mitchell* 377 (BM; BOL;
PRE; SRGH). **Malawi.** S: Shire Highlands, 1897, *Adamson* (K). **Mozambique.**
Z: Nauela, 12.x.1949, *Barbosa & Carvalho* 4464 (BOL). MS: Garuso, Jaegersberg,
11.vii.1955, *Schelpe* 5618 (B; BM; BOL; K; P).
Also in E. tropical Africa. Shaded humus-covered boulders in forest, 600–1000 m.

2. **Asplenium anisophyllum** Kunze in Linnaea, **10**: 511 (1836). Type from S. Africa.
    Asplenium geppii Carr., Cat. Afr. Pl. Welw. **2**, 2: 269 (1901). Type from Angola.

Rhizome c. 2 cm. in diam., erect, with tufted fronds and with pale-brown to
brown, concolorous, lanceolate, acute to acuminate scales up to 12×3.5 mm.

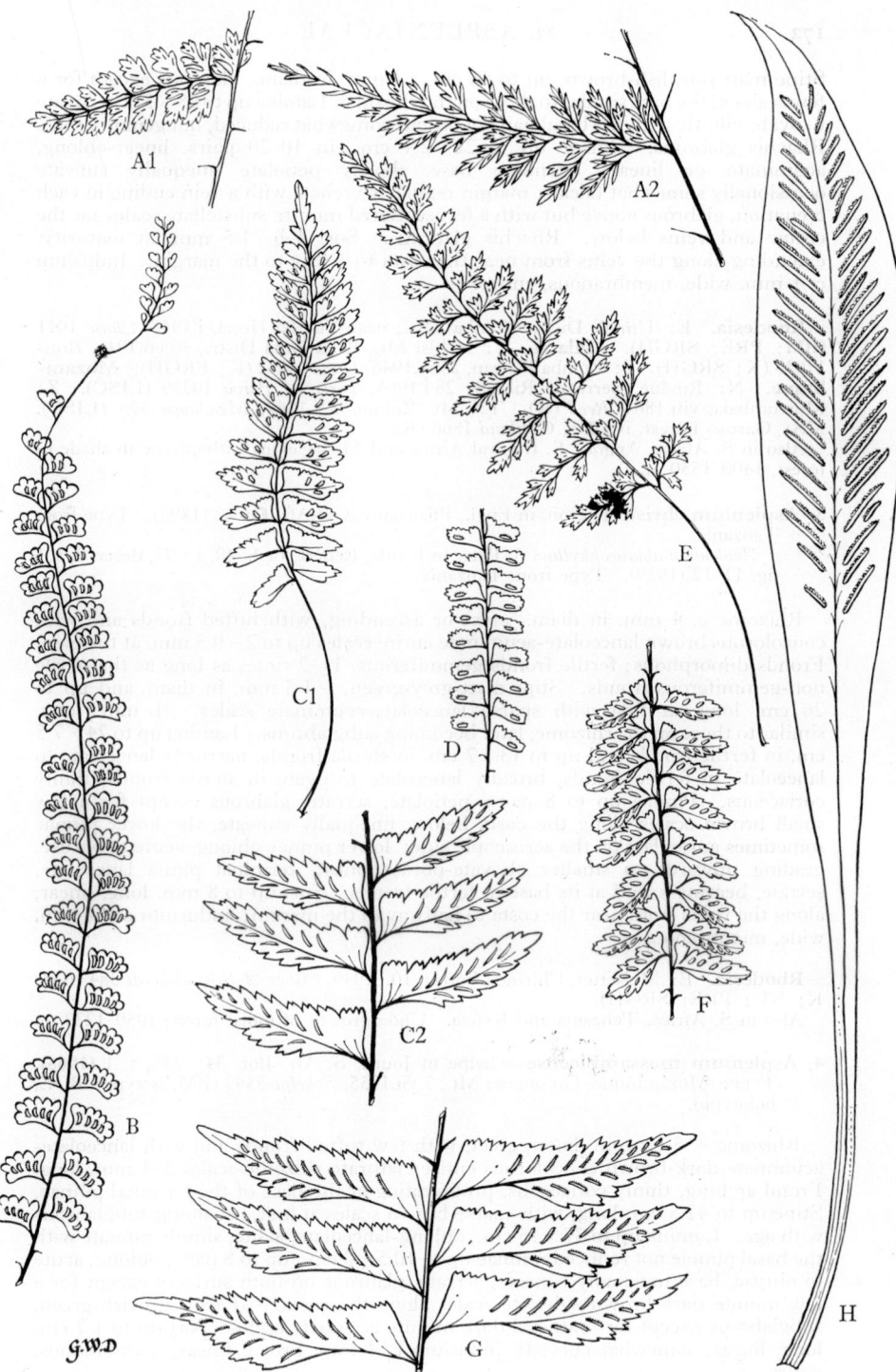

Tab. 53. A.—ASPLENIUM LOBATUM. A1, pinna of bipinnatifid form (× ⅔); A2, pinna of 3-pinnatifid form (× ⅔), both from *Chase* 7010. B—ASPLENIUM SANDERSONII. Fertile lamina (× ⅔) *Chase* 3396. C.—ASPLENIUM INAEQUILATERALE. C1, frond of small form (× ⅔) *Chase* 3516; C2, part of lamina of large (× ⅔) *Swynnerton* 840. D.—ASPLENIUM MONANTHES. Part of fertile lamina (× ⅔) *Chase* 4979. E.—ASPLENIUM VARIANS subsp. FIMBRIATUM. Frond (× ⅔) *Chase* 3319. F—ASPLENIUM ERECTUM var. USAMBARENSE. Part of lamina (× ⅔) from *Chase* 6693. G.—ASPLENIUM OBSCURUM. Part of lamina (× ⅔) *Chase* 6882. H.—ASPLENIUM HOLSTII. Fertile frond (× ⅔) *Holst* s.n.

Stipe matt-purplish-brown, up to 48 cm. × 4 mm. in diam., glabrous except for a
few scales at the base similar to those on the rhizome. Lamina up to 88 × 32 cm., ovate
to ovate-elliptic with the basal pairs of pinnae somewhat reduced, not gemmiferous.
Rhachis glabrous. Pinnae up to 22 × 2·4 cm., in 10–20 pairs, linear-oblong,
acuminate or linear, attenuate base, shortly petiolate unequally cuneate
occasionally somewhat falcate, margin regularly crenate with a vein ending in each
crenation, glabrous above but with a few scattered minute substellate scales on the
costae and veins below. Rhachis glabrous. Sori 4–6 × 1·5 mm. at maturity,
extending along the veins from near the costa to ½ way to the margin. Indusium
c. 1 mm. wide, membranous, entire.

Rhodesia. E: Umtali Distr., Vumba Mts., near Vumba Hotel, i.1948, *Chase* 1011
(BM; PRE; SRGH). **Malawi.** C: Nchisi Mt., Kota-Kota Distr., 30.vii.1946, *Brass*
17035 (K; SRGH). S: Zomba Plateau, 28.iv.1946, *Brass* 16062 (K; SRGH). **Mozam-
bique.** N: Ribáuè, Serra de Ribáuè, 28.i.1964, *Torre & Paiva* 10299 (LISC). Z:
Morrumbala, viii.1864, *Reed* (K). T: Mt. Zóbuè, 3.x.1942, *Mendonça* 573 (LISC).
MS: Garuso Forest, iv.1935, *Gilliland* 1866 (K).
Also in S. Africa, Angola, E. tropical Africa and Madagascar. Lithophytic in shade in
forest, 1400–1550 m.

3. **Asplenium christii** Hieron. in Engl., Pflanzenw. Ost-Afr. **C**: 82 (1895). Type from
   Tanzania.
       Asplenium amaurophyllum A. Peter in Fedde, Repert, Beih. **40**, 1: 73, descr. 5, t. 1
   fig. 11–12 (1929). Type from Tanzania.

Rhizome c. 4 mm. in diam., erect or ascending, with tufted fronds and with
concolorous brown lanceolate-acuminate entire scales up to 2 × 0·5 mm. at the base.
Fronds dimorphous; fertile fronds gemmiferous, 1½–2 times as long as the sterile
non-gemmiferous fronds. Stipe matt-grey-green, 1–1·5 mm. in diam. and up to
26 cm. long, at first with sparse lanceolate-acuminate scales ±1 mm. long,
similar to those on the rhizome, later becoming subglabrous. Lamina up to 24 × 7·5
cm. in fertile fronds and up to 13 × 7 cm. in sterile fronds, narrowly lanceolate to
lanceolate in fertile fronds, broadly lanceolate to ovate in sterile fronds, thinly
coriaceous. Pinnae up to 8 pairs, petiolate, serrate, glabrous except for a few
small brown scales along the costa below, unequally cuneate, the lower pinnae
sometimes auriculate on the acroscopic side; lower pinnae oblong, acute to obtuse,
grading upwards to smaller obovate-obtuse ones; terminal pinna lanceolate,
serrate, bearing a bud at its base in fertile fronds. Sori up to 8 mm. long, linear,
along the veins from near the costa to halfway to the margin; indusium c. 0·8 mm.
wide, membranous.

Rhodesia. E: Melsetter, Chirinda Forest, 10.i.1949, *Fisher & Schweickerdt* 384 (BM;
K; NU; PRE; SRGH).
Also in S. Africa, Tanzania and Kenya. Undergrowth of moist forest, 1050–1220 m.

4. **Asplenium mossambicense** Schelpe in Journ. S. Afr. Bot. **34**: 235, t. 1 (1968).
   Type: Mozambique, Gorongosa Mt., 7.vii.1955, *Schelpe* 5593 (BM, isotype; BOL,
   holotype).

Rhizome c. 5 mm. in diam., erect, with few tufted fronds and with lanceolate-
acuminate dark-brown concolorous entire clathrate rhizome-scales 3–4 mm. long.
Frond arching, thinly coriaceous, proliferating at the base of the terminal pinnae.
Stipe up to 42·5 cm. long, with sparse brown scales at first, becoming subglabrous
with age. Lamina up to 37 × 15 cm., oblong-lanceolate, acute, simply pinnate with
the basal pinnae not reduced, pinnae up to 10·5 × 3 cm., up to 8 pairs, oblong, acute
to obtuse, base unequally cuneate, serrate, glabrous on both surfaces except for a
few minute dark brown subulate scales along the costa. Rhachis greyish-green,
subglabrous except for scattered dark minute attenuate scales. Sori up to 1·7 cm.
long, linear, somewhat curved; indusium c. 1 mm. broad, linear, membranous,
entire.

Rhodesia. E: Chirinda Distr., 24.x.1947, *Wild* 2203 (K; SRGH). **Mozambique.**
MS: Serra da Gorongosa, Mt. Nhandore, 6.v.1964, *Torre & Paiva* 12291 (LISC).
Known only from this area. Deeply shaded boulders in forest, 1130–1220 m.

5. **Asplenium boltonii** Hook. ex Schelpe in Bol. Soc. Brot., Sér. 2, **41**: 204 (1967). Type from S. Africa.

Asplenium anisophyllum var. *elongatum* Mett., in Abh. Senckenb. Naturf. Ges. **3**: 143 (1859) reimpr. in Mett., Farngatt., **6**: 99 (1859). Syntypes from S. Africa and Réunion.

Asplenium anisophyllum sensu Sim, Ferns S. Afr. ed. 2: 151, t. 53 (1915).—Tardieu, Fl. Madag.: 203, t. 27 fig. 9–10 (1958).

Rhizome up to 2 cm. in diam., erect, with tufted fronds and with brown linear-lanceolate scales with pale margins, entire, up to 18 × 2 mm. at its base, composed of hyaline thin-walled cells. Stipe matt-grey-green, up to 14 cm. long and 4 mm. in diam., at first densely covered with reddish-brown hair-like scales up to 1·4 cm. long, later becoming subglabrous. Lamina up to 33 × 8 cm., ovate-lanceolate, with the lowest pinnae slightly reduced, bearing a bud or small plant at the base of the apical pinna. Pinnae up to 7·5 × 1·5 cm., in up to 28 pairs, lanceolate-attenuate, base broad and unequally cuneate, margin dentate with usually alternating deeper and shallower incisions, glabrous above but with scattered hair-like scales below when young. Rhachis matt-grey-green, with some hair-like scales similar to those on the stipe. Sori c. 4 mm. long, broadly elliptic when mature extending along the veins from near the costa to $\frac{1}{2}$ or less than the distance to the margin; indusium c. 1 mm. broad, entire, subelliptic.

Zambia. N: Isoka, 21.xi.1952, *Angus* 809 (K). **Rhodesia.** E: Inyanga, Romney-dale, 22.iii.1949, *Chase* 2038 (SRGH). **Malawi.** N: Mafinga Hills, Nyika Plateau, 27.viii.1962, *Tyrer* 588 (SRGH). S: Mlanje Distr., Luchenya Plateau-Chimbe, 13.vi.1962, *Richards* 16661 (K). **Mozambique.** N: Vila Cabral, 5.iii.1964, *Torre & Paiva* 11032 (LISC). MS: Manica, Serra Zuira, 11.xi.1965, *Torre & Paiva* 12874 (LISC).

Also in S. Africa, E. tropical Africa, Congo, Madagascar and Réunion. Lithophytic or rarely epiphytic in forest, 1800–1980 m.

6. **Asplenium gemmiferum** Schrad. in Gött. Gel. Anz. **1818**: 916 (1818). Type from S. Africa.

Asplenium macrolobium A. Peter in Fedde, Repert. Beih. **40, 1**: 77, descr. 6, t. 3 fig. 1–2 (1929). Type from Tanzania.

Rhizome to 3 cm. in diam., erect, with tufted fronds and with grey-brown, lanceolate acuminate irregularly fimbriate rhizome-scales, up to 9 × 2 mm., composed of thin-walled cells. Stipe matt-grey, up to 30 cm. long and 3 mm. in diam., at first densely clothed with scales similar to the rhizome scales gradually becoming subglabrous. Lamina up to 72 × 26 cm., oblong-lanceolate, pinnate, arching, bearing a gemma or small plant at the base of the apical pinna. Pinnae up to 15 × 3·6 cm., up to 12 pairs, lanceolate to oblong-lanceolate, petiolate, un-equally cuneate at the base, acuminate, margin minutely and regularly shallowly crenate, glabrous, carnose-coriaceous when fresh. Rhachis matt-grey-green, narrowly winged in the upper $\frac{1}{2}$, subglabrous as the stipe. Sori up to 1·5 cm. long, along the veins, usually extending from near the costa $\frac{1}{2}$-way to the margin; indusia linear membranous, entire.

Rhodesia. E: Umtali Distr., above Leopard Rock Hotel, 29.x.1956, *Chase* 6227 (BOL; LISC; SRGH). **Malawi.** C: Nchisi Mt., 28.vii.1946, *Brass* 17011 (K; SRGH). S: Cholo Mt., 23.ix.1946, *Brass* 17–753 (K; SRGH). **Mozambique.** MS: Garuso, Bandula Mt., 10.vii.1957, *Chase* 6592 (BM; BOL; LISC; PRE; SRGH).

Also in S. Africa, E. tropical Africa and E. Congo. Lithophytic on humus-covered boulders in deep shade in forest, 1200–1500 m.

7. **Asplenium atroviride** Schelpe in Bol. Soc. Brot., Sér. 2, **41**: 204 (1967). Type: Rhodesia, Vumba Mts., Witchwood, *Schelpe* 5446 (BM, isotype; BOL, holotype).

Rhizome c. 1 cm. in diam., erect, covered with old stipe bases with tufted fronds and with greyish-brown, lanceolate-subulate, subentire scales up to 10 × 1·5 mm. at its base, composed of clear thin-walled cells. Frond erect, not gemmiferous. Stipe matt-grey-green to brown, up to 5 cm. long and 3 mm. in diam., subglabrous, with a few scales towards the base similar to the rhizome scales. Lamina dark-green, up to 40 × 19 cm., oblong, erect, pinnate not gemmiferous. Pinnae in up to 6 sub-opposite pairs with a terminal pinna similar to the other pinnae, lanceolate to oblong, acute to acuminate, often abruptly so and often curved upwards, petiolate, base unequally truncate to unequally cuneate, margin irregularly and shallowly undulate-crenate, coriaceous. Rhachis matt-purplish-brown, at first set with dark irregularly fimbriate scales up to 2 mm. long, later becoming glabrous. Sori along the veins, extending from near the costa to more than ½-way to the margin; indusium c. 0·5 mm. wide, entire, membranous; spores reniform, winged.

Rhodesia. E: Umtali, Vumba, 27.vi.1955, *Chase* 5623 (BOL; LISC; SRGH). **Malawi.** S: Shire Highlands, 6.iv.1906, *Adamson* (K). **Mozambique.** Z: Namúli, 1887, *Last* (K). MS: Vila Gouveia, Serra de Chôa, 4.vii.1941, *Torre* 3006 (BM; COI; LISC).

Also in Kenya and Tanzania. Deeply shaded forest floors, 1220–1700 m.

8. **Asplenium obscurum** Bl., Enum. Pl. Jav.: 181 (1828).—Tardieu, Fl. Madag. Polypod. **1**: 190, t. 27 fig. 1–3 (1958). TAB. **53** fig. G. Type from Java.
    Asplenium serraeforme Mett. in Abh. Senckenb. Nat. Ges. **3**: 163 (1859) reimpr. in Mett. Farngatt. **6**: 119, n. 75. t. 4 fig. 1 (1859). Type of unknown origin.
    Hemiasplenium obscurum (Bl.) Tagawa in Act. Phytotax. Geobot. **7**: 83 (1938). Type as for *Asplenium obscurum.*

Rhizome 5 mm. in diam., creeping, with fronds spaced up to 4 cm. apart and with rather sparse brown clathrate subentire narrowly triangular acute rhizome-scales c. 2 × 1 mm. at its base. Frond erect, membranous, not proliferous. Stipe matt-brown, up to 17 cm. long, with very few short hairs when young, glabrous with age and with a few scales similar to those of the rhizome-scales near the base. Lamina pinnate, oblong to narrowly oblong-acuminate, basal pinnae not reduced, pinnae in 17–20 pairs, petiolate, rhombic, dimidiate for almost ½ the length of the costa, broadly to narrowly acute, very unequally cuneate at the base, serrate to mostly doubly serrate on the acroscopic margin and upper part of the basiscopic margin up to 5·5 × 1·5 cm., glabrous on both surfaces. Rhachis matt-brown with occasional multicellular hairs. Sori up to 7 mm. long, very narrowly oblong, set at about 45° to the costa about midway between the costa and margin; indusium narrowly oblong, membranous, entire.

Rhodesia. E: Melsetter, Chimanimani Mts., 16.ii.1958, *Mitchell* 405 (BM; BOL; SRGH). **Malawi.** N: Misuku Plateau, vii.1896, *Whyte* (K). S: Cholo Mt., 23.viii.1946, *Brass* 17754 (K; SRGH). **Mozambique.** MS: Bandula Mt., 24.iv.1958, *Chase* 6882 (BM; BOL; SRGH).

Also in Tanzania, Madagascar and Asia. Lithophytic on wet rocks and in stream beds in deep shade in forest, 1200–2140 m.

9. **Asplenium unilaterale** Lam., Encycl. Méth. Bot. **2**: 305 (1786).—Sim, Ferns S. Afr. ed. 2: 152, t. 54 (1915).—Tardieu in Mém. Inst. Fr. Afr. Noire, **28**: 191, t. 37 fig. 1–2 (1953); Fl. Camér. **3**, Ptérid.: 197, t. 29 fig. 1–2 (1964). Type from Mauritius.
    Asplenium resectum Sm., Pl. Icon. **3**: t. 72 (1794). Type from Réunion.
    Asplenium emarginato-dentatum Zenker ex Kunze in Linnaea, **24**: 263 (1851). Type from India.
    Asplenium amoenum C. Presl [Tent. Pterid.: 107 (1836) *nom. nud.*] ex Mett. in Abh. Senckenb. Naturf. Ges. **3**: 175 (1859) reimpr. in Mett., Farngatt. **6**: 131, n. 101 (1859). Type from India.

Rhizome 2 mm. in diam., creeping, with fronds spaced 1·5 cm. apart, with dark brown clathrate lanceolate entire rhizome-scales up to 2 mm. long. Frond erect, herbaceous. Stipe atrocastaneous, glabrous, shining, up to 13 cm. long. Lamina 23·5 × 8 cm. simply pinnate, narrowly oblong in outline with lowest pinnae not reduced, apex attenuate. Pinnae up to 4 × 1 cm., 12–20 pairs, rhombic, dimidiate for about ½ of their length, acute-serrate on the acroscopic margin, often doubly serrate towards the apex, glabrous on both surfaces, petiolate. Rhachis atrocastaneous, shiny, glabrous. Sori c. 3 mm. long, oblong to very narrowly oblong, set at an angle of about 30° to the costa in the upper ½ of the pinna; indusium narrowly oblong, membranous, entire.

Rhodesia. E: Melsetter Distr., Chipinga, Mt. Selinda, 27.viii.1951, *Taylor* 3380 (BOL). **Malawi.** S: Cholo Mt., 23.ix.1946, *Brass* 17755 (K; SRGH).

Widespread in tropical Africa and the Mascarene Is. Lithophytic in deep shade on wet rocks in forest, c. 1200 m.

10. **Asplenium trichomanes** L., Sp. Pl. **2**: 1080 (1753).—Sim, Ferns S. Afr. ed. 2: 140 44 fig. 1 (1915). Type from Europe.

Rhizome erect, with tufted fronds and with subulate attenuate clathrate dark-brown rhizome-scales up to 4 mm. long often with a dark central stripe and paler narrow clathrate borders. Frond erect to arching, firmly membranous. Stipe atrocastaneous, short, usually less than ⅙ the length of the lamina, glabrous. Lamina pinnate, narrowly linear in outline, somewhat tapering at the base. Pinnae up to 5 × 3·5 mm. (in Rhodesian specimen), up to 32 pairs, shortly petiolate, broadly oblong-circular, not dimidiate, becoming broadly crenate-flabellate towards the base, crenate-dentate, glabrous on both surfaces. Rhachis atrocastaneous, shallowly sulcate on the ventral surface, glabrous. Sori 1–3 per pinna, oblong at first, almost covering the dorsal surface of the pinna at maturity; indusium narrowly oblong, entire, membranous.

Rhodesia. E: Umtali Distr., Imbesa Valley, La Rochelle, 28.i.1962, *Chase* 7687 (K; SRGH).
Also in S. Africa, Kenya, NE. tropical Africa and Europe. Rock crevices and boulder bases, c. 1220 m.

11. **Asplenium monanthes** L., Mant. Pl. **1**: 130 (Oct. 1767).—Richter, Codex Bot. Linn.: 1030 (1835).—Sim, Ferns S. Afr. ed. 2: 141, t. 46 fig. 1 (1915). TAB. **53** fig. D. Type from S. Africa.
Asplenium monanthemum L., Syst. Nat. ed. 12, **2**: 690 (Oct. 1767). Type as above.

Rhizome erect to suberect with tufted fronds and with lanceolate-attenuate fine hair-pointed rhizome-scales up to 3·5 mm. long, with a black central stripe and paler clathrate entire borders. Frond erect, firmly membranous Stipe atrocastaneous, up to 11 cm. long, glabrous, nitid, terete. Lamina up to 30 × 2·5 cm., simply pinnate, narrowly linear in outline, acute, slightly decrescent. Pinnae up to 1·2 × 0·6 cm. at about the centre of the lamina, up to 45 pairs, shortly petiolate to subsessile, oblong-dimidiate becoming broadly cuneate-flabellate towards the base, prominently crenate-dentate on the acroscopic and outer margins, glabrous on both surfaces. Rhachis atrocastaneous, sulcate on the ventral surface. Sori usually solitary (occasionally 2), narrowly oblong, set along and near the basiscopic margin; indusium membranous, entire, linear.

Rhodesia. W: Victoria Falls, *Sim* (PRE). E: Melsetter Distr., Musapa Mt., 9.x.1950, *Chase* 3047 (BM; SRGH). **Malawi.** N: Nyika Plateau, Rufivi Stream, 28.x.1958, *Robson* 441 (K; LISC; SRGH). S: Mlanje Mt., Tuchila Plateau, 28.vii.1956, *Newman & Whitmore* 262 (SRGH).
Also in S. Africa, sporadic in tropical Africa and also in tropical America. Shaded forest floors, 1830–2400 m.

12. **Asplenium normale** D. Don, Prodr. Fl. Nepal.: 7 (1825). Type from Nepal.

Rhizome 5 mm. in diam., erect to creeping, with tufted fronds and lanceolate-attenuate subentire clathrate dark-brown rhizome-scales up to 2·5 mm. long, with a narrow ferrugineous margin up to 2·5 mm. long. Frond erect or arching, often proliferous near the apex, firmly membranous. Stipe atrocastaneous, up to 1·7 cm. long, terete, glabrous. Lamina up to 50 × 3 cm., simply pinnate, linear-acuminate in outline, hardly decrescent below. Pinnae up to 1·5 × 0·7 cm., up to 49 pairs, shortly petiolate, mostly oblong, slightly dimidiate, rounded, shallowly auriculate, acroscopically becoming increasingly deflexed and unequally triangular towards the base, shallowly crenate on the acroscopic and outer margins, glabrous on both surfaces. Rhachis castaneous to atrocastaneous, sulcate ventrally. Sori up to 3 mm. long, 2–7 per pinna, oblong; indusium linear, membranous, entire.

Malawi. S: Mlanje Mt., Luchenya Plateau, 30.vi.1946, *Brass* 16531 (K; SRGH). **Mozambique.** Z: Namúli, 1887, *Last* (K).
Also in E. tropical Africa and Asia. Shaded forest floors, c. 1890 m.

13. **Asplenium erectum** Bory ex Willd. in L., Sp. Pl., ed. 4, **5**: 328 (1810). Type from Réunion.

Rhizome 3–4 mm. in diam., erect, with tufted fronds and subulate dark-brown clathrate attenuate entire rhizome-scales 3 mm. long. Stipe matt-grey-brown, up to 9 cm. long, with narrow green wings in the upper ½, with sparse scales similar to the rhizome scales when young, becoming glabrous with age. Lamina up to 29 × 3·5 cm., erect, not proliferous, firmly membranous, pinnate to 2-pinnatifid, narrowly elliptic in outline. Pinnae up to 2×0.8 cm., up to 40 pairs, shortly petiolate, more or less auriculate acroscopically at the base with an acroscopic basal lobe separated to a greater or lesser degree, oblong, rounded to acute becoming flabellate towards the base of the frond, serrate except for the obtusely unequally cuneate base, glabrous on both surfaces. Rhachis dark matt-grey with 2 narrow green wings, glabrous. Sori 1·5–2 mm. long, up to 12 per pinna, narrowly oblong, mostly set at an acute angle on both sides of the costa; indusium narrowly oblong, entire, membranous.

Lamina very narrowly oblong; basal auricle of lower pinnae not free -   var. *erectum*
Lamina linear; basal auricle of lower pinnae free   -   -   -   -   var. *usambarense*

Var. erectum

Mozambique. MS: Vila Machado, Montes Chiluvo, 14.iv.1948, *Mendonça* 3939 (LISC).
Also in S. Africa and Mascarene Is. Forest floors in shade.

Var. usambarense (Hieron.) Schelpe in Bol. Soc. Brot., Sér. 2, **41**: 207 (1967). TAB. 53 fig. F. Type from Uganda.

Asplenium sphenolobium var. *usambarense* Hieron. in Deutsch. Z.-Afr. Exped., 1907–8, 2: 14 (1911). Type as above.
Asplenium usambarense (Hieron.) Hieron. in Hedwigia, **60**: 227 (1918) *nom. illegit.* Type as above.
Asplenium quintasii Gandog. in Bull. Soc. Bot. Fr. **66**: 305 (1919). Type from S. Tomé.

Zambia. N: Abercorn Distr., Chisau R. Gorge, Saisi Valley, 18.ii.1959, *Richards* 11804 (K). **Rhodesia.** E: Umtali Distr., Vumba Mts., Firwood, 11.viii.1957, *Chase* 6693 (BM; BOL; SRGH). **Malawi.** N: Misuku Hills, Walindi Forest, 12.ii.1958, *Robson* 594 (K; LISC; SRGH). S: Chiradzulu, 3.x.1859, *Kett* (K). **Mozambique.** MS: Garuso, Jaegersberg, 11.vii.1955, *Schelpe* 5629 (BM; BOL).
Also in S. Africa and tropical Africa. Shaded forest floors, 1500–2250 m.

14. Asplenium inaequilaterale Willd. in L., Sp. Pl., ed. 4, **5**: 322 (1810).—Tardieu in Mém. Inst. Fr. Afr. Noire, **28**: t. 34 fig. 4–5 (1953). TAB. 53 fig. C. Type from Réunion.

Asplenium brachyotus Kunze in Linnaea, **10**: 512 (1836). Type from S. Africa.
Asplenium erectum var. *brachyotus* (Kunze) Sim, Ferns S. Afr. ed. 2: 138, t. 66 (1892). Type as above.
Asplenium suppositum Hieron. in Engl., Bot. Jahrb. **46**: 353 (1911). Type from Angola.
Asplenium laetum sensu Sim, Ferns S. Afr. ed. 2: 150, t. 50 (1915) non Sw. (1806).
Asplenium laetum var. *brachyotus* (Kunze) Bonap., Not. Ptérid. **16**: 60 (1925). Type as for *Asplenium brachyotus*.

Rhizome up to 4 mm. in diam., erect, with tufted fronds and with dark-brown entire clathrate lanceolate attenuate rhizome-scales 2·5 mm. long. Frond erect, not proliferous, firmly membranous. Stipe matt-brown or greenish-brown, 19 cm. long, sparsely covered at first with minute hair-like scales, glabrous at maturity. Lamina up to 32×13 cm., pinnate, narrowly oblong-acuminate in outline with lowest pinnae not markedly reduced. Pinnae $1.9–6.3 \times 0.5–1.2$ cm., 12–20 pairs, narrowly to very narrowly oblong, obtuse to attenuate, unequally cuneate at the base, crenate or doubly crenate, shortly petiolate; apical segment attenuate-crenate, glabrous on both surfaces. Rhachis pale-brown, narrowly winged on the upper part of the lamina. Sori up to 5 mm. long, linear (to oblong in small fronds), set at an angle of about 45° to the costa; indusium linear to oblong, membranous, entire.

Zambia. N: Abercorn Distr., Chisau R. Gorge, 18.ii.1959, *Richards* 11802 (K). S: Palm Grove, Victoria Falls, 6.vii.1953, *Schelpe* 3940 (BM; BOL; K). **Rhodesia.** E: Chirinda Distr., 24.x.1947, *Wild* 2198 (K; SRGH). **Malawi.** N: Misuku Hills, 11.i.1959, *Robinson* 3150 (K; SRGH). C: Nchisi Mt., Kota Kota Distr., 28.vii.1946, *Brass* 17006 (K; SRGH). S: Mlanje Mt., 17.vi.1962, *Richards* 16755 (K). **Mozambique.** Z: Milange, 12.x.1942, *Torre* 4579 (COI; LISC). MS: Chimaio, R. Nhamissanguere, 19.ii.1948, *Garcia* 299 (BM; LISC).
Also in S. Africa and widespread in tropical Africa and Réunion. Deeply shaded floors of moist forest, 700–1680 m.

15. **Asplenium lobatum** Pappe & Raws., Syn. Fil. Afr. Austr.: 22 (1858). TAB. **53** fig. A. Type from S. Africa.

Asplenium gracile Pappe & Raws., loc. cit. non D. Don (1825). Type from S. Africa.

Asplenium lunulatum var. *gracile* Sim, Ferns S. Afr.: 139, t. 67 fig. a–b (1892); op. cit. ed. 2: 146, t. 149 (1915). Type as above.

Asplenium erectum var. *lobatum* (Pappe & Raws.) Alston & Schelpe in Journ. S. Afr. Bot. **18**: 161 (1952). Type as for *Asplenium lobatum*.

Asplenium erectum var. *gracile* (Pappe & Raws.) Tardieu, Fl. Madag., Polypod.: 222, t. 30 fig. 4–5 (1958). Type as for *Asplenium gracile*.

Rhizome up to 7 mm. in diam., erect, with tufted fronds and with pale-brown clathrate lanceolate-acuminate entire rhizome-scales up to 6 mm. long. Frond erect, usually not proliferous but occasionally proliferating at the base of the lamina, membranous. Stipe matt-greyish-brown, up to 23 cm., glabrous except at the base with scales similar to those on the rhizome. Lamina up to 35 × 12 cm., narrowly elliptic in outline, apex acuminate, lower pinnae variously reduced, 2-pinnate to 4-pinnatifid, pinnae up to 9 × 3 cm., narrowly oblong-acute to lanceolate-acuminate; pinnules rhombic in outline, coarsely serrate or divided into 3-fid to 2-fid or linear acute lobes, glabrous on both surfaces. Rhachis matt-grey-brown with narrow green wings most of the length. Sori up to 4 mm. long, linear to oblong, 3–6 per pinnule; indusium linear to narrowly oblong, membranous, entire.

Rhodesia. E: Umtali, Vumba Mts., 2.iv.1958, *Chase* 6865 (BM; BOL; K; SRGH). **Malawi.** N: Misuku Hills, 11.i.1959, *Robinson* 3170 (K; SRGH). C: Dedza, Mulum-duni, 31.iii.1961, *Chapman* 1201 (SRGH). S: Mlanje Mt., 21.vii.1956, *Newman & Whitmore* 102 (BM; SRGH). **Mozambique.** N: Mt. Massangulo, vi.1933, *Gomes e Sousa* 1490 (COI). MS: Báruè, Serra de Chôa, 24.viii.1949, *Pedro & Pedrógão* 8034 (BM).

Also in S. Africa and Madagascar. Deeply shaded floors of moist forest, 1460–1830 m.

Fragments of *A. abyssinicum* Fée, a species differing from *A. lobatum* by its very fragile fronds and shining castaneous rhachises, were reputedly collected by Sim at the Victoria Falls, but it has not been recorded from there since.

16. **Asplenium varians** Wall. ex Hook. & Grev., Ic. Fil.: t. 172 (1830). Type from India.

Subsp. **fimbriatum** (Kunze) Schelpe in Bol. Soc. Brot., Sér. 2, **41**: 211 (1967). TAB. **53** fig. E. Type from S. Africa.

Asplenium fimbriatum Kunze in Linnaea, **18**: 117 (1844). Type from S. Africa.

Rhizome 3 mm. in diam., erect, with tufted fronds and with dark-brown clath-rate-lanceolate attenuate subentire rhizome-scales up to 3 mm. long. Frond erect or arching, not proliferous, herbaceous. Stipe greyish-green when dried, becoming dark-brown at the base, 5·2 cm. long, with scattered subulate dark-brown weakly pseudoserrate scales up to 2 mm. long. Lamina up to 13 × 5 cm., 2-pinnate, narrowly elliptic in outline, acute with enlarged basal acroscopic pinnules, pinnae up to 10 pairs, pinnules and pinnule lobes obovate, up to 6 mm. long, sharply dentate on the outer margin, glabrous on both surfaces. Rhachis greyish-green when dried with narrow wings, glabrous. Sori up to 3 × 1 mm. at maturity, 2–5 per lobe; indusium linear, almost transparent, erose.

Rhodesia. E: Melsetter Distr., Black Mountain Inn, 14.xi.1953, *Chase* 5124 (BM; SRGH). **Mozambique.** MS: Manica Distr., Border Farm, 19.iii.1961, *Chase* 7439 (BOL; K; SRGH).

Also in S. Africa, E. Congo and east tropical Africa. Lithophytic on shaded boulders in forest, c. 1130 m.

17. **Asplenium auritum** Sw. in Schrad., Journ. Bot. **1800**, 2: 52 (1801). Type from Jamaica.

Rhizome up to 4 mm. in diam., erect, with tufted fronds and with shining entire dark-brown attenuate rhizome-scales up to 6 mm. long. Frond arching, not proliferous, thinly coriaceous, the sterile ½ as large as fertile fronds. Stipe matt-grey to greyish-green, often darker towards the base, up to 22 cm. long, glabrous. Lamina up to 26 × 10 cm., pinnate to 2-pinnate, triangular-lanceolate to narrowly oblong-lanceolate, acuminate, basal pinnae not reduced. Pinnae up to 5·5 × 2 cm., up to 15 pairs, petiolate, reduced upwards into a deeply pinnatifid apex, small pinnae narrowly oblong, crenate, larger fertile pinnae lanceolate, deeply pinnatifid or pinnate into cuneate-oblong irregularly crenate-serrate segments, petiolate. Rhachis matt-grey-green, glabrous except for scattered black hair-like scales c. 1 mm. long.   Sori 4 × 1·5 mm., oblong; indusium linear, entire, membranous.

Rhodesia. E: Umtali Distr., Vumba Mts., 18.vii.1957, *Chase* 6558 (BM; BOL; K; LISC; SRGH).   **Mozambique.**   Z: Gúruè, 28.vi.1943, *Torre* 5601 (BM; LISC).
Also in Madagascar, Mascarene Is., Congo and tropical America. Lithophytic or epiphytic in moist forest, c. 1220 m.
This species has been referred erroneously to *A. sulcatum* Lam.

18. **Asplenium pumilum** Sw., Nov. Gen. & Sp. Pl.: 129 (1788). Type from Jamaica.

Subsp. **hymenophylloides** (Fée) Schelpe in Bol. Soc. Brot., Sér. 2, **41**: 210 (1967). Type from Ethiopia.
  Asplenium pumilum var. *hymenophylloides* Fée, Mém. Fam. Foug. **7**: 54, t. 15 fig. 4 (1857). Type as above.
  Asplenium schimperanum Hochst. ex Fée, op. cit. **5**: 191 (1852) *nom. nud.*
  Asplenium eylesii Sim, Ferns S. Afr. ed. 2 : 147, t. 61 fig. 2 (1915). Type: Rhodesia, Mazoe Distr., Iron Mask Hills, *Eyles* 564 (PRE, holotype; SRGH, isotype).

Rhizome 3 mm. in diam., short, erect, with tufted fronds with subulate clathrate strongly pseudoserrate longly hair-pointed scales up to 4·5 mm. long usually with a black central stripe and margin of pale cells.   Fronds erect, softly herbaceous, not proliferous.   Stipe greenish above becoming dark-brown below, up to 6 cm. long, glabrous except for scattered pale hairs.   Lamina up to 6·5 × 4·5 cm., pinnate to 2-pinnatifid, oblong to triangular in outline, the basal pinnae not reduced, apex acute incised into obtuse oblong lobes.   Pinnae up to 3 × 1·6 cm., rhombic to narrowly triangular, deeply pinnatifid in larger pinnae into obtuse oblong deeply or shallowly crenate lobes with sparse transparent or whitish hairs along the veins of both surfaces.   Rhachis greyish-green with prominent pale-green wings and set with sparse transparent hair-scales.   Sori up to 4 mm. long, linear, slightly curved; indusium linear, transparent, erose.

Rhodesia. N: Sinoia, 21.iv.1946, *Rodin* 4360 (K; SRGH).   **Malawi.** C: Dedza Distr., Mua-Livulezi Forest Reserve, Namkokwe, 19.iii.1955, *E.M. & W.* 1062 (BM; LISC; SRGH).
Also in NE. and W. tropical Africa and NW. India. Lithophytic on boulders in forest, c. 1220 m.

19. **Asplenium friesiorum** C. Chr. in Notizbl. Bot. Gart. Berl. **9**: 181 (1924). Type from Kenya.
  Asplenium serra var. *natalensis* Bak., Syn. Fil., ed. 2: 485 (1883). Type from S. Africa.
  Asplenium pseudoserra Domin in Preslia, **8**: 6 (1929). Type from Tanzania.
  Asplenium monilisorum Domin, tom. cit.: 7 (1929). Type from S. Africa.
  Tarachia friesiorum (C. Chr.) Momose in Journ. Jap. Bot. **35**: 321, fig. 33–34 (1960). Type as for *Asplenium friesiorum*.

Rhizome up to 8 mm. in diam., widely creeping, with fronds widely spaced and with, lanceolate attenuate iridescent brown entire rhizome-scales up to 4 mm. long ending in a short hair-point.   Frond erect, not proliferous, thinly coriaceous. Stipe dull-brown or purplish-brown, up to 77 cm. long, set with brown lanceolate to ovate small scales similar to those on the rhizome.   Lamina up to 80 × 26 cm., simply pinnate, very narrowly oblong in outline, acuminate, basal pinnae somewhat reduced, apical segment deeply pinnatifid.   Pinnae up to 13 × 2 cm., up to 35 pairs, mostly shortly petiolate, linear-attenuate, unequally cuneate at the base tending to be auriculate acroscopically at the base, serrate to shallowly lobed with serrate lobes, glabrous on both surfaces except for scattered pale-brown ovate-acuminate scales c. 1 mm. long.   Rhachis dull-brown to purplish-brown, glabrous at maturity.   Sori up to 2 mm. wide at maturity, oblong, borne in 2 rows closely set along the costa; indusium linear, membranous, entire.

Zambia. N: Mpika Distr., 28.i.1952, *White* 3768 (K).   **Rhodesia.** E: Melsetter Distr., Chimanimani Mts., 1.ii.1958, *Mitchell* 266 (BM; BOL; PRE; SRGH).   **Malawi.** N: Nyika Plateau. 16.viii.1946, *Brass* 17273 (K; SRGH).   S: Mlanje Distr., Luchenya

Plateau, Mlanje Mt., 6.vi.1962, *Richards* 16557 (K). **Mozambique.** Z: Namúli, 1887, *Last* (K). MS: Gorongosa Mt., Gogogo Peak, 5.vii.1955, *Schelpe* 5510 (BM; BOL).
Also in S. Africa and widespread in tropical Africa. Forest undergrowth, 1700–1950 m.

20. **Asplenium protensum** Schrad. in Gött. Gel. Anz. **1818**: 916 (1818).—Sim, Ferns
S. Afr. ed. 2: 149, t. 51 (1915).—Tardieu in Mém. Inst. Fr. Afr. Noire, **28**: 183,
t. 37 fig. 5–6 (1953). Type from S. Africa.

Rhizome up to 5 mm. in diam., creeping, with fronds spaced c. 4 mm. apart and
with brown clathrate minutely pseudoserrate narrowly ovate rhizome-scales up to
2 mm. long. Frond arching, herbaceous, proliferous near the apex of the frond.
Stipe up to 18 cm. long, atrocastaneous, densely set with ovate to broadly ovate
clathrate pale scales of various sizes up to 1·5 mm. long together with multicellular
hairs. Lamina up to 65 × 10 cm., shallowly to deeply pinnatifid, very narrowly
elliptic in outline. Pinnae up to 6 × 1·6 cm., up to 46 pairs, lanceolate, rhombic,
attenuate, base unequally cuneate, pinnately divided into linear-oblong acute
2-fid or 3-fid or broadly cuneate and deeply incised lobes, acroscopic basal lobe
the largest and often overlapping the rhachis, thinly pubescent above and below
especially near the base, densely so along the costa, dorsally with clathrate ovate
brown scales up to 1 mm. long. Rhachis atrocastaneous, densely pubescent with
whitish hairs and with occasional ovate scales. Sori up to 5 mm. long, linear and
very narrowly oblong, set at 25° to the costa; indusium linear, membranous, entire.

Zambia. W: Mpongwe, 24.v.1960, *Robinson* 3722 (K; SRGH). **Rhodesia.** E:
Himalaya Distr., Tsetsera Range, 7.viii.1957, *Chase* 6690 (BM; BOL; SRGH). **Malawi.**
N: Nyika Plateau, Rufiri Stream, 28.x.1958, *Robson & Angus* 444 (K). S: Mlanje Distr.,
Luchenya Plateau, 13.vi.1962, *Richards* 16662 (K; SRGH). **Mozambique.** MS:
Gorongosa, Serra da Gorongosa, 22.x.1965, *Torre & Pereira* 12571 (LISC).
Also in S. Africa and widespread in tropical Africa. Lithophytic on shaded boulders in
forest, 1300–1800 m.

21. **Asplenium pellucidum** Lam., Encycl. Méth., Bot. **2**: 305 (1786). Type from
Mauritius.

Subsp. **pseudohorridum** (Hieron.) Schelpe in Bol. Soc. Brot., Sér. 2, **41**: 208 (1967).
Type from Tanzania.
 Asplenium protensum var. *pseudohorridum* Hieron. in Engl., Pflanzenw. Ost-Afr.
 C: 82 (1895). Type as above
 Asplenium pseudohorridum (Hieron.) Hieron. in Engl., Bot. Jahrb. **46**: 362 (1911).
 Type as above

Rhizome 7 mm. in diam., erect to suberect, with tufted fronds and with subulate
dark-brown distinctly clathrate subentire attenuate rhizome-scales up to 1·2 cm.
long. Frond arching, not proliferous, thinly coriaceous. Stipe matt-brown, up to
8 cm. long, densely covered with subulate hair-like scales similar to those on the
rhizome up to 5 mm. long. Lamina up to 22 × 9 cm., simply pinnate, elliptic in
outline, acute, basal pinnae reduced. Pinnae up to 5 × 0·8 cm., up to 36 pairs,
lanceolate to attenuate, base unequally cuneate, slightly auriculate acroscopically
shallowly incised into obtuse or shallowly crenate lobes, sparsely set with dark
hair-like and ovate longly acuminate scales up to 3 mm. long. Sori up to 5 mm.
long, linear set at about 15° to the costa; indusium linear, membranous, entire.

Rhodesia. E: Melsetter Distr., Chimanimani Mts., 13.ii.1958, *Mitchell* 380 (BM;
BOL; PRE; SRGH).
Also in Tanzania. Lithophytic and low-level epiphyte in forest, c. 600 m.

22. **Asplenium formosum** Willd. in L., Sp. Pl., ed. 4, **5**: 329 (1810). Type from
Venezuela.

Rhizome 3 mm. in diam., erect, with tufted fronds and with brown lanceolate
attenuate rhizome-scales up to 3 mm. long with a dark central stripe and pale entire
borders. Frond erect or slightly arching, not proliferous, chartaceous. Stipe
atrocastaneous, relatively short, up to 3 cm. long, shining, with 2 narrow pale-
brown wings. Lamina up to 31 × 3·5 cm., 2-pinnatifid, very narrowly oblong with
the lower pinnae gradually reduced, apex acute. Pinnae up to 1·8 × 0·7 cm., up
to 48 pairs, shortly petiolate, unequally narrowly oblong-rhombic, deeply incised
into linear obtuse segments c. 1 mm. wide, the acroscopic basal lobe sometimes

overlapping the rhachis, both surfaces glabrous. Rhachis atrocastaneous with a narrow pale wing. Sori up to 3 mm. long, 1–3 per segment, oblong, all facing acroscopically; indusium narrow-elliptic, membranous, entire.

Zambia. N: Kasama Distr., 18.ii.1961, *Robinson* 4396 (K; SRGH). W: Mwinilunga, 15.iv.1960, *Robinson* 3574 (K; SRGH). **Rhodesia.** E: Umtali, Norseland, Vumba Mt., 9.ii.1950, *Chase* 3517 (BOL; SRGH). **Malawi.** N: Livingstonia, Kagiwegswe R., 9.i.1959, *Robinson* 3121 (K; PRE; SRGH). S: Mlanje, *Whyte* (K; PRE). **Mozambique.** N: Ribáuè Mt., 19.vii.1962, *Schelpe & Leach* 6941 (BOL). Z: Gúruè, 28.vi.1943, *Torre* 5600 (BM; LISC). MS: Garuso, 27.i. 1949, *Chase* 3972 (SRGH).

Widespread in tropical Africa, tropical America, India and Ceylon. Lithophytic or epiphytic in forest, 600–1400 m.

23. **Asplenium ramlowii** Hieron. in Engl., Bot. Jahrb. **46**: 372 (1911). Type from Tanzania.

Rhizome c. 5 mm. in diam., creeping, with tufted fronds and with shining palebrown lanceolate acuminate entire scales 3–7 mm. long. Frond erect, not proliferous, coriaceous. Stipe dark-brown, up to 28 cm. long, at first densely set with scales similar to those of the rhizome, later becoming subglabrous. Lamina up to 28 × 4·3 cm., very narrowly oblong, simply pinnate to deeply 2-pinnatifid, acute; lower pinnae somewhat reduced towards the base. Pinnae up to 2·2 × 1·5 cm., petiolate, narrowly ovate-triangular in outline (to almost circular in small fronds), pinnately divided into a few obtuse serrate lobes up to 7 mm. broad, glabrous on both surfaces. Rhachis dark-brown at first densely set with shining scales similar to those on the stipe. Sori up to 8 mm. long, linear; indusium linear, membranous, erose.

Zambia. W: Luanshya Distr., 21.iii.1968, *Mutimushi* 2567 (BOL; SRGH). **Rhodesia.** W: Chesterfield, x.1958, *Miller* 5494 (BOL; K; SRGH). C: Rusape, 22.i.1949, *Fisher & Schweickerdt* 480 (K; NU; PRE; SRGH). E: Umtali Distr., 5.ii.1957, *Chase* 6316 (BOL; K; PRE; SRGH). S: Mushandike Dam, c. 26 km. W. of Fort Victoria, 20.ii.1963, *Mitchell* 782 (SRGH). **Mozambique.** N: Mt. Massangulo, xii.1932, *Gomes e Sousa* 1160 (COI). Z: Namúli Mt., 26.vii.1962, *Schelpe & Leach* 7021 (BOL). T: Moatize, 11.iii.1964, *Torre & Paiva* 11150 (LISC). MS: Chimanimani Mts., 2.iii.1907, *Johnson* 237 (K).

Also in E. tropical Africa. Rock crevices and around boulder bases in grassland, 910–1470 m.

24. **Asplenium torrei** Schelpe in Bol. Soc. Brot., Sér. 2, **41**: 209 (1967). Type: Mozambique, Gorongosa, *Torre & Paiva* 12563 (LISC, holotype).

Rhizome up to 4 mm. in diam., ascending, with tufted fronds and with linear lanceolate attenuate entire clathrate black rhizome-scales c. 3·5 mm. long. Frond apparently suberect, proliferous at the base of the apical segment, thinly coriaceous, darker above than below. Stipe matt-black, up to 30 cm. long, glabrous at maturity except for scattered minute linear-lanceolate dark scales. Lamina up to 35 × 10·5 cm., imparipinnate with a terminal segment resembling a subapical pinna. Pinnae up to 6·2 × 2 cm., elongate trapeziform, the lower slightly auricled acroscopically, sharply acute, slightly falcate, serrate, with the base cuneate, petiolate, glabrous on both surfaces. Rhachis matt-black, with scattered minute hair-like scales. Sori up to 1·5 × 0·1 cm. at maturity, linear, set at an acute angle to a fairly prominent costa; indusium membranous, entire.

Mozambique. MS: Gorongosa, Serra da Gorongosa, slopes of Mt. Nhandore, 22.x.1965, *Torre & Paiva* 12563 (LISC).

Not known from elsewhere. In forest at 1320 m.

25. **Asplenium megalura** Hieron. in Wiss. Ergebn. Deutsch. Z.-Afr. Exped. **2**: 17 (1910). Syntypes from Tanzania and Congo.

    Asplenium dimidiatum var. *longicaudatum* Hieron. in Engl., Pflanzenw. Afr. **2**: 28, t. 24 fig. c (1908) *nom. nud.*

Rhizome 3 mm. in diam., erect, with tufted fronds and with narrowly lanceolate-attenuate dark-brown entire rhizome-scales up to 6 mm. long with narrow palebrown borders. Frond arching, firmly membranous, not proliferous. Stipe brown, up to 19 cm., long, nitidous, glabrous. Lamina up to 43 × 12 cm., simply pinnate, narrowly to very narrowly oblong, lowest pinnae not reduced, apical segment often tricuspidate. Pinnae up to 7·4 × 2·9 cm., up to 11 pairs, trapeziform, very longly caudate-cuneate, the distal margin irregularly serrate-dentate and the cauda also serrate, glabrous on both surfaces, veins flabellate. Rhachis brown, shining, shallowly channelled ventrally, glabrous except for a few very small scales. Sori to 8 mm. long, linear; indusium linear, membranous, entire.

Zambia. N: Isoka Distr., Mafingi Mts., 21.xi.1952, *Angus* 817 (K; SRGH). E: Lundazi Distr., Nyika Plateau, Kangampande Mt., 8.v.1952, *White* 2791 (BM). **Malawi.** N: Nyika Plateau, 6.i.1959, *Robinson* 3097 (SRGH). C: Dowa Distr., Nchisi Mt., 5.v.1961, *Chapman* 1272 (SRGH). S: Mlanje Distr., slopes of Simpson Peak, 30.iii.1960, *Phipps* 2790 (K; SRGH). **Mozambique.** N: Vila Cabral, 25.ii.1964, *Torre & Paiva* 10806 (LISC). Z: Namúli Mt., NW. sector, 26.vii.1962, *Schelpe & Leach* 7043 (BOL). Widespread in tropical Africa. Lithophytic or epiphytic in forest, 1370–1980 m.

26. **Asplenium simii** Braithwaite & Schelpe in Bol. Soc. Brot., Sér. 2, **41**: 209 (1967). Type: Rhodesia: Vumba Mts., Elephant Forest, *Chase* 6274 (BOL, holotype; SRGH, isotype).
    Asplenium cuneatum var. *angustatum* Sim, Ferns S. Afr.: 152, t. 78 fig. 2 (1892); op. cit. ed. 2: 162, t. 63 fig. 2 (1915). Type from S. Africa.
    Asplenium splendens var. *angustatum* (Sim) C. Chr. in Dansk Bot. Ark. **7**: 100 (1932). Type as above.

Rhizome 5 mm. in diam., erect, with tufted fronds and with linear-attenuate black entire hair-pointed rhizome-scales 8 mm. long. Frond erect to arching, not proliferous, thinly coriaceous. Stipe matt-brown, up to 17 cm. long, glabrous except for scales, similar to those on the rhizome, at the extreme base. Lamina deeply 2-pinnatifid to 2-pinnate, narrowly oblong in outline, apex acute, lowest pinnae not reduced. Pinnae up to 3·8 × 2·8 cm., trapeziform, deeply incised into 3–5 obcuneate lobes, terminal lobe acuminate and caudate, glabrous on both surfaces, the upper surface darker than the lower, with occasional small scales similar to those on the rhachis along the veins, pinna segments irregularly crenate along the outer margin. Rhachis matt-brown, with dark clathrate lanceolate scales up to 2 mm. long. Sori up to 8 mm. long, linear; indusium linear, membranous, entire.

Rhodesia. E: Umtali Distr., Cloudlands, 18.i.1959, *Chase* 7253 (BOL; K). **Mozambique.** MS: Garuso, Jaegersberg, 11.vii.1955, *Schelpe* 5611 (BM; BOL). Also in S. Africa. Lithophytic or low-level epiphyte in moist forest, 1095–1700 m.

27. **Asplenium lividum** Mett. ex Kuhn in Linnaea, **36**: 100 (1869). Type from Venezuela.

Rhizome up to 7 mm. in diam., suberect, with tufted fronds and with brown clathrate entire lanceolate sharply acute rhizome-scales c. 3 mm. long. Frond arching, not proliferous, firmly herbaceous. Stipe matt-brown, often darker towards the base, up to 20 cm. long, with scattered scales similar to those on the rhizome, otherwise glabrous at maturity. Lamina up to 36·5 × 8·8 cm., narrowly to very narrowly oblong, acute, pinnatifid to weakly 3-pinnatifid, lowest pinnae not reduced. Pinnae up to 7 × 2·5 cm., petiolate, unequally rhombic-attenuate, very unequally cuneate at the base and deeply pinnatifid into narrowly oblanceolate cuneate to linear sharply serrate lobes, the basal acroscopic lobe often oblong and larger than the rest, glabrous on both surfaces; veins flabellate. Rhachis matt-brown, glabrous at maturity. Sori up to 7 mm. long, linear, 1–6 per lobe; indusium linear, membranous, entire.

Rhodesia. E: Umtali Distr., Vumba Mts., 3.vii.1957, *Chase* 6571 (BOL; K; SRGH). **Mozambique.** Z: Gúruè, Serra do Gúruè, 26.ii.1966, *Torre & Correia* 14927 (LISC). Also in S. Africa, tropical Africa and tropical America. Lithophytic on humus-covered boulders in forest, 1300–1530 m.

28. **Asplenium aethiopicum** (Burm. f.) Becherer in Candollea, **6**: 23 (1935). Type from S. Africa.
    Trichomanes aethiopicum Burm. f., Fl. Cap. Prodr. in Fl. Ind.: 32 (err. 28) (1768). Type as above.
    Asplenium adiantoides Lam., Encycl. Méth. Bot. **2**: 309 (1786) non (L.) C. Chr. (1905). Type from S. Africa.

Asplenium falsum Retz., Obs. Bot. **6**: 38 (1791). Type from S. Africa.

Asplenium furcatum Thunb., Prodr. Pl. Cap.: 172 (1800). Type from S. Africa.

Tarachia furcata (Thunb.) C. Presl, Epim. Bot.: 80 (1851) reimpr. in Abh. Königl. Böhm. Ges. Wiss., Ser. 5, **6**: 440 (1851). Type as above.

Asplenium gueinzianum Mett. ex Kuhn, Fil. Afr.: 103 (1868). Type from S. Africa.

Asplenium praemorsum sensu Sim, Ferns S. Afr. ed. 2: 163, t. 65, 66 (1915) non Sw. (1788).

Rhizome up to 7 mm. in diam., creeping or ascending, with tufted fronds and with linear-attenuate dark-reddish-brown clathrate entire to pseudoserrate scales up to 6 mm. long ending in a short hair-point. Frond usually arching, not proliferous, firmly herbaceous to thinly coriaceous. Stipe up to 28 cm. long, matt-brown, covered at first with a mixture of scales similar to those on the rhizome with numerous smaller almost hair-like scales when young often becoming subglabrous with age. Lamina up to 48 × 20 cm., 2-pinnate to 3-pinnate, oblong to lanceolate in outline, acute to broadly acute with the lower pinnae not significantly reduced. Pinnae up to 12 × 4 cm., lanceolate in outline, acute to caudate, divided into narrowly obcuneate to very narrowly oblong segments, the upper segments adnate and decurrent, irregularly incised and serrate towards their apices, glabrous ventrally when mature and often with copious hair-like scales on the dorsal surface. Rhachis dark-brown, covered with mostly hair-like scales but with some lanceolate scales similar to those on the stipe. Sori up to 8 mm. long, linear; indusium very narrowly linear, membranous, subentire.

Zambia. N: Kasama Distr., Malole Rocks, Kasama, 2.iii.1960, *Richards* 12701 (K). W: Mwinilunga, 15.iv.1960, *Robinson* 3571 (K). C: Mkushi, 2.i.1958, *Robinson* 2570 (K; SRGH). E: Lundazi, Kangampande Mt., Nyika Plateau, 6.v.1952, *White* 2734 (BOL; K). **Rhodesia.** N: Darwin, Umsengese Camp, *Whellan* 871 (LISC; PRE; SRGH). W: Besna Kobila, i.1955, *Miller* 2603 (BM; BOL; SRGH). C: Ngomakurira, 21.ii.1959, *Mitchell* 479 (BOL; SRGH). E: Umtali, Vumba Mts., 18.i.1959, *Chase* 7044 (BOL; K; SRGH). S: Zimbabwe, 8.vii.1953, *Schelpe* 3980 (BOL). **Malawi.** N: Nyika Plateau, Rufiri St., 28.x.1959, *Robson* 443 (K; SRGH). C: Dedza Mt., 10.ix.1929, *Burtt Davy* 154 (K). S: Zomba Plateau, 28.v.1946, *Brass* 16053 (K; SRGH). **Mozambique.** N: Delichinga Plateau, xii.1932, *Gomes e Sousa* 1051 (COI; K). Z: Milange Mt., 26.ii.1943, *Torre* 4864 (LISC). T: between Vila Mouzinho and Zóbuè, 19.vii.1949, *Barbosa & Carvalho* 3690 (LISC). MS: Gorongosa Mt., Gogogo Peak, 5.vii.1955, *Schelpe* 5514 (BM; BOL).

Also in S. Africa and widespread in tropical Africa. An extremely variable species, lithophytic or epiphytic in forest, 1060–2140 m.

A sterile depauperate form referable to the *A. aethiopicum* complex (*Wild* 6200 (BOL; SRGH)) was collected high on the main peak of Mlanje Mt. Its widely creeping rhizome with spaced fronds and shiny rhizome scales indicate that it may represent a depauperate form of either *A. volkensii* Hieron. or *A. goetzei* Hieron., or possibly an undescribed species.

29. **Asplenium buettneri** Hieron. in Wiss. Ergebn. Deutsch. Z. Afr. Exped. **2**: 23, t. 2 fig. 2 (1910). Syntypes from tropical Africa.

Rhizome up to 4 mm. in diam., creeping, with fronds up to 1 cm. apart, and with subulate dark-brown entire rhizome-scales up to 4 mm. long ending in a hair-point and with marginal cells having thinner walls than the central cells. Frond erect, not proliferous, thinly coriaceous to thinly herbaceous. Stipe up to 18 cm. long, dull brown at least at the base, paler above in younger specimens, set with dark clathrate lanceolate scales up to 1·5 mm. long, often ciliate at the base. Lamina up to 25 × 12 cm., narrowly oblong to triangular-ovate in outline, acute; basal pinnae not reduced. Pinnae up to 7 × 3·5 cm., elongate-obcuneate or rhombic above, gradually becoming dissected towards the base into cuneate irregularly serrate segments, the larger segments petiolate, pinnae glabrous on both surfaces, darker green ventrally than dorsally. Rhachis black dorsally, green and sulcate ventrally, with clathrate brown lanceolate scales ciliate at the base, up to 1·5 mm. long. Sori up to 8 mm. long, linear; indusium linear, membranous, entire.

Zambia. N: Mporokoso, 8 km. S. of Chienge, 5.xi.1952, *Angus* 719 (BM; BOL; K). S: Mapanza, 7.ii.1954, *Robinson* 514 (K). **Malawi.** N: Nkata Distr., Nkata Bay, slope in Kandoli Forest, 20.ii.1961, *Richards* 14414 (K; SRGH). **Mozambique.** MS: Mile

10, Trans-Zambesi Railway, iii.1921, *Honey* 723 (BOL; K; PRE). SS: R. Inhanombe near Inhambane, vii.1954, *Eccles* (BM; BOL).
Widespread in tropical Africa. Deeply shaded forest floors, 600–1070 m.

30. **Asplenium blastophorum** Hieron. in Engl., Bot. Jahrb. **46**: 378 (1911). Syntypes from W. Africa and southern Sudan.

Rhizome creeping, with tufted fronds and with dark-brown subulate entire rhizome-scales up to 5 mm. long. Frond arching, proliferating on the rhachis near the apex, thinly coriaceous. Stipe up to 35 cm. long, matt-black set with brown subulate or narrowly lanceolate scales similar to rhizome scales. Lamina up to 48 × 20 cm., 2-pinnate at least at the base, narrowly oblong to triangular, the lowest pinnae slightly larger than those above, apex acute. Pinnae up to 11 × 4·5 cm., lanceolate in outline, unequally cuneate at the base, attenuate, progressively deeply divided towards the base into oblong-cuneate or rhombic pinnae with the outer margins sharply serrate and irregularly shallowly incised (free pinnules with cuneate to narrowly cuneate bases), glabrous except for small lanceolate dark scales near the extreme base. Rhachis matt, with occasional dark subulate scales, sulcate ventrally. Sori up to 1·3 cm. long, linear; indusium linear, narrow, membranous, entire.

Zambia. S: Mumbwa Distr., Miukuikui I., 10.x.1965, *Mitchell* 3009 (SRGH). **Rhodesia.** E: Chirinda Forest, 19.vi.1906, *Swynnerton* 845a (K). **Malawi.** S: Blantyre, Shire Highlands, vi.1887, *Last* (K). **Mozambique.** Z: Namúli Mt., R. Licungo, 29.vii.1962, *Schelpe & Leach* 7090 (BOL). MS: Manica Distr., Bandula Forest, 1.iii.1958, *Chase* 6839 (K).
Also in S. Africa and widespread in tropical Africa. Lithophytic on wet rocks in deep shade in forest, 600–1220 m.

31. **Asplenium chaseanum** Schelpe in Bol. Soc. Brot., Sér. 2, **41**: 206 (1967). Type: Zambia, Fort Rosebery Distr., *White* 3163 (BOL, holotype; K, isotype).

Rhizome 4 mm. in diam., erect, with tufted fronds and with dark-brown subulate entire clathrate rhizome-scales up to 3 mm. long. Fronds arching, not proliferous, thinly coriaceous. Stipe black, up to 11 cm., set sparsely with attenuate clathrate brown scales c. 2 mm. long. Lamina up to 12 × 7 cm., deeply 2-pinnatifid, oblong-acute in outline with the basal pinnae slightly reduced. Pinnae up to 5 × 1·5 cm., elongate-trapeziform, deeply pinnatifid into oblong to narrowly oblong lobes irregularly serrate at their apices, glabrous on both surfaces, veins obscure. Rhachis black, with narrow greenish wings in the upper part, glabrous except for a few scales similar to those on the stipe. Sori up to 1 cm. long, linear; indusium linear, membranous, entire.

Zambia. N: Abercorn Distr., Chilongowelo Escarpment, 21.i.1963, *Richards* 17529 (K).
Not known from elsewhere. Shaded earth-banks in forest, c. 1500 m.

32. **Asplenium linckii** Kuhn, Fil. Deck.: 22 (1867). Type from Tanzania.
    Asplenium daubenbergii Rosenst. in Fedde, Repert. **4**: 2 (1907). Type from Tanzania.

Rhizome up to 5 mm. in diam., creeping, with loosely tufted fronds, shining brown entire lanceolate clathrate rhizome-scales up to 5 mm. long. Frond arching, not proliferous, firmly herbaceous. Stipe matt-brown, with appressed scales similar to those of the rhizome near the base, subglabrous above. Lamina up to 33 × 27 cm., 3–4-pinnate, ovate-triangular in outline, basal pinnae the largest. Pinnae up to 15 × 12 cm., triangular in outline, ultimate segments up to 8 × 2 mm., narrowly cuneate, sharply serrate at the apex, glabrous on both surfaces. Rhachis and secondary rhachises dark-brown, with sparse ovate-acuminate pale-brown clathrate scales c. 1 mm. long. Sori up to 4 mm. long, 1–2 per ultimate segment; indusium linear, membranous, entire.

Rhodesia. E: Vumba Mts., Umtali Distr., 3.vii.1957, *Chase* 6576 (BOL; K; PRE; SRGH). **Malawi.** S: Zomba Rock, 1896, *Whyte* (K).
Also in east tropical Africa. Deeply shaded and moist forest floors, c. 1500 m.

33. **Asplenium sandersonii** Hook., Sp. Fil. **3**: 147, t. 179 (1860).—Sim, Ferns S. Afr.

ed. 2: 139, t. 43 fig. 1 (1915).—Tardieu in Mém. Inst. Fr. Afr. Noire, **28**: 175, t. 33 fig. 4–5 (1953). TAB. **53** fig. B. Type from S. Africa.

Asplenium vagans Bak., Syn. Fil.: 195 (1867). Syntypes from S. Tomé and Madagascar.

Asplenium debile Mett. ex Kuhn, Fil. Afr.: 101 (1868) non Fée (1865). Type from Comoro Is.

Asplenium melleri Mett. ex Kuhn, tom. cit.: 106 (1868). Type from Madagascar.

Asplenium punctatum Mett. ex Kuhn, tom. cit.: 114 (1868). Type from S. Tomé.

Asplenium hanningtonii Bak. in Journ. of Bot. **21**: 245 (1883). Type from Tanzania.

Asplenium comorense C. Chr., Ind. Fil.: 105 (1906). Type from Comoro Is.

Rhizome c. 3 mm. in diam., erect to suberect, with tufted fronds and with brown lanceolate-attenuate entire clathrate rhizome-scales up to 3·5 mm. long. Fronds arching, carnose, coriaceous, proliferating with a bud at the end of a naked extension of the rhachis. Stipe stramineous, up to 7 cm., sulcate when dry, glabrous except for occasional subulate clathrate pseudoserrate scales up to 1·5 mm. long. Lamina up to 16·5 × 2·5 cm., simply pinnate, pinnae petiolate, rhombic-dimidiate to cuneate tending to lunate; larger pinnae up to 1·3 × 0·7 cm., shallowly lobed into up to 9 entire obtuse lobes on the acroscopic margins, subglabrous with occasional minute brown stellate scales c. 0·5 mm. in diam. on the dorsal surface. Rhachis stramineous, with narrow green wings, set with sparse subulate brown scales and minute paler substellate scales. Sori 2 mm. long, oblong, up to 5 per pinna; indusium oblong, longly ciliate, semi-transparent.

Rhodesia. E: Vumba Mts., 28.xii.1946, *Fisher* 1098 (BM; BOL; K; SRGH). **Malawi.** C: Kota Kota Distr., Mchinji Hills, 31.vii.1946, *Brass* 17065 (BM; K; SRGH). S: Gochi, Kirk Range, 30.i.1959, *Robson* 1372 (K; SRGH). **Mozambique.** N: Ribáuè, Serra de Ribáuè, 16.vi.1935, *Torre* 853 (COI; LISC). Z: Namúli Mt., 29.vii.1962, *Schelpe & Leach* 7089 (BOL). MS: Chimanimani Mts., 8.ii.1958, *Mitchell* 324 (BM; BOL; SRGH).

Also in S. Africa and widespread in tropical Africa, Madagascar and Comoro Is. Common gregarious epiphyte in moist forest, 970–1890 m.

34. **Asplenium dregeanum** Kunze in Linnaea, **10**: 517 (1836). Type from S. Africa.

Asplenium brachypteron Kunze [in Linnaea, **23**: 232 (1850) *nom. nud.*] ex Houlst. & Moore in Gard. Mag. Bot. **3**: 260 (1851). Type from Sierra Leone.

Asplenium gracile A. Peter in Fedde, Repert. Beih. **40**: 73, t. 5 fig. 2 (1929) non D. Don (1825). Type from Tanzania.

Rhizome 3 mm. in diam., erect, with tufted fronds and with lanceolate to narrowly ovate dark-brown clathrate entire rhizome-scales c. 1–5 mm. long. Frond arching, herbaceous, proliferating below the deeply pinnatifid apical segment with 3–5 lobes. Stipe greyish-brown with narrow green wings when fresh, up to 17 cm. long, usually less than ½ the length of the lamina, often uniform-grey-brown when dry with occasional clathrate scales similar to those of the rhizome becoming more frequent towards the base. Lamina up to 39 × 6 cm., deeply 2-pinnatifid, very narrowly oblong to narrowly oblong-elliptic in outline, apex acute to cuneate, lowest pinnae hardly reduced. Pinnae up to 3·5 × 1·1 cm., up to 32 pairs, petiolate, oblong or narrowly oblong in outline, variously acute-auriculate with the acroscopic basal lobe 2–4-fid, most of the other segments very narrowly oblong-obtuse (some larger segments 2-fid) up to 6 × 1·5 mm., subglabrous with a few minute substellate scales on the veins. Rhachis matt-greyish-green when dried, with occasional pale-brown minute substellate scales. Sori c. 2 mm. long, broadly elliptic at maturity; indusium elliptic, entire, membranous.

Zambia. N: Isoka, Mafingi Mts., 21.xi.1952, *Angus* 811 (BM; K). E: Nyika Plateau, 25.x.1958, *Robson & Angus* 345 (K; LISC; SRGH). **Rhodesia.** E: Umtali, Vumba Mts., Chinakwarimba, 8.xii.1948, *Chase* 1115 (BM; PRE; SRGH). **Malawi.** N: Misuku Hills, 11.i.1959, *Robinson* 3160 (K; SRGH). C: Nchisi Mt., Kota Kota Distr., 29.vii.1946, *Brass* 17028 (BM; K; SRGH). S: Cholo Mt., 20.ix.1946, *Brass* 17659 (BM; K; SRGH). **Mozambique.** N: Ribáuè, Serra de Ribáuè, 25.i.1964, *Torre & Paiva* 10239 (LISC). Z: Namúli, 1887, *Last* (K). MS: Serra da Gorongosa, Gogogo, 5.x.1946, *Simão* 958 (LM; SRGH).

Also in S. Africa and widespread in tropical Africa. Lithophytic on moist shaded boulders in forest, 1300–2150 m.

35. **Asplenium pseudoauriculatum** Schelpe in Bol. Soc. Brot., Sér. 2, **41**: 206 (1967).
TAB. **54** fig. B. Type: Mozambique, Manica e Sofala, Garuso, Jaegersberg, *Schelpe* 5626 (BM, isotype; BOL, holotype).

Rhizome 8 mm. in diam., erect, with tufted fronds and with dark-brown lanceo-late-attenuate subentire concolorous rhizome-scales 5–7 × 2 mm. Frond arching, thinly coriaceous, proliferous at the base of the deeply pinnatifid lanceolate apical segment. Stipe greyish-green when dried, c. 15 cm. long, with sparse dark-brown lanceolate scales up to 2 mm. long. Lamina up to 43 × 13 cm., deeply 2-pinnatifid, oblong lanceolate in outline, acuminate, basal pinnae hardly reduced. Pinnae up to 8·5 × 2 cm., petiolate, lanceolate-attenuate, unequally cuneate at the base, deeply pinnatifid into linear or very narrowly oblong-acute or oblanceolate 2-fid lobes up to 8 mm. long, basal acroscopic lobe broadly cuneate, shallowly incised in the upper ½ into 4–7 acute short lobes, glabrous except for a few substellate dark-brown minute scales on the lower surface. Rhachis matt-greyish-green with minute substellate to very narrowly lanceolate dark-brown scales. Sori up to 6 mm. long, linear, slightly curved; indusium very narrowly oblong, membranous, entire.

Rhodesia. E: Melsetter Distr., Chirinda Forest, 15.v.1953, *Chase* 4971 (BM; SRGH). **Malawi.** C: Kota Kota Distr., Nchisi Mt., 28.vii.1946, *Brass* 17007 (K; SRGH). S: Cholo Mt., 23.ix.1946, *Brass* 17762 (BM). **Mozambique.** Z: Namúli, 1887, *Last* (K). MS: Gorongosa, Serra da Gorongosa, 22.x.1965, *Torre & Pereira* 12534 (LISC). Also in E. Congo. Lithophytic on moist shaded boulders in forest, 1090–1650 m.

36. **Asplenium rutifolium** (Berg.) Kunze in Linnaea, **10**: 521 (1836). Type from S. Africa.
Caenopteris rutifolium Berg. in Act. Petropol. **1782**, 2: 249, t. 7 fig. 2 (1786). Type as above.
Adiantum achilleifolium Lam., Encycl. Méth., Bot. **1**: 43 (1783). Type from S. Africa.
Asplenium achilleifolium (Lam.) C. Chr., Ind. Fil.: 99 (1905) non Liebm. (1849). Type as above.
Asplenium bipinnatum sensu Sim, Ferns S. Afr. ed. 2: 169, t. 71 (1915) non (Forsk.) C. Chr. (1910).

Var. **bipinnatum** (Forsk.) Schelpe in Journ. S. Afr. Bot. **30**: 194 (1964). TAB. **54** fig. C. Type from Yemen.
Lonchitis bipinnata Forsk., Fl. Aegypt.-Arab.: CXXIV, 184 (1775). Type as above.
Caenopteris furcata Berg. in Act. Petropol. **1782**, 2: 249, t. 7 fig. 1 (1786). Type from Réunion.
Adiantum borbonicum Jacq., Collect. **3**: 286, t. 21 fig. 1 (1789). Type from Réunion.
Darea furcata (Berg.) Willd. in L., Sp. Pl., ed. 4, **5**: 297 (1810). Type as for *Caenopteris furcata*.
Darea disticha Kaulf., Enum. Fil.: 180 (1824) *nom. illegit.* Type from Yemen.
Caenopteris disticha Spreng. in L., Syst. Nat., ed. 16, **4**: 91 (1827). Type as above.
Darea stans Bory in Bélanger, Voy. Bot. **2**: 53 (1833). Type from Mascarene Is.
Asplenium stans (Bory) Kunze in Linnaea, **10**: 521 (1836). Type as above.
Asplenium borbonicum (Jacq.) Hook., Sp. Fil. **3**: 207 (1860). Type as for *Adiantum borbonicum*.
Asplenium distichum (Kaulf.) Mett. ex Salomon, Nomencl. Gefässkrypt.: 84 (1883). Type as for *Darea disticha*.
Asplenium bipinnatum (Forsk.) C. Chr. apud Hieron. in Wiss. Ergebn. Deutsch. Z.-Afr. Exped. **2**: 11 (1910). Type as for *Asplenium rutifolium* var. *bipinnatum*.
Asplenium linearilobum A. Peter in Fedde, Repert. Beih. **40**, 1: 80, t. 2 fig. 7–8 (1929). Type from Tanzania.

Rhizome erect, with tufted fronds and with lanceolate to narrowly ovate dark-brown slightly fimbriate acute rhizome-scales up to 4 mm. long with broad paler borders. Frond erect to arching, not proliferous, thinly to thickly coriaceous. Stipe pale brown to greyish-green when dry (green when fresh), up to 16 cm. long, glabrous except for scattered minute ovate acuminate dark-brown scales, becoming glabrous with age. Lamina up to 15 × 7·5 cm., deeply pinnatifid to 2-pinnatifid, narrowly oblong-acute in outline with basal pinnae hardly or not reduced, pinnate, apex acute; pinnae up to 5 × 1·2 cm., up to 15 pairs, petiolate, oblong attenuate or obtuse, deeply pinnatifid into mostly linear or very narrowly oblong-obtuse

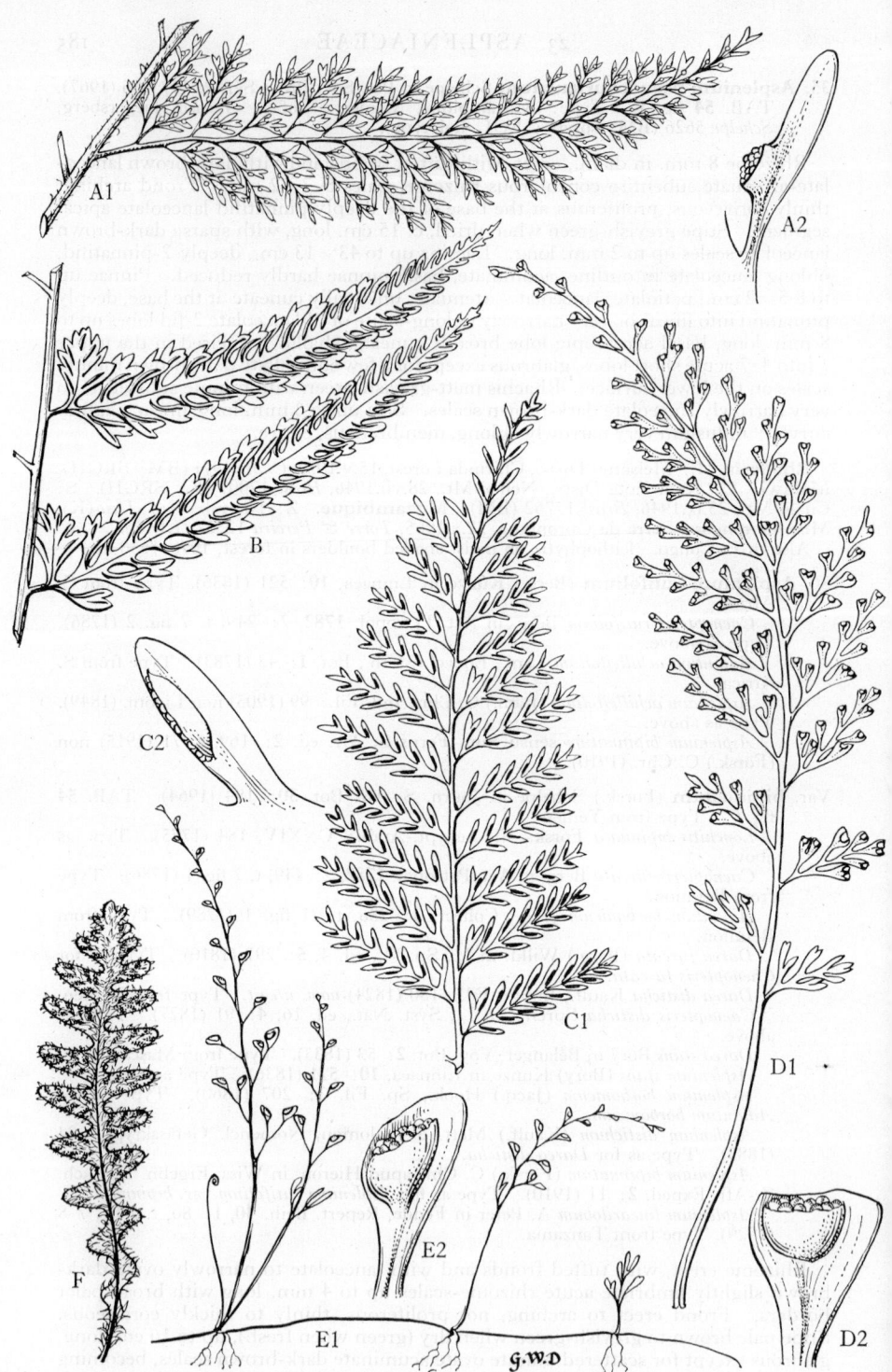

Tab. 54. A.—ASPLENIUM HYPOMELAS. A1, pinna (×⅔); A2, fertile segment (×5), both from *Fisher & Schweickerdt* 309. B.—ASPLENIUM PSEUDOAURICULATUM. Pinnae (×⅔) *Chase* 4971. C.—ASPLENIUM RUTIFOLIUM var. BIPINNATUM. C1, apex of lamina (×⅔); C2, fertile segment (×5), both from *Pedro & Pedrógão* 7289. D.—ASPLENIUM THECIFERUM var. CONCINNUM. D1, fertile frond (×⅔); D2, fertile segment (×5), both from *Chase* in GHS 43376. E.—ASPLENIUM MANNII. E1, habit (×⅔); E2, fertile segment (×5), both from *Fisher* 1240. F.—CETERACH CORDATUM. Fertile frond (×⅔) *Chase* 5519.

segments c. 1 mm. wide but with a 2–9 lobed acroscopic basal segment, or with some of the lower segments occasionally 2-fid, glabrous on both surfaces except for occasional minute dark narrowly ovate scales dorsally. Rhachis pale-brown when dry with occasional minute dark-brown ovate-lanceolate scales. Sori 2 mm. long, 1 per pinna lobe, linear, borne ½ way along the length of the lobe; indusium very narrowly oblong, membranous, entire.

Rhodesia. E: Melsetter, Chimanimani Mts., 2.ii.1958, *Mitchell* 282 (BM; BOL; SRGH). **Malawi.** N: Misuku Hills, Karonga Distr., 11.i.1959, *Robinson* 3156 (SRGH). C: Dedza Mt., 18.v.1963, *Chapman* 2062 (SRGH). S: Mlanje Distr., Luchenya Plateau, 2.vii.1946, *Brass* 16605 (K; PRE; SRGH). **Mozambique.** Z: Morrumbala, viii.1864, *Waller* (BM; K). T: Monte de Zóbuè, 3.x.1942, *Mendonça* 626 (BM; LISC). MS: Mossurize, Espungabera, Macuiana, 30.x.1944, *Mendonça* 2692 (BM; LISC). LM: Namaacha, Libombos, *Barbosa* 115 (BM; COI; LM).

Widespread in E. tropical Africa, Yemen, Madagascar and the Mascarene Is. Common lithophyte and epiphyte in forest, 880–2010 m.

36a. **Asplenium loxoscaphaides** Bak.
    Malawi N: *Simon, Williamson & Ball* 1802 (BOL; SRGH).

37. **Asplenium mannii** Hook., Sec. Cent. Ferns: t. 60 (1861).—Sim, Ferns S. Afr. ed. 2: 174, t. 61 (1915).—Tardieu in Mém. Inst. Fr. Afr. Noire, **28**: 198, t. 39 fig. 8–9 (1953). TAB. **54** fig. E. Type from Fernando Po.

Rhizome 1–3 mm. in diam., short, erect, with tufted fronds and with clathrate dark-brown triangular subentire acute rhizome-scales c. 1·7 mm. long. Frond erect, not proliferous, thinly coriaceous. New plants produced at intervals of 2–6 cm. along leafless green stolons. Stipe greyish-green when dried, up to 6·5 cm. long, glabrous. Lamina up to 5 × 2·5 cm., broadly lanceolate in outline attenuate mostly pinnate but with the lowest pinnae 2-fid and larger than those above. Pinnae up to 1·5 cm., up to 4 pairs, with linear terminal lobes mostly obliquely spathulate-obtuse, glabrous on both surfaces. Rhachis glabrous, narrowly winged. Sori up to 2·5 mm. long, solitary on each spathulate segment, oblong; indusium oblong, membranous, subentire.

Zambia. N: Abercorn Distr., Chisau R. Gorge, 18.ii.1959, *Richards* 1180 (K). E: Lundazi Distr., Nyika Plateau, 6.v.1952, *White* 2710 (BM; K). **Rhodesia.** E: Umtali, Cloudlands, Vumba, 3.vii.1948, *Fisher & Schweickerdt* 219 (BM; K; NU; PRE; SRGH). **Malawi.** N: Mafinga Hills, 12.iii.1961, *Robinson* 4453 (K; SRGH). C: Nchisi Mt., Kota Kota Distr., 31.vii.1946, *Brass* 17064 (BM; K; PRE; SRGH). S: Cholo Mt., 20.ix.1946, *Brass* 17677 (BM; K; PRE; SRGH). **Mozambique.** Z: Namúli, 1887, *Last* (K). MS: Chimanimani Mts., 7.ii.1958, *Mitchell* 314 (BM; COI; LISC; SRGH).

Widespread in tropical Africa. Common gregarious epiphyte in shade in moist forest, 1350–2140 m.

38. **Asplenium hypomelas** Kuhn, Fil. Afr.: 104 (1868). TAB. **54** fig. A. Type from Fernando Po.
    Davallia nigrescens Hook., Sec. Cent. Ferns: t. 93 (1861) non *Asplenium nigrescens* Bl. (1828). Type from Fernando Po.
    Loxoscaphe nigrescens (Hook.) Moore, Ind. Fil.: 297 (1861). Type as above.
    Davallia hollandii Sim in Trans. S. Afr. Phil. Soc. **16**: 274, t. 4 (1906). Type: Mozambique, Penhalonga, *Holland* (NBG, isotype; PRE, holotype).
    Asplenium floccigerum Rosenst. in Fedde, Repert. **4**: 3 (1907). Type from Tanzania.
    Asplenium hollandii (Sim) C. Chr., Ind. Fil., Suppl.: 11 (1913). Type as for *Davallia hollandii*.

Rhizome erect, often long, with tufted fronds and with narrowly lanceolate attenuate pale-brown sparsely fibrillose rhizome-scales up to 1·3 cm. long with short hair-points. Frond arching, not proliferous, herbaceous, dark-green when fresh. Stipe matt-brown, up to 15 cm. long, sulcate, covered with narrowly lanceolate hair-pointed fibrillose scales similar to the rhizome scales at first becoming glabrous with age. Lamina up to 90 × 46 cm., lanceolate to broadly lanceolate, deeply 4-pinnatifid, apex acuminate and with the basal pinnules somewhat reduced. Pinnae up to 26 × 9 cm., up to 25 pairs, lanceolate in outline, acuminate, 2-pinnatifid into oblanceolate acute ultimate segments c. 1·2 mm. broad, glabrous on both surfaces, unequally expanded around the sori. Rhachis and secondary rhachises matt-brown with brown hair-pointed scales; secondary rhachises narrowly winged for most of their length. Sori c. 1·3 mm. long, broadly oblong; indusium broadly elliptic to broadly oblong, membranous, entire.

Rhodesia. E: Umtali, Pioneer Farm, vii.1948, *Fisher & Schweickerdt* 309 (BM; K; NU; PRE; SRGH). S: Bikita, 28.viii.1922, *Nobbs* in *Eyles* 3643 (BOL; PRE; SAM; SRGH). **Malawi.** N: Nyika Plateau, 16.viii.1946, *Brass* 17274 (BM; K; SRGH). **Mozambique.** MS: N. of Penhalonga, 16.iv.1950, *Chase* 3164 (SRGH).

Widespread in tropical Africa. Low-level epiphyte in moist forest, often on *Cyathea manniana*, 1830–2250 m.

39. **Asplenium theciferum** (Kunth) Mett. in Ann. Sci. Nat., Sér. 5, **2**: 227 (1864).
Type from Venezuela.

Davallia thecifera Kunth, Nov. Gen. Sp. Pl. **1**: 23 (1816). Type as above.

Var. **concinnum** (Schrad.) Schelpe in Bol. Soc. Brot., Sér. 2, **41**: 210 (1967). TAB. **54** fig. D. Type from S. Africa.

Davallia concinna Schrad. in Gött. Gel. Anz. **1818**: 918 (1818). Type as above.

Davallia campyloptera Kunze in Linnaea, **10**: 544 (1836). Type from S. Africa.

Loxoscaphe concinnum (Schrad.) Moore in Hook. Journ. Bot. **5**: 227 (1853). Type for *Asplenium theciferum* var. *concinnum*.

Asplenium concinnum (Schrad.) Kuhn, Fil. Afr.: 99 (1868). Type as above.

Asplenium theciferum sensu Sim, Ferns S. Afr. ed. 2: 171, t. 72 (1915) non (Kunth) Mett. (1864).

Loxoscaphe theciferum var. *concinnum* (Schrad.) C. Chr. in Dansk Bot. Ark. **7**: 104 (1932). Type as for *Asplenium theciferum* var. *concinnum*.

Rhizome c. 5 mm. in diam., erect, with tufted fronds and with lanceolate attenuate irregularly ciliate dark-brown rhizome-scales 3–4 mm. long. Frond erect to arching, not proliferous, carnose-coriaceous. Stipe greyish-green when dry, up to 18 cm. long, with occasional lanceolate acuminate brown scales c. 1 mm. long. Lamina up to 18 × 4·5 cm., deeply 2-pinnatifid, narrowly oblong-lanceolate, acute with basal pinnae hardly reduced. Pinnae up to 3·3 × 1·7 cm., oblong to narrowly oblong-obtuse in outline, deeply pinnatifid with obliquely spathulate lobes, the basal acroscopic lobe usually 2-fid. Rhachis grey-green when dried, with narrow wings and occasional minute dark-brown lanceolate scales. Sori up to 2 mm. long, oblong to broadly oblong, solitary on the ends of the lobes, usually subtended on one side by a large or small triangular area of the lamina, indusium oblong to broadly oblong, membranous, entire.

Zambia. N: Abercorn Distr., Jembele Forest, 13.i.1965, *Richards* 19515 (K). W: Mwinilunga Distr., Matonchi, 6.xi.1937, *Milne-Redhead* 3130 (K). **Rhodesia.** E: Inyanga, Romneydale, 22.iii.1949, *Chase* 2046 (BM; BOL; SRGH). S: Belingwe Distr., Mt. Buhwe, 10.xii.1953, *Wild* 4313 (BM; SRGH). **Malawi.** N: Misuku Hills, Mugesse Forest, ix.1953, *Chapman* 233 (BM). C; Nchisi Mt., 3.ix.1929, *Burtt Davy* 1150 (K). S: Zomba Distr., Zomba Plateau, 28.v.1946, *Brass* 16048 (K; SRGH). **Mozambique.** N: Ribáuè, Serra de Ribáuè, Mepaluè, 25.i.1964, *Torre & Paiva* 10214 (LISC). Z: Gúruè, Serra do Gúruè, 24.ii.1966, *Torre & Correia* 14823 (LISC). T: Mt. Zóbuè, 3.x.1942, *Mendonça* 634 (BM; LISC). MS: Gorongosa, Gogogo Peak, 5.vii.1955, *Schelpe* 5512 (BM; BOL).

Also in S. Africa. Mid- to high-level epiphyte in forest, 850–2040 m.

2. CETERACH DC.

Ceterach DC. in Lam. & DC., Fl. Fr., ed. 3, **2**: 566 (1805).

Rhizome erect, short, with clathrate scales. Stipes short, tufted, densely scaly. Frond usually deeply pinnatifid to 2-pinnatifid, glabrous ventrally at maturity but densely paleaceous dorsally; veins usually anastomosing marginally. Sori elongate along the veins; indusium obsolete.

A genus of c. 5 spp. in Europe, Asia and Africa.

Ceterach cordatum (Thunb.) Desv. in Mém. Soc. Linn. Par. **6**, 2: 223 (1827).—Sim, Ferns S. Afr. ed. 2: 175, t. 73 (1915). TAB. **54** fig. F. Type from S. Africa.

Acrostichum cordatum Thunb., Prodr. Pl. Cap.: 171 (1800). Type as above.

Asplenium cordatum (Thunb.) Sw. in Schrad., Journ. Bot. **1800**, 2; 54 (1801). Type as above.

Grammitis cordata (Thunb.) Sw., Syn. Fil.: 23, 217 (1806). Type as above.

Cincinalis cordata (Thunb.) Desv. in Mag. Ges. Naturf. Fr. Berl. **5**: 311 (1811). Type as above.

Notholaena cordata (Thunb.) Desv. in Journ. de Bot. **1**, App.: 92 (1813). Type as above.

Ceterach crenata Kaulf., Enum. Fil.: 85 (1824) *nom. illegit.* Type from S. Africa.
Gymnogramma cordata (Thunb.) Schlechtend., Adumbr.: 16 (1825). Type as for *Ceterach cordatum.*
Ceterach capense Kunze in Linnaea, **10**: 496 (1836). Type from S. Africa.
Grammitis capensis (Kunze) Moore, Ind. Fil.: 232 (1857). Type as above.

Rhizome up to 4 mm. in diam., erect, with tufted fronds and with clathrate lanceolate acuminate dark-brown pseudoserrate rhizome-scales up to 4·5 mm. long, frequently with a hair-point. Fronds suberect, not proliferous, thinly carnose coriaceous, inrolled when dry. Stipe atrocastaneous, relatively short, up to 4 mm. long, with dense broadly lanceolate acuminate scales up to 5 mm. long. Lamina up to 14·5 × 5 cm., elliptic to narrowly elliptic, pinnatifid to 2-pinnatifid, glabrous ventrally, densely set, with ovate-acuminate to lanceolate imbricate pale-brown scales dorsally; the lower pinnae gradually reduced; pinnae up to 2·5 × 0·9 cm., narrowly oblong, adnate to auriculate, weakly undulate to pinnatifid with broadly oblong crenate segments. Sori up to 2 mm. long, linear.

Rhodesia. N: Mazoe Distr., Umvukwes Range, Mtoroshanga Pass, ii.1957, *Mitchell* 123 (BOL). W: Matobo, Besna Kobila, x.1958, *Miller* 5501 (BOL; K; SRGH). C: Salisbury, Ruwa Rocks, 4.iii.1961, *Richards* 14524 (K). E: Umtali Distr., Umtali Commonage, 2.x.1960, *Chase* 7383 (K; SRGH). S: Victoria Distr., Zimbabwe, 1.vii.1930, *Hutchinson & Gillett* 3331 (K).
Also in S. Africa and sporadic in tropical Africa. Rock crevices and around bases of boulders, 1090–1500 m.

24. THELYPTERIDACEAE

Terrestrial plants. Rhizome creeping or erect, with brown non-peltate rhizome scales. Stipe not articulated, with 2 vascular strands at the base which fuse upwards to form a single U-shaped strand. Lamina pinnate to 2-pinnate (rarely 3-pinnatifid), usually narrowly oblong in outline, glabrous, pubescent or pilose; veins free, or few to many pairs of veins arising from adjoining costules anastomosing into a vein which runs to the sinus between the pinna lobes. Sori round, with or without a reniform indusium, to linear and exindusiate. Spores monolete, with perispore.

Fronds not proliferous or only proliferous near the apex; soral paraphyses absent
1. **Thelypteris**
Fronds proliferous anywhere on the rhachis; soral paraphyses present
2. **Ampelopteris**

1. THELYPTERIS Schmidel

Thelypteris Schmidel, Ic. Pl., ed. Kneller: 45, t. 11, 13 (1763).

Rhizome creeping or erect, with tufted or spaced fronds and with non-peltate brown rhizome-scales. Stipe not articulated. Lamina pinnate to deeply 2-pinnatifid (rarely 3-pinnatifid); veins free, or with 1 or more pairs of veins from adjoining costules anastomosing into a vein which runs to the sinus between the pinna lobes. Sori round to elongate, with or without an indusium; paraphyses absent.
A large cosmopolitan genus with c. 30 spp. in continental Africa.
Among species with round sori, those with anastomosing veins have previously been referred to the genus *Cyclosorus* Link while those with free veins have been referred to *Thelypteris* in a narrow sense. This distinction is untenable since both conditions may be found on a single frond. The separation of *Leptogramma* J. Sm., with elongate exindusiate sori, is similarly untenable when the variation in sorus shape exhibited by the tropical American species is taken into consideration. These generic names can, however, be useful at subgeneric level.
Sori circular, with or without indusia:
    Veins of the pinna lobes not anastomosing or the basal vein meeting at the sinus or in a membrane extending from the sinus towards the costa:

Fronds glabrous and without glands but with small ovate scales borne along the dorsal
surface of the costae   -   -   -   -   -   -   -   -   -   -   1. *confluens*
Fronds hairy or glandular at least on the costae and veins dorsally:
  Fronds abruptly decrescent with a series of much reduced pinnae at the base:
    Indusia glandular; rhizome erect   -   -   -   -   -   -   2. *longicuspis*
    Indusia pilose; rhizome creeping   -   -   -   -   -   -   3. *friesii*
  Fronds not decrescent or if abruptly decrescent with a long series of much reduced
    pinnae at the base:
    Basal veins quite free:
      Lamina with at least a few reddish glands dorsally; lamina bases long-tapering
                                                                    4. *strigosa*
      Lamina without red glands; lamina bases shortly decrescent   5. *bergiana*
    Basal veins meeting in the sinus between the pinna lobes or in a membrane extend-
      ing from the sinus:
      Rhizome erect; frond gradually decrescent   -   -   -   6. *gueinziana*
      Rhizome creeping; frond with 2 basal pairs of pinnae somewhat reduced and
        deflexed   -   -   -   -   -   -   -   -   -   -   7. *chaseana*
Veins of the pinna lobes with 1 or more pairs anastomosing below the sinus between the
  pinna lobes:
  1 pair of veins anastomosing below the sinus:
    Yellow glands present on the veins dorsally   -   -   -   -   8. *extensa*
    Hairs present on the veins dorsally:
      Lamina gradually decrescent; rhizome creeping   -   -   -   12. *dentata*
      Lamina not decrescent; rhizome erect   -   -   -   -   9. *quadrangularis*
  2 or more pairs of veins anastomosing below the sinus:
    Lamina proliferous, rhizome erect   -   -   -   -   10. *madagascariensis*
    Lamina not proliferous:
      Lamina decrescent:
        Costae and costules hairy   -   -   -   -   -   -   -   12. *dentata*
        Costae and costules glabrous   -   -   -   -   -   11. *prismatica*
      Lamina not decrescent:
        Veins 6–13 per pinna lobe; pinnae incised ½-way to the costa -   13. *totta*
        Veins 15–24 per pinna lobe; pinnae incised more than ½-way to the costa
                                                                    14. *striata*
Sori elongate, without indusia   -   -   -   -   -   -   -   -   15. *pozoi*

1. **Thelypteris confluens** (Thunb.) Morton in Contr. US. Nat. Herb. **38**: 71 (1967).
   TAB. **55** fig. E.   Type from S. Africa.
   Pteris confluens Thunb., Prodr. Pl. Cap.: 171 (1800).   Type as above.
   Aspidium thelypteris var. *squamigerum* Schlechtend., Adumbr.: 23, t. 11 (1825).
   Type from S. Africa.
   Lastrea squamulosa C. Presl, Tent. Pterid.: 76 (1836) reimpr. in Abh. Königl.
   Böhm. Ges. Wiss., Ser. 4, **5**: 76 (1837) *nom. nud.*   Type from S. Africa.
   Nephrodium squamulosum Hook. f., Fl. Nov. Zeyl. **2**, 2: 39 (1854).   Type from
   New Zealand.
   Aspidium squamigerum (Schlechtend.) Fée, Mém. Fam. Foug. **8**: 104 (1857).
   Type from S. Africa.
   Lastrea thelypteris var. *squamigera* (Schlechtend.) Bedd., Handb. Ferns Brit.
   Ind., Suppl.: 54 (1892).   Type as above.
   Dryopteris thelypteris var. *squamigera* (Schlechtend.) C. Chr., Ind. Fil.: 297
   (1905).   Type as above.
   Thelypteris palustris var. *squamigera* (Schlechtend.) Weath. in Contr. Gray Herb.
   n.s., **73**: 40 (1924).   Type as above.
   Thelypteris squamulosa (Hook. f.) Ching in Bull. Fan Mem. Inst. Biol. Bot. **6**:
   5, 329 (1936).   Type as for *Naphrodium squamulosum*.
   Dryopteris thelypteris sensu Sim, Ferns S. Afr. ed. 2: 101, t. 16 (1915).

Rhizome up to 3 mm. in diam., widely creeping, with fronds borne at intervals
up to 5 cm., and with dark-brown ovate somewhat undulate rhizome scales up to
2 mm. long, becoming black and subglabrous with age.   Frond erect (or somewhat
arching in shade forms), non-proliferous, very firmly herbaceous (or thinly herbaceous in shade forms).   Stipe up to 45 cm. long, pale-brown for most of its length
but black at the base, glabrous.   Lamina up to 50 × 21 cm., deeply 2-pinnatifid,
lanceolate to broadly lanceolate in outline, apex acute, basal pinnae slightly reduced.
Pinnae up to 11 × 2 cm., very narrowly oblong, acute, deeply pinnatifid into oblong,
obtuse to acute, entire or slightly undulate; lobes up to 1·2 × 0·4 cm., glabrous
ventrally, glabrous, glandular or thinly pilose and with pale-brown ovate scales
along the costa dorsally.   Rhachis pale-brown, thinly pilose to glabrous.   Sori

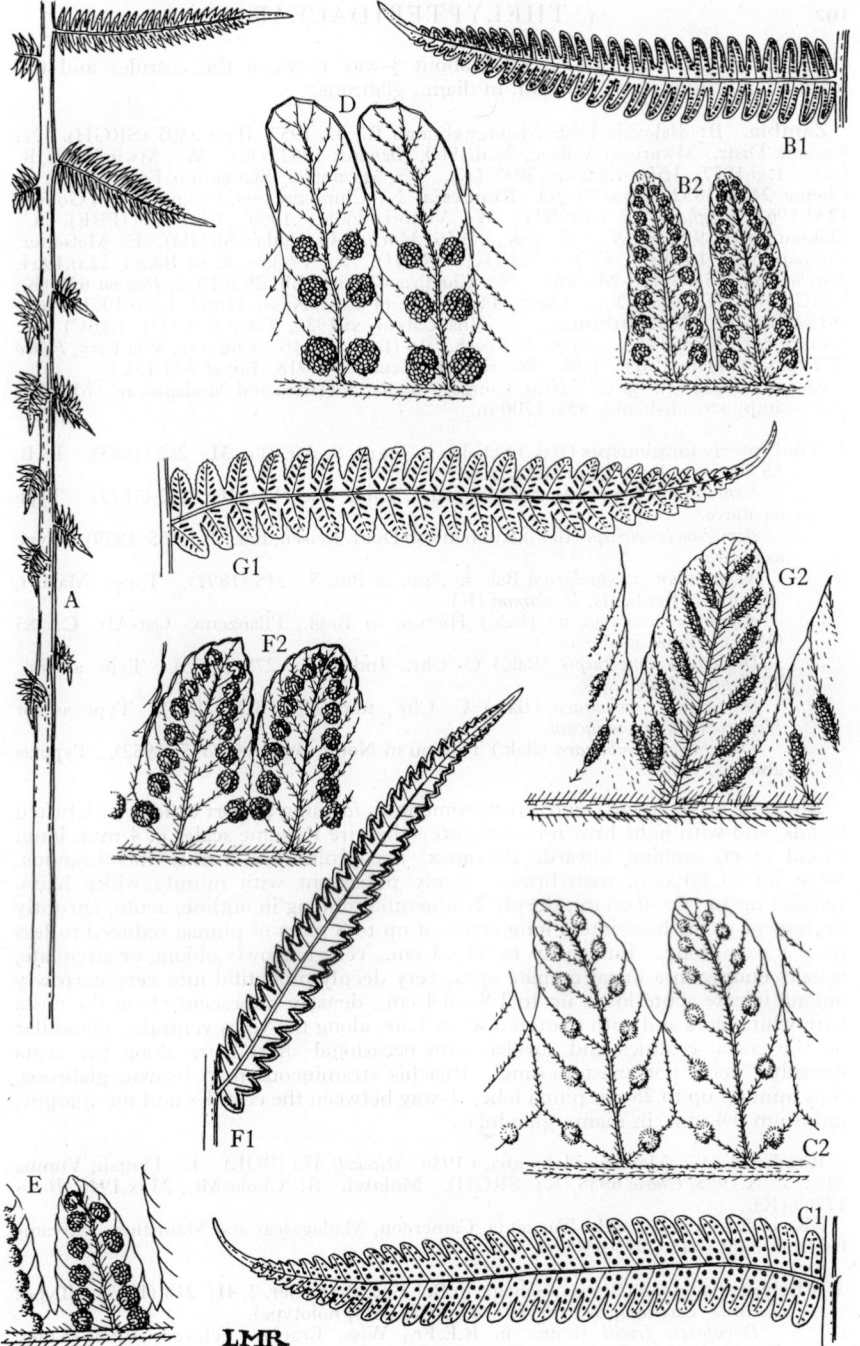

Tab. 55. A.—THELYPTERIS LONGICUSPIS. Base of frond (×⅔) *Chase* 3390. B.—THELYPTERIS
BERGIANA. B1, pinna (×⅔); B2, pinnule (×4), both from *Fisher* 1546. C.—
THELYPTERIS DENTATA var. DENTATA. C1, pinna (×⅔); C2, pinnule (×4). both
from *Schelpe* 4072. D.—THELYPTERIS MADAGASCARIENSIS. Pinnule (×4) *Chase*
13919. E.—THELYPTERIS CONFLUENS. Pinnule (×4) *Chase* 3915. F.—THELYP-
TERIS TOTTA. F1, pinna (×⅔); F2, pinnule (×4), both from *Torre* 5543. G.—
THELYPTERIS POZOI. G1, pinna (×⅓); G2, pinnule (×4), both from *Chase* 3038.

circular, up to 18 per pinna lobe, about ½-way between the costules and the margin; indusium up to 0·5 mm. in diam., glabrous.

Zambia. B: Balovale Distr., Lungwebungu R., 4.x.1957, *West* 3505 (SRGH). N: Kasama Distr., Mwaruski Village, 27.ii.1960, *Richards* 12627 (K). W: Mwinilunga, R. Luao, 1.xi.1937, *Milne-Redhead* 3047 (K). S: Mazabuka, Siamambo Forest Reserve, Choma, 24.vii.1952, *Angus* 32 (K). **Rhodesia.** N: Chimvuri Vlei, 3·2 km. SW. of Gokwe, 12.xi.1963, *Bingham* 905 (SRGH). W: Victoria Falls, i.1906, *Allen* 229 (PRE). C: Makoni Distr., 9·6 km. W. of Rusape, i.1956, *Mitchell* 44 (BOL; SRGH). E: Melsetter, Chimanimani, 1.ii.1957, *Phipps* 373 (K; SRGH). S: 1·6 km. E. of Bikita, 22.ii.1964, *Mitchell* 813 (SRGH). **Malawi.** N: Chadiza Chipiri Hill, 29.ii.1958, *Robson* 629 (K; LISC; SRGH). C: Dowa Distr., 4·8 km. W. of Lake Nyasa Hotel, 1.viii.1951, *Chase* 3915 (SRGH). **Mozambique.** N: Vila Cabral, xi.1933, *Torre* 6 (COI; LISC). Z: Namúli Mt., 26.vii.1962, *Schelpe & Leach* 7016 (BOL). MS: Chimoio, Vila Pery, *Pedro & Pedrógão* 6095 (BM). LM: Rikatla, Marracuene, i.1918, *Junod* 7 (LISC).

Also in Sudan, E. tropical Africa, Congo, Angola, S. Africa and Madagascar. Marshes and swampy stream-banks, 855–1700 m.

2. **Thelypteris longicuspis** (Bak.) Schelpe in Journ. S. Afr. Bot. **31**; 262 (1965). TAB. 55 fig. A. Type from Madagascar.
   Nephrodium longicuspe Bak. in Journ. Linn. Soc., Bot. **16**: 202 (1877). Type as above.
   Aspidium longicuspe (Bak.) Kuhn in von Deck. Reisen, Bot., **3**, 3: 65 (1879). Type as above.
   Nephrodium zambesiacum Bak. in Ann. of Bot. **5**: 318 (1891). Type: Malawi, Zambezi Highlands, *Buchanan* (K).
   Aspidium zambesiacum (Bak.) Hieron. in Engl., Pflanzenw. Ost-Afr. **C**: 85 (1895). Type as above.
   Dryopteris longicuspis (Bak.) C. Chr., Ind. Fil.: 275 (1905). Type as for *Thelypteris longicuspis*.
   Dryopteris zambesiaca (Bak.) C. Chr., tom. cit.: 318 (1905). Type as for *Nephrodium zambesiacum.*
   Thelypteris zambesiaca (Bak.) Tardieu in Notul. Syst. **14**: 345 (1952). Type as above.

Rhizome up to 30 cm. high, erect, sometimes forming a short caudex with tufted fronds, and with light-brown ovate acute subentire rhizome scales c. 4 mm. long. Frond erect, arching towards the apex, non-proliferous, firmly membranous. Stipe up to 60 cm., matt-brown, thinly pubescent with minute white hairs. Lamina up to 1·5 × 0·66 m., deeply 2-pinnatifid, oblong in outline, acute, abruptly decrescent at the base into a long series of up to 8 pairs of pinnae reduced to less than 2·5 cm. long. Pinnae up to 33 × 3 cm., very narrowly oblong or attenuate, usually ending in a linear crenate apex, very deeply pinnatifid into very narrowly oblong falcate acute lobes up to 1·8 × 0·4 cm., densely pubescent along the costa with white hairs and with scattered white hairs along the veins ventrally, glandular on the costa, costules and surface with occasional short hairs along the costa dorsally; veins not anastomosing. Rhachis stramineous matt-brown, glabrous. Sori minute, up to 25 per pinna lobe, ½-way between the costules and the margin; indusium 0·4 mm. in diam., glandular.

Rhodesia. C: Makoni, Mahamba, i.1956, *Mitchell* 37a (BOL). E: Umtali, Vumba Mts., 25.ix.1958, *Chase* 6988 (K; SRGH). **Malawi.** S: Cholo Mt., 23.ix.1946, *Brass* 17760 (K).

Also in Ethiopia, Uganda, Tanzania, Cameroon, Madagascar and Mauritius. Stream-banks in light shade, 975–1900 m.

3. **Thelypteris friesii** (Brause) Schelpe in Bol. Soc. Brot., Sér. 2, **41**: 216 (1967). Type: Zambia, Luvingo, 24.x.1911, *Fries* 1107 (UPS, holotype).
   Dryopteris friesii Brause in R.E.Fr., Wiss. Ergebn. Schwed. Rhod.-Kong. Exped. Bot. **1**: 1 (1914). Type as above.

Rhizome creeping, with fronds spaced 1–4 cm. apart, and with lanceolate or acute entire dark-brown rhizome-scales up to 4·5 mm. long. Frond erect, non-proliferous, firmly membranous. Stipe up to 38 cm. long, greyish-brown, thinly pubescent with minute white hairs. Lamina up to 1·7–0·48 m., deeply 2-pinnatifid, narrowly oblong in outline, acute with the base suddenly decrescent into a widely spaced series (up to 6 pairs) of much reduced pinnae less than 2 cm. long. Pinnae up to 26 × 2·5 cm., very narrowly oblong attenuate and deeply pinnatifid into very narrowly oblong falcate acute lobes; pinna lobes up to 1·5 × 0·3 cm., costa with dense white hairs, more scattered along the veins, ventrally; costae, costules, veins and lamina pilose with soft white hairs dorsally; veins not anastomosing with up to 13 pairs of veins per lobe. Rhachis stramineous to pale-brown, pilose with soft white hairs. Sori up to 20 per lobe, ½-way between the costules and the margin; indusium 0·5 mm. in diam., glandular, with few or many long white hairs.

Zambia. N: Abercorn Distr., Abercorn, 23.iii.1960, *Richards* 12789 (K). W: Mwinilunga Distr., Matonchi R., 6.xi.1937, *Milne-Redhead* 3141 (K). **Rhodesia.** E: Umtali, Eastlands, 10.viii.1952, *Chase* 4627 (BOL). **Malawi.** N: Mafinga foothills above Chiuagu 11.i.1958, *Robson & Fanshawe* 568 (K; SRGH). C: Dedza, 2.ii.1959, *Robson* 1423 pro parte (K; LISC; SRGH).
Not yet found elsewhere. Stream-banks in light shade, 1130–1705 m.

4. **Thelypteris strigosa** (Willd.) Tardieu, Fl. Madag., Polypod. **1**: 274 (1958). Type from Mauritius.
    Aspidium strigosum Willd. in L., Sp. Pl., ed. 4, **5**: 249 (1810). Type as above.
    Aspidium pulchrum Bory ex Willd. in L., tom. cit.; 253 (1810). Type from Réunion.
    Nephrodium strigosum (Willd.) Desv. in Mém. Soc. Linn. Par. **6**, 2: 256 (1857). Type as for *Thelypteris strigosa*.
    Nephrodium pulchrum (Bory ex Willd.) Desv., loc. cit. Type as for *Aspidium pulchrum*.
    Lastrea strigosa (Willd.) C. Presl, Tent. Pterid.: 75 (1836) reimpr. in Abh. Königl. Böhm. Ges. Wiss., Ser. 4, **5**: 75 (1837). Type as for *Thelypteris strigosa*.
    Lastrea pulchra (Bory ex Willd.) C. Presl, loc. cit., reimpr. loc. cit. Type as for *Aspidium pulchrum*.
    Dryopteris strigosa (Willd.) C. Chr., Ind. Fil.: 295 (1905). Type as for *Thelypteris strigosa*.

Rhizome c. 6 mm. in diam., erect with tufted fronds, and with narrowly ovate acute concolorous entire brown rhizome scales up to 4 mm. long. Frond suberect to arching, non-proliferous, herbaceous. Stipe up to 12 cm. long, matt-brown to dark-brown, thinly pubescent with minute white hairs. Lamina up to 36 × 11 cm., 2-pinnatifid, oblanceolate in outline, acuminate, with a long tapering base with numerous reduced pinnae. Pinnae up to 6 × 1·2 cm., narrowly oblong-lanceolate attenuate (the reduced basal pinnae deflexed, auriculate), deeply pinnatifid into acute oblong lobes, sparsely set with minute pale hairs on both surfaces along the costa, costules and veins and with reddish glands on the dorsal surface; up to 7 pairs of veins per lobe, not anastomosing. Rhachis pale-brown, thinly pubescent with minute stiff hairs. Sori up to 16 per lobe, exindusiate.

Rhodesia. E: Pungwe Rest Huts, *Schelpe* 5670 (BOL). **Malawi.** N: Nyika Plateau, 14.viii.1946, *Brass* 17219 (K; SRGH).
Also in Madagascar, Mauritius and Réunion. Moist stream-banks in light shade, 700–1615 m.

5. **Thelypteris bergiana** (Schlechtend.) Ching in Bull. Fan Mem. Inst. Biol., Bot. **10**: 251 (1941). TAB. **55** fig. B. Type from S. Africa.
    Polypodium bergianum Schlechtend., Adumbr.: 20, t. 9 (1825). Type as above.
    Aspidium bergianum (Schlechtend.) Mett. in Abh. Senckenb. Naturf. Ges. **2**: 363 (1858) reimpr. in Mett., Farngatt., Pheg. u. Aspl.: 78, n. 188 (1858). Type as above.
    Lastrea bergiana (Schlechtend.) Moore, Ind. Fil.: 86 (1858). Type as above.
    Nephrodium bergianum (Schlechtend.) Bak., Syn. Fil.: 269 (1867). Type as above.
    Nephrodium sewellii Bak. in Journ. Linn. Soc. Bot. **15**: 418 (1876). Type from Madagascar.
    Nephrodium anateinophlebium Bak., op. cit. **16**: 202 (1877). Type from Madagascar.
    Dryopteris bergiana (Schlechtend.) Kuntze, Rev. Gen. Pl. **2**: 812 (1891).—Sim, Ferns S. Afr. ed. 2: 93, t. 10 excl. fig. b & c (1915). Type as for *Thelypteris bergiana*.
    Aspidium maranguense Hieron. in Engl., Pflanzenw. Ost-Afr. **C**: 85 (1895). Type from E. Africa.
    Dryopteris sewellii (Bak.) C. Chr., Ind. Fil.: 292 (1905). Type as for *Nephrodium sewellii*.

G

Dryopteris anateinophlebia (Bak.) C. Chr., tom. cit.: 252 (1905). Type as for *Nephrodium anateinophlebium.*

Dryopteris maranguensis (Hieron.) C. Chr., tom. cit.: 276 (1905). Type as for *Aspidium maranguense.*

Dryopteris palmii C. Chr. in Ark. för Bot., Ser. 2, **14**: 1, t. 2 fig. 6 (1916). Type from Madagascar.

Dryopteris prolixa var. *bergiana* (Schlechtend.) Alston apud Gilliland in Journ. S. Afr. Bot. **4**: 149 (1938). Type as for *Thelypteris bergiana.*

Lastrea maranguensis (Hieron.) Copel., Gen. Fil.: 139 (1947). Type as for *Aspidium maranguense.*

Rhizome erect with tufted fronds and pale-brown linear, or acute to acuminate subentire sparsely ciliate rhizome scales up to 8 mm. long. Frond arching, non-proliferous, herbaceous. Stipe up to 17 cm. long, greyish-brown, thinly pubescent with minute white hairs and with scales similar to those of the rhizome about the base. Lamina up to 95 × 25 cm., narrowly elliptic in outline, deeply 2-pinnatifid, apex acuminate with deeply pinnatifid terminal segment; lower pinnae gradually decrescent; middle pinnae up to 12·5 × 2 cm., very narrowly oblong-acuminate deeply pinnatifid into oblong obtuse entire lobes up to 10 × 4 mm., pubescent along the costa and with long scattered hairs on the costules and veins ventrally; pilose on the costa, costules and lamina surface with short whitish hairs dorsally; veins not anastomosing. Rhachis light-brown, pubescent with minute white hairs. Sori up to 14 per pinna lobe, more or less submarginal; indusium minute.

Zambia. N: Dulamidi Kloof, Abercorn Distr., 28.i.1952, *Richards* 593 (K). E: Lundazi, Nyika Plateau, 6.v.1952, *White* 2727 (K). **Rhodesia.** W: Victoria Falls, *Allen* 9 (K). E: Umtali Distr., Vumba Mts., 24.ii.1957, *Chase* 6340 (BOL; K; SRGH). **Malawi.** S: Mlanje Mt., 7.vii.1946, *Brass* 16712 (SRGH). **Mozambique.** N: Mt. Massungulo, iii.1933, *Gomes e Sousa* 1262 (COI).

Also in Ethiopia, Sudan, E. tropical Africa, Cameroon, Fernando Po, Congo and Madagascar. Streambanks in forest and moist areas along forest margins, 1160–2135 m.

6. **Thelypteris gueinziana** (Mett.) Schelpe in Journ. S. Afr. Bot. **31**: 262, 264, t. 1 fig. a (1965). Type from S. Africa.

Aspidium gueinzianum Mett. in Abh. Senckenb. Naturf. Ges. **2**: 367 (1857) ("*gueintzianum*") reimpr. Mett., Farngatt., Pheg. u. Asp.: 83 n. 201 (1858). Type as above.

Lastrea gueinziana (Mett.) Moore, Ind. Fil.: 93 (1858). Type as above.

Nephrodium gueinzianum (Mett). Hieron. in Engl., Bot. Jahrb. **28**: 341 (1900). Type as above.

Rhizome erect, with tufted fronds, and with brown ovate entire acute rhizome-scales up to 4 mm. long. Frond arching, non-proliferous, firmly herbaceous. Stipe stramineous, thinly pubescent with minute white hairs. Lamina up to 34 × 18 cm., deeply 2-pinnatifid, elliptic in outline with an acute deeply pinnatifid apex with the basal pair of pinnae somewhat reduced and deflexed. Pinnae up to 11 × 1·4 cm., narrowly oblong or attenuate, caudate with a long narrow crenate segment at the apex, deeply pinnatifid into narrowly oblong falcate acute lobes up to 9 × 2 mm. with up to 11 pairs of veins per lobe, the basal pair free or meeting in the sinus, pilose on both surfaces. Rhachis stramineous, pilose with white hairs. Sori up to 16 per lobe, ½-way between the costules and the margin; indusium 0·7 mm. in diam., glandular and pilose with white hairs.

Zambia. E: Petauke, 15.xii.1958, *Robson* 955 (K; LISC; SRGH). **Rhodesia.** N: Mrewa, i.1959, *Boltt* (SRGH). W: Victoria Falls rain forest, *Taylor* 3106 (NBG). E: Umtali, 10.viii.1956, *Chase* 6167 (BOL; K; SRGH). S; Fort Victoria, Msali R., 29. viii.1952, *Chase* 5737 (BOL; SRGH). **Malawi.** S: Shire Highlands, xii.1893, *Scott Elliot* 8579 (K). **Mozambique.** MS: Gorongosa Mt., Morambodzi Falls, 2.vii.1955, *Schelpe* 5458 (BOL).

Also in Sudan, E. tropical Africa, Congo, Angola, S. Africa and Madagascar. Shaded or exposed stream-banks and river-banks, 610–1830 m.

7. **Thelypteris chaseana** Schelpe in Journ. S. Afr. Bot. **31**: 264, t. 1 fig. f.(1965). Type from SW. Africa.

Rhizome creeping, with fronds spaced up to 1 cm. apart, and with brown anceolate acuminate entire slightly pilose rhizome-scales up to 11 mm. long. Frond erect to slightly arching, non-proliferous, herbaceous. Stipe up to 63 cm.

long, pale-brown, glabrous except for dark-brown scales about the base similar to those on the rhizome. Lamina up to 70 × 20 cm., deeply 2-pinnatifid, narrowly elliptic to lanceolate in outline, acuminate with a deeply pinnatifid terminal segment and with up to 3 pairs of basal pinnae reduced and deflexed; middle pinnae up to 11·5 × 1·8 cm., very narrowly oblong-acuminate and with the basal acroscopic pinna lobes enlarged and frequently overlapping the rhachis and with the other pinna lobes narrowly oblong, somewhat falcate, obtuse, up to 10 × 3·5 mm., pilose dorsally with stiff white hairs on the costae and costules; veins free or anastomosing, meeting the sinus or in a membrane extending down from the sinus. Rhachis pale-brown, pilose. Sori up to 16 per pinnule lobe, circular, placed ½-way between the costule and the margin; indusium c. 1 mm. in diam., pilose with long white hairs.

Zambia. N: Abercorn Distr., 24.iv.1959, *Richards* 11329b (K; SRGH). W: Mwinilunga Distr., 6.xi.1937, *Milne-Redhead* 3127 (K). C: Kashitu R., vii.1909, *Rogers* 8300 (K). **Rhodesia.** N: Sebungwe Distr., Kariangwe Spring, 12.viii.1951, *Whellan* 535 (SRGH). C: Goromonzi Distr., 61 km. NE. of Salisbury, *Mitchell* 584 (BOL; SRGH). **Malawi.** N: Chisenga, 26.viii.1962, *Tyrer* 609 (SRGH). C: Dedza, 2.ii.1959, *Robson* 1423 pro parte (K). S: Manganje Hills, 1861, *Miller* (K).
Also in E. tropical Africa, Congo, Angola and SW. Africa. Forest floors and shaded or exposed streambanks, 1220–1700 m.

8. **Thelypteris extensa** (Bl.) Morton in Amer. Fern Journ. **49**: 113 (1959). Type from Java.
    Aspidium extensum Bl., Enum. Pl. Jav.: 156 (1828). Type as above.
    Dryopteris extensa (Bl.) Kuntze, Rev. Gen. Pl. **2**: 812 (1891). Type as above.
    Nephrodium wakefieldii Bak. in Ann. of Bot. **5**: 326 (1891). Type from Kenya.
    Aspidium wakefieldii (Bak.) Hieron. in Engl., Pflanzenw. Ost-Afr. **C**: 85 (1895). Type as above.
    Dryopteris wakefieldii (Bak.) C. Chr., Ind. Fil.: 301 (1905). Type as above.
    Cyclosorus extensus (Bl.) H. Ito in Bot. Mag. Tokyo, **51**: 713, fig. 6 (1937). Type as for *Thelypteris extensa*.
    Cyclosorus wakefieldii (Bak.) Ching in Bull. Fan Mem. Inst. Biol. Bot. **10**: 250 (1941). Type as for *Nephrodium wakefieldii*.

Rhizome erect, with tufted fronds and with dark-brown attenuate, minutely ciliate rhizome-scales up to 9 mm. long. Frond arching, non-proliferous, membranous. Stipe up to 40 cm. long, brown nitid, darker at the base, sparsely and minutely pubescent with short white hairs and with dark-brown scales similar to those on the rhizome, towards the base. Lamina up to 1 × 0·60 m., deeply 2-pinnatifid, broadly oblong-acute in outline with the basal pinnae not reduced. Pinnae up to 35 × 2·9 cm., very narrowly oblong in outline in upper pinnae, lanceolate in lower pinnae, attenuate-caudate, the basal pinnae narrowed towards the base, pinnatifid into somewhat falcate narrowly oblong acute entire lobes, up to 1·2 × 0·5 cm., glabrous on both surfaces except for curved white hairs on the ventral costa and margin, and sparsely pubescent on the costa dorsally and with yellow glands along the veins; 1 pair of veins anastomosing. Sori up to 16 per lobe; indusium up to 0·6 mm. in diam., with yellow glands.

Mozambique. MS: Cheringoma, Furnas de Inhaminga, 25.v.1948, *Mendonça* 4377 (BM; LISC).
Also in E. tropical Africa, Pemba, the Chagos Archipelago and from Ceylon to the Philippines. Known in our area only from a shaded moist locality in calcareous caves.

9. **Thelypteris quadrangularis** (Fée) Schelpe in Journ. S. Afr. Bot. **30**: 196, t. 1 fig. b (1964). Type from Guyana.
    Nephrodium quadrangulare Fée, Mém. Fam. Foug. **5**: 308 (1852). Type as above.
    Dryopteris contigua Rosenst. in Meded. Rijksherb. Leiden, no. **31**: 8 (1917) Type from Borneo.
    Dryopteris quadrangularis (Fée) Alston in Journ. of Bot. **75**: 253 (1937). Type as for *Thelypteris quadrangularis*.
    Cyclosorus contiguus (Rosenst.) Copel., Gen. Fil.: 142 (1947). Type as for *Dryopteris contigua*.
    Cyclosorus quadrangularis (Fée) Tardieu in Notul. Syst. **14**: 345 (1952). Type as for *Thelypteris quadrangularis*.

Rhizome erect, with tufted fronds, and with brown to dark-brown lanceolate or attenuate, entire to ciliate, often thinly pilose rhizome-scales up to 6 mm. long. Frond non-proliferous, softly herbaceous. Stipe up to 46 cm., pale-brown, dark at the base, thinly pubescent with minute whitish hairs, and with dark-brown scales similar to those of the rhizome at the base. Lamina up to 59 × 30 cm., deeply 2-pinnatifid, lanceolate in outline, apex acuminate with a deeply pinnatifid terminal segment; the lowest 2 pairs of pinnae somewhat reduced, deflexed; middle pinnae up to 16 × 1·7 cm., very narrowly oblong or attenuate, deeply pinnatifid into narrowly oblong somewhat falcate, obtuse entire lobes up to 3 mm. broad (the basal acroscopic lobe somewhat larger than the rest), thinly pilose ventrally, thinly pilose with long white hairs up to 0·7 mm. long, especially on the costa and costules dorsally; the basal pair of veins anastomosing at a very obtuse angle below the sinus. Rhachis pale-brown, pilose with curved whitish hairs especially along the ventral furrow. Sori up to 13 per pinna lobe, circular, borne ½-way between the costules and the margin; indusium 1 mm. in diam., densely pilose with long white silky hairs.

Rhodesia. E: Chirinda, 25.x.1947, *Wild* 2188 (K; SRGH). **Malawi.** C: Nchisi Mt., Kota Kota Distr., 28.vii.1946, *Brass* 16999 (K; SRGH). **Mozambique.** N: Ribáuè Mt., 19.vii.1962, *Schelpe & Leach* 6925 (BOL). Z: Namúli Mt., Chá Moçambique, 29.vii.1962, *Schelpe & Leach* 7077 (BOL). MS: Bandula Mt., 23.x.1957, *Chase* 6725 (BM; BOL; K; SRGH).

Also in Kenya and W. tropical Africa from Sierra Leone to Angola, in tropical S. America and E. tropical Asia. Deeply shaded stream-banks and undergrowth in forest, 700–1500 m.

10. **Thelypteris madagascariensis** (Fée) Schelpe in Journ. S. Afr. Bot. **31**: 267 (1965). TAB. **55** fig. D. Type from Madagascar.

Gymnogramme unita Kunze in Linnaea, **18**: 115 (1844) non *Thelypteris unita* (L.) Morton (1959). Type from S. Africa.

Goniopteris madagascariensis Fée, Gen. Fil.: 251 (1852). Type as for *Thelypteris madagascariensis*.

Goniopteris patens Fée, tom. cit.: 253 (1852). Type from S. Africa.

Phegopteris unita (Kunze) Mett. in Abh. Senckenb. Naturf. Ges. **2**: 306 (1858) reimpr. in Mett., Farngatt., Pheg. u. Asp.: no. 42 (1858). Type as for *Gymnogramme unita*.

Goniopteris silvatica Pappe & Raws., Syn. Fil. Afr. Austr. : 39 (1858) *nom. illegit.* Type from S. Africa.

Polypodium unitum (Kunze) Hook., Sp. Fil. **5**: 5 (1863) non L. (1759). Type as for *Gymnogramme unita*.

Goniopteris unita (Kunze) J. Sm., Hist. Fil.: 192 (1875). Type as above.

Nephrodium patens (Fée) J. Sm., tom. cit.: 208 (1875) non (Sw.) Desv. (1827). Type as for *Goniopteris patens*.

Nephrodium costulare Bak. in Journ. Linn. Soc. Bot., **16**: 203 (1877). Type from Madagascar.

Dryopteris madagascariensis (Fée) C. Chr., Ind. Fil.: 276 (1905). Type as for *Thelypteris madagascariensis*.

Dryopteris silvatica C. Chr., tom. cit.: 292 (1905).—Sim, Ferns S. Afr. ed. 2: 100, t. 15 (1915). Type as for *Goniopteris silvatica*.

Dryopteris costularis (Bak.) C. Chr., tom. cit.: 258 (1905). Type as for *Nephrodium costulare*.

Dryopteris gladiata C. Chr. in Ark. för Bot., Ser. 2, **14**, 19: 4, t. 1 (1916). Type from Madagascar.

Cyclosorus silvatica Ching in Bull. Fan Mem. Inst. Biol. Bot. **10**: 249 (1941). Type as for *Goniopteris silvatica*.

Cyclosorus patens (Fée) Copel., Gen. Fil.: 143 (1947). Type as for *Goniopteris patens*.

Cyclosorus costularis (Bak.) Adams & Alston in Bull. Brit. Mus. (Nat. Hist.), Bot., **1**: 158 (1955). Type as for *Nephrodium costulare*.

Rhizome erect, with tufted fronds. Fronds arching, proliferating by gemmae borne in an angle of the rhachis near the frond apex, herbaceous. Stipe up to 0·5 m., pale-brown, glabrous. Lamina up to 1·5 × 0·5 m., lanceolate in outline, pinnate. Pinnae acute with basal ones hardly reduced but somewhat deflexed; middle pinnae up to 25 × 2·2 cm., very narrowly oblong -or attenuate with the base truncate (with the acroscopic basal lobe often somewhat enlarged), shallowly incised forming broadly oblong somewhat falcate obtuse lobes up to 4 mm. broad; glabrous except for scattered short white hairs on the costa, costules and veins dorsally; about 4 pairs of veins anastomosing below the sinus either in a membrane extending from the sinus or below it. Rhachis pale-brown, glabrous. Sori circular, up to 20 per pinna lobe, borne close to the costules, exindusiate.

Rhodesia. E: Chirinda Forest, 17.vi.1906, *Swynnerton* 864 (BM; K). **Malawi.** S: Shire Highlands, *Adamson* 159 (K). **Mozambique.** MS: Garuso, Jaegersberg, 11.vii. 1955, *Schelpe* 5622 (BM; BOL).

Also in Sudan, E. tropical Africa, W. tropical Africa, S. Africa and Madagascar. Deeply shaded moist forest floors, 1095–1585 m.

11. **Thelypteris prismatica** (Desv.) Schelpe in Bol. Soc. Brot., Sér. 2, **41**: 217 (1967).
Type from Mauritius.
Nephrodium prismaticum Desv. in Mém. Soc. Linn. Par. **6**, 2: 256 (1827). Type as above.
Cyclosorus prismaticus (Desv.) Tardieu, Fl. Madag., Polypod. **1**: 294 (1958). Type as above.

Rhizome not seen. Frond arching, non-proliferous, herbaceous. Stipe pale-brown, glabrous, nitidus. Lamina up to 1 × 0·48 m., shallowly 2-pinnatifid, narrowly ovate-elliptic in outline, acute with c. 3 pairs of reduced basal pinnae; pinnae up to 24 × 2·2 cm., broadly linear or attenuate, base truncate, shallowly pinnatifid into short oblong broadly acute entire lobes up to 4 mm.; glabrous ventrally except for few white curved hairs along the costa, glabrous dorsally, the lamina with a rather warted appearance; 8 pairs of veins per lobe with 2 pairs of veins anastomosing below the sinus and with the fifth vein sometimes running into the sinus. Rhachis pale-brown, glabrous. Sori circular, up to 8 per pinna lobe borne nearer the costule than the sinus membrane; indusium 0·5 mm. in diam., glabrous.

Malawi. N: Misuku Hills, 11.i.1959, *Robinson* 3171 (K).
Also in Madagascar, the Comoro Is. and Mauritius. Probably in forest, c. 1675 m.

12. **Thelypteris dentata** (Forsk.) E. St. John in Amer. Fern Journ. **26**: 44 (1936).
Type from Yemen.
Polypodium dentatum Forsk., Fl. Aegypt.-Arab.: CXXV, 185 (1775). Type as above.
Dryopteris dentata (Forsk.) C. Chr. in Kongel. Dansk. Vid. Selsk., Afd. 8, **6**: 24 (1920). Type as above.
Aspidium natalense Fée, Mém. Fam. Foug. **8**: 102 (1857). Type from S. Africa.
Cyclosorus hispidulus A. Peter in Fedde, Repert, Beih. **40**, 1: 58 descr. 10, t. 4 fig. 1–2 (1929). Type : Rhodesia, Victoria Falls, *Peter* 30794 (not found).
Cyclosorus dentatus (Forsk.) Ching in Bull. Fan Mem. Inst. Biol. Bot. **8**: 206 (1938). Type as for *Thelypteris dentata*.

Rhizome c. 7 mm. in diam., creeping, with fronds closely spaced and with dark-brown ovate to lanceolate acuminate entire thinly pilose rhizome-scales up to 6 mm. long. Frond arching, non-proliferous, herbaceous. Stipe up to 20 cm. long, pale-brown to greyish-brown, glabrous, thinly pubescent with minute white hairs and with lanceolate scales about the base similar to those on the rhizome. Lamina up to 1·3 × 0·4 m., pinnate, elliptic to narrowly elliptic in outline, apex acuminate with a deeply pinnatifid terminal segment, lower pinnae gradually decrescent; middle pinnae up to 21 × 2·2 cm., very narrowly oblong or attenuate with a long linear acuminate shallowly crenate apex, deeply pinnatifid into oblong slightly falcate obtuse to acute entire lobes, up to 5 mm. broad, pilose along the costa with a few scattered hairs on the costules and veins ventrally, thinly pubescent with minute white hairs on the costae and costules dorsally and along the margins and with even smaller hairs on the lamina; 1 or more pairs of veins (usually a pair and a single vein) anastomosing at and below the sinus. Rhachis pale-brown, pilose with stiff often curved hairs and thinly pubescent with minute white hairs. Sori circular, up to 14 per segment, borne ½-way between the costule and the margin; indusium up to 1 mm. in diam., pilose with short white hairs.

1 pair of veins anastomosing below the sinus - - - - var. *dentata*
2 or more pairs of veins anastomosing below the sinus - - var. *buchananii*

Var. **dentata** TAB. 55 fig. C.

Zambia. W: Solwezi Distr., Nyalisonga R., 1930, *Milne-Redhead* 798 (K). C: Luangwa Valley, tributary of Mutinsasa R., 11.xi.1965, *Astle* 4071 (SRGH). **Rhodesia.** N: Sinoia, 21.iv.1948, *Rodin* 4362 (K; SRGH). W: Victoria Falls, 6.vii.1959, *Schelpe* 3929 (BM; BOL). C: Makoni Distr., Mahamba, 9·6 km. W. of Rusape, i.1956, *Mitchell* 39 (BOL; SRGH). E: Umtali, Lowlands E. of Umtali, 10.viii.1956, *Chase* 6165 (BOL; K; SRGH). S: 1·6 km. E. of Bikita, 22.ii.1964, *Mitchell* 815 (SRGH). **Malawi.** S: Cholo Mt., 23.ix.1946, *Brass* 17764 (K). **Mozambique.** N: Ribáuè Mt., 19.vii.1962, *Schelpe & Leach* 6939 (BOL). Z: Namúli Mt., Chá Mozambique, 29.vii.1962, *Schelpe & Leach* 7076 (BOL).

Widespread in tropical and subtropical Africa and America, in Yemen, the Seychelles and Comoro Is. and Madagascar. Forest floors and margins and a weed in tea plantations. 700–1525 m.

Var. **buchananii** Schelpe in Journ. S. Afr. Bot. **31**: 265, t. 1 fig. d (1965). Type: Mozambique, Manica e Sofala, Garuso, Jaegersberg, *Schelpe* 5599 (BOL).

Rhodesia. E: Melsetter, Chimanimani Mts., E. bank of Haroni Gorge, 16.ii.1958, *Mitchell* 401 (BOL; SRGH). **Mozambique.** Z: Metolola, 9.ix.1949, *Barbosa & Carvalho* 3999 (BOL). MS: Chimoio, Gondola, R. Nhamissanguere, 3.ii.1948, *Garcia* 28 (LISC).

Also in E. tropical Africa and subtropical S. Africa. Shaded stream-banks and moist shaded forest floors, 610–915 m.

13. **Thelypteris totta** (Thunb.) Schelpe in Journ. S. Afr. Bot. **29**: 91 (1963). TAB. **55** fig. F. Type from S. Africa.

> *Polypodium tottum* Thunb., Prodr. Pl. Cap.: 172 (1800). Type as above.
> *Aspidium goggilodus* Schkuhr, Krypt. Gew. **1**: 193, t. 33c (1809). Type from Guyana.
> *Polystichum goggilodus* (Schkuhr) Gaud. in Freyc., Voy. Bot.: 326 (1827). Type as above.
> *Cyclosorus goggilodus* (Schkuhr) Link, Hort. Berol. **2**: 128 (1833). Type as above.
> *Aspidium ecklonii* Kunze in Linnaea, **10**: 546 (1836). Type from S. Africa.
> *Nephrodium plantianum* Pappe & Raws., Syn. Fil. Afr. Austr.: 14 (1858). Type from S. Africa.
> *Aspidium plantianum* (Pappe & Raws.) Kuhn, Fil. Afr.: 139 (1868). Type as above.
> *Dryopteris goggilodus* (Schkuhr) Kuntze, Rev. Gen. Pl. **2**: 811 (1891) (" gongylodes ").—Sim, Ferns S. Afr. ed. 2: 97, t. 13 (1915). Type as for *Aspidium goggilodus*.
> *Thelypteris goggilodus* (Schkuhr) Small, Ferns S.E. States: 248 (1938) (" gongylodes "). Type as above.

Rhizome up to 6 mm. in diam., widely creeping with fronds spaced up to 12 cm. apart, and with sparse black narrowly ovate acute entire scales up to 1·3 mm. long. Frond erect, non-proliferous, firmly herbaceous to thinly coriaceous. Stipe up to 60 cm. long, pale-brown or greyish-brown, darker at the base, glabrous except for a few black scales about the base similar to those on the rhizome. Lamina up to 84 × 30 cm., 2-pinnatifid, oblong-lanceolate in outline, apex shortly acuminate, basal pinnae not reduced. Pinnae up to 19·5 × 2 cm., very narrowly oblong acute to attenuate, shallowly incised into quadrate acute to obtuse lobes up to 7 mm. broad, subglabrous ventrally, pilose with white hairs on the costae, costules and veins and sometimes on the lamina dorsally, the basal pair of veins anastomosing well below the sinus. Rhachis pale-brown, thinly pubescent to subglabrous. Sori up to 18 per lobe, circular; indusia 1 mm. in diam., densely pilose with white hairs.

Botswana. N: Okavango R., Sepopa, 17.iii.1965, *Wild & Drummond* 7104 (BOL; K; LISC; SRGH). **Zambia.** B: Kataba, 12.vi.1963, *Fanshawe* 7829 (SRGH). N: Samfya, 28.xii.1961, *Symoens* 9085 (K). C: Chiwefwe, 19.ix.1954, *Mutimushi* 1056 (SRGH). E: Kenyuma R., 23.viii.1929, *Burtt Davy* 867 (K). S: Livingstone Distr., Katambora, 26. viii.1947, *Greenway & Brenan* 7988 (K). **Rhodesia.** W: Wankie, Victoria Falls, *Scott & Priestley* (BM). E: Umtali, Eastlands Farm, 10.viii.1952, *Chase* 4625 (BM; SRGH). **Malawi.** N: Mzimba Distr., Marymount, 16.iv.1967, *Pawek* 964 (SRGH). C: Dowa Distr., W. of Lake Nyasa Hotel, 10.viii.1951, *Chase* 3916 (SRGH). S: Elephant Marsh, Lower Shire R., 1956, *Robertson* 10 (K). **Mozambique.** N: Montepuez, 25.xii.1963, *Torre & Paiva* 9667 (LISC). Z: Zambezia, Quelimane, Madal Plantation, 25.viii.1962, *Wild & Pedro* 5885 (K; SRGH). T: Zóbuè, 11.i.1966, *Correia* 372 (LISC). MS: Chimoio, Vila Pery, *Pedro & Pedrógão* 6096 (BM). SS: Inhambane, Mongue, 20.i.1954, *Schelpe* 4475 (BM; BOL). LM: Lourenço Marques, *Borle* 30 (PRE; S–PA).

Also in Ethiopia, Sudan, E. tropical Africa, W. tropical Africa, Angola, S. Africa and Madagascar. Marshes, *Papyrus* swamps and wet ditches, 3–1130 m.

14. **Thelypteris striata** (Schumach.) Schelpe in Journ. S. Afr. Bot. **31**: 268 (1965). Type from W. Africa.
    Aspidium striatum Schumach. in Kongel. Dansk. Vid. Selsk., Afh. **4**: 230 (1829). Type as above.
    Polypodium pallidivenium Hook., Sp. Fil. **5**: 8 (1863). Type from W. Africa.
    Nephrodium pallidivenium (Hook.) Bak. in Hook. & Bak., Syn. Fil.: 290 (1867). Type as above.
    Dryopteris striata (Schumach.) C. Chr., Ind. Fil.: 294 (1905). Type as for *Thelypteris striata*.
    Dryopteris hemitelioides Christ in Ann. Mus. Cong. **5**: 26 (1909). Type from the Congo.
    Cyclosorus striatus (Schumach.) Ching in Bull. Fan Mem. Biol., Bot. **10**: 249 (1941). Type as for *Thelypteris striata*.

Rhizome up to 5 mm. in diam., widely creeping, with fronds spaced up to 15 cm. apart, and with dark-brown ovate acuminate entire scales up to 2 mm. long. Frond erect, non-proliferous, firmly herbaceous. Stipe up to 90 cm. long, pale-brown, becoming very dark at the base, glabrous. Lamina up to 77 × 25 cm., 2-pinnatifid, narrowly oblong-ovate in outline, acute, the basal pinnae not reduced; pinnae up to 21 × 2·5 cm., narrowly oblong or attenuate often somewhat contracted towards the base, incised more than ½-way to the costa into oblong to narrowly oblong often slightly falcate subacute lobes up to 6 mm. broad, usually with more than 18 pairs of veins per lobe, the basal pair anastomosing below the sinus, the sub-basal pair reaching the sinus, glabrous ventrally except for a pilose costa, densely to thinly pilose on the costa, costules, veins and lamina dorsally. Rhachis mostly glabrous with age but thinly pilose at first with stiff white hairs. Sori circular, up to 35 per pinna lobe, contiguous at maturity, borne ½-way between the costule and margin; indusium c. 0·5 mm. in diam., thinly pilose.

Zambia. B: Senanga, 27.vii.1962, *Fanshawe*, 6981 (SRGH). N: Mporokoso Distr., L. Chisi, Mweru-wa-Ntipa, 13.xii.1960, *Richards* 13672 (K; SRGH). **Malawi.** S: W. shore of Lake Nyasa, ix.1861, *Kirk* (K).
Widespread in tropical Africa. *Papyrus* swamps and swamp forest, c. 1000 m.

15. **Thelypteris pozoi** (Lagasca) Morton in Bull. Soc. Bot. Fr. **106**: 234 (1959). TAB. 55 fig. G. Type from Spain.
    Polypodium tottum Willd. in L., Sp. Pl., ed. 4, **5**: 201 (1810) non Thunb. (1800). Type from S. Africa.
    Hemionitis pozoi Lagasca, Nov. Gen. et Sp.: 33 (1816). Type as for *Thelypteris pozoi*.
    Gymnogramma totta Schlechtend., Adumbr.: 15, t. 6 (1825). Type as for *Polypodium tottum*.
    Acrostichum pilosiusculum Wikstr. in Kongl. Vet. Acad. Handl. **1825**: 439 (1826). Type from Madeira.
    Polypodium africanum Desv. in Mém. Soc. Linn. Par. **6**, 2: 239 (1827). Type from S. Africa.
    Gymnogramma pozoi (Lagasca) Desv., tom. cit.: 216 (1827). Type as for *Thelypteris pozoi*.
    Grammitis totta (Schlechtend.) C. Presl, Tent. Pterid.: 209, t. 9 fig. 4 (1836) reimpr. in Abh. Königl. Böhm. Ges. Wiss., Ser. 4, **5**: 209, t. 9 fig. 4 (1837). Type as for *Polypodium tottum*.
    Leptogramma totta (Schlechtend.) J. Sm. in Hook., Journ. of Bot. **4**: 52 (1841). Type as above.
    Phegopteris totta (Schlechtend.) Mett. in Abh. Senckenb. Naturf. Ges. **2**: 302 (1858) reimpr. Mett., Farngatt., Pheg. u. Asp. : 18, n.31 (1858). Type as above.
    Pleurosorus pozoi (Lagasca) Trevisan in Att. Soc. Ital. Sci. Nat. **17**: 256 (1875). Type as for *Thelypteris pozoi*.
    Ceterach pozoi (Lagasca) A. Braun ex Milde in Bot. Zeit. **1886**: 310 (1886). Type as above.
    Aspidium tottum (Schlechtend.) Engl., Hochgebirgsfl. Trop. Afr.: 99 (1892). Type as for *Polypodium tottum*.
    Nephrodium tottum (Schlechtend.) Diels in Engl. & Prantl, Nat. Pflanzenfam. **1**, **4**: 170 (1899). Type as above.
    Dryopteris africana (Desv.) C. Chr., Ind. Fil.: 250 (1905).—Sim, Ferns S. Afr. ed. 2: 102, t. 23 (1915). Type as for *Polypodium africanum*.

Polypodium eliasii Sennen & Pau in Bull. Soc. Geogr. Bot. Mans, **1910**: 94 (1910). Type from Spain.

Aspidium africanum (Desv.) Aschers. & Graebn., Syn. Mitteleur. Fl., ed. 2, **1**: 28, 153 (1912). Type as for *Polypodium africanum.*

Leptogramma africana (Desv.) Nakai ex Mori, Enum. Pl. Corea: 13 (1922). Type as above.

Lastrea africana (Desv.) Copel., Gen. Fil.: 138 (1947). Type as above.

Lastrea totta (Schlechtend.) Ohwi in Bull. Nat. Sci. Mus. Tokyo, ser. 2, **3**: 98 (1956). Type as for *Polypodium tottum.*

Leptogramma pilosiuscula (Wikstr.) Alston in Bol. Soc. Brot., Sér. 2. **30**: 17 (1956). Type as for *Acrostichum pilosiusculum.*

Leptogramma pozoi (Lagasca) Heywood in Fedde, Repert. **64**: 19 (1961). Type as for *Thelypteris pozoi.*

Stenogramma pozoi (Lagasca) Iwatsuki, Act. Phytotax. et Geobot. **19**: 124 (1963). Type as above.

Rhizome c. 3 mm. in diam., erect, with tufted fronds and with dark-brown lanceolate ciliate rhizome-scales c. 1 mm. long. Frond arching, non-proliferous, softly herbaceous. Stipe up to 23 cm. long, pale-brown becoming darker at the base, thinly pubescent with minute whitish hairs. Lamina up to 34 × 14 cm., 2-pinnatifid, lanceolate to narrowly elliptic in outline, acute, with the lower pinnae somewhat reduced and deflexed. Pinnae up to 7·5 × 1·2 cm., sessile, very narrowly oblong or attenuate, base truncate, incised about ½-way to the costa into quadrate rounded undulate lobes c. 4 mm. broad; pilose with short white hairs along the costa and set with longer white hairs along the costules and veins on both surfaces; Lamina thinly pilose dorsally; c. 6 pairs of veins per lobe, not anastomosing. Rhachis pale-brown, pilose with short and long white hairs. Sori up to 1·5 m. long, linear along the veins, exindusiate.

Rhodesia. E: Inyanga, Pungwe Gorge, 15.vii.1955, *Schelpe* 5708 (BOL). **Malawi.** N: Nyika Plateau, 24.x.1958, *Robson* 311 (K; LISC; SRGH).

Widespread in tropical Africa, S. Africa, Madeira, Azores and Spain. Moist shaded forest floors, c. 1000–1740 m.

2. AMPELOPTERIS Kunze

Ampelopteris Kunze in Bot. Zeit. **6**: 114 (1848).

Rhizome creeping, with tufted fronds and with non-peltate black rhizome-scales. Stipe not articulated. Lamina pinnate, freely proliferous along the rhachis. Veins from adjoining costules anastomosing to form an excurrent vein. Sori circular to elongate, exindusiate, with capitate paraphyses.

A monotypic genus distributed through the tropics and subtropics of the Old World.

Ampelopteris prolifera (Retz.) Copel., Gen. Fil.: 144 (1947). TAB. **56**. Type from ?Java.

Hemionitis prolifera Retz., Obs. Bot.: 38 (1791). Type as above.

Meniscium proliferum (Retz.) Sw., Syn. Fil.: 19, 207 (1806). Type as above.

Goniopteris prolifera (Retz.) C. Presl, Tent. Pterid.: 183 (1836) reimpr. in Abh. Königl. Böhm. Ges. Wiss., Ser. 4, **5**: 183 (1837). Type as above.

Ampelopteris elegans Kunze in Bot. Zeit. **6**: 114 (1848). Origin unknown.

Polypodium luxurians Kunze in Linnaea, **23**: 280 (1850). Type from ? E. Indies.

Phegopteris luxurians (Kunze) Mett. in Abh. Senckenb. Naturf. Ges. **2**: 309 (1858) reimpr. Mett., Farngatt., Pheg. u. Asp.: 25, n. 51 (1858). Type as above.

Nephrodium proliferum (Retz.) Keys., Polpod. Cyath. Herb. Bung.: 49 (1873). Type as for *Ampelopteris prolifera.*

Phegopteris prolifera (Retz.) Kuhn, Von Deck. Reisen, Bot. **3**: 5, 44 (1879). Type as above.

Dryopteris prolifera (Retz.) C. Chr., Ind. Fil.: 286 (1905).—Sim, Ferns S. Afr. ed. 2: 99, t. 14 (1915). Type as above.

Cyclosorus proliferus (Retz.) Tardieu & C. Chr. in Notul. Syst. **14**: 346 (1952). Type as above.

Rhizome up to 1 cm. in diam., creeping with closely spaced fronds and black entire triangular acuminate rhizome-scales up to 2 mm. long. Frond arching, proliferating at intervals along the rhachis of the large fronds, firmly membranous. Stipe up to 40 cm. long, pale-brown, nitidous. Lamina up to 1 × 0·26 m., very

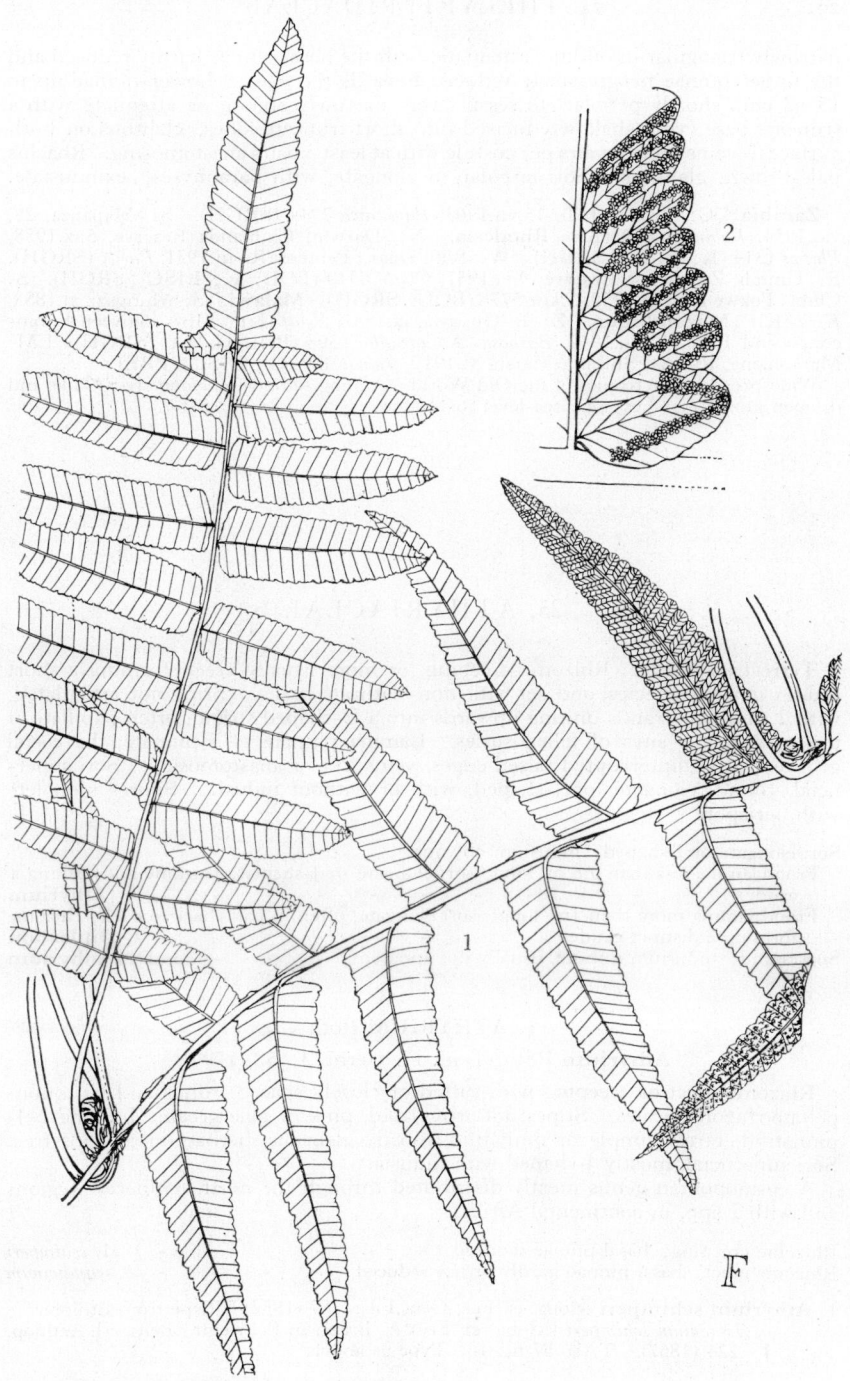

Tab. 56. AMPELOPTERIS PROLIFERA. 1, portion of frond (×⅔); 2, portion of fertile pinna
(×4) *Welwitsch* s.n.

narrowly triangular in outline, attenuate, with the basal pinnae hardly reduced and the upper pinnae progressively reduced towards the apex. Lower pinnae up to 15 × 2 cm., shortly petiolate to sessile, very narrowly oblong or attenuate with a truncate base, very shallowly incised into short truncate lobes, glabrous on both surfaces; veins up to 8 pairs per costule with at least 5 pairs anastomosing. Rhachis pale-brown, glabrous. Sori circular to elongate, with paraphyses, exindusiate.

Zambia. C: Broken Hill, 15.vii.1963, *Fanshawe* 7944 (SRGH). S: Mapanza, 29. viii.1954, *Robinson* 879 (K). **Rhodesia.** N: Darwin, Chimanda Reserve, 5.ix.1958, *Phipps* 1314 (K; LISC; SRGH). W: Bubi Distr., Bembesi R., iii.1931, *Pardy* (SRGH). E: Umtali, Zimunya's Reserve, 9.vi.1957, *Chase* 6540 (BOL; K; LISC; SRGH). S: Chibi, Tokwe R., 5.ix.1955, *Chase* 5776 (BOL; SRGH). **Malawi.** S: Mungazi, xi.1883, *Kirk* (K). **Mozambique.** Z: R. Quaqwa, ix.1887, *Scott* (K). MS: Between Furancungo and Bene, 13.vii.1949, *Barbosa & Carvalho* 3598 (BOL; LMA; SRGH). LM: Marracuene, Rikatla, Shibidji Marsh, xi.1917, *Junod* 389 (G; LISC; PRE).

Widespread in the tropics of the Old World. Among *Phragmites* along river banks and in open grassy marshes, near sea-level to 1100 m.

25. ATHYRIACEAE

Terrestrial plants. Rhizome creeping, or more usually erect, forming a short caudex in some species, and set with non-clathrate scales. Stipe not articulated, with 2 vascular strands uniting upwards into a U-shaped strand, often papillate at the base at the sites of fallen scales. Lamina pinnate to 3-pinnate, the costal grooves with uninterrupted raised edges, veins free or anastomosing. Sori superficial, round, elongate, or J-shaped, with or without indusia. Spores monolete with perispore.

Sori elongate or J-shaped; indusium obvious:
  Frond lamina less than 0·8 m. long; sori elongate or J-shaped; rhizome not forming a
    caudex - - - - - - - - - - - - - 1. **Athyrium**
  Frond lamina more than 1m. long; sori elongate, often in pairs back to back; rhizome
    forming a distinct caudex - - - - - - - - 2. **Diplazium**
Sori round; indusium minute, usually not apparent - - - 3. **Dryoathyrium**

1. ATHYRIUM Roth

Athyrium Roth, Tent. Fl. Germ. **3**: 58 (1799).

Rhizome erect or creeping with tufted or closely spaced fronds and with non-peltate rhizome-scales. Stipes not articulated, pink or pale green. Lamina 2–4-pinnatifid, rarely simple or pinnatifid, mostly glabrous, herbaceous; veins free. Sori superficial, mostly J-shaped with indusia.

A cosmopolitan genus mostly distributed through the north temperate regions and with 2 spp. in continental Africa.

Rhizome creeping; basal pinnae reduced - - - - - - 1. *schimperi*
Rhizome erect; basal pinnae hardly or not reduced - - - - 2. *scandicinum*

1. **Athyrium schimperi** Moug. ex Fée, Gen. Fil.: 187 (1852). Type from Ethiopia.
    Asplenium schimperi (Moug. ex Fée) A. Braun in Schweinf., Beitr. Fl. Aethiop.
  1: 224 (1867). TAB. **57** fig. B. Type as above.

Rhizome up to 6 mm. in diam., creeping, with closely spaced fronds and with reddish-brown lanceolate-acuminate rhizome-scales up to 7 mm. long. Frond erect, herbaceous. Stipe pale-brown with a very dark-brown base, up to 38 cm. long, glabrous, except for dark-brown attenuate scales up to 7 mm. long about the base similar to those of the rhizome. Lamina up to 70 × 28 cm., deeply 3-pinnatifid, oblong-lanceolate, acute to acuminate, with the basal pair of pinnae usually

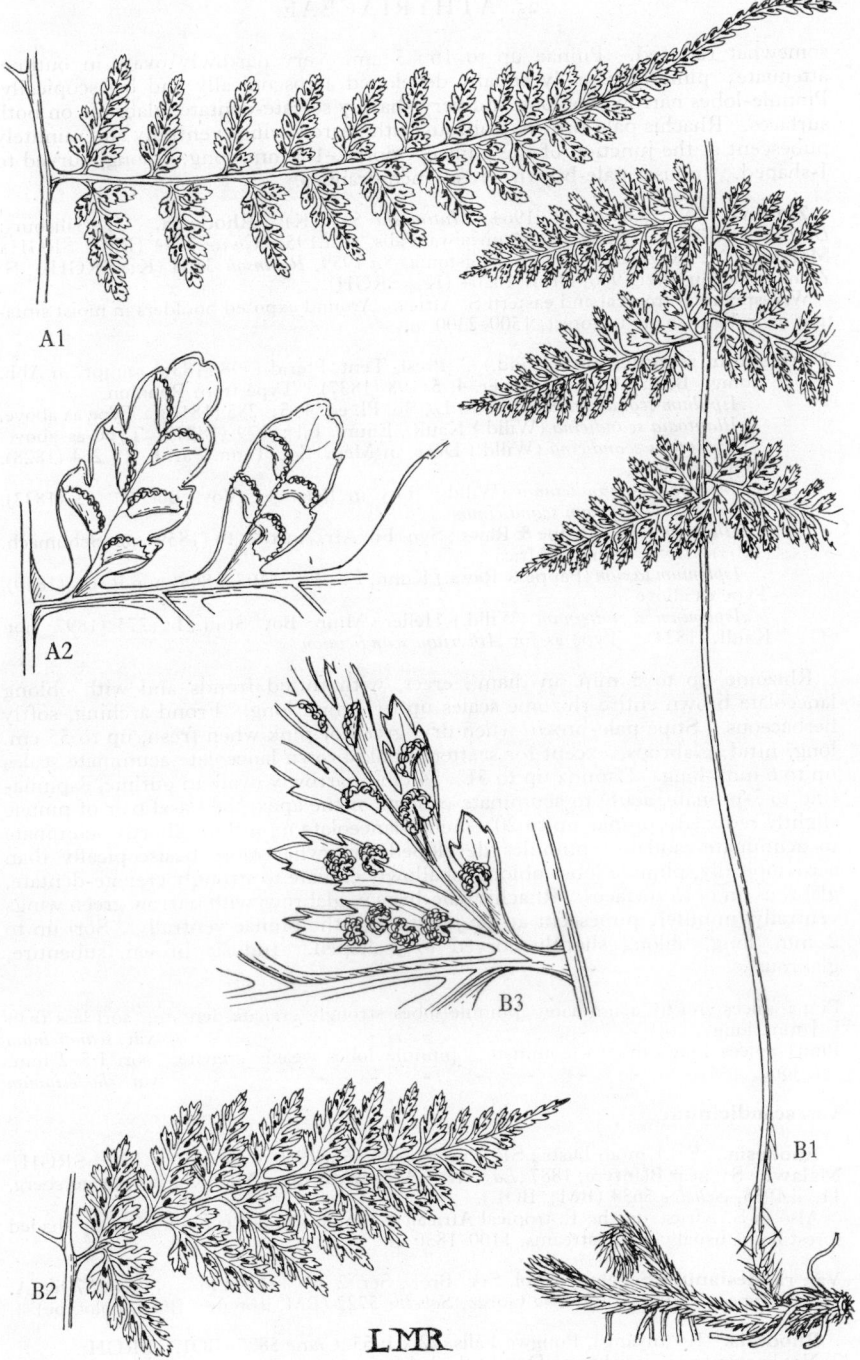

LMR

Tab. 57. A.—ATHYRIUM SCANDICINUM var. RHODESIANUM. A1, pinna (×⅔); A2, lower
surface of pinnule showing sori (×4), both from *Schelpe* 5722. B.—ATHYRIUM SCHIM-
PERI. B1, rhizome, stipe and lower part of lamina (×⅔); B2, pinna (×⅔); B3,
lower surface of pinnule showing sori (×4), all from *Chase* 3774.

somewhat reduced. Pinnae up to 16×5 cm., very narrowly ovate in outline, attenuate; pinnules usually equally developed acroscopically and basiscopically. Pinnule-lobes narrowly oblong and very sharply serrate-dentate, glabrous on both surfaces. Rhachis pale-brown, glabrous with narrow wings ventrally and minutely pubescent at the junction of the pinnae. Sori 1–1·5 mm. long, oblong, curved to J-shaped. Indusia pale-brown, erose, membranous.

Zambia. W: Kitwe, 22.i.1964, *Mutimushi* 547 (K). **Rhodesia.** C: Salisbury, *Darling* (PRE). E: Inyanga, Nyamaziwa Falls, 11.i.1957, *Chase* 3774 (BM; SRGH). **Malawi.** N: Kaziweziwe R., Livingstonia, 8.i.1959, *Robinson* 3105 (K; SRGH). S: Zomba Plateau, 4.vi.1946, *Brass* 16204 (K; SRGH).
Widespread in tropical and eastern S. Africa. Around exposed boulders in moist situation or in light shade in forest, 1500–2300 m.

2. **Athyrium scandicinum** (Willd.) C. Presl, Tent. Pterid.: 98 (1836) reimpr. in Abh. Königl. Böhm. Ges. Wiss. Ser. 4, **5**: 98 (1837). Type from Réunion.
    Aspidium scandicinum Willd. in L., Sp. Pl. ed. 4, **5**: 285 (1810). Type as above.
    Allantodia scandicina (Willd.) Kaulf., Enum. Fil.: 179 (1824). Type as above.
    Cystopteris scandicina (Willd.) Desv. in Mém. Soc. Linn. Par. **6**, 2: 264 (1828). Type as above.
    Nephrodium scandicinum (Willd.) Bory in Bélanger, Voy. Bot. **2**: 63 (1833). Type as for *Athyrium scandicinum*.
    Athyrium laxum Pappe & Raws., Syn. Fil. Afr. Austr.: 16 (1859) non Schumach. (1803). Type from S. Africa.
    Asplenium laxum (Pappe & Raws.) Kuhn, Fil. Afr.: 105 (1868) non R. Br. (1810). Type as above.
    Asplenium scandicinum (Willd.) Heller, Minn. Bot. Stud. **1**: 775 (1897) non Kaulf. (1824). Type as for *Athyrium scandicinum*.

Rhizome up to 5 mm. in diam., erect, with tufted fronds and with oblong lanceolate brown entire rhizome scales up to 7 mm. long. Frond arching, softly herbaceous. Stipe pale-brown when dry, green or pink when fresh, up to 55 cm. long, nitid, glabrous, except for scattered pale-brown lanceolate acuminate scales up to 6 mm. long. Lamina up to 51×34 cm., narrowly ovate in outline, 3-pinnatifid to 3-pinnate, acute to acuminate-caudate at the apex, the basal pair of pinnae slightly reduced; pinnae up to 20×6 cm., lanceolate in outline, shortly acuminate to acuminate-caudate; pinnules developed somewhat more basiscopically than acroscopically, pinnule lobes oblong, shallowly crenate to strongly crenate-dentate, glabrous on both surfaces. Rhachis pale-brown, glabrous with narrow green wings ventrally, minutely pubescent at the junction of the pinnae ventrally. Sori up to 2 mm. long, oblong slightly curved to J-shaped. Indusia brown, subentire, glabrous.

Pinna apices shortly acuminate; pinnule lobes strongly crenate-dentate; sori less than
  1 mm. long - - - - - - - - - var. *scandicinum*
Pinna apices long-caudate-acuminate; pinnule lobes weakly crenate; sori 1·5–2 mm.
  long. - - - - - - - - var. *rhodesianum*

Var. **scandicinum**

Rhodesia. E: Umtali Distr., SE. Vumba Mts., 5.ii.1950, *Chase* 2199 (BM; SRGH). **Malawi.** S: near Blantyre, 1887 ,*Last* (K). **Mozambique.** MS: Garuso, Jaegersberg, 11.vii.1955, *Schelpe* 5634 (BM; BOL).
Also in S. Africa, on the E. tropical African mountains and Réunion. Deeply shaded forest floors usually along streams, 1100–1830 m.

Var. **rhodesianum** Schelpe in Bol. Soc. Brot., Sér. 2, **41**: 211 (1967). TAB. **57** fig. A.
    Type: Rhodesia, Pungwe Gorge, *Schelpe* 5722 (BM, isotype; BOL, holotype).

Rhodesia. E: Inyanga, Pungwe Falls, 22.x.1955, *Chase* 5827 (BOL; SRGH).
Not known from elsewhere. Deeply shaded forest stream-banks, 1740–2075 m.

2. **DIPLAZIUM** Sw.
    Diplazium Sw. in Schrad., Journ. Bot. **1800**, 2: 61 (1801).

Rhizome mostly erect, often forming a short caudex, with non-peltate rhizome-scales and frequently bearing strong black roots. Stipe not articulated. Lamina usually large, 3–4-pinnate, mostly glabrous, firmly herbaceous to coriaceous; veins free. Sori superficial, elongate, with the indusia, at least those near the costule, in pairs set back to back.

A pantropic genus with 5 spp. in tropical Africa.

Sori 2–3 mm. long - - - - - - - - - -   **1.** *nemorale*
Sori 1–1·5 mm. long - - - - - - - -   - **2.** *zanzibaricum*

1. **Diplazium nemorale** (Bak.) Schelpe in Bol. Soc. Brot., Sér. 2, **41**: 212 (1967). TAB. **58** fig. A. Type from ? Madagascar.

> *Asplenium nemorale* Bak. in Journ. Linn. Soc., Bot. **15**: 417 (1876). Type as above.
> *Asplenium madagascariense* Bak., loc. cit. Type from Madagascar.
> *Asplenium hyophilum* Hieron. in Engl., Pflanzenw. Ost-Afr. **C**: 84 (1895). Type from Tanzania.
> *Diplazium hyophilum* (Hieron.) C. Chr., Ind. Fil.: 233 (1905). Type as above
> *Diplazium stolzii* Brause in Engl., Bot. Jahrb. **53**: 381 (1915). Type from Tanzania.
> *Diplazium arborescens* var. *nemorale* (Bak.) C. Chr. in Perrier, Cat. Pl. Madag., Ptérid.: 36 (1932). Type from Madagascar.

Rhizome erect. Frond arching, firmly herbaceous. Stipe smooth, green when fresh, up to 1 m. long and 1 cm. in diam., sulcate. Lamina up to 1·5 × 0·9 m., 3-pinnatifid, ovate in outline, acute, the 2 basal pairs of pinnae somewhat reduced. Pinnae up to 45 × 14 cm., oblong-lanceolate, acuminate; pinnules mostly petiolate, up to 1·8–8·2 cm., incised more than ½-way to the costa into broadly oblong serrate obtuse lobes up to 7 mm. broad, glabrous on both surfaces. Rhachis pale-greenish-brown when dry, sulcate, glabrous. Second degree rhachises pale-brown with narrow wings ventrally. Sori 2–3 mm. long, linear, often back to back. Indusium brown, membranous, entire.

Rhodesia. E: Umtali, Vumba Mts., 5.viii.1955, *Chase* 5700 (BM; BOL; SRGH). **Mozambique.** MS: Garuso, Jaegersberg, 11.vii.1955, *Schelpe* 5623 (BM; BOL). Also in Tanzania (Kyimbila), Madagascar and Comoro Is. Deeply shaded stream-banks in forest, 1100–1770 m.

2. **Diplazium zanzibaricum** (Bak.) C. Chr., Ind. Fil.: 241 (1905). TAB. **58** fig. B. Type from Zanzibar.

> *Asplenium zanzibaricum* Bak. in Ann. of Bot. **5**: 311 (1891). Type as above.
> *Aspidium sulcinervium* Hieron. in Engl., Pflanzenw. Ost-Afr. **C**: 85 (1895). Type from Tanzania.
> *Dryopteris sulcinervia* (Hieron.) C. Chr., tom. cit.: 296 (1905). Type as above.
> *Cornopteris sulcinervia* (Hieron.) Tardieu in Mém. Inst. Sci. Madag. **7**: t. 2 fig. 5–7 (1956); Fl. Madag. Polypod. **1**: 257, t. 35 fig. 5–7 (1958). Type as for *Aspidium sulcinervium*.

Rhizome up to 40 cm. high and 15 cm. in diam., erect. Frond arching. Stipe pale-brown when dry, up to 1 m. long and 1 cm. in diam., subglabrous above but with numerous attenuate shortly ciliate brown scales up to 1·1 cm. long, among a very short dense brown tomentum about the base. Lamina up to 1·5 × 1·4 m., broadly ovate in outline, acute, with the basal pair of pinnae slightly reduced, very deeply 3-pinnatifid, up to 68 × 25 cm., oblong-ovate acute with the basal pinnules slightly reduced. Pinnules up to 13 × 2·5 cm., oblong acute-acuminate to acuminate-cordate cut almost to the costules into narrow oblong pinnatifid obtuse lobes up to 10·3 × 5 mm., glabrous on both surfaces except for minute whitish hairs on the costules. Rhachis pale-brown, sulcate, glabrous; second degree rhachises pale-brown with narrow wings ventrally. Sori 1–1·5 mm. long, narrowly oblong, often back to back. Indusium brown, membranous, erose.

Zambia. E: Lundazi, 7.v.1952, *White* 2767 (BOL). **Rhodesia.** E: Umtali, Vumba Mts., NE. of Castle Beacon, 24.viii.1956, *Chase* 6181 (BM; BOL; K; LISC; SRGH). **Mozambique.** MS: Garuso, Jaegersberg, *Schelpe* 5625 (BM; BOL; K; P). Also in the Transvaal and widespread in tropical Africa. Stream-banks in deep shade in forest, 1100–2135 m.

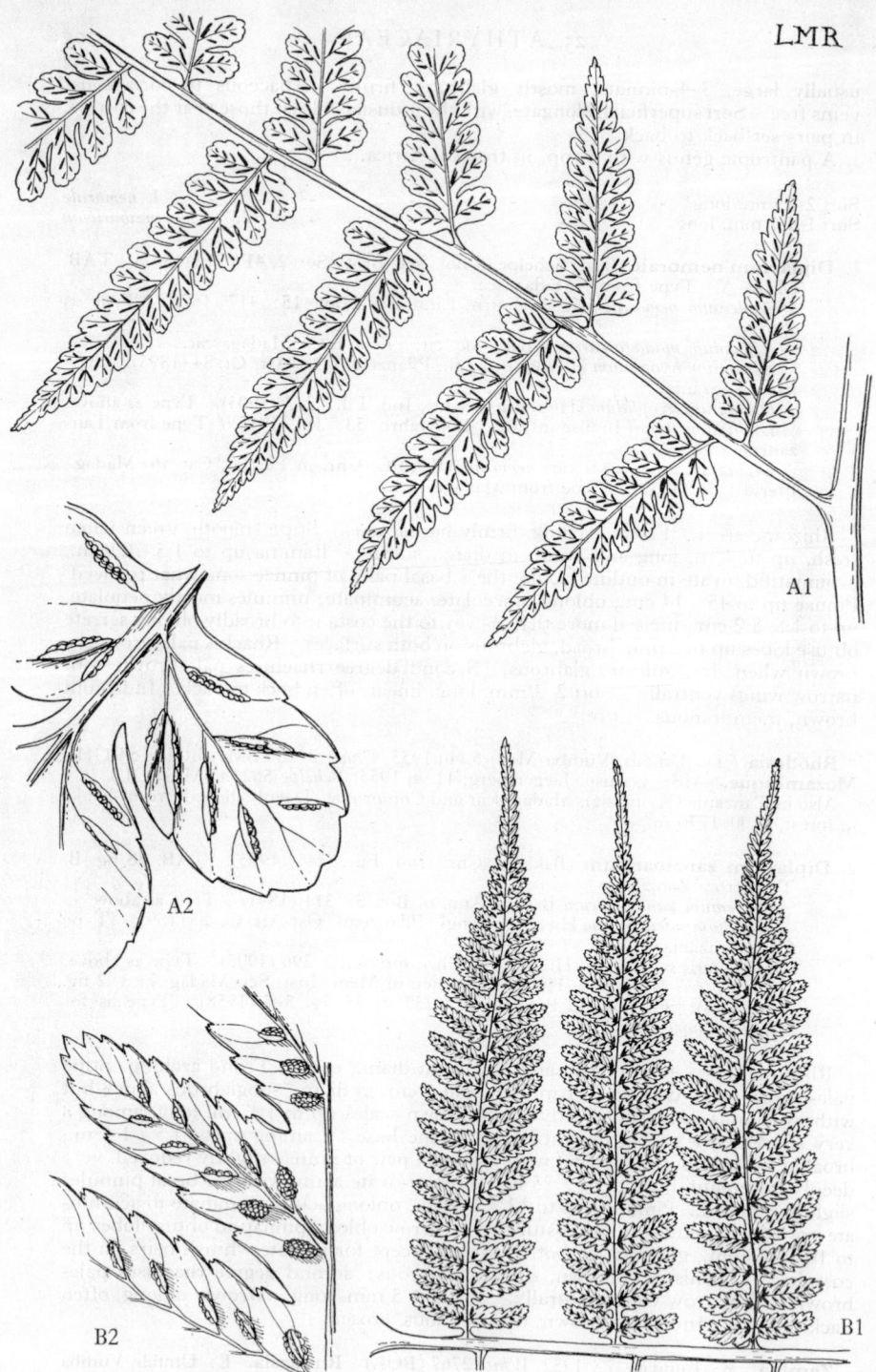

Tab. 58. A.—DIPLAZIUM NEMORALE. A1, part of pinna (× ⅔); A2, pinnule lobe (×4),
both from *Schelpe* 5623. B.—DIPLAZIUM ZANZIBARICUM. B1, part of pinna (× ⅔)
Schelpe 5625; B2, pinnule lobe (×4) *Chase* 5701.

3. DRYOATHYRIUM Ching

Dryoathyrium Ching in Bull. Fan Mem. Inst. Biol. Bot. **11**: 79 (1941).

Rhizome erect, forming a short caudex, with non-peltate rhizome-scales. Stipes not articulated. Lamina large 4-pinnatifid, glabrous except for minute short hairs along the narrowly and evenly winged costules, herbaceous; veins free. Sori superficial, round, with minute reniform indusia not apparent in mature sori.

A tropical genus of c. 10 spp. distributed from tropical Africa and Madagascar to China and Japan. Only 1 sp. occurs in continental Africa.

Dryoathyrium boryanum (Willd.) Ching in Bull. Fan Mem. Inst. Biol., Bot. **11**: 79 (1941). TAB. **59**. Type from Réunion.

Aspidium boryanum Willd. in L., Sp. Pl. ed. 4, **5**: 285 (1810). Type as above.

Lastrea boryana (Willd.) Moore, Ind. Fil.: 86 (1858). Type as above.

Nephrodium boryanum (Willd.) Bak., Syn. Fil.: 284 (1867). Type as above.

Aspidium glabratum Mett. ex Kuhn, Fil. Afr.: 133 (1868). Type from Fernando Po.

Nephrodium glabratum (Mett. ex Kuhn) Bak., Syn. Fil. ed. 2: 500 (1874). Type as above.

Dryopteris glabrata (Mett. ex Kuhn) Kuntze, Rev. Gen. Pl. **2**: 812 (1891). Type as above.

Aspidium kiboschense Hieron. in Engl., Pflanzenw. Ost-Afr. **C**: 85 (1895). Type from Tanzania.

Dryopteris kiboschensis (Hieron.) C. Chr., Ind. Fil.: 273 (1906). Type as above.

Dryopteris boryana (Willd.) C. Chr., tom. cit.: 255 (1906). Type as for *Dryoathyrium boryanum*.

Athyrium boryanum (Willd.) Tagawa in Act. Phytotax. Geobot. **4**: 144 (1935). Type as above.

Ctenitis boryana (Willd.) Copel., Gen. Fil.: 123 (1947). Type as above.

Thelypteris glabrata (Mett. ex Kuhn) Tardieu in Notul. Syst. **14**: 344 (1952). Type as for *Aspidium glabratum*.

Thelypteris glabrata var. *hirsuta* Tardieu, loc. cit. Type from Ivory Coast.

Athyrium glabratum (Mett. ex Kuhn) Alston in Bol. Soc. Brot., Sér. 2, **30**: 11 (1956). Type as for *Aspidium glabratum*.

Cornopteris boryana (Willd.) Tardieu in Amer. Fern Journ. **48**: 32 (1958). Type as for *Dryoathyrium boryanum*.

Parathyrium boryanum (Willd.) Holttum in Kew Bull. **13**: 449 (1958). Type as above.

Rhizome erect, forming a short caudex and with tufted fronds and pale-brown concolorous subentire ovate to lanceolate rhizome-scales up to 5 mm. long. Frond arching, herbaceous. Stipe matt-greenish-brown, up to 1 m. long, glabrous except for pale-brown scales about the base similar to those on the rhizome. Lamina up to 1 × 0·64 m., narrowly ovate in outline, acute, basal pinnae hardly reduced. Pinnae up to 34 × 15 cm., attenuate, shortly petiolate, deeply 3-pinnatifid, oblong-lanceolate-acuminate in outline, developed equally acroscopically and basiscopically, the basal pinnules somewhat reduced; pinnules up to 7·5 × 2·2 mm., oblong-acuminate, base truncate, deeply pinnatifid into narrowly oblong obtuse deeply crenate-serrate lobes up to 9 × 3 mm. with an angular sinus between them, glabrous on both surfaces except for minute short blunt brownish hairs along the costules and veins; veins free. Rhachis pale-brown, glabrous; secondary rhachises pale-brown, with scattered minute blunt hairs and with a narrow laminar green wing at least about the insertion of the pinnules. Sori circular up to 10 per pinnule lobe c. 0·6 mm. in diam. Indusium membranous, minute, subentire.

Rhodesia.. E: Vumba Mts., NE. of Castle Beacon, 24.viii.1956, *Chase* 6182 (BM; K; LISC; SRGH). **Mozambique.** MS: Penhalonga, 24.vi.1955, *Schelpe* 5347 (BOL). S. Nigeria, Cameroon, Fernando Po, Uganda, Kenya, Madagascar, Réunion and widespread in tropical Asia. Wet shaded situations in forest, 1190–1700 m.

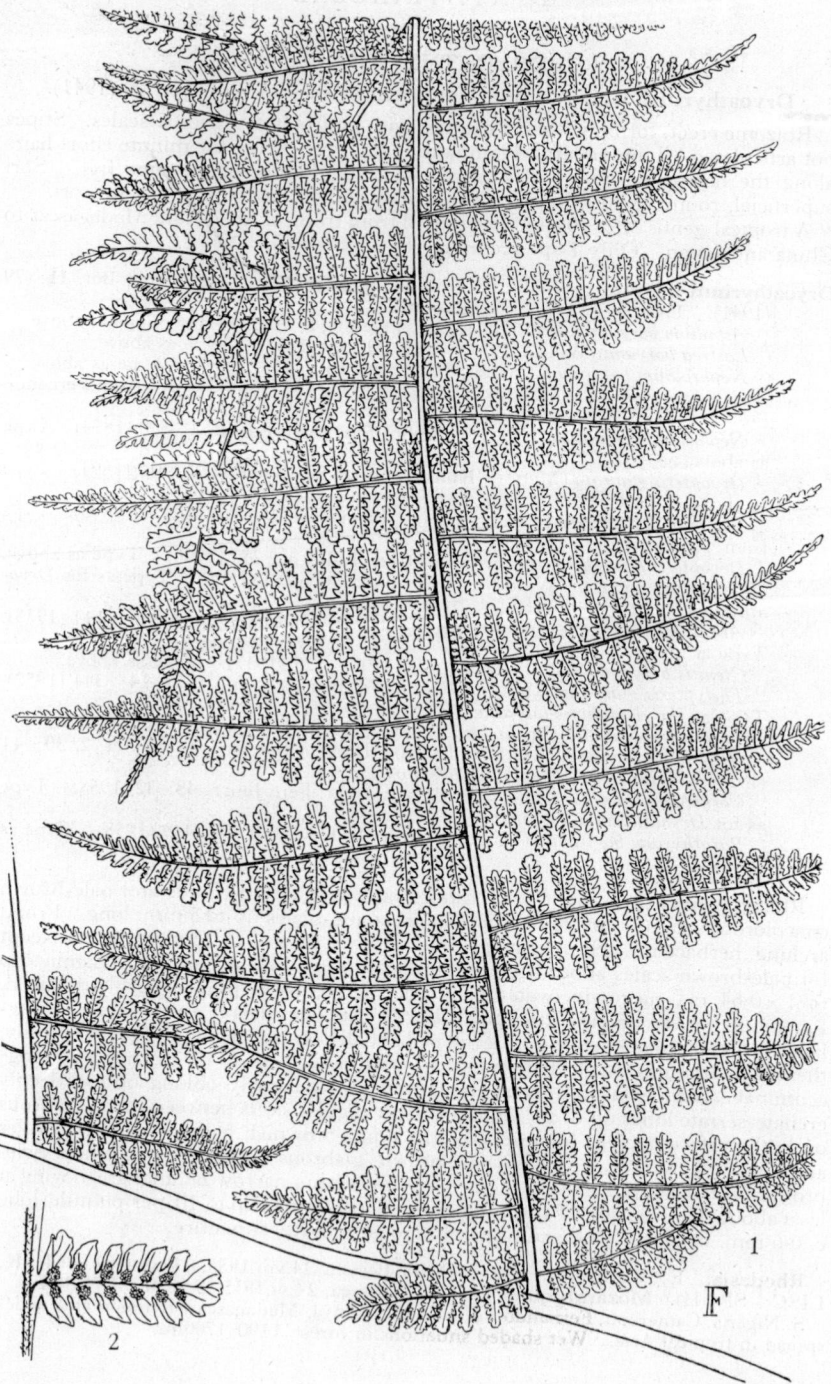

Tab. 59. DRYOATHYRIUM BORYANUM. 1, pinna (× ⅔); 2, fertile pinnule segment (× 2), both from *Schelpe* 5663.

26. LOMARIOPSIDACEAE

Epiphytic, lithophytic or terrestrial plants. Rhizome creeping or scandent, dorsiventral with non-peltate scales. Stipe articulated or not. Fronds simple, or pinnate with the basiscopic pinna margin decurrent on the rhachis; veins free or anastomosing, without included veinlets; fertile fronds acrostichoid (i.e. the dorsal surface of the lamina covered with sporangia without indusia). Spores monolete with perispores.

Veins free:
  Fronds simple; stipes articulated - - - - - - 1. **Elaphoglossum**
  Fronds pinnate; stipes not articulated - - - - - 2. **Lomariopsis**
Veins anastomosing - - - - - - - - - 3. **Bolbitis**

1. ELAPHOGLOSSUM Schott

Elaphoglossum Schott, Gen. Fil.: t. 14 adn. (1834).

Rhizome creeping, with tufted or spaced fronds and with brown or black rhizome-scales. Stipes articulated at the base. Fronds strongly to weakly dimorphous, the fertile fronds acrostichoid. Lamina simple, glabrous, or thinly to densely beset with entire or fimbriate or stellate scales; veins parallel and free (in all African species).

A genus of over 400 spp. with the greatest concentration in tropical S. America.

Sterile lamina glabrous or with minute inconspicuous appressed peltate or substellate scales:
  Sterile lamina base very long tapering:
    Sterile lamina less than 5 × 1 cm.; rhizome long-creeping, c. 1 mm. in diam.
      3. *marojejyense*
    Sterile lamina more than 10 × 1·5 cm.; rhizome short-creeping, fronds clustered:
      Stipe scales ovate:
        Sterile lamina more than 60 cm. long, narrowly oblanceolate 6. *rhodesianum*
        Sterile lamina less than 25 cm. long, oblanceolate to elliptic 1. *acrostichoides*
      Stipe scales linear - - - - - - - - 2. *zambesiacum*
  Sterile lamina base cuneate not long tapering:
    Sterile lamina with the apex rounded; rhizome-scales ferrugineous, appressed
      5. *lastii*
    Sterile lamina with the apex acute to acuminate, caudate; rhizome-scales brown:
      Sterile lamina very narrowly oblong acute; rhizome-scales less than 7 mm. long
        1. *acrostichoides*
      Sterile lamina elliptic to broadly elliptic; rhizome-scales 8–12 mm. long
        4. *macropodium*
Sterile lamina with obvious acicular substellate or ciliate scales 1 mm. or more in diam.
  (including cilia) at least on the under surface:
  Lamina-scales lanceolate, more or less inrolled:
    Lamina-scales very dark brown, borne mainly along the midrib and margin
      7. *hybridum*
    Lamina-scales pale-brown or reddish-brown, scattered over the surface of the lamina:
      Sterile frond linear; scales on lamina surface much smaller than those along midrib
        or on stipe - - - - - - - - - - 8. *aubertii*
      Sterile frond spathulate, oblanceolate, oblong or elliptic:
        Sterile lamina spathulate or oblanceolate, rounded, less than 5 cm. long
          9. *spathulatum*
        Sterile lamina oblong or elliptic, acute to caudate, more than 5 cm. long:
          Apex of sterile lamina caudate, or if acute then the fertile frond less than ⅔ as
            long as the sterile frond - - - - - - 10. *mildbraedii*
          Apex of sterile lamina acute or rounded - - - - 11. *chevalieri*
  Lamina-scales ciliate or substellate:
    Cilia of substellate lamina-scales longer than the diameter of the scale; rhizome-
      scales castaneous - - - - - - - - - 12. *petiolatum*
    Cilia of lamina-scales as long as or shorter than the length of the scale:
      Rhizome-scales all dark-castaneous - - - - - - 13. *deckenii*

Rhizome-scales brown, with or without dark-castaneous margins or cilia:
  Lamina-scales closely imbricate, mostly oblong, 1–2 mm. long; sterile lamina
    tapering abruptly at the base - - - - - - 14. *kuhnii*
  Lamina-scales not imbricate, mostly round, 1 mm. or less in diam.; sterile
    lamina long-tapering at the base - - - - - 15. *welwitschii*

1. **Elaphoglossum acrostichoides** (Hook. & Grev.) Schelpe in Journ. S. Afr. Bot. **30**:
   196 (1964). Type from S. Africa.
   Vittaria acrostichoides Hook. & Grev., Icon. Fil.: t. 186 (1830). Type as above.
   Drymoglossum acrostichoides (Hook. & Grev.) Moore, Ind. Fil.: 31 (1857). Type
   as above.
   Acrostichum lineatum Kuhn ex Christ in Denkschr. Schweiz. Naturf. Ges. **36**:
   146 (1899) non Cav. (1799). Type from Madagascar.
   Elaphoglossum preussii Hieron. in Engl., Bot. Jahrb. **46**: 402 (1911). Type from
   West Africa.
   Elaphoglossum conforme var. *lineatum* (Kuhn ex Christ) C. Chr. in Dansk Bot.
   Ark. **7**: 166, t. 64 fig. 1–2 (1932); in Perrier, Acad. Malgache Cat. Pl. Madag.
   Ptérid.: 61 fig. 1–2 (1932). Type from Madagascar.
   Elaphoglossum petiolatum sensu Sim, Ferns S. Afr. ed. 2: 287, t. 149 fig. 2 (1915)
   non (Sw.) Urban (1903).

Rhizome c. 3 mm. in diam., creeping, with fronds spaced 5–10 mm. apart, and
with lanceolate acuminate subentire rhizome-scales up to 6 × 1·2 mm., concolorous
pale-brown or with a castaneous upper ½ with occasional filamentous outgrowths
along the margin. Frond erect to arching, thinly coriaceous. Stipe up to 22 cm.
long, pale-brown stramineous, densely to sparsely set with pale-brown oblong
lacerate to ovate entire scales up to 4 mm. long. Stipe of fertile frond usually
longer than that of the sterile frond. Sterile lamina up to 35 × 3·5 cm., very
narrowly elliptic-acuminate, base cuneate shortly to longly decurrent, margin
entire to shallowly irregularly undulate, glabrous on both surfaces except for small
entire to lacerate scales along the midrib and minute substellate brown to black
scales scattered on the dorsal surface of the lamina; midrib shallowly sulcate above;
veins obscure about 0·8 mm. apart. Fertile lamina up to 17 × 1·5 cm., linear, acute
to acuminate, base narrowly cuneate, shortly or long-tapering.

Rhodesia. E: Inyanga Distr., Nyamziwa Falls, 11.i.1951, *Chase* 3792 (BM; SRGH).
Malawi. S. Mlanje Mt., Chambe to Lichenya path, 8.v.1958, *Chapman* 603 (BOL; K;
SRGH). **Mozambique.** Z: Namúli Mt., 26.vii.1962, *Schelpe & Leach* 7048 (BOL).
MS: Gorongosa Mt., near Gogogo Peak, 6.vii.1955, *Schelpe* 5543 (BM; BOL).
Also in S. Africa and widespread at higher altitudes in tropical Africa and Madagascar.
Epiphytic or lithophytic in deep shade in forest, 750–1980 m.

2. **Elaphoglossum zambesiacum** Schelpe in Bol. Soc. Brot., Sér. 2, **41**: 214 (1967).
   Type: Zambia, Luwingu, *Fanshawe* 8708 (SRGH, holotype).

Rhizome c. 4 mm. in diam., creeping, with closely spaced fronds and with pale-
brown concolorous lanceolate acuminate rhizome-scales 5 × 1 mm., mostly entire
but with occasional hair-like marginal outgrowths. Stipe up to 11 cm. long, pale
greenish-grey or pale brown, at first with dense patent linear, irregularly ciliate
pale-brown scales up to 5 mm. long, later becoming subglabrous. Sterile lamina
up to 10 × 3 cm., coriaceous oblanceolate to very narrowly elliptic-acute, base
narrowly cuneate-decurrent, glabrous ventrally, with sparse minute dark-brown
stellate scales less than 0·4 mm. long, margin with a pale cartilaginous border.
Fertile lamina very narrowly oblanceolate, base decurrent, glabrous.

Zambia. N: Abercorn Distr., Kambole Escarpment, 18.ii.1961, *Robinson* 4398 (K).
Also in Katanga. Streamside rocks in shade.

3. **Elaphoglossum marojejyense** Tardieu in Mém. Inst. Sci. Madag., Sér. B, **6**: 238,
   t. 8 fig. 1–4 (1955). Type from Madagascar.

Rhizome c. 1 mm. in diam., widely creeping, with spaced fronds and with brown
concolorous ovate-oblong to narrowly ovate subentire rhizome-scales up to 2 × 0·8
mm. Frond erect, coriaceous. Stipe up to 1 cm. long, glabrous except for a few
slightly fimbriate brown scales similar to those on the rhizome. Sterile lamina up
to 2·5 × 7 cm., oblong-elliptic, obovate or spathulate, acute to obtuse, very longly
decurrent, glabrous. Fertile lamina up to 1·4 × 0·6 cm., oblong-elliptic to broadly
elliptic, long-decurrent, glabrous.

Rhodesia. E: Chimanimani Mts., 11.ii.1958, *Mitchell* 367 (BOL; K; SRGH).
Also in Madagascar. Sheltered mossy rock crevices at c. 1830 m.

4. **Elaphoglossum macropodium** (Fée) Moore, Ind. Fil.: 11 (1857). Type from
Réunion.
Acrostichum macropodium Fée, Mém. Fam. Foug. **2**: 30, t. 6 fig. 2 (1845). Type
as above.
Elaphoglossum conforme var. *latifolium* sensu Sim, Ferns S. Afr. ed. 2: 286, t.
148 (1915).

Rhizome c. 5 mm. in diam., creeping, with closely spaced fronds and with semi-patent pale-brown concolorous narrowly lanceolate attenuate irregularly fimbriate rhizome-scales up to 12 × 2 mm. Frond suberect, coriaceous. Stipe up to 24 cm. long, darker brown at the extreme base, nitidous, glabrous except for a few scales at the extreme base similar to the rhizome scales. Sterile lamina up to 36 × 8 cm., acuminate (broadly acute to broadly obtuse in smaller fronds) broadly cuneate and scarcely decurrent at the base; veins c. 1 mm. apart; midrib shallowly sulcate ventrally, convex dorsally, glabrous except for minute very scattered substellate scales less than 0·5 mm. long. Fertile lamina up to 26 × 4 cm., oblong-lanceolate, acute, shortly decurrent, glabrous ventrally.

Rhodesia. E: Stapleford, 16.vii.1955, *Schelpe* 5742 (BM; BOL). **Malawi.** S:
Mlanje Mt., xii.1898, *Mahon* (K). **Mozambique.** Z: Namúli Mt., 26.vii.1962, *Schelpe
& Leach* 7046 (BOL). MS: Chimanimani Mts., 2.iii.1907, *Johnson* 227 (K).
Also in S. Africa, Tanzania and Réunion. Lithophyte or low-level epiphyte in deep
shade in forest, 1600–1770 m.

5. **Elaphoglossum lastii** (Bak.) C. Chr., Ind. Fil.: 309 (1905). Type probably from
Tanzania.
Acrostichum lastii Bak. in Ann. of Bot. **5**: 491 (1891). Type as above.
Acrostichum volkensii Hieron. in Engl., Pflanzenw. Ost-Afr. **C**: 81 (1895). Type
from Tanzania.
Elaphoglossum volkensii (Hieron.) C. Chr., tom. cit.: 318 (1905). Type as above.
Elaphoglossum obtusum A. Peter in Fedde, Repert. Beih. **40**, 1: 2 (1929). Type
from Tanzania.

Rhizome up to 6 mm. in diam., creeping, with fronds spaced 1 cm. apart, and with adpressed ferrugineous linear-lanceolate-attenuate sometimes subciliate scales up to 5 × 0·9 mm. Stipe up to 17 cm. long, stramineous, dark-brown at the base, glabrous except for reddish-brown scales similar to those on the rhizome, which are seldom persistent except at the base. Frond erect or arching, thinly to thickly coriaceous. Sterile lamina up to 33 × 5·2 cm., narrowly oblong, always obtuse, base cuneate hardly decurrent, with an extremely narrow pale recurved margin and with minute substellate brown scales less than 0·5 mm. in diam. Fertile lamina up to 6 × 1·8 cm., very narrowly oblong-acute to obtuse, base narrowly cuneate, scarcely decrescent, glabrous ventrally.

Rhodesia. E: Umtali, Pioneer Farm, vii.1948, *Fisher & Schweickerdt* 326 (K; PRE).
Malawi. S: Luchenya Plateau, Mlanje Mt., 30.vi.1946, *Brass* 16550 (K). **Mozambique.**
MS: Gorongosa Mt., 4.vii.1955, *Schelpe* 5489 (BOL).
Also in Tanzania and Kenya. Mid-level to high-level epiphyte in forest, 1370–1830 m.

6. **Elaphoglossum rhodesianum** Schelpe in Bol. Soc. Brot., Sér. 2, **41**: 213 (1967).
TAB. 60. Type : Zambia, Shiwa Ngandu, *Greenway* 5449 (EA, isotype; K, holo-type).

Rhizome c. 5 mm. in diam., creeping, with tufted fronds and with pale-brown concolorous lanceolate subentire acuminate rhizome-scales up to 9 × 2 mm. Frond erect, thinly coriaceous. Stipe up to 56 cm. long on fertile frond, up to 32 cm. on sterile frond, 3 mm. in diam., stramineous above the articulation, dark-brown below, nitidous with scattered narrowly ovate brown slightly fimbriate scales. Sterile lamina up to 43 × 3·5 cm., narrowly oblanceolate-acuminate longly decurrent below, glabrous on both surfaces, veins c. 1 mm. apart, midrib shallowly sulcate ventrally, convex dorsally. Fertile lamina up to 25 × 1·3 cm., very narrowly linear-elliptic, attenuate, long-tapering below, glabrous.

Zambia. N: Isoka Distr., 48 km. S. of Shiwa Ngandu, 28.xi.1952, *Angus* 861a (K).
Also in SE. Congo (Kundelungu Plateau). Undergrowth of swamp forest, c. 1640 m.

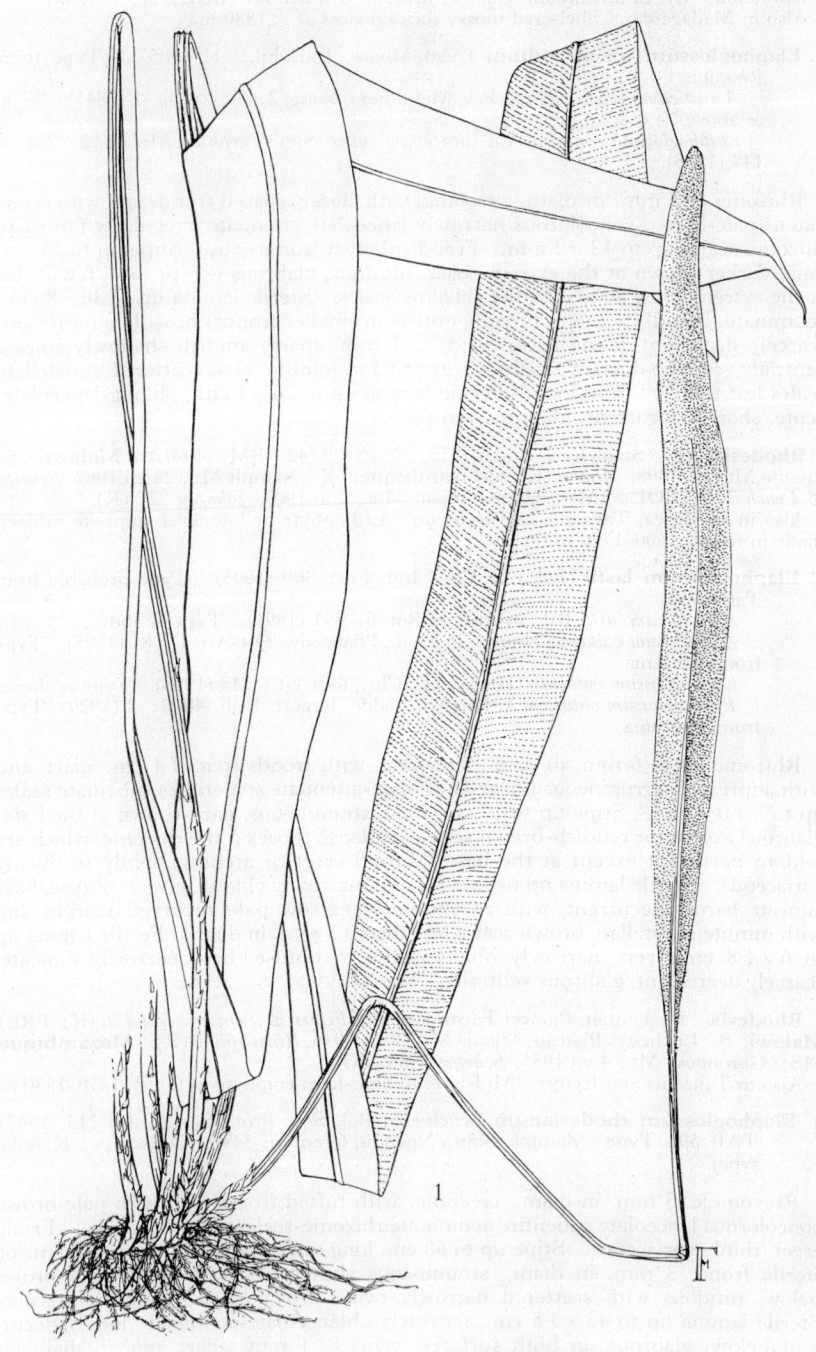

Tab. 60. ELAPHOGLOSSUM RHODESIANUM. 1, habit (× ½) *Greenway* 5449.

7. **Elaphoglossum hybridum** (Bory) Brack. in U.S. Expl. Exped. **16**: 69 (1854). Type from Réunion.

Acrostichum hybridum Bory, Voy. Quatre Princ. Iles **3**: 95 (1804). Type as above.

Acrostichum ciliatum Desv. in Mag. Ges. Naturf. Fr. Berl. **5**: 310 (1811.) Type from Réunion.

Acrostichum ciliare Carm. in Trans. Linn. Soc. Lond. **12**: 510 (1818) non Thouars (1804). Type from ? Tristan da Cunha.

Olfersia hybrida (Bory) C. Presl, Tent. Pterid.: 235 (1836) reimpr. in Abh. Königl. Böhm. Ges. Wiss., Ser. 4, **5**: 235 (1837). Type as for *Elaphoglossum hybridum*.

Acrostichum tricholepis Bak. in Journ. of Bot. **29**: 5 (1891). Type from Madagascar.

Elaphoglossum tricholepis (Bak.) C. Chr., Ind. Fil.: 317 (1905). Type as above.

Rhizome up to 4 mm. in diam., creeping, with clustered fronds and with dark-brown lanceolate attenuate entire rhizome-scales up to 6 × 1 mm. Fronds arching, firmly herbaceous. Stipe stramineous to pale-brown fairly densely set with dark-brown lanceolate attenuate dark-brown squarrose entire scales up to 4 mm. long, becoming subglabrous with age. Sterile lamina up to 40 × 5·5 cm., very narrowly oblong to very narrowly elliptic (or narrowly elliptic in small fronds), acuminate base narrowly to broadly cuneate, abruptly decrescent or not, set mainly along the margin and the midrib with very-dark-brown lanceolate attenuate entire scales up to 4 mm. long, smaller along the margin, midrib shallowly sulcate ventrally convex dorsally, veins about 1·5 mm. apart. Fertile lamina up to 12·5 × 3 cm., narrowly oblong, acute, narrowly ovate in small fronds, base broadly cuneate, hardly decurrent.

Rhodesia. E. Inyanga, Inyangani, 25.vii.1957, *Chase* 6666 (BM; SRGH). **Malawi.** S: Zomba Distr., Zomba Plateau, 7.vi.1946, *Brass* 16302 (K). **Mozambique.** MS: Gorongosa Mt., Gogogo Peak, 5.vii.1953, *Schelpe* 5523 (BM; BOL).

Also in S. Africa, the high eastern tropical African mountains, Madagascar, Comoro Is., the Mascarene Is., Tristan da Cunha and Gough Is. Lithophyte or low-level epiphyte in shade, 1700–1980 m.

8. **Elaphoglossum aubertii** (Desv.) Moore, Ind. Fil.: 5 (1857). Type from Réunion.

Acrostichum aubertii Desv. in Mag. Ges. Naturf. Fr. Berl. **5**: 309 (1811). Type as above.

Acrostichum boivinii Mett. ex Kuhn, Fil. Afr.: 43 (1868). Type from Mauritius.

Rhizome short, creeping, with clustered fronds and with brown lanceolate - acuminate entire hair-pointed rhizome-scales up to 6·5 × 0·7 cm. at the base. Frond arching, firmly membranous. Stipe stramineous to pale-brown with persistent rather squarrose brown linear-lanceolate-acuminate entire persistent scales up to 5 mm. long. Stipe of fertile fronds much longer than in sterile fronds; up to 18 cm. long (up to 7 cm. in sterile fronds). Sterile lamina up to 29 × 1·7 cm., linear, acuminate, base rounded cuneate not decurrent, margin entire or irregularly and shallowly undulate, with brown inrolled lanceolate scales similar to those on the stipe along the midrib and with smaller ovate to lanceolate inrolled scales scattered on both surfaces and margins. Fertile lamina up to 10 × 1·3 cm., very narrowly oblong, acute, base broadly cuneate to cordate-truncate, ventral surface subglabrous.

Rhodesia. E: Umtali Distr., Stapleford Forest Reserve, 1.v.1952, *Chase* 4509 (BM; BOL; SRGH). **Malawi.** N: Nyika Plateau, 16.viii.1946, *Brass* 17243 (K). S: Mlanje Distr., Mlanje Mt , 12.vi.1962, *Richards* 16647 (K). **Mozambique.** MS: Gorongosa Mt., Gogogo Peak, 5.vii.1955, *Schelpe* 5522 (B; BM; BOL; K; P; US).

Also in S. Africa and on the higher mountains of tropical Africa, and in Madagascar, the Comoro Is. and the Mascarene Is. Lithophyte, or occasionally low-level epiphyte, in deep shade in forest, 1370–1830 m.

9. **Elaphoglossum spathulatum** (Bory) Moore, Ind. Fil.: 14 (1857). Type from Réunion.

Acrostichum spathulatum Bory, Voy. Quatre Princ. Iles, **1**: 363, t. 20 fig. 1 (1804). Type as above.

Acrostichum piloselloides C. Presl, Rel. Haenk. **1**: 14, t. 2 fig. 1 (1825). Type from S. America (Peru).

Olfersia spathulata (Bory) C. Presl, Tent. Pterid.: 233 (1836) reimpr. in Abh.

Königl. Böhm. Ges. Wiss., Ser. 4, 5: 233 (1837). Type as for *Elaphoglossum spathu-latum*.
 Elaphoglossum piloselloides (C. Presl) Moore, Ind. Fil.: 13 (1857). Type as for *Acrostichum piloselloides*.
 Elaphoglossum uluguruense Reim. in Notizbl. Bot. Gart. Berl. **12**: 80 (1934). Type from Tanzania.
 Elaphoglossum spathulatum var. *uluguruense* (Reim.) Schelpe in Journ. S. Afr. Bot. **30**: 197 (1964). Type as above.

Rhizome 1·5 mm. in diam., short, creeping, with clustered fronds and with pale-brown lanceolate attenuate entire rhizome-scales up to 4 mm. long. Frond erect, firmly membranous. Stipe stramineous, up to 4 cm. in fertile fronds, up to 1·7 cm. in sterile fronds, fairly densely beset with squarrose lanceolate longly acuminate somewhat inrolled subentire pale-brown scales up to 2 mm. long. Sterile lamina up to 4·5 × 0·9 cm., narrowly oblanceolate to spathulate, rounded with a longly decurrent base fairly densely set on both surfaces with pale-brown inrolled entire scales up to 2 mm. long. Fertile lamina up to 1·8 × 1 cm. broadly elliptic to circular, apex rounded, hardly decurrent, with scales similar to those on the sterile lamina but not persistent except along the midrib.

Zambia. W: Lukera Falls N. of Mwinilunga, 25.i.1938, *Milne-Redhead* 4334 (K). **Rhodesia.** N: Mazoe, 12.iv.1959, *Drummond* 6049 (K; SRGH). E: Inyanga, 25.vii.1957, *Chase* 6662 (K; SRGH). **Malawi.** S: Mt. Mlanje, 10.iii.1963, *Wild* 6192 (BOL; SRGH). **Mozambique.** Z; Namúli Mt., R. Licungo, 29.vii.1962, *Schelpe & Leach* 7094 (BOL). MS: Serra da Gorongosa, Mt. Nhandore, 6.v.1964, *Torre & Paiva* 12280 (LISC).
 Also in S. Africa, Tanzania, Madagascar and Réunion, and tropical America. Litho-phytic on stream-bank boulders in shade, 970–2135 m.

10. Elaphoglossum mildbraedii Hieron. in Deutsch. Z.-Afr. Exped. 1907–1908, **2**: 34 (1910). Type from Ruwenzori.

Rhizome c. 4 mm. in diam., shortly creeping, with clustered fronds and with castaneous narrowly lanceolate attenuate rhizome-scales up to 5 mm. long, minutely serrulate on the upper ½. Frond arching, membranous. Stipe up to 23 cm. long, greyish-green to pale-brown, set at first with narrowly lanceolate sub-entire squarrose ferrugineous inrolled scales up to 2 mm. long, becoming sub-glabrous with age. Sterile lamina 12–15 × 2·5 cm., narrowly elliptic to narrowly oblong, acute to acuminate-cordate, base cuneate hardly decrescent; set on both surfaces with scattered ferrugineous narrowly lanceolate inrolled scales up to 2·5 mm. long and sometimes with scattered minute white hairs between the scales. Fertile lamina up to 5 × 2·2 cm. elliptic-oblong to ovate-triangular with hairs on the ventral surface similar to those on the sterile lamina.

Malawi. S: Mlanje Mt., 11.vi.1962, *Richards* 16630a (K).
 Also in E. tropical African high mountains. Lithophytic in deep shade, 2010 m.

11. Elaphoglossum chevalieri Christ in Journ. de Bot., Sér. 2, **2**: 23 (1909). Type from S. Tomé.

Rhizome up to 4 mm. in diam., creeping, with clustered fronds and with appressed brown lanceolate acuminate concolorous rhizome-scales up to 5 × 1 mm. at the base. Frond erect to arching, firmly to thinly membranous. Stipe pale-greenish-brown, up to 16 cm. long, densely set with subulate subentire inrolled pale-reddish-brown scales up to 6 mm. long. Sterile lamina up to 12 × 3·2 cm., narrowly oblong-elliptic (narrowly oblong to narrowly elliptic in depauperate forms), acute, base cuneate, shortly decurrent, sparsely beset on both surfaces, more thickly so along the margin and the midrib, dorsally with reddish-brown sub-ulate inrolled subentire scales up to 4 mm. long. Fertile lamina up to 5 × 1·2 cm., narrowly oblong, elliptic (to subcircular in depauperate forms) obtuse, rounded or broadly truncate, shortly decurrent, sometimes with a narrow sterile margin and with scattered subulate scales on the ventral surface.

Rhodesia. E: Inyanga, Pungwe Gorge, 25.vii.1957, *Chase* 6667 (BM; BOL; COI; LISC; SRGH). **Malawi.** S: Mlanje Distr., Luchenya Plateau, 5.vii.1946, *Brass* 16671 (K). **Mozambique.** Z: Namúli, 1887, *Last* (K). MS: Gorongosa Mt., near Gogogo Peak, 6.vii.1955, *Schelpe* 5539 (BM; BOL; K).

Also in W. tropical Africa. Lithophytic on boulders in deep shade in forest, 1220–1830 m.

12. **Elaphoglossum petiolatum** (Sw.) Urb., Symb. Ant. **4**: 61 (1903). Type from W. Indies.
 Acrostichum petiolatum Sw., Nov. Gen. Sp. Pl.: 128 (1788). Type as above.

Subsp. **salicifolium** (Willd. ex Kaulf.) Schelpe in Contr. Bol. Herb. **1**: 34 (1969). Type from Réunion.
 Acrostichum lancifolium Desv. in Ges. Naturf. Fr. Berl. Mag. **5**: 310 (1811). Type from Mauritius.
 Acrostichum salicifolium Willd. ex Kaulf., Enum. Fil.: 58 (1824). Type as for *Elaphoglossum petiolatum* subsp. *salicifolium*.
 Elaphoglossum viscosum var. *salicifolium* (Willd. ex Kaulf.) Hieron. in Mildbr., Wiss. Ergebn. Deutsch. Z.-Afr. Exped. 1907–08, **2**: 35 (1910–1914). Type as above.
 Elaphoglossum petiolatum var. *salicifolium* (Willd. ex Kaulf.) C. Chr. in Dansk Bot. Ark. **7**: 168 (1932). Type as above.
 Elaphoglossum salicifolium (Willd. ex Kaulf.) Alston in Exell, Cat. Vasc. Pl. S. Tomé: 92 (1944). Type as above.
 Elaphoglossum lancifolium (Desv.) Morton in Contr. U.S. Nat. Herb. **38**: 32 (1967). Type as for *Acrostichum lancifolium*.

Rhizome c. 2·5 cm. in diam., creeping, with tufted fronds, and with lanceolate-attenuate subentire rhizome-scales up to 4·5 × 0·7 mm., concolorous, dark-brown, often castaneous. Frond erect or arching, firmly membranous to thinly coriaceous. Stipe up to 13 cm. long, densely covered at first with long ciliate oblong to lanceolate scales up to 2 mm. long. Sterile lamina up to 22 × 1·5 cm., thinly covered on both surfaces with very longly ciliate scales, the cilia much longer than the scale-diameter, often almost stellate, with some oblong longly ciliate scales along the midrib on the ventral surface. Fertile lamina up to 18 × 0·6 cm., linear-acute, base cuneate not decurrent.

Rhodesia. E: Umtali, Norseland, 7.iii.1949, *Chase* 3308 (BM; BOL; SRGH). **Malawi.** N: Nyika Plateau, Nthonjera Mt., 27.xi.1963, *Lemon* 1012 (SRGH). S: Mlanje Mt., Great Ruo Gorge, 17.vi.1962, *Richards* 16725 (K). **Mozambique.** Z: Namúli Mt., 26.vii.1962, *Schelpe & Leach* 7047 (BOL).
 Also in western tropical Africa, Madagascar, Comoro Is. and Mascarene Is. Lithophyte and low-level epiphyte in shade in forest, 970–1370 m.

13. **Elaphoglossum deckenii** (Kuhn) C. Chr., Ind. Fil.: 305 (1906). Type from Tanzania.
 Acrostichum deckenii Kuhn in Fil. Deck.: 16 (1867). Type as above.

Rhizome up to 2·5 mm. in diam., creeping, with clustered fronds, and with castaneous lanceolate ciliate rhizome-scales up to 3 × 0·8 mm. Frond erect to arching, firmly membranous. Stipe up to 6 cm. long, densely beset with lanceolate to narrowly ovate pale-brown ciliate scales up to 3 × 1·2 mm. Sterile lamina up to 14·5 × 2·5 cm., beset on both surfaces with pale-reddish-brown narrowly ovate to oblong ciliate scales up to 1·5 mm. long. Fertile lamina up to 10 × 1 cm., linear-lanceolate to ovate, with pale-reddish-brown longly ciliate scales on the ventral surface and along the midrib of the dorsal surface.

Rhodesia. E: Inyanga, Chipungu Waterfall, 4.iv.1949, *Chase* 2055 (BM; SRGH). **Malawi.** S: West Peak, Mlanje Mt., 11.vi.1962, *Richards* 16630b (K).
 Also on the high E. tropical African mountains. Lithophytic on moist deeply shaded boulders, c. 1850 m.
 Specimens from our area have none or very few of the stipe scales with the black cilia usually seen in the E. African specimens.

14. **Elaphoglossum kuhnii** Hieron. in Engl., Bot. Jahrb. **46**: 399 (1911). Type from W. Africa.

Rhizome creeping, with clustered fronds and with lanceolate acuminate serrate to subentire pale-brown concolorous rhizome-scales up to 6 × 0·8 mm., some with black margins and black cilia. Frond coriaceous to firmly membranous, erect. Stipe up to 12·5 cm., with dense narrowly ovate to lanceolate long-ciliate pale-brown scales up to 5·5 × 1 mm. Sterile lamina up to 12·5 × 2·1 cm., very narrowly oblong, apex rounded to acute, base cuneate, hardly decrescent, covered on both surfaces with oblong to oblong-lanceolate pale-brown scales with cilia as long as the width of the scales. Fertile lamina up to 7 × 0·8 cm., very narrowly oblong, rounded; base broadly cuneate hardly decurrent, with ciliate scales on the ventral surface similar to those on the sterile lamina.

Rhodesia. C: Chindamora Reserve, Mkumbe R., 29.iv.1956, *Wild* 4810 (BM; BOL; SRGH). E: Umtali, Odzani R. Falls, 14.vii.1955, *Schelpe* 5647 (BM; BOL; K). Also in W. tropical Africa. Lithophytic on shaded rocks, 1220–1400 m.

15. **Elaphoglossum welwitschii** (Bak.) C. Chr., Ind. Fil.: 318 (1905). Type from Angola.
    Acrostichum welwitschii Bak., Syn. Fil., ed. 2: 521 (1874). Type as above.

Rhizome up to 6 mm. in diam., creeping, with tufted fronds and with ferrugineous lanceolate acuminate concolorous rhizome-scales up to 3·5 × 0·7 mm. Stipe up to 10 cm. long, pale brown densely set with pale-reddish-brown ovate, long-ciliate scales c. 1 mm. long. Sterile fronds arching, firmly membranous; fertile fronds erect. Sterile lamina up to 12 × 1·7 cm., very narrowly elliptic, attenuate, base tapered, margins usually recurved, both surfaces with almost contiguous but not imbricate ciliate ovate to oblong scales up to 0·8 mm. long, the cilia about as long as the width of the scales. Fertile lamina linear with a tapered base.

Zambia. N: Mpika Distr., near Serenje, 5.iv.1961, *Richards* 14969 (K). **Malawi.** N: Vipya, round Luwawa Dam, 5.i.1962, *Chapman* 1542 (SRGH).
Also in Angola, Congo and Tanzania. *Brachystegia* woodland at about 1600 m.

2. LOMARIOPSIS Fée

Lomariopsis Fée, Mém. Fam. Foug. **2**: 10, 66 (1845).

Rhizome climbing, with spaced fronds and with dark rhizome-scales. Stipes not articulated. Fronds dimorphous. Lamina pinnate with articulated pinnae except for the terminal pinna; sterile pinnae oblong to narrowly lanceolate; veins parallel and free; fertile pinnae narrowly linear; sori acrostichoid.

A tropical genus of over 20 spp., with 8 spp. in tropical Africa, but only 1 of these occurring in our area. Fertile fronds are usually infrequently produced here.

Lomariopsis warneckei (Hieron.) Alston in Journ. of Bot. **72**, Suppl. 2: 6 (1934). TAB. **61**. Type from Tanzania.
    Stenochlaena warneckei Hieron. in Engl., Bot. Jahrb. **46**: 383 (1911). Type as above.
    Lomariopsis nigrescens Holttum in Kew Bull. **1939**: 627, fig. 14 (1939). Type from Fernando Po.

Rhizome c. 8 mm. in diam., creeping, with fronds spaced 2–4 cm. apart and with dark-brown narrowly lanceolate non-clathrate concolorous subentire rhizome scales up to 7 × 1 mm. Fronds arching, thinly coriaceous, strongly dimorphous. Stipes pale-brown, up to 14 × 0·2 cm., set at first with scales similar to the rhizome-scales, becoming glabrous later except at the base. Sterile lamina up to 53 × 23 cm., narrowly ovate-deltate in outline, imparipinnate, pinnae linear to lanceolate, attenuate, glabrous, base cuneate, apex gradually acuminate, margin minutely and irregularly crenate, up to 12 × 1·6 cm., rhachis pale-brown, glabrous or with a few small dark brown scales, narrowly winged in the upper part. Fertile lamina up to 30 × 25 cm., narrowly lanceolate-triangular in outline, pinnate; pinnae up to 7·5 × 0·4 cm., linear attenuate, shortly petiolate, base unequally cuneate, glabrous; rhachis pale-brown, glabrous at maturity. Sori covering the entire under surface of the fertile lamina.

Malawi. N: Misuku Hills, 11.i.1959, *Robinson* 3164 (K; SRGH). **Mozambique.** Z: Namúli, 1887, *Last* (K). MS: Garuso, Jaegersberg, Bandula Mt., 11.vii.1955, *Schelpe* 5628 (B; BM; BOL; K; P).
Also in Tanzania, Kenya and Fernando Po. Low-level epiphyte in dense shade in forest, c. 1100 m.

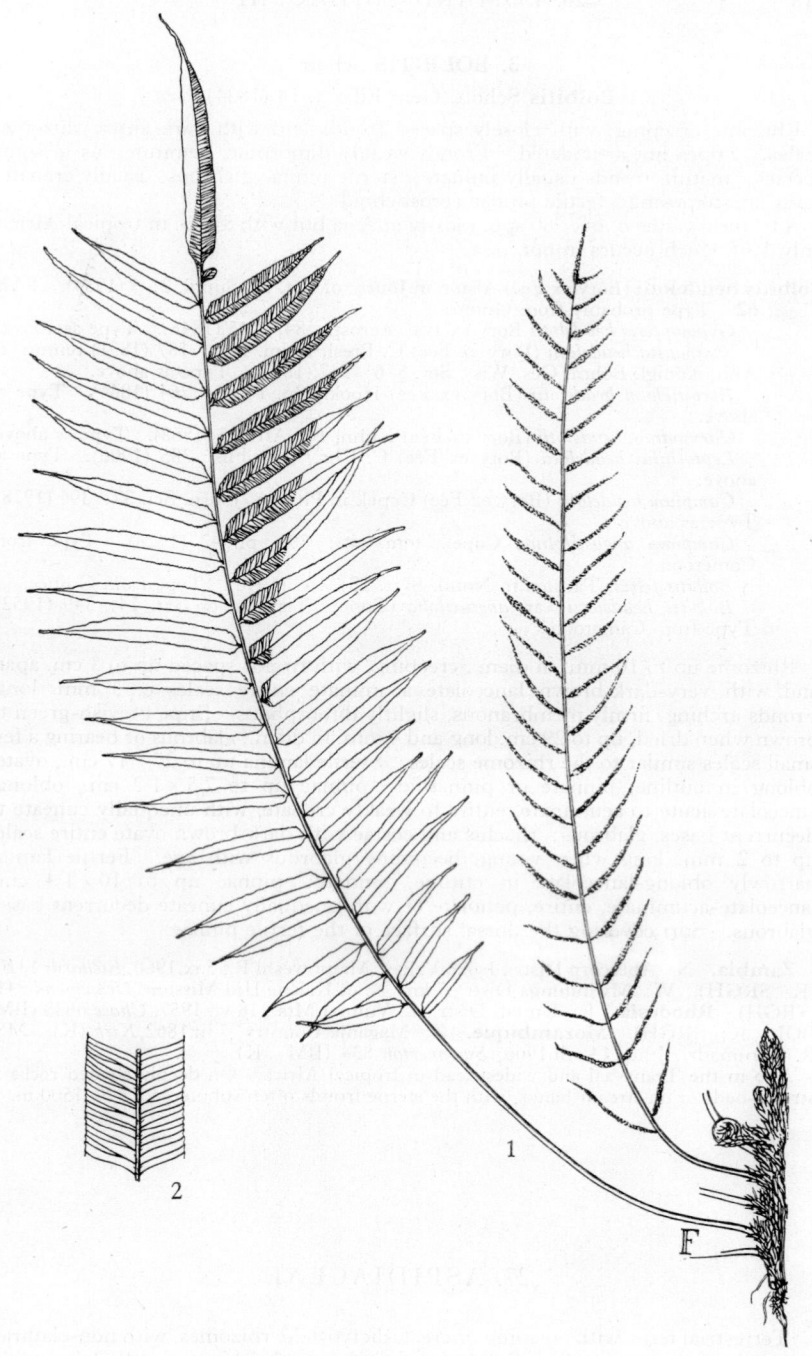

Tab. 61. LOMARIOPSIS WARNECKEI. 1, habit (× ⅓); 2, part of sterile pinna (× 1), both from *Longfield* 22.

3. BOLBITIS Schott

Bolbitis Schott, Gen. Fil.: t. 14 (1834).

Rhizome creeping, with closely spaced fronds and with dark entire rhizome-scales. Stipes not articulated. Fronds weakly dimorphic, gemmiferous in some species; mature fronds usually pinnate; sterile pinnae glabrous, usually crenate; veins anastomosing; fertile pinnae acrostichoid.

A tropical genus of over 80 spp. mostly in Asia but with 8 spp. in tropical Africa, only 1 of which occurs in our area.

Bolbitis heudelotii (Bory ex Fée) Alston in Journ. of Bot. **72**, Suppl. 2: 3 (1934). TAB. 62. Type probably from Guinée.

> *Gymnopteris heudelotii* Bory ex Fée, Acrost.: 84, t. 45 (1845). Type as above.
> *Anapausia heudelotii* (Bory ex Fée) C. Presl, Epim. Bot.: 187 (1851) reimpr. in Abh. Königl. Böhm. Ges. Wiss. Ser. 5, **6**: 547 (1851). Type as above.
> *Acrostichum heudelotii* (Bory ex Fée) Hook., Sp. Fil. **5**: 264 (1864). Type as above.
> *Chrysodium heudelotii* (Bory ex Fée) Kuhn, Fil. Afr.: 5 (1868). Type as above.
> *Leptochilus heudelotii* (Bory ex Fée) C. Chr., Ind. Fil.: 385 (1906). Type as above.
> *Campium heudelotii* (Bory ex Fée) Copel. in Philipp. Journ. Sci. **37**: 396 (1928). Type as above.
> *Campium angustifolium* Copel., tom. cit.: 396, pl. 47 (1928). Type from Cameroon.
> *Bolbitis felixii* Tardieu in Notul. Syst. **13**: 169 (1948). Type from Guinée.
> *Bolbitis heudelotti* var. *angustifolia* (Copel.) Tardieu, op. cit. **14**: 349 (1952). Type from Cameroon.

Rhizome up to 10 mm. in diam., creeping, with fronds spaced up to 3 cm. apart and with very-dark-brown lanceolate acuminate entire scales c. 3 mm. long. Fronds arching, firmly membranous, slightly dimorphous. Stipe greyish-green to brown when dried, up to 29 cm. long and 4 mm. in diam., glabrous or bearing a few small scales similar to the rhizome scales. Sterile lamina up to 27 × 17 cm., ovate-oblong in outline, pinnate or pinnatifid; pinnae up to 7·5 × 1·2 cm., oblong-lanceolate acute to acuminate, entire or weakly crenate, with unequally cuneate to decurrent bases, glabrous; rhachis and costae with dark-brown ovate entire scales up to 2 mm. long when young, becoming glabrous with age. Fertile lamina narrowly oblong-lanceolate in outline, pinnate; pinnae up to 10 × 1·4 cm., lanceolate-acuminate, entire, petiolate or with unequally cuneate decurrent bases, glabrous. Sori covering the dorsal surface of the fertile pinnae.

Zambia. N: Abercorn Distr., Isoko Valley, Mambweshi R., 5.ix.1960, *Richards* 13201 (K; SRGH). W: Mwinilunga Distr., 7 km. N. of Kalene Hill Mission, *Drummond* 8418 (SRGH). **Rhodesia.** E: Umtali Distr., E. Vumba Mts., 18.vii.1957, *Chase* 6638 (BM; BOL; K; SRGH). **Mozambique.** Z: Maganja Country, 7.iii.1862, *Kirk* (K). MS: R. Curumadzi, Jihu, 14.viii.1906, *Swynnerton* 834 (BM; K).

Also in the Transvaal and widespread in tropical Africa. On deeply shaded rocks in stream-beds or on stream-banks, with the sterile fronds often submerged, 820–1500 m.

27. ASPIDIACEAE

Terrestrial ferns with creeping or erect, dictyostelic rhizomes, with non-clathrate non-peltate rhizome-scales. Stipes not articulate to the rhizome, with 2–7 vascular bundles. Lamina mostly 2-pinnatifid to 4-pinnatifid, often with the lower pinnae basiscopically developed, glabrous or with scales or unicellular or multicellular hairs. Pinnae not articulated to the rhachis. Rhachis groove glabrous or with scales but not with multicellular hairs. Veins free or anastomosing. Sori superficial, circular with or without peltate or reniform indusia. Spores monolete, usually with a perispore.

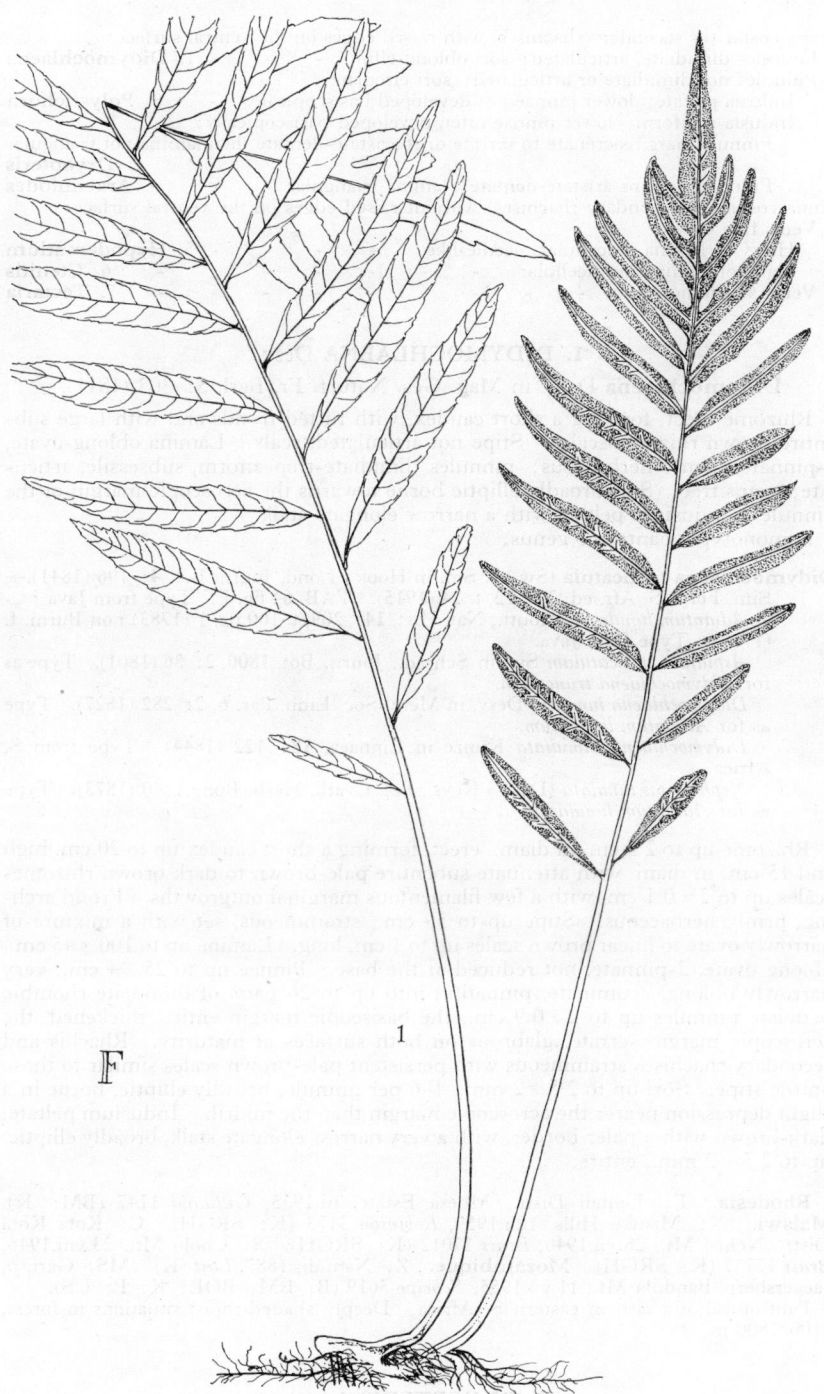

Tab. 62. BOLBITIS HEUDELOTII. 1, habit (× ⅓) *Chase* 6638.

Pinna costae (or secondary rhachises) with raised edges on the ventral surface:
  Pinnules dimidiate, articulated; sori oblong-elliptic -   -   - 1. **Didymochlaena**
  Pinnules not dimidiate or articulated; sori circular:
    Indusia peltate; lower pinnae not developed basiscopically   -   3. **Polystichum**
    Indusia reniform; lower pinnae often developed basiscopically:
      Pinnule margins crenate to serrate or if aristate-dentate then lamina not triangular
                                                            2. **Dryopteris**
      Pinnule margins aristate-dentate; lamina triangular -   -   4. **Arachniodes**
Pinna costae (or secondary rhachises) without raised edges on the ventral surface:
  Veins free:
    Hairs on lamina unicellular, needle-like   -   -   -   - 5. **Hypodematium**
    Hairs on lamina multicellular - - - - - -   6. **Ctenitis**
  Veins anastomosing   - - - - - - -   7. **Tectaria**

1. DIDYMOCHLAENA Desv.

Didymochlaena Desv. in Mag. Ges. Naturf. Fr. Berl. **5**: 303 (1811).

Rhizome erect, forming a short caudex, with tufted fronds and with large sub-entire brown rhizome-scales. Stipe not articulated, scaly. Lamina oblong-ovate, 2-pinnate, firmly herbaceous; pinnules dimidiate-trapeziform, subsessile, articulate; veins free. Sori broadly elliptic borne towards the acroscopic margin of the pinnules; indusium peltate with a narrow elongate stalk.

A monotypic pantropic genus.

Didymochlaena truncatula (Sw.) J. Sm. in Hook., Lond. Journ. Bot. **4**: 196 (1841).—
  Sim, Ferns S. Afr. ed. 2: 112, t. 24 (1915). TAB. **64** fig. E. Type from Java ?
    Adiantum lunulatum Houtt., Nat. Hist. **14**: 209, t. 100 fig. 1 (1783) non Burm. f.
  (1768). Type from Java.
    Aspidium truncatulum Sw. in Schrad., Journ. Bot. **1800**, 2: 36 (1801). Type as
  for *Didymochlaena truncatula.*
    Didymochlaena lunulata Desv. in Mém. Soc. Linn. Par. **6**, 2: 282 (1827). Type
  as for *Adiantum lunulatum.*
    Didymochlaena dimidiata Kunze in Linnaea, **18**: 122 (1844). Type from S.
  Africa.
    Nephrolepis lunulata (Desv.) Keys., Pol. Cyath. Herb. Bung.: 40 (1873). Type
  as for *Adiantum lunulatum.*

Rhizome up to 2·5 cm. in diam., erect, forming a short caudex up to 20 cm. high and 15 cm. in diam. with attenuate subentire pale-brown to dark brown rhizome-scales up to 2 × 0·1 cm. with a few filamentous marginal outgrowths. Frond arching, firmly herbaceous. Stipe up to 50 cm., stramineous, set with a mixture of narrowly ovate to linear brown scales up to 1 cm. long. Lamina up to 100 × 45 cm., oblong ovate, 2-pinnate, not reduced at the base. Pinnae up to 25 × 4 cm.; very narrowly oblong, acuminate, pinnatifid into up to 26 pairs of dimidiate rhombic petiolate pinnules up to 2 × 0·9 cm., the basiscopic margin entire, thickened, the acroscopic margin serrate, glabrous on both surfaces at maturity. Rhachis and secondary rhachises stramineous with persistent pale-brown scales similar to those on the stipe. Sori up to 2·5 × 2 mm., 1–6 per pinnule, broadly elliptic, borne in a slight depression nearer the acroscopic margin than the midrib. Indusium peltate, dark-brown with a paler border, with a very narrow elongate stalk, broadly elliptic, up to 2·5 × 2 mm., entire.

Rhodesia. E: Umtali Distr., Mbesa Estate, iii.1935, *Gilliland* 1747 (BM; K).
Malawi. N: Misuku Hills, 12.i.1959, *Robinson* 3173 (K; SRGH). C: Kota Kota Distr., Nchisi Mt., 28.vii.1946, *Brass* 17012 (K; SRGH). S: Cholo Mt., 23.viii.1946, *Brass* 17752 (K; SRGH). **Mozambique.** Z: Namúli, 1887, *Last* (K). MS: Garuso, Jaegersberg, Bandula Mt., 11.vii.1955, *Schelpe* 5619 (B; BM; BOL; K; P; US).
  Pantropical and also in eastern S. Africa. Deeply shaded moist situations in forest, 1100–1800 m.

2. DRYOPTERIS Adans.

Dryopteris Adans., Fam. Pl. **2**: 20, 551 (1763) *nom. conserv.*

Rhizome erect or creeping, with tufted fronds, and usually with broad con-colorous entire or subentire rhizome-scales. Stipe not articulated, often with

broad to narrow persistent scales. Lamina oblong-lanceolate to broadly triangular in outline, herbaceous, 2-pinnatifid to 4-pinnate, with the basal pair of pinnae often developed basiscopically, the ultimate segments crenate to serrate, rarely aristate-dentate; veins free. Sori circular with reniform indusia.
A cosmopolitan genus of c. 150 spp.

Lamina with the basal pinnae not, or hardly, reduced, 3–4-pinnatifid:
  Rhachis and costae with numerous patent dark-brown persistent hair-like scales
                                            6. *squamiseta*
  Rhachis and costae glabrous or with pale-brown scales:
    Costae with scattered ovate-acuminate scales; lamina broadly triangular
                                            3. *kilemensis*
    Costae glabrous or with linear scales; lamina oblong to narrowly triangular:
      Frond proliferous - - - - - - - - - 5. *manniana*
      Frond not proliferous:
        Pinnule lobes aristate-serrate - - - - - - 4. *callolepis*
        Pinnule lobes crenate to serrate, not aristate:
          Pinnae set at about 70° to the rhachis - - - - 2. *inaequalis*
          Pinnae set at about 45° to the rhachis - - - 1. *athamantica*
Lamina with a gradually tapering base, deeply 2-pinnatifid into oblong–truncate lobes
                                          7. *wallichiana*

1. **Dryopteris athamantica** (Kunze) Kuntze, Rev. Gen. Pl. **2**: 812 (1891).—Sim, Ferns S. Afr. ed. 2: 112, t. 24 (1915).—Tardieu in Mém. Inst. Fr. Afr. Noire, **28**: 153, t. 29 fig. 7 (1953). Type from S. Africa.
    Aspidium athamanticum Kunze in Linnaea, **18**: 123 (1844). Type as above.
    Lastrea athamantica (Kunze) Moore in Hook., Journ. Bot. **5**: 311 (1853). Type as above.
    Lastrea plantii Moore, tom. cit.: 227 (1853). Type from S. Africa.
    Nephrodium athamanticum (Kunze) Hook., Sp. Fil. **4**: 125, t. 258 (1862). Type as for *Dryopteris athamantica*.
    Nephrodium eurylepium A. Peter in Fedde, Repert., Beih. **40**, 1: 57 (1929). Syntypes from Tanzania, Burundi and Rhodesia: Vumba Mts., *Peter* 30573 (not found).

Rhizome up to 2·5 cm. in diam., creeping, with tufted fronds and with ferrugineous narrowly lanceolate attenuate to linear acuminate rhizome-scales up to 2·2 × 0·1 cm., with or without marginal filamentous outgrowths. Frond erect, thinly coriaceous. Stipe up to 37 cm. long, stramineous, subglabrous above, densely paleaceous with dark reddish-brown linear-attenuate to lanceolate-attenuate scales up to 2 cm. long towards the base. Lamina up to 68 × 26 cm., narrowly ovate-oblong in outline, acute with the basal pinnae hardly reduced, 3-pinnatifid. Pinnae up to 23 × 12 cm., suberect, forming an angle of less than 50° to the rhachis, unequally narrowly triangular to oblong, acute or acuminate, unequally cuneate, outline of the base a right angle, more developed acroscopically than basiscopically, glabrous on both surfaces. Rhachis stramineous with scattered reddish-brown hair-like scales. Sori up to 1·5 mm. in diam., circular, borne about ½-way between the costules and the margin. Indusium c. 1 mm. in diam., glabrous, membranous, erose.

Zambia. N: Abercorn Distr., Ndundu, Abercorn, 6.iii.1963, *Richards* 18061 (K). W: Solwezi Distr., Mutando Bridge Camp, 24.vi.1930, *Milne-Redhead* 591 (K). **Rhodesia.** C. Makoni Distr., Timaru, i.1956, *Mitchell* 45 (BOL; SRGH). E: Umtali near Odzani R. Bridge, 14.vii.1955, *Schelpe* 5653 (BM; BOL). S: Bikita Distr., Hunyana Mt., 22.ii.1964, *Mitchell* 802 (SRGH). **Malawi.** N: S. Nyika Mts., vi.1896, *Whyte* (K). C: Chongoni Forestry School, Chiwao Hill, 4.ii.1959, *Robson* 1442 (K; LISC; SRGH). S: between Lake Chilwa and Lake Chiuta, x.1898, *Cunningham* 13 (K). **Mozambique.** N: Planalto de Maniamba, i.1932, *Sousa* 1219 (K). Z: Namúli Mt., 25.vii.1962, *Schelpe & Leach* 7000 (BOL). T: Angónia, Metengo Balane, 11.v.1948, *Mendonça* 4160 (BM; LISC). MS: Manica, 19 km. N. of Macequece, *Pedro & Pedrógão* 7044 (BM).
Also in S. Africa and widespread in tropical Africa. Around boulder-bases in grassland and about forest margins, 1200–2135 m.

2. **Dryopteris inaequalis** (Schlechtend.) Kuntze, Rev. Gen. Pl. **2**: 813 (1891). Type from S. Africa.
    Aspidium inaequale Schlechtend., Adumbr.: 23, t. 12 (1825). Type as above.
    Lastrea inaequalis (Schlechtend.) C. Presl, Tent. Pterid.: 77 (1836) reimpr. in Abh. Königl. Böhm. Ges. Wiss., Ser. 4, **5**: 77 (1837). Type as above.

Aspidium inaequale var. *montanum* Kunze in Linnaea, **10**: 549 (1836). Type from S. Africa.

Lastrea pentagona Moore in Hook., Journ. Bot. **5**: 227 (1853). Type from S. Africa.

Nephrodium inaequale (Schlechtend.) Hook., Sp. Fil. **4**: 125 (1862) non Schrad. (1824). Type as for *Dryopteris inaequalis*.

Polystichum inaequale (Schlechtend.) Keys., Polypod. Cyath. Herb. Bung.: 44 (1873). Type as above.

Nephrodium pentheri Krasser in Ann. Naturh. Hofmus. Wien, **15**: 5, t. 2 fig. 1–5 (1900). Type from S. Africa (Orange Free State).

Dryopteris pentheri (Krasser) C. Chr., Ind. Fil.: 284 (1905). Type as above.

Dryopteris pentheri var. *montana* (Kunze) Alston in Bol. Soc. Brot., Sér. 2, **30**: 14 (1956). Type from S. Africa.

Rhizome up to 3·5 cm. in diam., creeping, with tufted fronds and linear attenuate pale-brown rhizome-scales up to 2 cm. long, with filamentous marginal outgrowths. Frond arching, thinly or firmly membranous. Stipe up to 80 cm. long, stramineous, subglabrous or thinly set with pale-brown scales in the upper part, densely scaly towards the base. Lamina up to 90 × 33 cm., broadly to narrowly ovate, triangular 2-pinnatifid, apex acuminate with the basal pinnae hardly or not reduced. Pinnae up to 33 × 16 cm., not suberect forming an angle of more than 50° to the rhachis, oblong to narrowly oblong or ovate-acuminate, not conspicuously developed acroscopically. Pinnules oblong to very narrowly oblong in outline, crenate to deeply pinnatifid into oblong crenate-serrate (rarely pinnatifid) obtuse to truncate lobes glabrous on both surfaces except for scattered pale-brown hair-like scales along the costules. Rhachis stramineous, glabrous at maturity. Sori up to 1·5 mm. in diam., circular; indusium up to 1·5 mm. in diam., glabrous, entire.

Rhizome and stipe scales brown - - - - - - var. *inaequalis*
Rhizome and stipe scales black - - - - - - var. *atropaleacea*

Var. inaequalis

Zambia. N: Abercorn, Kalambo Falls, 30.iii.1952, *Richards* 1241 (K). E: Lundazi, 6.v.1952, *White* 2725 (K). **Rhodesia.** N: Mazoe, Umvukwes Ridge, Mtoroshanga **Pass,** 5.iii.1961, *Vesey-FitzGerald* 3089 (BOL). C: Makoni, Silverbow, 15.vi.1957, *Chase* 6531 (BOL; K; LISC; SRGH). E: Umtali, Rowa Township, Zimunya's Reserve, 5.ii.1957, *Chase* 6318 (BOL; K; SRGH). S: Bikita Distr., Hunyana Mt., 22.ii.1964, *Mitchell* 808 (SRGH). **Malawi.** N: Rumpi Distr., Kaziwiziwe R., 8.i.1959, *Richards* 10561 (K). S: Zomba Distr., Zomba Plateau, 28.v.1946, *Brass* 16055 (K). **Mozambique.** N: Ribáuè Mt., 19.vii.1952, *Schelpe & Leach* 6926 (BOL). Z: Namúli Mt., 25.vii.1962, *Schelpe & Leach* 7002 (BOL). T: Macanga, Mt. Furancungo, 15.iii.1966, *Pereira, Sarmento & Marques* 1692 (LMU). MS: Manica, Mavita Rotanda, 29.i.1948, *Barbosa* 908 (LISC).

Also in S. Africa and widely found in tropical Africa. Forest floors and margins, 760–2140 m.

Var. atropaleacea Schelpe in Bol. Soc. Brot., Sér. 2, **41**: 213 (1967). Type from Tanzania.

Zambia. N: Abercorn Distr., Itembwa Gorge, 15.i.1964, *Richards* 18774 (K). Also in Tanzania. Known from our region only in a cave in a steep cliff, 1500 m.

3. **Dryopteris kilemensis** (Kuhn) Kuntze, Rev. Gen. Pl. **2**: 813 (1891) ("kilmensis"). Type from Tanzania.

Aspidium kilemense Kuhn, Fil. Deck.: 24 (1867) ("kilimensis"). Type as above.

Nephrodium kilemense (Kuhn) Bak., Syn, Fil.: 498 (1874). Type as above.

Nephrodium lastii Bak. in Ann. of Bot. **5**: 324 (1891). Type: Mozambique, Namúli, *Last* (K).

Aspidium lastii (Bak.) Hieron. in Engl., Pflanzenw. Ost-Afr. C: 85 (1895). Type as above.

Dryopteris lastii (Bak.) C. Chr., Ind. Fil.: 274 (1905). Type as above.

Rhizome up to 1 cm. in diam., creeping, with tufted fronds and with brown linear to narrowly lanceolate acuminate entire rhizome-scales up to 2·5 × 0·3 cm. Frond arching, membranous. Stipe up to 1·2 m. long, glabrous at maturity above but set towards the base with persistent large and small brown scales with irregular filamentous marginal outgrowths. Lamina 1·2 × 0·8 m., ovate in outline, 4-pinnatifid to 4-pinnate, acute, with the basal pinnae the largest and much developed basiscopically. Middle pinnae up to 32 × 15 cm., oblong ovate in outline, acuminate, developed somewhat basiscopically with petiolate 2-pinnatifid oblong acuminate pinnules; the ultimate segments mostly oblong, rather truncate, serrate glabrous above and below. The basal pinnae up to 34 × 26 cm., unequally ovate-triangular with the basal basiscopic pinnule up to 18 cm. long. Rhachis pale-brown, glabrous at maturity; secondary rhachises stramineous with scattered pale-brown ovate entire acuminate scales up to 2 cm. long, narrowly winged towards the apices. Sori up to 1·5 mm. in diam., circular; indusia up to 1 mm. in diam., entire, membranous.

Zambia. N: Fort Rosebery, Lake Bangweulu, 30.viii.1952, *White* 3171 (BOL; K). **Rhodesia.** E: Umtali, Vumba Mts., Elephant Forest, 25.vi.1955, *Schelpe* 5390 (BM; BOL). **Malawi.** N: Nyika Plateau, 17.viii.1946, *Brass* 17285 (K; SRGH). S: Mlanje Mt., Luchenya Plateau, 7.vii.1946, *Brass* 16707 (K; SRGH). **Mozambique.** Z: Namúli. 1887, *Last* (K). MS: Gorongosa Mt., Gogogo Peak, 5.vii.1955, *Schelpe* 5519 BM; BOL).
Also in E. tropical Africa. Shaded forest undergrowth, 1890–2350 m.

4. **Dryopteris callolepis** C. Chr. in Notizbl. Bot. Gart. Berl. **9**: 177 (1924). Type from Kenya.

Rhizome creeping, with tufted fronds and with pale-brown concolorous lanceolate acuminate entire scales up to 1·7 × 0·4 cm. Frond arching, herbaceous. Stipe up to 31 cm. long, stramineous, becoming darker towards the base, set with lanceolate acuminate entire pale-brown scales with or without a dark central stripe, up to 1·3 × 0·3 cm. Lamina up to 67 × 36 cm., oblong-lanceolate in outline, deeply 3-pinnatifid, broadly acute and with the basal pair of pinnae slightly reduced but conspicuously developed basiscopically. Upper pinnae narrowly oblong, acute, the basal pinnae unequally triangular, acute; pinnules narrowly oblong acute to obtuse, pinnately divided into oblong obtuse aristate-dentate, often decurrent lobes c. 2 mm. broad, glabrous on both surfaces at maturity. Rhachis stramineous, sparsely set with pale-brown concolorous linear-lanceolate scales up to 5 mm. long. Sori 1–4 per pinnule lobe, circular, c. 1 mm. in diam.; indusium brown membranous, erose.

Rhodesia. E: Inyanga Distr., source of Pungwe R., 25.v.1954, *Chase* 5262 (BM; BOL; SRGH).
Also in S. Africa and on the higher E. tropical African mountains. Shaded forest floors c. 1980 m.

5. **Dryopteris manniana** (Hook.) C. Chr., Ind. Fil.: 276 (1905).—Tardieu in Mém. Inst. Fr. Afr. Noire, **28**: 149, t. 29 fig. 4–5 (1950). TAB. **63**. Type from Fernando Po.
Polypodium mannianum Hook., Sp. Fil. **4**: 253 (1862). Type as above.
Phegopteris manniana (Hook.) Kuhn, Fil. Afr.: 123 (1868). Type as above.

Rhizome up to 7 mm. in diam., suberect, with tufted fronds and with brown subentire narrowly lanceolate attenuate rhizome-scales up to 1·4 × 0·15 cm. with irregular short marginal outgrowths. Frond arching, herbaceous, proliferous in the axils of the uppermost pinnae. Stipe stramineous, up to 30 cm. long, glabrous except for pale-brown scales similar to those of the rhizome about the base. Lamina up to 39 × 26 cm., ovate to ovate-oblong in outline, acuminate, 2-pinnate to 2-pinnatifid, the basal pinnae about the same length as the subbasal pinnae but weakly developed basiscopically. Central pinnae narrowly ovate-triangular, sharply acute, petiolate, broadly cuneate at the base, pinnately divided into slightly petiolulate to adnate crenate to pinnatifid segments; the upper oblong to narrowly oblong or rotund, the lower narrowly ovate-triangular to rotund up to 3 × 1·5 cm., glabrous on both surfaces at maturity. Rhachis stramineous, glabrous except for scattered brown linear to hair-like scales up to 4 mm. long, especially about the junction of the pinnae with the rhachis. Sori c. 1 mm. in diam., circular, exindusiate.

Rhodesia. E: Vumba Mts., 3.vii.1957, *Chase* 6572 (BM; BOL; SRGH). **Malawi.** S: Mt. Mlanje, Lower Ruo Plateau, 17.viii.1956, *Newman & Whitmore* 477 (BM; SRGH). **Mozambique.** Z: Namúli, 1887, *Last* (K).
Also in Kenya and W. tropical Africa. Shaded undergrowth of forest, 1520–1770 m.

6. **Dryopteris squamiseta** (Hook.) Kuntze, Rev. Gen. Pl. **2**: 813 (1891).—Tardieu in

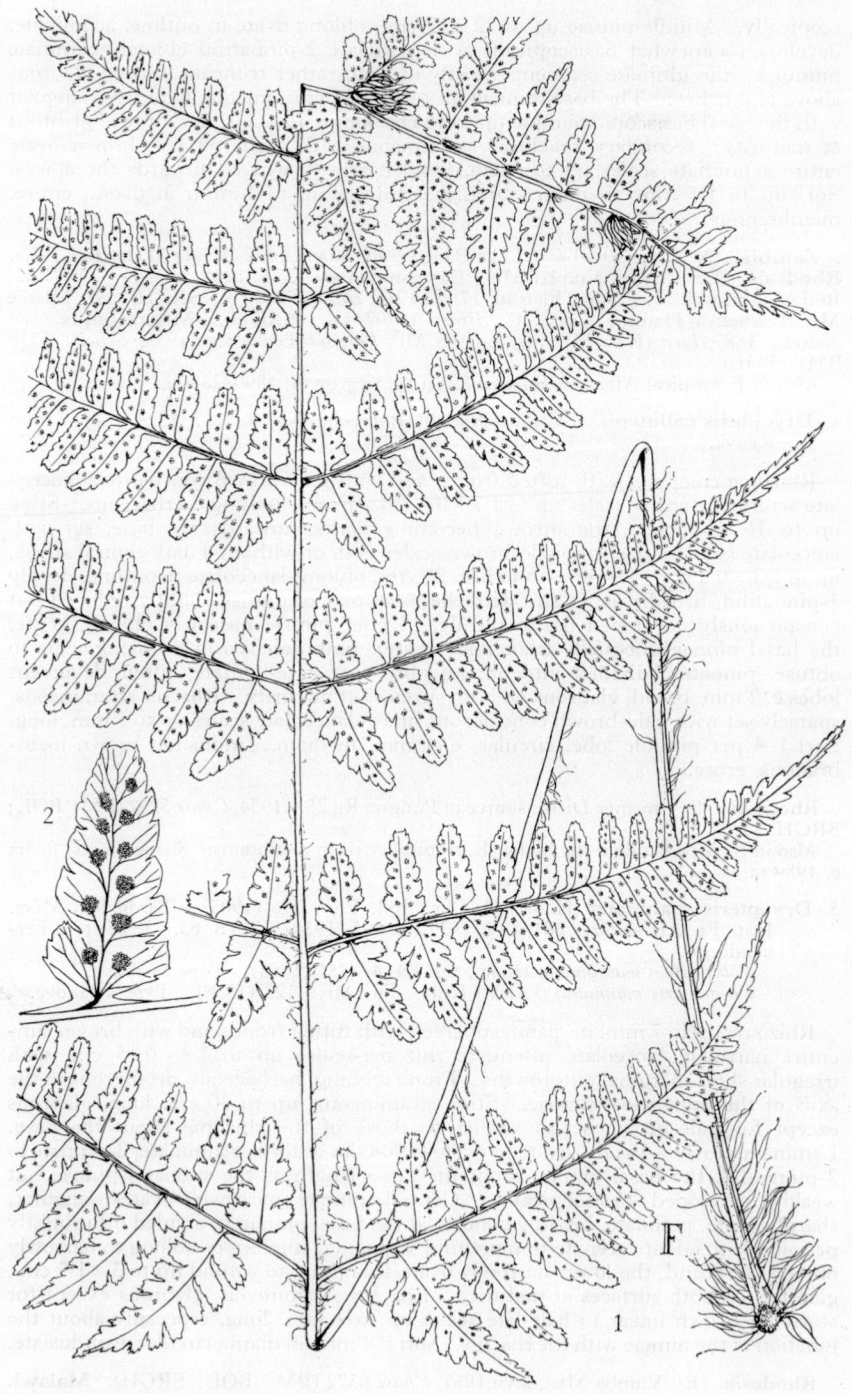

Tab. 63. DRYOPTERIS MANNIANA. 1, frond (× ⅔); 2, fertile pinnule (× 2), both from *Chase* 6572.

Mém. Inst. Fr. Afr. Noire, **28**: 149, t. 29 fig. 1–2 (1953). Type from Fernando Po.
Nephrodium squamisetum Hook., Sp. Fil. **4**: 140, t. 268 (1858). Type as above.
Aspidium squamisetum (Hook.) Kuhn, Fil. Deck.: 24 (1867). Type as above.
Nephrodium buchananii Bak. in Hook. & Bak., Syn. Fil.: 498 (1874). Type from
S. Africa.
  Lastrea buchananii (Bak.) Bedd., Handb. Ferns Brit. Ind.: 255 (1883). Type
as above.
  Dryopteris buchananii (Bak.) Kuntze, tom. cit.: 812 (1891).—Sim, Ferns S. Afr.
ed. 2: 108, t. 20 (1915). Type as above.

Rhizome up to 7 mm. in diam., creeping, with tufted fronds and with brown
lanceolate attenuate entire rhizome-scales up to 9 × 1·2 mm. Frond arching, herbaceous. Stipe up to 53 cm. long, castaneous at the base, stramineous above,
with dense dark-brown subulate spreading entire scales up to 5 mm. long together
with a few broader lighter-brown scales towards the base. Lamina up to 44 × 42
cm., ovate-triangular in outline, 3-pinnate to 4-pinnatifid, apex acute, with
the basal pair of pinnae the largest and conspicuously developed basiscopically.
Upper pinnae narrowly oblong-attenuate, the basal pinnae unequally triangular-
attenuate; pinnules oblong acute mostly deeply pinnatifid to pinnately divided into
narrowly oblong obtuse weakly crenate often decurrent lobes c. 8 × 4 mm., the larger pinnules of the basal pinnae oblong-acuminate in outline, pinnately divided
into deeply pinnatifid oblong broadly acute segments. Rhachis stramineous, set
with subulate dark-brown spreading scales up to 4 mm. long. Sori up to 1·2 mm. in
diam., circular; indusium brown, membranous, entire.

Rhodesia. E. Umtali, Stapleford Forest Reserve, 25.ix.1952, *Chase* 4640 (BM; SRGH).
Sporadic in S. Africa and W. tropical Africa and also in Fernando Po, Madagascar and
the Mascarene Is. Deeply shaded wet undergrowth of forest, c. 1580–2070 m.

7. **Dryopteris wallichiana** (Spreng.) Hylander in Bot. Notis. **1952**: 352 (1953). Type
from Nepal.
  Aspidium paleaceum D. Don, Prodr. Fl. Nepal.: 4 (1825) non Sw. (1806). Type
from Nepal.
  Aspidium wallichianum Spreng. in L., Syst. Veg. ed. 16, **4**: 104 (1827). Type as
for *Dryopteris wallichiana.*
  Aspidium donianum Spreng. in L., tom. cit., Suppl.: 320 (1827). Type from
Nepal.
  Dryopteris paleacea Hand.-Mazz. in Verh. Zool.-Bot. Ges. Wien, **58**: 100 (1908).
Type as for *Aspidium paleaceum.*
  Dryopteris paleacea var. *madagascariensis* C. Chr. [in Acad. Malgache, Cat. Pl.
Madag. Ptérid.: 27 (Feb. 1932) *nom. nud.*] in Dansk Bot. Ark. **7**: 53, t. 11 fig. 5–9
(June 1932). Type from Madagascar.
  Dryopteris doniana (Spreng.) Ching in Sunyatsenia, **6**: 3 (1941). Type as for
Aspidium donianum.

Rhizome up to 16 cm. high and 8 cm. in diam., erect, forming a short caudex
with fronds arranged in cycad fashion and with dark-brown linear attenuate
rhizome-scales up to 3 × 0·3 cm. with some filamentous marginal outgrowths.
Frond suberect, arching, firmly membranacous. Stipe up to 19 cm. long and 1 cm.
in diam. at the base, matt-brown, with dense dark-brown subentire attenuate
scales up to 1 × 0·36 m., narrowly elliptic in outline, very deeply 2-pinnatifid, acute-
acuminate, gradually decrescent at the base, pinnae subsessile. Pinnae up to
18 × 2·7 cm. at the base, narrowly oblong-attenuate, widest at the base, very deeply
pinnatifid with oblong truncate serrate lobes up to 1·2 × 0·5 cm.; the basal pair of
lobes auriculate, darker green above than below, glabrous on both surfaces except
for scattered linear attenuate pale-brown scales up to 2·5 mm. long on the costae
and sometime costules on both surfaces. Rhachis matt-brown, densely set with
persistent dark-brown linear attenuate scales up to 4 mm. long. Sori up to 1·2
mm. in diam., circular, up to 10 per pinna lobe set about ½-way between the
costule and the margin; indusium c. 1 mm. in diam., glabrous, entire, membranous.

Rhodesia. E: Inyangani Mt., source of Pungwe R., 1.v.1960, *Chase* 7337 (BOL; K;
LISC; SRGH).
  Also in Madagascar and Asia. Known only from a single locality in continental Africa,
in shade of *Widdringtonia*, 1980 m.

H

3. POLYSTICHUM Roth

Polystichum Roth, Tent. Fl. Germ. **3**: 31, 69 (1799) *nom. conserv.*

Rhizome erect to creeping with tufted fronds and densely set with entire to variously lacerate pale to very-dark-brown scales of different shapes and sizes. Stipe not articulated, with large persistent brown to black scales around the base. Lamina narrowly elliptic to oblong-lanceolate in outline, herbaceous to coriaceous, pinnate to 4-pinnatifid with the lower pinnae not basiscopically developed, the ultimate segments crenate to aristate-dentate, veins free. Sori circular with peltate indusium, rarely exindusiate.

A cosmopolitan mainly temperate genus of over 200 spp.

Pinnule margins strongly dentate-aristate:
Rhachis scales brown - - - - - - - - - 2. *setiferum*
Rhachis scales black - - - - - - - - - 3. *luctuosum*
Pinnule margins bluntly crenate-serrate not aristate - - - 1. *zambesiacum*

1. **Polystichum zambesiacum** Schelpe in Bol. Soc. Brot., Sér. 2, **41**: 215 (1967). TAB. **64** fig. A. Type: Rhodesia, Umtali Distr., Stapleford, *Schelpe* 5751 (BOL, holotype).

Rhizome up to 2·5 cm. in diam., creeping, with tufted fronds and with ferrugineous narrowly lanceolate acuminate subentire rhizome-scales up to 2 × 0·2 cm. Fronds arching, herbaceous. Stipe pale-brown, up to 73 cm. long, set mainly towards the base with narrowly ovate-acuminate castaneous scales up to 8 mm. long with paler borders and mixed with paler narrower scales. Lamina up to 66 × 52 cm., ovate in outline, acute with the lowest pair of pinnae somewhat reduced and deflexed, shallowly to deeply 3-pinnatifid, 2-pinnate in small specimens. Pinnae very narrowly oblong and conspicuously attenuate up to 28 × 5·2 cm. Pinnules petiolate narrowly ovate-triangular in outline, broadly acute, base unequally cuneate, obtuse, deeply pinnatifid into crenate-serrate lobes without aristae, with the acroscopic basal lobe the largest, glabrous on both surfaces except for scattered minute pale-brown scales along the costules dorsally. Rhachis stramineous with scattered dark-brown subulate hair-pointed scales together with smaller paler scales. Sori c. 1 mm. in diam., circular; indusiam c. 0·5 mm., evanescent.

Rhodesia. E: Umtali Distr., Stapleford Forest Reserve, 3.ii.1961, *Chase* 7429 (K; SRGH). **Malawi.** N: Nyika Plateau, 16.viii.1946, *Brass* 17255 (K; SRGH). S: Mlanje Mt., 1.vii.1946, *Brass* 16566 (K; SRGH). **Mozambique.** Z: Namúli, 1887, *Last* (K). MS: Báruè, Chôa Mt., 9.xii.1965, *Torre & Correia* 13498 (LISC).
Known only from our area. Frequent on forest floors, occasional in shade near forest margins, 1770–1830 m.

2. **Polystichum setiferum** (Forsk.) Moore ex Woynar in Mitt. Naturw. Ver. Steiermark. **49**: 181 (1913). Type from Yemen.
Polypodium setiferum Forsk, Fl. Aegypt.–Arab.: 185 (1775). Type as above.

Var. **fuscopaleaceum** (Alston) Schelpe in Bol. Soc. Brot., Sér 2, **41**: 216 (1967). TAB. **64** fig. B. Type from Cameroon.
Polystichum fuscopaleaceum Alston in Bol. Soc. Brot., Sér. 2, **30**: 22 (1956). Type as above.

Rhizome creeping, with tufted fronds. Frond arching, softly herbaceous, not proliferous. Stipe up to 43 cm. long, brown below becoming stramineous above, with large oblong-lanceolate acuminate dark-brown scales up to 1·3 × 0·3 cm. together with smaller and paler linear acuminate scales. Lamina up to 50 × 25 cm., narrowly ovate-oblong in outline, acute, with the basal pinnae hardly reduced, 2-pinnate. Pinnae up to 13 × 2·9 cm., narrowly oblong, attenuate, with a basal acroscopic pinnule slightly larger. Pinnules closely spaced, usually overlapping, slightly petiolate up to 1·5 × 0·7 cm. at the base, the base forming a right angle, unequally cuneate, auriculate acroscopically, narrowly oblong, very broadly acute, strongly serrate-aristate, glabrous ventrally at maturity, thinly pilose with soft hairs along the veins dorsally. Rhachis stramineous, set with numerous brown broadly acuminate-lanceolate to linear and hair-like brown scales up to 5 mm. long, many with filamentous outgrowths around the base. Sori c. 1 mm. in diam.. circular; indusium c. 1 mm. in diam., membranous, erose.

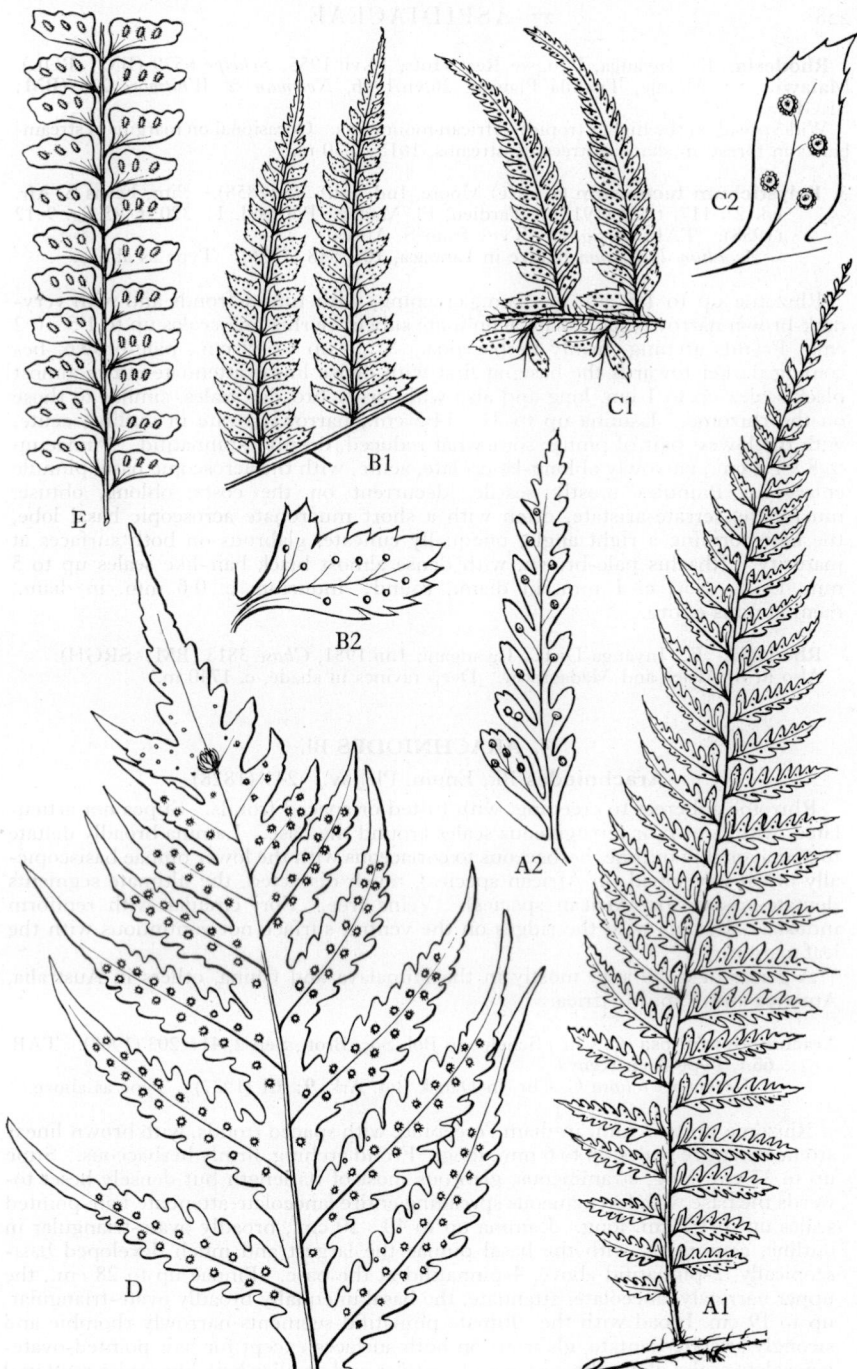

Tab. 64. A.—POLYSTICHUM ZAMBESIACUM. A1, pinna (× ⅔); A2, pinnule (× 2), both from
Schelpe 5699. B.—POLYSTICHUM SETIFERUM var. FUSCOPALEACEUM. B1, pinnae
(× ⅔); B2, pinnule (× 3), both from *Schelpe* 5679. C.—POLYSTICHUM LUCTUOSUM.
C1, pinnae (× ⅔); C2, pinnule (× 3), both from *Chase* 1083. D.—TECTARIA GEMMIFERA.
Part of frond (× ⅓) *Chase* 3276. E.—DIDYMOCHLAENA TRUNCATULA. Part of pinna
(× ⅔) *Chase* 5856.

Rhodesia. E: Inyanga, Pungwe Rest Huts, 14.vii.1955, *Schelpe* 6579 (BM; BOL).
Malawi. S: Mlanje, Tuchila Plateau, 26.vii.1956, *Newman & Whitmore* 214 (BM; SRGH).
Widespread on the higher tropical African mountains. Occasional on margins of stream-banks in forest, in shade of trees by streams, 1615–1830 m.

3. **Polystichum luctuosum** (Kunze) Moore, Ind. Fil.: 95 (1858).—Sim, Ferns S. Afr.
   ed. 2: 117, t. 28 (1915).—Tardieu, Fl. Madag., Polypod. **1**: 320, t. 45 fig. 9–12
   (1958). TAB. **64** fig. C. Type from S. Africa.
   Aspidium luctuosum Kunze in Linnaea, **10**: 548 (1836). Type as above.

Rhizome up to 1·5 cm. in diam., creeping, with tufted fronds and with very-dark-brown narrowly lanceolate acuminate subentire rhizome-scales up to 1·5 × 0·2 cm. Fronds arching, firmly herbaceous. Stipe up to 19 cm., pale-brown, becoming darker towards the base, at first with dense linear attenuate entire almost black scales up to 1 cm. long and also with larger broader scales, similar to those on the rhizome. Lamina up to 31 × 14·5 cm., narrowly ovate in outline, acute, with the lowest pair of pinnae somewhat reduced, deeply 2-pinnatifid. Pinnae up to 8 × 1·7 cm., narrowly oblong-lanceolate, acute, with the acroscopic basal pinnule enlarged. Pinnules mostly sessile, decurrent on the costa, oblong, obtuse, mucronate, serrate-aristate, often with a short mucronate acroscopic basal lobe, the base forming a right angle, unequally cuneate, glabrous on both surfaces at maturity. Rhachis pale-brown, with dense almost black hair-like scales up to 5 mm. long. Sori c. 1 mm. in diam., round; indusium c. 0·6 mm. in diam., membranous entire.

Rhodesia. E: Inyanga Distr., Inyangani, 1.iii.1951, *Chase* 3813 (BM; SRGH).
Also in S. Africa and Madagascar. Deep ravines in shade, c. 1740 m.

4. ARACHNIODES Bl.

Arachniodes Bl., Enum. Pl. Jav.: 241 (1828).

Rhizome suberect to creeping, with tufted or spaced fronds. Stipes not articulated, with brown or ferrugineous scales around the base. Lamina broadly deltate to pentagonal in outline, herbaceous to coriaceous with the lower pinnae basiscopically much developed (in African species), much dissected, the ultimate segments dentate-aristate (in African species). Veins free. Sori circular with reniform indusia. Rhachis with the ridges on the ventral surface not continuous with the leaf margin.
A genus of c. 50 spp. mostly in the Himalaya and China, others in Australia, America and 1 sp. in Africa.

Arachniodes foliosa (C. Chr.) Schelpe in Bol. Soc. Brot., Sér. 2, **41**: 203 (1967). TAB.
   65. Type from Kenya.
   Dryopteris foliosa C. Chr. in Dansk Bot. Ark. **9**: 61 (1937). Type as above.

Rhizome up to 7 mm. in diam., creeping, with spaced fronds, with brown linear attenuate entire scales up to 6 mm. long. Frond arching, firmly herbaceous. Stipe up to 37 cm. long, stramineous, glabrous most of its length but densely beset towards the base with ferrugineous spreading entire lanceolate attenuate hair-pointed scales up to 1·2 cm. long. Lamina up to 40 × 30 cm., broadly ovate-triangular in outline, acuminate, with the basal pinnae the largest and much developed basiscopically, 2-pinnatifid above, 4-pinnatifid at the base. Pinnae up to 28 cm., the upper narrowly lanceolate, attenuate, the basal unequally broadly ovate-triangular, up to 19 cm. broad with the ultimate pinnatifid segments narrowly rhombic and strongly aristate-dentate, glabrous on both surfaces except for hair pointed-ovate-lanceolate scales along the costae and costules and smaller hair-like scales scattered on the veins. Rhachis stramineous with scattered hair-pointed brown to pale-brown scales. Sori up to 1 mm. in diam.; indusium minutely papillate, entire.

Rhodesia. E: Umtali, Banti N. Forest, W. aspect, 3.x.1962, *Chase* 7818 (K; SRGH).
Also in S. Africa and Kenya. Deeply shaded stream-banks in forest at about 1850 m.

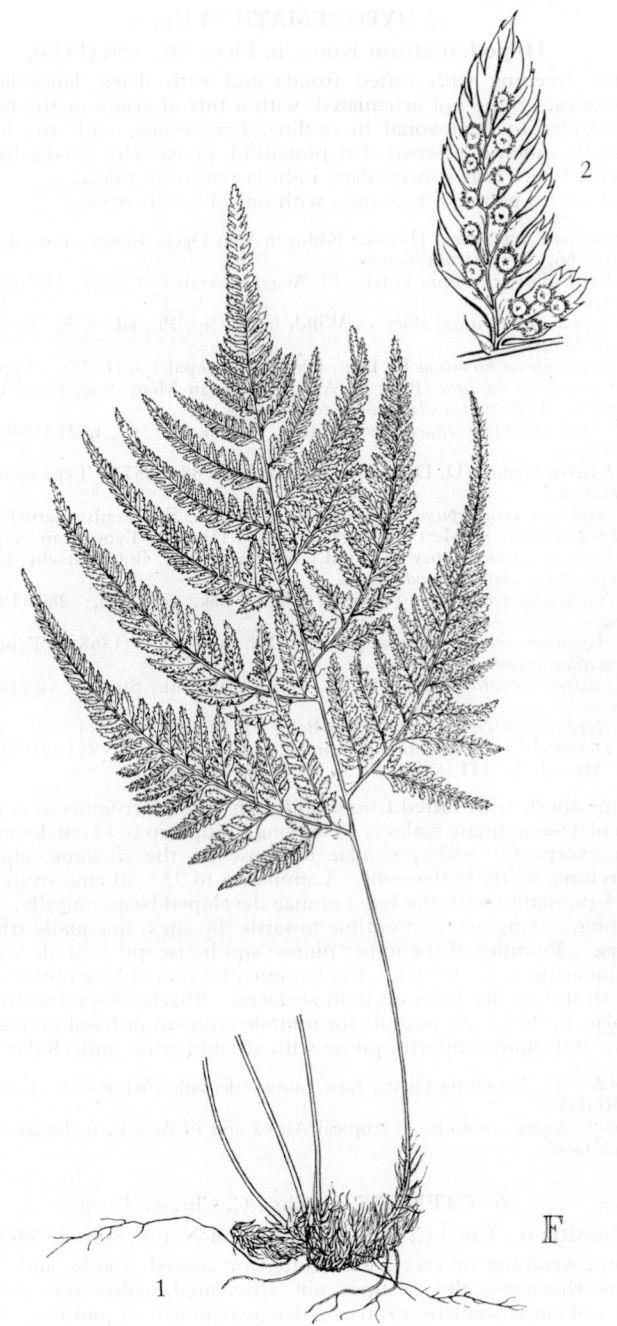

Tab. 65. ARACHNIODES FOLIOSA. 1, habit (× ⅓); 2, pinna (× 2), both from *Chase* 7818.

5. HYPODEMATIUM Kunze

Hypodematium Kunze in Flora, **16**: 690 (1833).

Rhizome creeping with tufted fronds and with dense lanceolate-acuminate rhizome-scales. Stipe not articulated, with a tuft of scales at the base. Lamina ovate-triangular to pentagonal in outline, herbaceous, with the lowest pinnae basiscopically much developed, 3–4-pinnatifid, pilose with needle-like unicellular hairs; veins free. Sori subcircular; indusia reniform, pilose.

An Old World genus of c. 3 spp., with only 1 sp. in Africa.

Hypodematium crenatum (Forsk.) Kuhn in Von Deck. Reisen, Bot., **3**, 3: 37 (1879). TAB. **66**. Type from Yemen.

> *Polypodium crenatum* Forsk., Fl. Aegypt.–Arab.: CXXV, 185 (1775). Type as above.
> *Aspidium odoratum* Bory ex Willd. in L., Sp. Pl., ed. 4, **5**: 286 (1810). Type from Mauritius.
> *Nephrodium hirsutum* D. Don, Prodr. Fl. Nepal.: 6 (1825). Type from Nepal.
> *Cystopteris odorata* (Bory ex Willd.) Desv. in Mém. Soc. Linn. Par. **6**, 2: 264 (1827). Type as for *Aspidium odoratum*.
> *Hypodematium rueppellianum* Kunze, Farrnkr. **1**: 41, t. 21 (1840). Type from Ethiopia.
> *Lastrea hirsuta* (D. Don) Moore, Ind. Fil.: 88 (1857). Type as for *Nephrodium hirsutum*.
> *Aspidium eriocarpum* Wall. ex Mett. in Abh. Senckenb. Naturf. Ges. **2**: 344 (1858) reimpr. in Mett., Farngatt. **4**: 344 (1858). Type from Nepal.
> *Lastrea odorata* (Bory ex Willd.) Wawra in Oest. Bot. Zeitschr. **13**: 144 (1863). Type as for *Aspidium odoratum*.
> *Nephrodium odoratum* (Bory ex Willd.) Bak., Syn. Fil.: 280 (1868). Type as above.
> *Aspidium crenatum* (Forsk.) Kuhn, Fil. Afr.: 129 (1868). Type as for *Hypodematium crenatum*.
> *Lastrea crenata* (Forsk.) Bedd., Ferns Brit. Ind., Suppl.: 18 (1876). Type as above.
> *Nephrodium crenatum* (Forsk.) Bak., Fl. Maurit.: 49 (1877). Type as above.
> *Dryopteris crenata* (Forsk.) Kuntze, Rev. Gen. Pl. **2**: 811 (1891).—Sim, Ferns S. Afr. ed. 2: 111, t. 22 (1915). Type as above.

Rhizome short, with tufted fronds and with dense ferrugineous concolorous entire lanceolate-acuminate scales c. 1 cm. long. Stipe up to 12 cm. long, stramineous, glabrous except for scales, similar to those on the rhizome, about the base. Frond arching, softly herbaceous. Lamina up to 33 × 30 cm., ovate-triangular in outline, 4-pinnatifid with the basal pinnae developed basiscopically. Pinnae up to 22 cm. long, oblong-acute in outline towards the apex, unequally triangular-ovate at the base. Pinnules of the upper pinnae and basiscopic pinnule segments of the lowest pinnae up to 2 × 0·9 cm., deeply pinnatifid into oblong obtuse crenate lobes, pilose with unicellular hairs on both surfaces. Rhachis stramineous pilose. Sori subcircular, up to 12 per pinnule (or pinnule segment of basal pinnae), 1–1·5 mm. in diam.; indusium reniform, pilose with straight white unicellular hairs.

Zambia. W: Kasempa Distr., East Lunga Sleeping Sickness Area, x.1963, *Mitchell* 24/24 (SRGH).

Also in S. Africa, sporadic in tropical Africa and in Asia E. to Japan. In crevices in schist rock face.

6. CTENITIS (C. Chr.) C. Chr. ex Tardieu

Ctenitis (C. Chr.) C. Chr. ex Tardieu in Notul. Syst. **7**: 86 (1938).

Rhizome creeping or erect, with tufted or spaced fronds, and with linear to lanceolate rhizome-scales. Stipes not articulated, paleaceous about the base. Lamina oblong-lanceolate to triangular-pentagonal in outline, 2–4-pinnatifid, herbaceous with multicellular hairs along the costae and costules; veins free. Sori circular with reniform indusia.

A pantropic genus of c. 150 spp.

Lamina oblong-lanceolate; basal pinnae not developed basiscopically   -   1. *cirrhosa*
Lamina broadly triangular; basal pinnae prominently developed basiscopically
                                                                    2. *lanuginosa*

Tab. 66. HYPODEMATIUM CRENATUM. 1, rhizome and base of stipes ($\times\frac{1}{2}$); 2, frond ($\times\frac{1}{2}$); 3, pinnule (lower surface) ($\times10$); 4a, enlargement of pinnule-lobe showing sorus ($\times18$); 4b, hairs on the indusium ($\times18$), all from *Schimper* 617.

1. **Ctenitis cirrhosa** (Schum.) Ching in Sunyatsenia, **5**: 250 (1940).   TAB. **67** fig. A.
Type from Guinée.
   Aspidium cirrhosum Schumach. in Kongel. Dansk. Vid. Selsk., Nat. & Math. Afh.
   4: 231 (1829).   Type as above.
   Aspidium comorense Kuhn, Fil. Afr.: 128 (1828) *nom. nud.*   Type from Comoro
   Is.
   Nephrodium crinobulbon Hook., Sp. Fil. **4**: 92, t. 224 (1862).   Type from S.
   Tomé.
   Nephrodium cirrhosum (Schumach.) Bak. in Hook. & Bak., Syn. Fil.: 472 (1868).
   Type as for *Ctenitis cirrhosa.*
   Nephrodium welwitschii Bak., tom. cit.: 274 (1868).   Type from Angola.
   Nephrodium spekei Bak., tom. cit.: 263 (1868).   Type from Comoro Is.
   Lastrea crinobulbon (Hook.) J. Sm., Hist. Fil.: 214 (1875).   Type as for *Nephro-
   dium crinobulbon.*
   Dryopteris cirrhosa (Schumach.) Kuntze, Rev. Gen. Pl. **2**: 812 (1891).   Type as
   for *Ctenitis cirrhosa.*
   Dryopteris spekei (Bak.) Kuntze, tom. cit.: 813 (1891).   Type as for *Nephrodium
   spekei.*
   Dryopteris pulvinata C. Chr. in Bonap., Not. Ptérid. **16**: 177, t. 6 fig. a (1925).
   Type from Madagascar.
   Dryopteris crinobulbon (Hook.) C. Chr., Ind. Fil., Suppl. **3**: 84 (1934).   Type as
   for *Nephrodium crinobulbon.*
   Ctenitis crinobulbon (Hook.) Ching in Sunyatsenia, **5**: 250 (1940).   Type as above.
   Thelypteris spekei (Bak.) Ching in Bull. Fan Mem. Inst. Biol. Bot. **10**: 254 (1941).
   Type as for *Nephrodium spekei.*
   Dryopteris nimbaensis Tardieu in Notul. Syst. **13**: 370 (1948).   Type from
   Guinée.

Rhizome erect, with tufted fronds and with linear attenuate golden-brown
entire rhizome-scales up to 1 cm. long.   Frond arching, thinly herbaceous.   Stipe
pale-brown, up to 43 cm. long, sparsely set with dark-brown hair-like scales above,
densely set with linear attenuate entire scales up to 9 mm. long towards the base.
Lamina up to 80 × 40 cm., deeply 2-pinnatifid, oblong-lanceolate in outline, acute
with the basal pinnae not reduced.   Pinnae up to 20 × 3·2 cm., very narrowly
oblong, attenuate, slightly narrowed towards the base in the lowest pinnae but with
a slightly enlarged acroscopic basal lobe in the upper pinnae; pinnatifid
nearly to the costa into narrowly oblong obtuse to subacute subentire lobes up to
1·6 × 0·35 cm., widened at the extreme base, mostly glabrous but with lanceolate
acuminate brown scales along the costa dorsally and with numerous pale-brown
multicellular hairs along the costa ventrally together with smaller scales on the
costules and minute pale hairs along the veins dorsally and with scattered marginal
hairs.   Rhachis stramineous, be set with dark subulate scales up to 3 mm. long
with paler margins.   Sori c. 0·6 mm. in diam., circular, about midway between
the costule and the margin; indusium c. 0·6 mm. in diam., pilose and ciliate with
minute hairs.

Zambia. N: Mpongwe, 11.x.1962, *Fanshawe* 7087 (SRGH). **Rhodesia.** E: Mel-
setter Distr., Umvumvumvu R., 15.xi.1953, *Chase* 5127 (BM; BOL; SRGH). **Malawi.**
N: Karonga Distr., Misuku Hills, 11.i.1959, *Robinson* 3171 (SRGH). **Mozambique.**
MS: Chimoio, Garuso Mt., 19.x.1944, *Mendonça* 2533 (BM; LISC).
   Widespread in tropical Africa, Madagascar and Comoro Is.   Shaded streambanks in
forest, 700–1280 m.

2. **Ctenitis lanuginosa** (Willd. ex Kaulf.) Copel. in Gen. Fil.: 124 (1947).   TAB. **67**
fig. B.   Type from Mauritius.
   Aspidium lanuginosum Willd. ex Kaulf., Enum. Fil.: 244 (1814).   Type as above.
   Nephrodium lanuginosum (Willd. ex Kaulf.) Desv. in Mém. Soc. Linn. Par. **6**, 2:
   262 (1827).   Type as above.
   Aspidium catopteron Kunze in Linnaea, **10**: 550 (1836).   Type from S. Africa
   (Natal).
   Lastrea odorata C. Presl, Tent. Pterid.: 77 (1836) reimpr. in Abh. Königl.
   Böhm. Ges. Wiss. **4**, 5: 77 (1837) *nom. nud.*
   Lastrea lanuginosa (Willd. ex Kaulf.) Moore, Ind. Fil.: 87 (1858).   Type as for
   Ctenitis lanuginosa.
   Lastrea catoptera (Kunze) Pappe & Raws., Syn. Fil. Afr. Austr.: 12 (1858).
   Type as for *Aspidium catopteron.*
   Aspidium odoratum Sieb. ex Mett. in Abh. Senckenb. Naturf. Ges. **2**: 399 (1858)

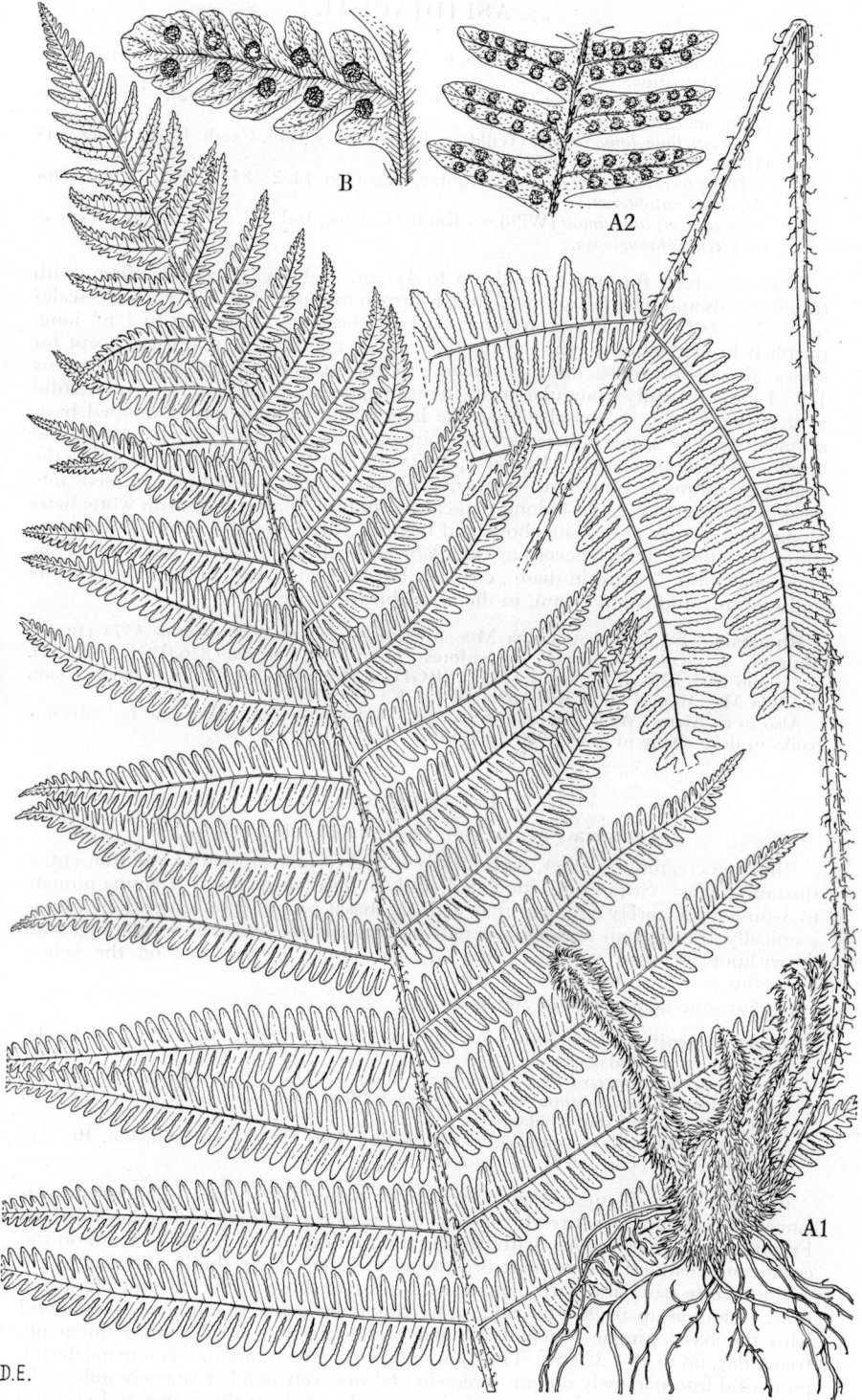

D.E.

Tab. 67. A.—CTENITIS CIRRHOSA. A1, habit (× ½); A2, section of pinna showing lower surface of pinnules (× 1), both from *Pedro & Pedrógão* 7477. B.—CTENITIS LANU-GINOSA. Pinnule (× 2) *Chase* 6596.

reimpr. in Mett., Farngatt. **4**: 115, n. 273 (1858) non Bory ex Willd. (1810).   Type from Mauritius.

Nephrodium catopteron (Kunze) Hook., Sp. Fil. **4**: 137 (1862).   Type as for *Aspidium catopteron.*

Polystichum lanuginosum (Willd. ex Kaulf.) Keys., Pol. Cyath. Herb. Bung.: 45 (1873).   Type as for *Ctenitis lanuginosa.*

Dryopteris catoptera (Kunze) Kuntze, Rev. Gen. Pl. **2**: 812 (1891).   Type as for *Aspidium catopteron.*

Dryopteris lanuginosa (Willd. ex Kaulf.) C. Chr., Ind. Fil.: 273 (1906).   Type as for *Ctenitis lanuginosa.*

Rhizome erect, forming a trunk up to 45 cm. high and 15 cm. in diam. with tufted fronds and linear acuminate golden-brown minutely serrulate rhizome-scales up to 2·7 × 0·25 cm.   Frond arching, softly herbaceous.   Stipe up to 1 m. long, purplish-brown when fresh, drying to brown; glabrous at maturity except for scales similar to those on the rhizome borne at the extreme base.   Lamina 1·3 × 1·2 m., broadly triangular-ovate in outline, deeply 2-pinnatifid to 4-pinnatifid at the base, acute, the basal pinnae the largest and prominently developed basi-scopically.   The upper pinnae narrowly oblong-acute, the basal pinnae broadly and unequally triangular-acute.   Pinnae up to 75 cm., set at an angle of about 45° to the rhachis; ultimate pinnatifid segments oblong, acute to obtuse, incised into weakly crenate to crenate-oblong truncate to obtuse lobes, pilose with white hairs along the costules and veins above and below.   Rhachis and secondary rhachises minutely pilose at first becoming subglabrous with age towards the base of the rhachis.   Sori c. 1 mm. in diam., circular, borne about ½-way between the costule and margin; indusium 1 mm. in diam., glabrous, erose.

Rhodesia. E: Umtali, Vumba Mts., Cloudlands, 25.vi.1955, *Schelpe* 5375 (BOL). **Malawi.** C: Kota Kota Distr., Nchisi forest, 11.vii.1960, *Chapman* 836 (BOL; SRGH). S: Cholo Mt., 23.ix.1946, *Brass* 17757 (SRGH).   **Mozambique.** MS: Cement Dam, Bandula Mt., 10.vii.1957, *Chase* 6596 (BOL; K; SRGH).

Also in S. Africa, West tropical African Is., Madagascar and Mascarene Is.   Stream-banks in deep shade in forest, 900–1600 m.

7. TECTARIA Cav.

Tectaria Cav. in Ann. Hist. Nat. **1**: 115 (1799).

Rhizome creeping to erect, with mostly tufted fronds and with large thin entire rhizome-scales.   Stipes not articulated, scaly at least at the base.   Lamina pinnate to 3-pinnatifid, rarely simple, often with the basal pair of pinnae developed basi-scopically, the ultimate segments crenate but not aristate; veins anastomosing, with or without included veinlets.   Sori circular, dorsal or terminal on the veins; indusium peltate, reniform or absent.

A pantropic genus of over 200 spp.

Tectaria gemmifera (Fée) Alston in Journ. of Bot. **77**: 288 (1939).   TAB. **64** fig. D. Type from Madagascar.

Sagenia gemmifera Fée, Mém. Fam. Foug. **5**: 313 (1852).   Type as above.

Aspidium coadunatum var. *gemmiferum* (Fée) Mett. ex Kuhn, Fil. Afr.: 128 (1868).   Type from Madagascar.

Aspidium gemmiferum (Fée) Ching in Bull. Fan Mem. Inst. Biol., Bot. **10**: 237 (1941).   Type as for *Tectaria gemmifera.*

Rhizome up to 2 cm. in diam., erect, with tufted fronds and very-dark-brown lanceolate acuminate entire rhizome-scales up to 8 mm. long with paler borders. Frond arching, herbaceous, with proliferating bulbils up to 1 cm. in diam. on the costae and costules.   Stipe matt-brown, up to 75 cm. long, thinly pubescent with minute white hairs and with dark scales similar to those of the rhizome about the base.   Lamina up to 90 × 60 cm., ovate-triangular-acute in outline, 3-pinnatifid with the basal pinnae much developed basiscopically; basal pinnae unequally triangular, up to 48 × 38 cm.   The upper pinnae narrowly oblong-acuminate, deeply pinnatifid into narrowly oblong somewhat falcate crenate lobes sparsely pubescent on both surfaces with short white hairs, densely so along the costae and costules and veins of the ventral surface; veins anastomosing usually without included veinlets.   Rhachis brown or straw-coloured, pubescent with minute pale-brown hairs, with decurrent lamina-wings near the apex. Sori up to 2 mm. in diam., circular; indusium c. 1 mm. in diam., membranous, minutely ciliate, reniform.

Zambia. N: Abercorn Distr., Chisau R. Gorge, 18.xi.1959, *Richards* 11803 (K; SRGH). **Rhodesia.** E: Vumba Mts., Hawkhurst, 14.viii.1955, *Chase* 5715 (BM; BOL; SRGH). **Malawi.** N: Misuku Distr., Misuku Hills, 11.i.1959, *Richards* 10561 (K). C: Nchisi Mt., 19.ii.1959, *Robson & Steele* 1661 (LISC; SRGH). S: Cholo Mt., 23.viii.1946, *Brass* 17751 (K). **Mozambique.** N: Monte Massungulo, vi.1933, *Gomes e Sousa* 1494 (COI). Z: Namúli, R. Licungo, 29.vii.1962, *Schelpe & Leach* 7032 (BOL). T: Mt. Zóbuè, 3.x.1942, *Mendonça* 574 (LISC). MS: Mavita, Messambusi, 14.vii.1949, *Pedro & Pedrógão* 6471 (BM).

Also in the Transvaal and widespread in E. tropical Africa, Angola and Madagascar. Shaded floors of forests, 700–1525 m.

28. BLECHNACEAE

Terrestrial, lithophytic or epiphytic plants. Rhizome creeping to erect, sometimes forming a caudex, dictyostelic, and with non-peltate, non-clathrate scales. Stipe not articulated, with many vascular strands. Fronds pinnate or pinnatifid (infrequently 2-pinnatifid) with the basal pinnae reduced or not, dimorphous to a greater or lesser extent; veins free or anastomosing. Sori linear (less frequently disrupted) usually borne on a secondary vein parallel to the costa, between the costa and the margin; indusia linear, opening towards the costa, or absent; spores monolete with or without a perispore.

Indusium present; fertile frond usually pinnate; rhizome erect or creeping, rarely ascending trees - - - - - - - - - 1. **Blechnum**
Indusium absent; fertile frond usually 2-pinnate; rhizome ascending trees
2. **Stenochlaena**

1. BLECHNUM L.

Blechnum L., Sp. Pl. **2**: 1077 (1753); Gen. Pl. ed. 5: 485 (1754).

Rhizome creeping or erect, sometimes forming a short caudex, with tufted fronds and brown rhizome-scales. Stipe not articulated. Lamina pinnatifid to pinnate (rarely 2-pinnatifid); veins free in the sterile fronds. Sori linear, borne between the costa and the margin; indusium linear opening towards the costa.

A cosmopolitan genus of over 200 spp., mostly in the southern hemisphere, with 7 spp. in continental Africa.

Sterile pinnae all sessile with the bases adnate to the rhachis:
 Frond thickly coriaceous, the base of the lamina not or hardly decrescent; sterile pinnae unequally cuneate - - - - - - - - - 4. *tabulare*
 Frond softly membranous to thinly coriaceous, the base of the lamina tapered (decrescent); sterile pinnae equally dilated at the base:
  Pinna apices obtuse to acute - - - - - - - - 2. *inflexum*
  Sterile pinna apices attenuate to acute-acuminate:
   Sterile pinnae entire - - - - - - - - 1. *attenuatum*
   Sterile pinnae minutely serrate - - - - - 3. *ivohibense*
Sterile pinnae with at least the lower ones petiolate:
 Sterile pinna margin serrate - - - - - - - 5. *capense*
 Sterile pinna margin entire or shallowly undulate:
  Sterile pinna apex mucronate - - - - - - 7. *australe*
  Sterile pinna apex attenuate - - - - - - 6. *punctulatum*

1. **Blechnum attenuatum** (Sw.) Mett. in Fil. Hort. Bot. Lips.: 64, t. 3 fig. 1–6 (1856). Type from Mauritius.
  Onoclea attenuata Sw. in Schrad., Journ. Bot. **1800**, 2: 73 (1801). Type as above.
  Lomaria attenuata (Sw.) Willd., Sp. Pl., ed. 4, **5**: 290 (1810). Type as above.
  Lomaria acuminata Desv., Mag. Ges. Naturf. Fr. Berl. **5**: 326 (1811). Type from Réunion.
  Spicanta attenuata (Sw.) Kuntze, Rev. Gen. Pl. **2**: 821 (1891). Type as for *Blechnum attenuatum*.

Blechnum polypodioides var. *holstii* Hieron. in Engl., Pflanzenw. Ost-Afr., **C**: 81 (1895). Type from Tanzania.
Blechnum attenuatum var. *holstii* (Hieron.) Schelpe in Journ. Linn. Soc. Bot. **53**: 493 (1952). Type as above.

Rhizome 1–3 cm. in diam., creeping or ascending, with tufted fronds and with narrowly linear attenuate mostly entire rhizome-scales up to 2·5 cm. long, sometimes with a black central stripe. Frond erect or arching, thinly to firmly coriaceous Stipe pale-brown becoming castaneous towards the base, up to 10 cm. long, glabrous except for brown linear attenuate scales up to 10 mm. long similar to those on the rhizome. Lamina up to 92 × 28 cm., narrowly elliptic in outline, pinnate acuminate with an oblong-lanceolate attenuate terminal segment, base with up to 16 pairs of lower pinnae gradually reduced, the lowermost rounded and rudimentary. Sterile pinnae up to 23 × 1·4 cm., linear attenuate, entire, frequently with revolute margins, the bases dilated and adnate to the rhachis, glabrous on both surfaces, the ventral surface darker; fertile pinnae up to 10 × 0·15 cm., very narrowly linear, bluntly apiculate with a broadened base adnate to the rhachis. Rhachis stramineous, sulcate ventrally. Sori linear, extending almost to the length of the fertile pinna. Indusium c. 1 mm. broad, linear, brown at maturity.

Plant epiphytic; rhizome creeping, less than 2 cm. in diam.; fronds arching
                                                                        var. *attenuatum*
Plant not epiphytic, terrestrial or lithophytic; rhizome more than 2 cm. in diam.; fronds
erect to suberect -   -   -   -   -   -   -   -   -   -   var. *giganteum*

Var. attenuatum
Zambia. N: Isoka Distr., Mafingi Mts., W. of Chisenga, 21.xi.1952, *White* 3743 (BM; BOL; K). **Rhodesia.** E: Inyanga, 23.vii.1957, *Chase* 6645 (BOL; K; SRGH).
Malawi. S: Zomba Distr., Zomba Plateau, 3.vi.1946, *Brass* 16172 (K).
Also in the Mascarene Is. Usually epiphytic in forest, c. 1830 m.

Var. **giganteum** (Kaulf.) Bonap. in Sarasin & Roux, Nova Caledonia, **1**: 43 (1914). Type from S. Africa.
        Lomaria heterophylla Desv. in Mag. Ges. Naturf. Fr. Berl. **5**: 330 (1811). Type from S. Africa.
        Lomaria gigantea Kaulf., Enum. Fil.: 150 (1824). Type as for *Blechnum attenuatum* var. *giganteum*.
        Lomaria hamata Kaulf., loc. cit. Type from S. Africa.
        Blechnum giganteum (Kaulf.) Schlechtend., Adumbr.: 36, t. 20 & 22 fig. 1 (1827). Type as for *Blechnum attenuatum* var. *giganteum*.
        Blechnum heterophyllum (Desv.) Schlechtend., tom. cit.: 37 (1827). Type as for *Lomaria heterophylla*.
        Lomaria decipiens Pappe & Raws., Syn. Fil. Afr. Austr.: 29 (1858). Type from S. Africa.

Rhodesia. E: Chimanimani Mts., 26.ix.1906, *Swynnerton* 835a (K). **Malawi.** S: Mlanje Mt., Tuchila Plateau, 23.vii.1956, *Newman & Whitmore* 153 (BM; SRGH).
Mozambique. Z: Namúli Mt., 26.vii.1962, *Schelpe & Leach* 7032 (BOL). MS: Manica, Mavita, Rotanda, 26.x.1944, *Mendonça* 2669 (BM; LISC).
Also in S. Africa. Lithophytic or terrestrial in deep shade in forest, 880–1770 m.

2. **Blechnum inflexum** (Kunze) Kuhn, Fil. Afr.: 92 (1868).—Sim, Ferns S. Afr. ed. 2: 178, t. 74 (1915). Type from S. Africa.
        Lomaria inflexa Kunze, Farrnkr. **1**, 7: 150, t. 65 (1844). Type as above.
        Lomaria discolor var. *natalensis* Bak., Syn. Fil. ed. 2: 481 (1874). Type from S. Africa.
        Struthiopteris inflexa (Kunze) Ching in Sunyatsenia, **5**: 243 (1940). Type as for *Blechnum inflexum*.

Rhizome up to 4 mm. in diam., procumbent to erect, with tufted fronds and with brown linear attenuate entire rhizome-scales up to 1·2 cm. long. Frond erect, thinly coriaceous. Stipe up to 9 cm. long, brown, becoming darker towards the base and densely set about the base with scales similar to the rhizome scales. Sterile lamina up to 30 × 5·5 cm., very deeply pinnatifid to pinnate, very narrowly oblong-elliptic in outline, acute, the basal pinnae gradually reduced; sterile pinnae narrowly oblong, entire, acute-obtuse, base sessile, slightly auriculate acroscopically, glabrous on both surfaces. Fertile lamina conspicuously shorter than the sterile lamina, up to 11·5 × 2·2 cm., very narrowly oblong in outline, acute, the basal pinnae not conspicuously reduced; fertile pinnae up to 1·7 × 0·3 cm., narrowly oblong, obtuse, petiolate. Rhachis brown, glabrous at maturity except for a few scattered small pale-brown scales. Sori linear, extending the length of the fertile pinnae; indusium up to 0·8 mm. broad, dark-brown, lacerate.

Rhodesia. E: Umtali Distr., Bideford, 30.i.1948, *Fisher* 1428 (NU). Also in S. Africa. In dense mats on stream-banks above 1675 m.

3. **Blechnum ivohibense** C. Chr. in Arch. Bot. (Caen) **2**, Bull. Mens.: 211 (1928). TAB. **68**. Type from Madagascar.
    Blechnum bakeri var. *glabra* Bonap., Not. Ptérid. **16**: 72 (1925) *nom. nud.*— Tardieu, Fl. Madag., Polypod. **49**: t. 2 fig. 1–3 (1960). Type from Madagascar.
    Blechnum umbrosum A. Peter in Fedde, Repert. Beih. **40**, 2: 82, 9, t. 3 fig. 5–8 (1929). Type from Tanzania.

Rhizome up to 10 × 0·5 cm., erect, with tufted fronds and with black linear entire rhizome-scales up to 3 mm. long. Frond suberect, firmly membranous. Stipe matt-grey-brown, black at the base, up to 20 cm. long. Lamina up to 40 × 7·5 cm., pinnate to very deeply pinnatifid, very narrowly elliptic in outline, acuminate with up to 8 pairs of pinnae, reduced at the base; sterile pinnae up to 4 × 1 cm., narrowly oblong sometimes somewhat falcate, base sessile, shallowly undulate-crenate towards the apex, glabrous on both surfaces; fertile pinnae very narrowly linear, up to 2·8 × 0·1 cm., bluntly apiculate, sessile. Rhachis stramineous, glabrous except for scattered minute dark lanceolate scales. Sori linear, extending for almost the length of the fertile pinnae; indusium vestigial.

Rhodesia. E: Inyanga Distr., Syfret's Place, 23.vii.1957, *Chase* 6659 (BOL; K; SRGH). **Mozambique.** Z: Serra de Gúruè, Marrequelo, R. Namiroé, *Mendonça* 2162 (BM; LISC).
Also in Tanzania and Madagascar. Shaded stream-banks in forest in deep shade, c. 1830 m.

4. **Blechnum tabulare** (Thunb.) Kuhn, Fil. Afr.: 94 (1868). Type from S. Africa (Cape Prov.).
    Pteris tabularis Thunb., Prodr. Pl. Cap.: 171 (1800). Type as above.
    Lomaria coriacea Schrad. in Gött. Gel. Anz. **1818**: 916 (1818). Type from S. Africa.
    Lomaria gueinzii Moug. ex Fée, Gen. Fil.: 69 (1852). Type from S. Africa.
    Lomaria cycadoides Pappe & Raws., Syn. Fil. Afr. Austr.: 28 (1858). Type from S. Africa.
    Lomaria dalgairnsiae Pappe & Raws., tom. cit.: 27 (1858). Type from S. Africa.
    Blechnum cycadoides (Pappe & Raws.) Kuhn, tom. cit.: 91 (1868). Type as for *Lomaria cycadoides*.
    Blechnum dalgairnsiae (Pappe & Raws.) Kuhn, tom. cit.: 92 (1868). Type as for *Lomaria dalgairnsiae*.
    Lomaria tabularis (Thunb.) Mett. ex Bak. in Mart., Fl. Bras. **1**, 2: 418 (1870). Type as for *Blechnum tabulare*.

Rhizome up to 90 cm. long and 10 cm. in diam., erect or procumbent, with tufted fronds and with very narrowly linear attenuate entire rhizome-scales up to 3·5 cm. long with a dark-brown central stripe and paler borders. Frond erect to suberect. Stipe pale-brown, up to 20 cm. long, glabrous except for the base which is densely set with scales similar to the rhizome scales. Lamina up to 1·4 × 0·36 m., narrowly oblong in outline, acute with a narrowly oblong acute terminal segment, base with up to 9 pairs of basal pinnae gradually reduced, the lowermost rudimentary. Sterile pinnae up to 22 × 2·4 cm., very narrowly oblong acute, subsessile with an unequally cuneate base, entire, usually revolute, glabrous on both surfaces except for minute brown lanceolate scales up to 1·5 mm. long along the costae; fertile pinnae up to 22 × 0·3 cm., linear, acute to rotund at apex, subsessile to very shortly petiolate. Rhachis pale-brown, glabrous at maturity. Sori linear, extending almost the whole length of the fertile pinnae; indusium pale-brown, c. 1·5 mm. broad, linear, lacerate at maturity.

Zambia. N: N. Shiwa Ngandu, 19.ix.1938, *Greenway & Trapnell* 5713 (K). W: Mwinilunga Distr., R. Dobeka S. of Dobeka Bridge, 17.xii.1937, *Milne-Redhead* 3706 (K). **Rhodesia.** C: Rusape, 1922, *Eyles* 3488 (SRGH). E: Umtali Distr., Vumba Mts., Nimbus Farm, 24.v.1959, *Chase* 7134 (BOL; K; SRGH). **Malawi.** N: Rumpi Distr., Lake Kaulime, Nyika Plateau, 4.i.1959, *Richards* 10465 (K). S: Mlanje Mt., Luchenya Plateau, 27.vi.1946, *Brass* 16480 (K; SRGH). **Mozambique** Z: Serra de Gúruè, 20.

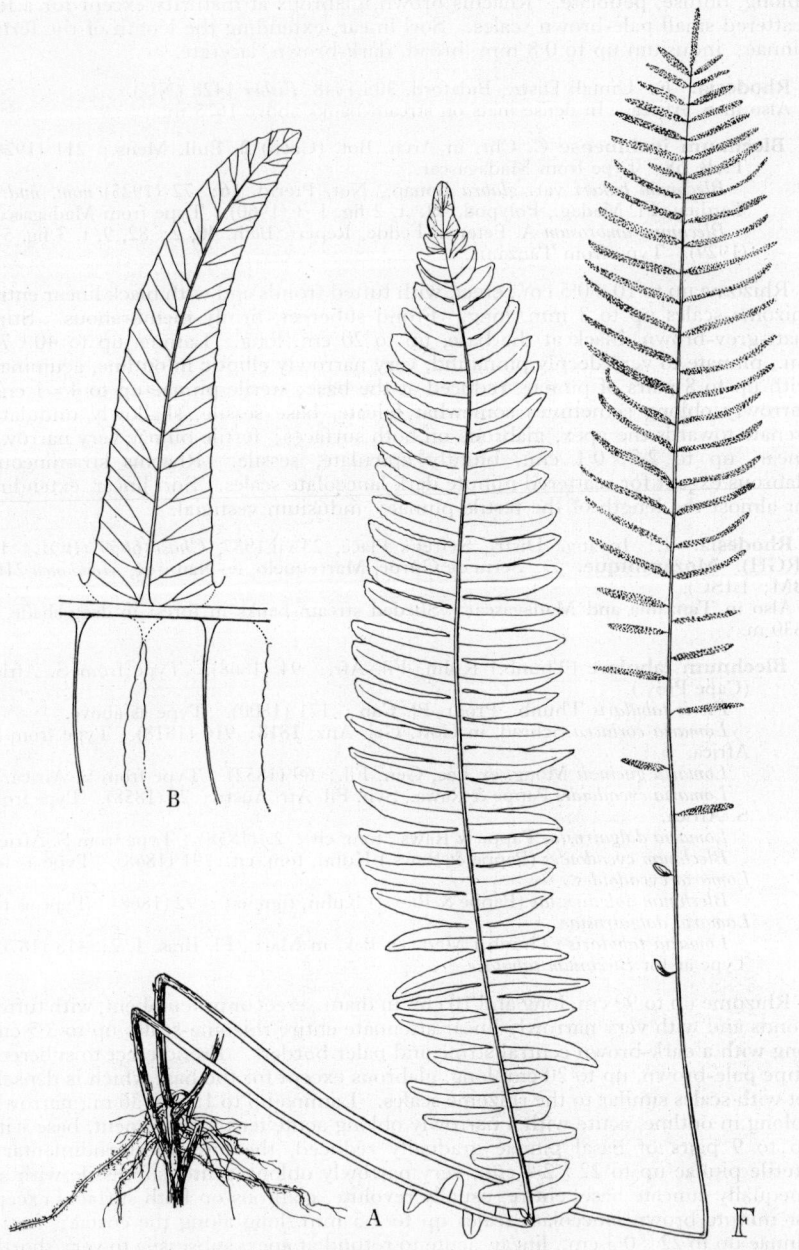

Tab. 68. BLECHNUM IVOHIBENSE. 1, habit (× ⅓); 2, enlargement of sterile pinna (×2), both from *Chase* 6659.

ix.1944, *Mendonça* 2174 (BM; LISC; LMA). MS: Gorongosa Mt., Gogogo Peak, 6.vii.1955, *Schelpe* 5534 (BOL).

Also in S. Africa, Madagascar, Tanzania, Uganda, Congo and Angola. Moist depressions or around boulders in grassland, occasional in forest, 1370–1830 m.

5. **Blechnum capense** (L.) Schlechtend., Adumbr.: 34, t. 18 (1825). Type from S. Africa (Cape Prov.).

   Osmunda capensis L., Mant. Pl. Alt.: 306 (1771). Type as above.
   Onoclea capensis (L.) Sw., Syn. Fil.: 111 (1806). Type as above.
   Lomaria capensis (L.) Willd. in L., Sp. Pl., ed. 4, **5**: 291 (1810). Type as above.
   Spicanta capensis (L.) Kuntze, Rev. Gen. Pl. **2**: 821 (1891). Type as above.

Rhizome up to 2 cm. in diam., creeping, with tufted fronds and with brown ovate-acuminate entire rhizome-scales up to 1 cm. long. Frond arching, firmly herbaceous. Stipe up to 47 cm. long, nitid, castaneous at least in the lower ½, glabrous at maturity. Sterile lamina up to 63 × 21 cm., imparipinnate, oblong-acute in outline, lower pinnae not reduced; pinnae up to 13·5 × 1·7 cm., minutely serrate, glabrous on both surfaces except for scattered brown ovate-lanceolate scales up to 3 mm. long along the costa dorsally. Fertile lamina up to 40 × 13 cm., narrowly oblong-acute in outline, lower pinnae not reduced; pinnae up to 12 × 0·4 cm., undulate, narrowly linear-acuminate, petiolate, slightly auricled at the base, glabrous except for brown scales along the costa dorsally. Sori linear, almost marginal; indusium brown, c. 1 mm. broad.

Rhodesia. E: Umtali Distr., Stapleford Forest Reserve, 29.xi.1951, *Chase* 4268 (BM; SRGH). **Malawi.** S: Mlanje, 1891, *Whyte* (BM). **Mozambique.** MS: Gorongosa Mt., Gogogo Peak, 6.vii.1955, *Schelpe* 5533 (BM; BOL).

Also in S. Africa. Streambanks in forest, 1525–1770 m.

6. **Blechnum punctulatum** Sw. in Schrad., Journ. Bot. **1800**, 2: 74 (1801). Type from S. Africa.

   Blechnum rigidum Sw. in Schrad., tom. cit.: 75 (1801). Type from S. Africa.
   Lomaria auriculata Desv., in Mag. Ges. Naturf. Fr. Berl. **5**: 330 (1811). Type from S. Africa.
   Lomaria densa Kaulf., Enum. Fil.: 151 (1824). Type from S. Africa.
   Lomaria punctulata (Sw.) Kunze in Linnaea, **10**: 507 (1836). Type as for *Blechnum punctulatum*.
   Mesothema punctulata (Sw.) C. Presl, Epim. Bot.: 113 (1851). Type as above.
   Mesothema rigidum (Sw.) C. Presl, loc. cit. Type as for *Blechnum rigidum*.
   Blechnopteris punctulata (Sw.) Trevisan in Att. Ist. Veneto, **2**, 2: 166 (1851). Type as for *Blechnum punctulatum*.
   Lomaria rigida (Sw.) Fée, Mém. Fam. Foug. **5**: 68 (1852). Type as for *Mesothema rigidum*.
   Lomaria dregeana Fée, op. cit. **10**: 9 (1865). Type from S. Africa.
   Struthiopteris punctulata (Sw.) Trevisan, op. cit. **3**, 14: 573 (1869). Type as for *Blechnum punctulatum*.
   Struthiopteris dregeana (Fée) Trevisan, tom. cit.: 572 (1869). Type as for *Lomaria dregeana*.
   Struthiopteris rigida (Sw.) Trevisan, loc. cit. Type as for *Mesothema rigidum*.
   Spicanta punctulata (Sw.) Kuntze, Rev. Gen. Pl. **2**: 822 (1891). Type as for *Blechnum punctulatum*.

Rhizome creeping, up to 8 cm. long and c. 1 cm. in diam., with tufted fronds and with narrowly linear attenuate entire dark-brown rhizome-scales up to 2 cm. long with paler borders. Frond erect, coriaceous. Stipe pale-brown, up to 5 cm. long, glabrous except for tufted scales similar to those on the rhizome about the base. Lamina up to 76 × 10 cm., very narrowly oblong in outline, acute, with a long tapering base, the lower pinnae up to 5 × 0·8 cm. at the base, attenuate from a shallowly auriculate base, subsessile to shortly petiolate, entire, glabrous or thinly glandular on both surfaces but with prominent submarginal hydathodes on the ventral surface, fertile pinnae up to 5·6 × 0·4 cm., linear from a weakly auriculate base, acute. Rhachis brown, glabrous at maturity. Sori linear extending for most of the length of the fertile pinnae; indusium up to 0·6 mm. broad, subentire.

Rhodesia. N: Mazoe, Iron Mask Hill, iv.1906, *Eyles* 342 (BM; SRGH). **Malawi.** S: Mlanje Distr., Mlanje Mt., 11.vi.1962, *Richards* 16631 (K).

Also in S. Africa and Madagascar. Lithophytic or terrestrial in shade of forest.

7. **Blechnum australe** L., Mant. Pl. Alt.: 130 (1767).—Sim, Ferns S. Afr. ed. 2: 188, t. 84 (1915).—Launert in Prodr. Fl. SW. Afr. **8**: 1 (1969). Type from S. Africa.
Lomaria pumila Kaulf., Enum. Fil.: 151 (1824). Type from S. Africa.
Lomaria australis (L.) Link, Fil. Hort. Berol.: 75 (1841). Type as for *Blechnum australe*.
Mesothema australe (L.) C. Presl, Epim. Bot.: 111 (1851) reimpr. in Abh. Königl. Böhm. Ges. Wiss., Ser. 5, **6**: 472 (1851). Type as above.
Blechnopteris australis (L.) Trevisan in Att. Ist. Veneto, **2**, 2: 166 (1851). Type as above.
Struthiopteris australis (L.) Trevisan, op. cit. **3**, 14: 572 (1869). Type as above.
Spicanta australis (L.) Kuntze, Rev. Gen. Pl. **2**: 821 (1891). Type as above.

Rhizome up to 5 mm. in diam., creeping, with tufted fronds and with dark-brown ovate-lanceolate acuminate entire rhizome-scales up to 5·5 × 1 mm. Frond erect or arching, firmly membranous. Stipe up to 14 cm. long, pale-brown, darker and with scales similar to the rhizome scales about the base. Lamina up to 47 × 12·5 cm. (usually smaller), pinnate, narrowly elliptic to elliptic-oblong in outline, acute or acuminate and with several pairs of basal pinnae reduced. Sterile pinnae up to 6 × 1·1 cm., narrowly to very narrowly oblong, obtuse to acute, mucronate, base somewhat auriculate, subsessile, glabrous or thinly glandular. Fertile pinnae up to 27 × 0·2 mm., linear, mucronate with a dilated auriculate subsessile base. Rhachis stramineous, glabrous or glandular. Sori linear extending almost the length of the fertile pinnae; indusium c. 0·6 mm. broad, brown, membranous, erose.

Rhodesia. E: Melsetter Distr., Lessops Pass, 11.i.1947, *Fisher* 1257 (BM; SRGH). Also in S. Africa, Madagascar and the S. Atlantic Is. Deeply shaded stream-banks and waterfalls.

2. STENOCHLAENA J. Sm.

Stenochlaena J. Sm. in Journ. of Bot. **3**: 401 (1841).

Large ferns with rhizomes creeping along the ground but eventually becoming scandent epiphytes; rhizome-scales sparse. Fronds remote, dimorphic, pinnate or 2-pinnate. Sterile pinnae articulate with basal glands, firmly membranous or chartaceous with sharply cartilaginous serrate margins. Fertile pinnae linear or divided into linear segments, almost entirely covered with sporangia below; paraphyses absent.

A small genus of the Old World tropics and sub-tropics which has been referred to the Polypodiaceae (sens. strict.) by Alston (1959). The present author agrees with Copeland (1947) that this genus is better referred to the Blechnaceae on the venation and spore characters.

Stenochlaena tenuifolia (Desv.) Moore in Gard. Chron. **1856**: 193 (1856).—Sim, Ferns S. Afr. ed. 2: 192, t. 85–86 (1915). TAB. **69**. Type from Madagascar.
Lomaria tenuifolia Desv. in Mag. Ges. Naturf. Fr. Berl. **5**: 326 (1811). Type as above.
Lomaria meyerana Kunze in Linnaea, **10**: 509 (1836). Type from S. Africa.
Stenochlaena meyerana (Kunze) C. Presl, Epim. Bot.: 166 (1851) reimpr. in Abh. Königl. Böhm. Ges. Wiss., Ser. 5, **6**: 526 (1851). Type as above.
Lomariobotrys tenuifolia (Desv.) Fée, Mém. Fam. Foug. **5**: 46 (1852). Type as for *Stenochlaena tenuifolia*.
Lomariobotrys meyerana (Kunze) Fée, tom. cit.: 46, t. 5 fig. a (1852). Type as for *Lomaria meyerana*.
Polybotrya meyerana (Kunze) Mett., Fil. Hort. Bot. Lips.: 24, t. 1 fig. 4 & 7 (1856). Type as above.
Acrostichum meyeranum (Kunze) Hook., Garden Ferns: t. 16 (1862). Type as above.
Polybotrya tenuifolia (Desv.) Kuhn, Fil. Afr.: 52 (1868). Type as for *Stenochlaena tenuifolia*.
Acrostichum tenuifolium (Desv.) Bak., Syn Fil.: 412 (1868). Type as above.
Lomariopsis tenuifolia (Desv.) Christ, Farnkr.: 42 (1897). Type as above.

Rhizome up to 20 m. long and 1–1·5 cm. in diam., creeping along the ground or ascending trees, sparsely clothed with dark-brown subulate rhizome-scales up to 5 mm. long, becoming glabrous with age. Sterile fronds erect, pinnate, fertile

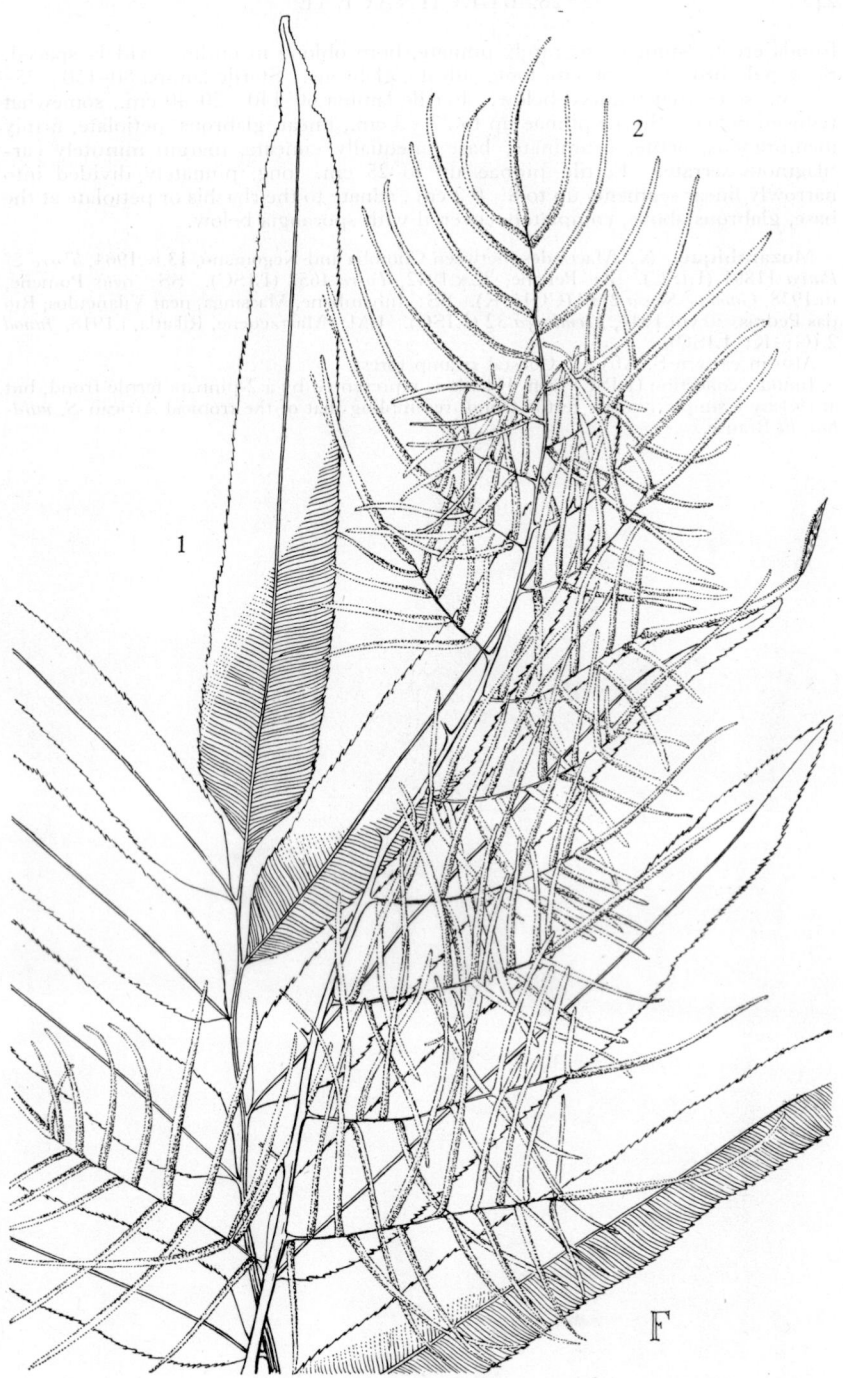

Tab. 69. STENOCHLAENA TENUIFOLIA. 1, apical part of sterile frond (× ⅔) *Junod* 2 (K);
2, apical part of fertile frond (× ⅔) *Boivin* s.n. (Madagascar).

fronds erect, 2-pinnate or rarely pinnate, both oblong in outline, widely spaced. Stipe pale-brown, 30–50 cm. long, sulcate, glabrous.   Sterile lamina 80–150 × 25–40 cm., somewhat reduced below.   Fertile lamina 60–140 × 20–40 cm., somewhat reduced below.   Sterile pinnae up to 27 × 3 cm., linear, glabrous, petiolate, firmly membranous, acute, acuminate, base unequally cuneate, margin minutely cartilaginous-serrate.   Fertile pinnae up to 25 cm. long, pinnately divided into narrowly linear segments up to 8 × 0·2 cm., adnate to the rhachis or petiolate at the base, glabrous above, completely covered with sporangia below.

Mozambique. N: Macondes, between Chomba and Negomano, 13.iv.1964, *Torre & Paiva* 11891 (LISC).   Z: Pebane, 22.x.1942, *Torre* 4652 (LISC).   SS: near Pomene, iii.1938, *Gomes e Sousa* 2117 (COI; K).   SS: Inhambane, Massinga, near Vilanculos, Rio das Pedras, 30.viii.1944, *Mendonça* 32 (LISC).   LM: Marracuene, Rikatla, i.1918, *Junod* 2 (G; K; LISC).

Also in eastern S. Africa.   Coastal swamp forest.

Junod's collection (LISC) from Rikatla is represented by a 2-pinnate fertile frond, but at (K) by a simply pinnate fertile frond, resembling that of the tropical African *S. mildbraedii* Brause.

INDEX TO BOTANICAL NAMES

Acropteris radiata, 138
ACROSTICHUM, 42, **98**
  alcicorne, 145
  aubertii, 213
  aureum, **99,** tab. **31**
  barbarum, 46
  boivinii, 213
  calomelanos, 107
  ciliare, 213
  ciliatum, 213
  cordatum, 188
  deckenii, 215
  ebeneum, 107
  guineense, 99
  heudelotii, 218
  hybridum, 213
  inaequale, 99
  lanceolatum, 146
  lancifolium, 215
  lastii, 211
  leptophyllum, 99
  lineatum, 210
  macropodium, 211
  meyeranum, 240
  pectinatum, 52
  petiolatum, 215
  piloselloides, 213
  pilosiusculum, 199
  polypodioides, 158
  punctatum, 156
  radiatum, 138
  salicifolium, 215
  serrulatum, 144
  siliquosum, 102
  spathulatum, 213
  spicatum, 159
  tenuifolium, 240
  thalictroides, 102
  tricholepis, 213
  volkensii, 211
  welwitschii, 216
Acrosticum, 98
ACTINIOPTERIS, 42, 98, **136**
  australis, 138
    var. *radiata,* 138
  dimorpha, **136,** 138, tab. **42** fig. A
  pauciloba, 136, **138,** tab. **42** fig. B
  radiata, 136, **138,** tab. **42** fig. C
ADIANTACEAE, 98
ADIANTUM, 42, 43, 98, **108**
  achilleifolium, 185
  aethiopicum, 112
  alatum, 110
  altissimum, 92
  arcuatum, 110
  balansae, 110
  borbonicum, 185
  caffrorum, 56, 124
  capillus, 112
  capillus-gorgonis, 108
  capillus-veneris, 108, **112,** tab. **36** fig. B
  caudatum, 108
  coriandrifolium, 112
  cuneatum, 113
  cycloides, 112

  fontanum, 112
  globatum, 123
  hirtum, 124
  hispidulum, 108, **111,** tab. **36** fig. A
  incisum, **108,** tab. **36** fig. C
  lindsaea, 111
  lobulatum, 111
  lunatum, 110
  lunulatum, 110, 220
  marginatum, 112
  mendoncae, 108, **111,** tab. **36** fig. E
  mettenii, 108, **110,** tab. **36** fig. F
  multifidum, 123
  oatesii, 111
  paradiseae, 112
  patens, **111**
    subsp. *oatesii,* 108, **111**
    var. *oatesii,* 111
  pedatum, 111
  pellucidum, 112
  philippense, 108, **110,** tab. **36** fig. G
  poiretii, 108, **112**
    var. poiretii, **113,** tab. **36** fig. D
    var. sulphureum, **113**
  pseudocapillus, 112
  raddianum, 108, **113**
  soboliferum, 110
  sulphureum, 113
  thalictroides, 112
  viride, 135
  williamsii, 113
Alcicornium angolense, 145
Aleuritopteris farinosa, 122
Allantodia scandicina, 204
Allosorus adiantoides, 135
  angulosus, 131
  aquilinus, 88
  boivinii, 131
  calomelanos, 132
  capensis, 88
  consobrinus, 133
  coriifolius, 88
  doniana, 129
  durus, 132
  farinosus, 122
  lanuginosus, 88
  pectiniformis, 128
  quadripinnatus, 133
  viridis, 135
Alsophila capensis, 75
AMPELOPTERIS, 42, 189, **200**
  elegans, 200
  prolifera, **200,** tab. **56**
Amphicosmia capensis, 75
  riparia, 75
Anapausia heudelotii, 218
ANEMIA, 41, **52**
  angolensis, **54**
  anthriscifolia, 54
  schimperana var. *angustiloba,* 54
  simii, **54,** tab. **15** fig. A
    var. *angustiloba,* 54
Anisosorus occidentalis, 86
ANOGRAMMA, 43, 98, **99**
  conspersa, 105

ANOGRAMMA (*contd.*)
  leptophylla, **99**, tab. **32**
  rosea, 105
ANTROPHYUM, 41, 94, **96**
  lanceolatum, 151
  mannianum, **96**, tab. **30**
ARACHNIODES, 44, 220, **228**
  foliosa, **228**, tab. **65**
ARTHROPTERIS, 44, 159, **162**
  monocarpa, **163**, tab. **50**
  orientalis, **163**
ASPIDIACEAE, 218
Aspidium acutum, 160
  africanum, 200
  albopunctatum, 163
  articulatum, 165
  athamanticum, 221
  bergianum, 193
  biserratum, 160
  boryanum, 207
  capense, 74
  catopteron, 232
  cirrhosum, 232
  coadunatum var. *gemmiferum*, 234
  comorense, 232
  crenatum, 230
  donianum, 225
  ecklonii, 198
  eriocarpum, 230
  extensum, 195
  gemmiferum, 234
  glabratum, 207
  goggilodus, 198
  gueinzianum, 194
  gueinztianum, 194
  guineense, 160
  imbricatum, 162
  inaequale, 221
    var. *montanum*, 222
  kiboschense, 207
  kilemense, 222
  lanuginosum, 232
  lastii, 222
  leucosticton, 163
  longiscupe, 192
  luctuosum, 228
  maranguense, 193
  natalense, 197
  odoratum, 230, 232
  paleaceum, 225
  plantianum, 198
  pulchrum, 193
  scandicinum, 204
  speluncae, 89
  splendens, 160
  squamigerum, 190
  squamisetum, 225
  striatum, 199
  strigosum, 193
  sulcinervium, 205
  thelypteris var. *squamigerum*, 190
  thonningii, 163
  tottum, 199
  truncatulum, 220
  undulatum, 162
  wakefieldii, 195
  wallichianum, 225
  zambesiacum, 192
ASPIDOTIS, 43, 98, **113**

  schimperi, **113**, tab. **37** fig. A
ASPLENIACEAE, 167
ASPLENIUM, 41, 42, 43, **167**
  abyssinicum, 177
  achilleifolium, 185
  adiantoides, 181
  adscensionis, 117
  aethiopicum, 170, **181**
  africanum var. *holstii*, 170
  amaurophyllum, 172
  amoenum, 174
  anisophyllum, 169, **170**
    var. *elongatum*, 173
  anisophyllum, 173
  aquilinum, 88
  atroviride, 169, **173**
  auritum, 170, **177**
  bipinnatum, 185
  blastophorum, 170, **183**
  boltonii, 169, **173**
  borbonicum, 185
  brachyotus, 176
  brachypteron, 184
  buettneri, 170, **182**
  chaseanum, 169, **183**
  christii, 169, **172**
  comorense, 184
  concinnum, 188
  cordatum, 188
  cuneatum var. *angustatum*, 181
  daubenbergii, 183
  debile, 184
  dimidiatum var. *longicaudatum*, 180
  distichum, 185
  dregeanum, 169, **184**
  emarginato-dentatum, 174
  erectum, 170, **175**
    var. *brachyotus*, 176
    var. erectum, **176**
    var. *gracile*, 177
    var. *lobatum*, 177
    var. usambarense, **176**, tab. **53** fig. F
  eylesii, 178
  falsum, 182
  fimbriatum, 177
  floccigerum, 187
  formosum, 169, **179**
  friesiorum, 169, **178**
  furcatum, 182
  gemmiferum, 169, **173**
  geppii, 170
  goetzii, 182
  gracile, 177, 184
  gueinzianum, 182
  hanningtonii, 184
  hollandii, 187
  holstii, 169, **170**, tab. **53** fig. H
  hyophilum, 205
  hypomelas, 170, **187**, tab. **54** fig. A
  inaequilaterale, 170, **176**, tab. **53** fig. C
  laetum, 176
    var. *brachyotus*, 176
  laxum, 204
  leptophyllum, 99
  linckii, 170, **183**
  linearilobum, 185
  lividum, 170, **181**
  lobatum, 170, **177**, tab. **53** fig. A
  loxoscaphoides, **187**

lunulatun var. *gracile*, 177
macrolobium, 173
madagascariense, 205
mannii, 169, **187,** tab. **54** fig. E
mary-annae, 102
megalura, 169, **180**
melleri, 184
monanthemum, 175
monanthes, 169, **175,** tab. **53** fig. D
monilisorum, 178
mossambicense, 169, **172**
nemorale, 205
nigrescens, 187
normale, 169, **175**
obscurum, 169, **174**
pellucidum, 169, **179**
   subsp. pseudohorridum, **179**
polydactylon, 138
praemorsum, 182
protensum, 169, **179**
   var. *pseudohorridum*, 179
pseudoauriculatum, 169, **185,** tab. **54**
   fig. B
pseudohorridum, 179
pseudoserra, 178
pumilum, 170, **178**
   subsp. hymenophylloides, **178**
   var. *hymenohpylloides*, 178
punctatum, 184
quintasii, 176
radiatum, 138
ramlowii, 169, **180**
resectum, 174
rutifolium, 170, **185**
   var. bipinnatum, **185,** tab. **54** fig. C
sandersonii, 169, **183,** tab. **53** fig. B
scandicinum, 204
schimperanum, 178
schimperi, 202
serra var. *natalensis*, 178
serraeforme, 174
serrulatum, 144
simii, 170, **181**
sphenolobum var. *usambarense*, 176
splendens var. *angustatum*, 181
stans, 185
sulcatum, 178
suppositum, 176
theciferum, 170, **188**
   var. concinnum, **188,** tab. **54** fig. D
theciferum, 188
torrei, 169, **180**
trichomanes, 169, **174**
unilaterale, 169, **174**
usambarense, 176
vagans, 184
varians, 170, **177**
   subsp. fimbriatum, **177,** tab. **53**
   fig. E
volkensii, 182
zanzibaricum, 205
ATHYRIACEAE, 202
ATHYRIUM, 43, **202**
   boryanum, 207
   glabratum, 207
   laxum, 204
   scandicinum, 202, **204**
      var. rhodesianum, **204,** tab. **57** fig. A
      var. scandicinum, **204**

schimperi, **202,** tab. **57** fig. B
AZOLLA, 40, **69**
   africana, 69
   guineensis, 69
   nilotica, 69, **70,** tab. **20**
   pinnata, **69**
      var. africana, **69**
AZOLLACEAE, 69

BELVISIA, 41, 145, **159**
   spicata, **159,** tab. **48** fig. D
Bernhardia capensis, 15
   mascarenica, 15
BLECHNACEAE, 235
Blechnopteris australis, 240
   punctulata, 239
BLECHNUM, 42, **235**
   attenuatum, **235**
      var. attenuatum, **236**
      var. giganteum, **236**
      var. *holstii,* 236
   australe, 235, **240**
   bakeri var. *glabra,* 237
   capense, 235, **239**
   cycadoides, 237
   dalgairnsiae, 237
   giganteum, 236
   heterophyllum, 236
   inflexum, 235, **236**
   ivohibense, 235, **237,** tab. **68**
   polypodioides var. *holstii,* 236
   punctulatum, 235, **239**
   rigidum, 239
   tabulare, 235, **237**
   umbrosum, 237
BLOTIELLA, 43, **81,** 86
   crenata, 81, **82,** tab. **23**
   currorii, 82, **84**
   glabra, 81, **82**
   natalensis, **82**
BOLBITIS, 40, 42, 209, **218**
   felixii, 218
   heudelotii, **218,** tab. **62**
      var. angustifolia, 218

Caenopteris disticha, 185
   furcata, 185
   rutifolium, 185
Calymella polypodioides, 48
Campium angustifolium, 218
   heudelotii, 218
Candollea lanceolata, 146
Cassebeera farinosa, 122
CERATOPTERIS, 40, 42, 98, **102**
   siliquosa, 105
   thalictroides, **102,** tab. **34**
Ceropteris argentea, 105
   calomelanos, 107
CETERACH, 42, 43, 167, **188**
   capense, 189
   cordatum, **188,** tab. **54** fig. F
   crenata, 189
   pozoi, 199
CHEILANTHES, 43, 56, 98, **122,** 128
   angustifrondosa, 122, **123**
   aspera, 92
   atherstonei, 133

CHEILANTHES (*contd.*)
  bergiana, 122, **124**
  bolusii, 123
  commutata, 92
  cornuta, 135
  eckloniana, 122, **125**
  elata, 124
  farinosa, **122**
  firma, 133
  fuscata, 56
  glandulosa, 125
  hastata var. *stenophylla*, 135
  heterophylla, 124
  hirta, 122, **124**
    var. *intermedia*, 125
    var. *laxa*, 125
  inaequalis, 122, **126**
    var. *buchananii*, **126,** tab. **39** fig. B
    var. *inaequalis*, **126,** tab. **39** fig. A
  kirkii, 121
  leachii, 122, **125**
  linearis, 133
  mossambicensis, 122, **123**
  multifida, 122, **123**
  quadripinnata, 133
  schimperi, 113
  similis, 122, **124**
  sparsisora, 92
  streetiae, 124
  triangula, 133
  viridis, 135
Chrysodium aureum, 99
  heudelotii, 218
  inaequale, 99
  vulgare, 99
Chrysopteris phymatodes, 153
Cincinalis aquilina, 88
  cordata, 188
Colina caffrorum, 56
Colysis irioides, 156
CONIOGRAMME, 42, 98, **102**
  africana, **102,** tab. **33**
Cormophyllum capense, 75
Cornopteris boryana, 207
  sulcinervia, 205
CTENITIS, 44, 220, **230**
  boryana, 207
  cirrhosa, 230, **232,** tab. **67** fig. A
  crinobulbon, 232
  lanuginosa, 230, **232,** tab. **67** fig. B
Ctenopteris cultrata, 143
  elastica, 143
  rigescens, 143
  villosissima, 142
CYATHEA, 41, **70**
  angolensis, 74
  burkei, 74
  capensis, 70, 72, **74,** tab. **21** fig. D
  deckenii, 72
  dregei, 72, **74,** tab. **21** fig. E
  holstii, 72
  humilis, 72
  laurentiorum, 74
  manniana, **72,** tab. **21** fig B
  mossambicensis, 70, **72,** tab. **21** fig. A
  riparia, 75
  sellae, 74
  thomsonii, **72,** tab. **21** fig. C

  usambarensis, 74
  zambesiaca, 72
CYATHEACEAE, **70**
Cyclophorus giesenhagenii, 146
  lanceolatus, 146
  mechowii, 147
  rhodesianus, 147
  schimperanus, 147
  spissus, 146
    var. *continentalis*, 146
  stoltzii, 147
  stolzii, 147
  tener, 146
  vittarioides, 146
Cyclosorus, 189
  contiguus, 195
  costularis, 196
  dentatus, 197
  extensus, 195
  goggilodes, 198
  hispidulus, 197
  patens, 196
  prismaticus, 197
  proliferus, 200
  quadrangularis, 195
  silvatica, 196
  striatus, 199
  wakefieldii, 195
Cystopteris odorata, 230
  scandicina, 204

Darea disticha, 185
  furcata, 185
  stans, 185
DAVALLIA, 43, 159, **167**
  campyloptera, 188
  chaerophylloides, **167,** tab. **52**
  concinna, 188
  denticulata, var. *intermedia*, 167
  hollandii, 187
  nigrescens, 187
  nitidula, 167
  schnellii, 167
  speluncae, 89
  thecifera, 188
  vogelii, 167
DAVALLIACEAE, **159**
DENNSTAEDTIACEAE, **81**
Dicranodium leptophyllum, 99
DICRANOPTERIS, 41, 46, **50**
  linearis, **50,** tab. **13**
DIDYMOCHLAENA, 43, **220**
  dimidiata, 220
  lunulata, 220
  truncatula, 220, tab. **64** fig. E
DIPLAZIUM, 43, 202, **204**
  arborescens var. *nemorale*, 205
  hyophilum, 205
  nemorale, **205,** tab. **58** fig. A
  stolzii, 205
  zanzibaricum, **205,** tab. **58** fig. B
DORYOPTERIS, 43, 98, **121,** 128
  articulata, 131
  concolor, **121**
    var. *kirkii*, **121,** tab. **37** fig. C
    var. *nicklesii*, **121,** tab. **37** fig. B
  kirkii, 121

nicklesii, 121
Drymoglossum acrostichoides, 210
DRYNARIA, 42, 144, **149**
  excavata, 151
  irioides, 156
  lepidota, 153
  macrocarpa, 153
  phymatodes, 155
  polycarpa, 156
  volkensii, **149**, tab. **46**
  vulgaris, 155
DRYOATHYRIUM, 43, 202, **207**
  boryanum, **207**, tab. **59**
DRYOPTERIS, 43, 44, **220**
  africana, 199
  anateinophlebia, 194
  athamantica, **221**
  bergiana, 193
  boryana, 207
  buchananii, 225
  callolepis, 221, **223**
  catoptera, 234
  cirrhosa, 232
  contigua, 195
  costularis, 196
  crenata, 230
  crinobulbon, 232
  dentata, 197
  doniana, 225
  extensa, 195
  foliosa, 228
  friesii, 192
  glabrata, 207
  gladiata, 196
  goggilodus, 198
  gongylodes, 198
  hemitelioides, 199
  inaequalis, **221**
    var. *atropaleacea*, **222**
    var. *inaequalis*, **222**
  kiboschensis, 207
  kilmensis, 222
  kilemensis, 221, **222**
  lanuginosa, 234
  lastii, 222
  longicuspis, 192
  madagascariensis, 196
  manniana, 221, **223**, tab. **63**
  maranguensis, 194
  nimbaensis, 232
  orientalis, 163
  paleacea, 225
    var. *madagascariensis*, 225
  palmii, 194
  pentheri, 222
    var. *montana*, 222
  prolifera, 200
  prolixa var. *bergiana*, 194
  pulvinata, 232
  quadrangularis, 195
  sewellii, 193
  silvatica, 196
  spekei, 232
  squamiseta, 221, **223**
  striata, 199
  strigosa, 193
  sulcinervia, 205
  thelypteris, 190
    var. *squamigera*, 190

wakefieldii, 195
wallichiana, 221, **225**
zambesiaca, 192

ELAPHOGLOSSUM, 41, **209**
  acrostichoides, 209, **210**
  aubertii, 209, **213**
  chevalieri, 209, **214**
  conforme var. *latifolium*, 211
    var. *lineatum*, 210
  deckenii, 209, **215**
  hybridum, 209, **213**
  kuhnii, 210, **215**
  lancifolium, 215
  lastii, 209, **211**
  macropodium, 209, **211**
  marojejyense, 209, **210**
  mildbraedii, 209, **214**
  obtusum, 211
  petiolatum, 209, **215**
    subsp. salicifolium, **215**
    var. *salicifolium*, 215
  petiolatum, 210
  piloselloides, 214
  preussii, 210
  rhodesianum, 209, **211**, tab. **60**
  salicifolium, 215
  spathulatum, 209, **213**
    var. *uluguruense*, 214
  tricholepis, 213
  uluguruense, 214
  viscosum var. *salicifolium*, 215
  volkensii, 211
  welwitschii, 210, **216**
  zambesiacum, 209, **210**
Ellobocarpus oleraceus, 102
EQUISETACEAE, **32**
EQUISETALES, 13, **32**
EQUISETUM, **32**
  azoricum, 34
  burchellii, 32
  campanulatum, 32
  elongatum, 32
  ephedroides, 32
  hungaricum, 32
  incanum, 32
  multiforme, 32
  pannonicum, 32
  procerum, 32
  ramosissimum, **32**, tab. **7**
  ramosum, 32
  sieboldii, 32
  thunbergii, 32
Eupteris aquilina, 88

FILICALES, 13, **40**
Furcaria thalictroides, 105

GLEICHENIA, 41, 46, **48**
  argentea, 48
  boryi, 50
  elongata, **48**, tab. **12**
  glauca, 48
  linearis, 50
  polypodioides, **48**
  ruwenzoriensis, 48
  umbraculifera, 48, **50**

GLEICHENIACEAE, 41, **46**
Goniopteris madagascariensis, 196
  patens, 196
  prolifera, 200
  silvatica, 196
  unita, 196
GRAMMITIDACEAE, 141
GRAMMITIS, 41, **141**
  capensis, 189
  cordata, 188
  coriacea, 149
  flabelliformis, 143
  lanceolata, 149
  leptophylla, 99
  nanodes, **141**, tab. **44** fig. A
  serrulata, 144
  totta, 199
Gymnogramma abyssinica, 151
  argentea, 105
  aurantiaca, 107
  calomelanos, 107
    var. *aureoflava*, 108
  conspersa, 105
  cordata, 189
  distans, 107
  lanceolata, 151
  leptophylla, 99
  novae-zeylandiae, 99
  pozoi, 199
  rosea, 105
  schwackeana, 99
  thiebautii, 115
  totta, 199
  unita, 196
Gymnopteris heudelotii, 218

Hemianemia schimperana var. *angustiloba*,
  54
Hemiasplenium obscurum, 174
Hemionitis argentea, 105
  leptophylla, 99
  pozoi, 199
  prolifera, 200
Hemitelia, 70
  capensis, 75
  riparia, 75
Hippochaete ramosissimum, 34
HISTIOPTERIS, 43, 81, **84**
  incisa, **84**, tab. **24**
  vespertilionis, 86
Humata chaerophylloides, 167
Huperzia gnidioides, 18
  ophioglossoides, 18
  phlegmaria, 20
  saururus, 17
  verticillata, 17
Hymenolepis spicata, 159
  var. *occultivenia*, 159
  var. *usambarensis*, 159
HYMENOPHYLLACEAE, 41, **75**
HYMENOPHYLLUM, 41, 75, **78**
  capense, **79**
  capillare, 79, **80**, tab. **22** fig. H
  dregeanum, 80
  fulvum, 80
  fumarioides, 79
  henkelii, 80
  holotrichum, 80

  kuhnii, 80
  lineare, 80
  meyeri, 80
  natalense, 79
  parvum, 79
  pendulum, 80
  polyanthos, **79**
    var. *kuhnii*, **80**, tab. **22** fig. G
    var. mossambicense, **80**, tab. **22**
     fig. F
  sibthorpioides, **79**, tab. **22** fig. I
  tabulare, 79
  thomassetii, 80
  tunbridgense, 79, **80**, tab. **22** fig. E
  zeyheri, 79
HYPODEMATIUM, 44, 220, **230**
  crenatum, **230**, tab. **66**
  rueppellianum, 230
HYPOLEPIS, 43, 81, **92**
  aspera, 92
  bergiana, 124
  pteridioides, 117
  punctata, 92
  rugulosa var. *africana*, 92
  schimperi, 113
  sparsisora, **92**, tab. **28**
Hypopeltis biserrata, 160

ISOETACEAE, 29
ISOETALES, 13, 29
ISOETES, **30**
  aequinoctialis, **30**
  alstonii, **30**, tab. **6**
  nigritiana, 30
  rhodesiana, **30**
Isoloma lanuginosa, 160

Lastrea africana, 200
  athamantica, 221
  bergiana, 193
  boryana, 207
  buchananii, 225
  catoptera, 232
  crenata, 230
  crinobulbon, 232
  gueinziana, 194
  hirsuta, 230
  inaequalis, 221
  lanuginosa, 232
  maranguensis, 194
  odorata, 230, 232
  pentagona, 222
  plantii, 221
  pulchra, 193
  squamulosa, 190
  strigosa, 193
  thelypteris var. *squamigera*, 190
  totta, 200
Lepidoneuron biserratum, 160
Lepidotis cernua, 20
  funiculosa, 18
  inflexa, 20
  longifolia, 18
  gueintzii, 152
Lepisorus
  schraderi, 152

Leptochilus heudelotii, 218
Leptogramma, 189
  africana, 200
  pilosiuscula, 200
  pozoi, 200
  totta, 199
LINDSAEA, 42, **139**
  acutifolia, 160
  ensifolia, **139**, tab. **43**
  lanuginosa, 160
  loherana, 139
  membranacea, 139
  odorata, **139**
LINDSAEACEAE, **139**
Litobrochia articulata, 131
  dura, 132
  incisa, 86
  vespertilionis, 86
Lomaria acuminata, 235
  attenuata, 235
  auriculata, 239
  australis, 240
  capensis, 239
  coriacea, 237
  cycadoides, 237
  dalgairnsiae, 237
  decipiens, 236
  densa, 239
  discolor var. *natalensis*, 236
  dregeana, 239
  gigantea, 236
  gueinzii, 237
  hamata, 236
  heterophylla, 236
  inflexa, 236
  meyerana, 240
  pumila, 240
  punctulata, 239
  rigida, 239
  tabularis, 237
  tenuifolia, 240
Lomariobotrys meyerana, 240
  tenuifolia, 240
LOMARIOPSIDACEAE, 209
LOMARIOPSIS, 42, 209, **216**
  nigrescens, 216
  tenuifolia, 240
  warnckei, **216**, tab. **61**
LONCHITIS, 43, 81, **86**
  adscensionis, 117
  bipinnata, 188
  caffrorum, 56
  crenata, 82
  currorii, 84
  friesii, 86
  glabra, 82
  gracilis, 82
  mannii, 84
  natalensis, 82
  occidentalis, **86**, tab. **25**
  pubescens, 82
  stenochlamys, 82
LOXOGRAMME, 41, 144, **149**
  africana, 151
  lanceolata, **149**, tab. **48** fig. E
  suberosa, 151
Loxoscaphe concinnum, 188
  nigrescens, 187
  theciferum var. *concinnum*, 188

LYCOPODIACEAE, 15
LYCOPODIALES, 13, **15**
LYCOPODIUM, **17**
  aberdaricum, 21
  acerosum, 17
  affine, 21
  ambiguum, 18
  boryanum, 20
  carolinianum, 17, **21**
    var. affine, **21**
    var. carolinianum, 21
    var. grandifolium, **21**
    var. tuberosum, **21**
  cernuum, 17, **20**
  clavatum, 17, **20**
    var. inflexum, **20**
    var. natalense, 21
  clavatum, 20
  cymosum, 20
  dacrydioides, 17, **18**
  dregei, 23
  ericetorum, 21
  flagelliforme, 18
  funiculosum, 18
  gnidioides, 17, **18**
    var. *strictum*, 18
  heeschii, 20
  hupeanum, 20
  imbricatum, 25
  inflexum, 20
  kraussianum, 26
  lehmannii, 20
  longifolium, 18
  marianum, 20
  moritzii, 20
  nudum, 15
  ophioglossoides, 17, **18**, tab. **3**
  phlegmaria, 17, **20**
  pinifolium, 18
  sarcocaulon, 21
  sanguin., 25
  sanguinolentum, 25
  saururus, **17**
  secundum, 20
  setaceum, 17
  sikkimense, 20
  strictum, 18
  trianae, 20
  tuberosum, 21
  verticillatum, **17**
  yemense, 25
LYGODIQM, 41, **57**
  brycei, 57
  kerstenii, **57**, tab. **16** fig. A
  microphyllum, 57, tab. **16** fig. B
  scandens, 57
  subalatum, 57

MARATTIA, **38**
  dregeana, 40
  fraxinea, **40**, tab. **9**
    var. salicifolia, **40**
  fraxinea, 40
  natalensis, 40
  salicifolia, 40
MARATTIACEAE, 38
MARATTIALES, 13, **38**
MARSILEA, 40, **59**

aegyptiaca, 59, **62**, tab. **18** fig. E
MARSILEA (*contd.*)
  apposita, 59, 60, **65**
  biloba, 65, 66
  burchellii, 60, **66,** tab. **18** fig. C
  capensis, 60, **66,** tab. **18** fig. D
  capensis, 65
  coromandelina, 59
  crenulata, 60
  diffusa, 60
  dregeana, 64
  ephippiocarpa, 59, **62**, tab. **17** fig. A
  farinosa, 59, **65**, tab. **17** fig. C
  fischeri, 64
  macrocarpa, 60, **64,** tab. **17** fig. D
  macrocarpa, 65
  minuta, 59, **60**, tab. **17** fig. B
  rotundata, 60, 64
  tenax, 60
  vera, 59, **65**, tab. **18** fig. A
  villifolia, 60, **64**, tab. **18** fig. B
  villosa, 64
MARSILEACEAE, 59
Mecodium kuhnii, 80
Meniscium proliferum, 200
Mertensia caeruleo-glauca, 48
  linearis, 50
  umbraculifera, 50
Mesothema australe, 240
  punctulata, 239
  rigidum, 239
MICROGRAMMA, 41, 145, **155**
  lycopodioides, **155**, tab. **48** fig. C
  mauritiana, 155
  owariensis, 155
MICROLEPIA, 43, 81, **89**
  setosa, 89
  speluncae, **89**, tab. **27**
Micropteris orientalis, 144
  serrulata, 144
MICROSORIUM, 41, 42, 145, 153, **156**
  iroides, 156
  irregulare, 156
  pappei, 156, **158**, tab. **48** fig. B
  polycarpon, 156
  punctatum, **156**, tab. **48** fig. A
  scandens, 156
Microtrichomanes parvulum, 79
MOHRIA, 41, **54**
  achilleifolia, 56
  caffrorum, **56,** tab. **15** fig. B
    var. *multisquamosa*, 56
  crenata, 56
  lepigera, **56**, tab. **15** fig. C
  thurifraga, 56
  vestita, 56
Myriopteris hirta, 125
  intermedia, 125

Nephrodium acutum, 160
  albopunctatum, 163
  anateinophlebium, 193
  athamanticum, 221
  bergianum, 193
  biserratum, 160
  boryanum, 207
  buchananii, 225
  catopteron, 234

cirrhosum, 232
costulare, 196
crenatum, 230
crinobulbon, 232
eurylepium, 221
glabratum, 207
gueinzianum, 194
hirsutum, 230
inaequale, 222
kilemense, 222
lanuginosum, 232
lastii, 222
longicuspe, 192
monocarpum, 163
odoratum, 230
pallidivenium, 199
patens, 196
pentheri, 222
plantianum, 198
prismaticum, 197
proliferum, 200
pulchrum, 193
quadrangulare, 195
scandicinum, 204
sewellii, 193
spekei, 232
splendens, 160
squamisetum, 225
squamulosum, 190
strigosum, 193
tottum, 199
wakefieldii, 195
welwitschii, 232
zambesiacum, 192
NEPHROLEPIS, 42, **159**
  acuta, 160
  acutifolia, **160,** tab. **49** fig. B
  biserrata, 160
  caudata, 160
  cordifolia var. *compacta*, 162
  duffei, 160
  filipes, 162
  imbricata, 162
  lunulata, 220
  pluma, 162
  punctulata var. *hirsuta*, 160
  splendens, 160
  undulata, 160, **162**, tab. **49** fig. A
Niphobolus adnascens, 146
  fissus, 146
  giesenhagenii, 146
  lanceolatus, 146
  polycarpus, 156
  schimperanus, 147
  schraderi, 152
  spissus, 146
  stolzii, 147
  tener, 146
Notholaena, 122
  bipinnata, 126
  buchananii, 126
  calomelanos, 132
  capensis, 124
  cordata, 188
  eckloniana, 125
  hirta, 125
  inaequalis, 126
  krebsiana, 125
  leachii, 125

lepigera, 56
  tricholepis, 126
Oetosis elongata, 96
  isoetifolia, 94
OLEANDRA, 41, 159, **165**
  africana, 165
  articulata, var. *welwitschii*, 165
  densifrons, 165
  distenta, **165,** tab. **51**
    var. *hirsuta*, 165
    var. *villosa*, 165
Olfersia hybrida, 213
  spathulata, 213
Onoclea attenuata, 235
  capensis, 239
  polypodioides, 48
OPHIOGLOSSACEAE, 34
OPHIOGLOSSALES, 13, 34
OPHIOGLOSSUM, **34**
  aitchisonii, 37
  aphrodisiacum, 38
  brevipes, 38
  capense, 37
    var. *regulare*, 37
  costatum, 34, **38,** tab. **8** fig. D
  cuspidatum, 37
  felixii, 38
  fibrosum, 38
  gomezianum, 34, **35,** tab. **8** fig. A
  lancifolium, **35,** tab. **8** fig. C
  lusoafricanum, 35
  pedunculosum, 38
  petiolatum, 38
  polyphyllum, 34, **37**
  regulare, 37
  reticulatum, 35, **37**
  rubellum, 34, **35,** tab. **8** fig. E
  tapinum, 37
  thomasii, 34, **35,** tab. **8** fig. B
  vulgatum, 35, **37**
    var. *aitchisonii*, 37
    var. *kilimandscharicum*, 37
    var. *polyphyllum*, 37
  wightii, 38
Ornithopteris, 54
  aquilina, 88
OSMUNDA, 41, **44**
  capensis, 44, 239
  leptophylla, 99
  marginalis, 56
  regalis, **44,** tab. **10**
    var. *capensis*, 44
  thurifera, 56
  thurifraga, 56
  totta, 46
OSMUNDACEAE, 41, 44

Paesia aquilina, 88
Parathyrium boryanum, 207
Parkeriaceae, 102,
PELLAEA, 42, 43, 98, **128**
  adiantoides 131, 135
  angulosa, 128, **131,** tab. **40** fig. A
  articulata, 131
  boivinii, 128, **131**
  burkeana, 132
  calomelanos, 128, **132**
    var. calomelanos, **132**

  var. swynnertoniana, 132, **133,** tab.
    40 fig. B
  consobrina, 133
  doniana, 128, **129,** tab. **40** fig. C
  dura, 128, **132,**
  goudotii, 128
  hastata, 132
    var. *glauca*, 135
  involuta, 136
  longipilosa, 128, **129,** tab. **40** fig. D
  pectiniformis, **128**
  quadripinnata, 128, **133,** tab. **41**
  swynnertoniana, 133
  viridis, 128, **133**
    var. glauca, **135**
    var. involuta, 135, **136**
    var. viridis, **135**
Pellaeopsis articulata, 131
  burkeana, 132
Phegopteris incisa, 86
  luxurians, 200
  manniana, 223
  prolifera, 200
  sparsisora, 92
  totta, 199
  unita, 196
PHYMATODES, 42, 144, **153**
  elongata, 152
  excavata, 151
  grossa, 155
  irioides, 156
  phymatodes, 155
  polycarpa, 156
  scolopendria, **153,** tab. **47**
  simplex, 151
  vulgaris, 153
PITYROGRAMMA, 43, 98, **105**
  argentea, **105,** tab. **35**
  aurantiaca, 105, **107**
  austroamericana, 108
  calomelanos, 105, **107**
    var. aureoflava, 107, **108**
    var. *austroamericana*, 108
    var. calomelanos, **107**
  chamaesorbus, 107
  insularis, 107
Plananthus gnidioides, 18
  saururus, 17
PLATYCERIUM, 41, 42, 144, **145**
  alcicorne, **145,** frontisp.
  angolense, 145
  bifurcatum, 145
  elephantotis, **145,** frontisp.
  vassei, 145
  velutinum, 145
Platyloma adiantoides, 135
  calomelanos, 132
PLEOPELTIS, 42, 144, **151**
  corradii, 152
  ensifolia, 153
  excavata, **151,** tab. **45** fig. C
  irioides, 156
  kaulfussiana, 153
  lanceolata, 152, 153
  lepidota, 153
  macrocarpa, 151, **152,** tab. **45** fig. B
  marginalis, 153
  mildbraedii, 152
  phymatodes, 155

PLEOPELTIS (*contd.*)
  polycarpa, 156
  preussii, 152
  punctata, 156
  rotundum, 152
  schraderi, 151, **152**
  simplex, 151
  stolzii, 152
  vesciculari-palacea, 152
Pleurosorus pozoi, 199
Polybotrya meyerana, 240
  tenuifolia, 240
POLYPODIACEAE, 41, **144**
POLYPODIUM, 42, 145, **158**
  adnascens, 146
  adspersum, 153
  africanum, 199
  anguinum, 155
  astrosorum, 155
  bergianum, 193
  caffrorum, 56
  capense, 74
  coriaceum, 151
  crassinerve, 156
  crenatum, 230
  cultratum, 143
    var. *elasticum*, 143
  dentatum, 197
  duale, 144
  ecklonii, 158
  elasticum, 143
  eliasii, 200
  elongatum, 152
  excavatum, 151
  flabelliforme, 143
  grossum, 153
  gueintzii, 152
  irioides, 156
  lanceolatum, 152
  lepidotum, 153
  leptophyllum, 99
  ligustrifolium, 155
  lineare, 50
    var. *schraderi*, 152
  lingulatum, 156
  loxogramme, 151
  luxurians, 200
  lycopodioides, 155
    var. *mackenii*, 155
    var. *myrtillifolium*, 155
  mackenii, 155
  macrocarpum, 153
  mannianum, 223
  marginale, 153
  mauritianum, 155
  myrtillifolium, 155
  nanodes, 141
  newtonii, 143
  normale var. *madagascariense*, 158
  oosorum, 143
  orientale, 163
  ovariense, 155
  pallidivenium, 199
  pappei, 158
  pectinatum, 163
  pertusum, 146
  phymatodes, 153
  polycarpon, 156
  polypodioides, **158**

  subsp. ecklonii, **158,** tab. **45** fig. D
  punctatum, 156
  rigescens, 143
  scolopendria, 153
  schimperanum, 147
  schraderi, 152
  serrulatum, 144
  setiferum, 226
  simplex, 151
  speluncae, 89
  spissum, 146
  suberosum, 151
  tottum, 198, 199
  trapezoides, 115
  unitum, 196
  villosissimum, 142
    var. *majus*, 142
  vittarioides, 146
POLYSTICHUM, 43, 220, **226**
  capense, 75
  fuscopaleaceum, 226
  goggilodus, 198
  inaequale, 222
  lanuginosum, 234
  luctuosum, 226, **228,** tab. **64** fig. C
  setiferum, **226**
    var. fuscopaleaceum, **226,** tab. **64** fig
      B
  zambesiacum, 226, tab. **64** fig. A
PSILOTACEAE, 15
PSILOTALES, 13, **15**
PSILOTUM, **15**
  natalense, 15
  nudum, **15,** tab. **2**
  triquetrum, 15
Pteridella adiantoides, 131
  angulosa, 131
  doniana, 129
  dura, 132
  involuta, 136
  pectiniformis, 128
  quadripinnata, 133
PTERIDIUM, 43, 81, **88**
  aquilinum, **88**
    var. *africanum*, 89
  subsp. aquilinum, **88**
    subsp. *capense*, 88
    subsp. *caudatum* var. *africanum*, 89
    subsp. centrali-africanum, 88, **89,**
      tab. **26**
  capense, 88
  centrali-africanum, 89
PTERIS, 42, 43, 98, **115**
  abrahamii, 118
  acanthoneura, 120
  acuminatissima, 116
  adamii, 117
  adiantoides, 135
  adscensionis, 117
  aequalis, 116
  alpinii, 116
  angulosa, 131
  aquilina, 88
    var. *lanuginosa*, 88
  arguta var. *flabellata*, 117
  articulata, 131
  biaurita, 118
  boivinii, 131
  brevisora, 117

brunoniana, 84
buchananii, 115, **120**
burkeana, 132
calomelanos, 132
capensis, 88
catoptera, 115, **118**
   var. catoptera, **118**
   var. horridula, 118, tab. **38**
concolor, 121
confluens, 190
consobrina, 133
coriifolia, 88
costata, 115
cretica, 115, **116**
currorii, 84
dentata, 115, **117**
  subsp. flabellata, **117**
diversifolia, 115
doniana, 129
dura, 132
ensifolia, 115
farinosa, 122
flabellata, 117
friesii, 115, **118**
glabra, 82
guichenotiana, 116
hamulosa, 115, **120**
hastifolia, 135
inaequilateralis, 116
incisa, 84
intricata, 115, **117**
involuta, 136
lanceolata, 115
lanuginosa, 88
lunulata, 110
mannii, 84
microdonta, 116
montana, 86
nervosa, 116
obliqua, 115
pectiniformis, 128
pentaphylla, 116
polymorpha, 135
pteridioides, 115, **117**
quadriaurita, 120
quadripinnata, 133
radiata, 138
semiserrata, 116, 117
serraria, 116
siliquosa, 102
subquadripinnata, 120
succulenta, 105
tabularis, 237
tenuifolia, 116
thalictroides, 102
vespertilionis, 84
viridis, 135
vittata, **115**
vulcanica, 116
Pteropsis angustifolia, 94
Pycnodoria cretica, 116
  vittata, 116
PYRROSIA, 41, 144, **146**
  adnascens, 146
  lanceolata, **146**
  mechowii, 147
  rhodesiana, 146, **147**
  schimperana, 146, **147**, tab. **45** fig. A
    var. *mechowii*, 147

  stolzii, 146, **147**
Ramondia, 57

Sagenia gemmifera, 234
SALVINIA, 40, **67**
  adnata, 67
  auriculata, **67**, tab. **19** fig. B
  hastata, 67, tab. **19** fig. A
  hildebrandtii, 67
  mollis, 67
SALVINIACEAE, 67
SCHIZAEA, 41, **52**
  pectinata, **52**, tab. **14**
SCHIZAEACEAE, 41, **52**
Schizolegnia, 139
  ensifolia, 139
Schizoloma, 139
  ensifolia, 139
SELAGINELLA, 22
  abyssinica, 23, **27**, tab. **5**
  bueensis, 27
  cathedrifolia, 26
  cooperi, 26
  depressa, 26
  dregei, 22, **23**, tab. **4**
  eublepharis, 22, **27**
  goetzei, 27
  grisea, 23
  hortensis, 26
  imbricata, 22, **25**
  kirkii, 27
  kraussiana, 22, **26**
  mackenii, 26
  mittenii, 22, **26**
  nivea, 22, **25**
  njam-njamensis, 22, **23**
  perpusilla, 23, **29**
  pervillei, 27
  preussii, 27
  rupestris forma *dregei*, 23
    var. *recurva* forma *dregeana*, 23
  subgen. SELAGINELLA, 22
  subgen. STACHYGYNANDRUM, 22
  subisophylla, 23, **25**
  tectissima, 26
  tenerrima, 23, **29**
  vogelii, 27
  welwitschii, 26
  whytei, 27
SELAGINELLACEAE, 22
SELAGINELLALES, 13, **22**
Selenodesmium cupressoides, 78
  rigidum, 78
Selliguea coriacea, 151
  lanceolata, 151
Sphaerocionium pendulum, 80
Spicanta attenuata, 235
  australis, 240
  capensis, 239
  punctulata, 239
STENOCHLAENA, 42, 235, **240**
  meyerana, 240
  mildbraedii, 242
  tenuifolia, **240**, tab. **69**
  warneckii, 216
Stenogramma pozoi, 200
Sticherus umbraculifera, 50
Struthiopteris australis, 240

Struthiopteris (*contd.*)
  dregeana, 239
  inflexa, 236
  punctulata, 239
  regalis, 44
  rigida, 239
STYLITES, 30

Tarachia friesiorum, 178
  furcata, 182
TECTARIA, 43, 220, **234**
  gemmifera, **234,** tab. **64** fig. D
THELYPTERIDACEAE, 189
THELYPTERIS, 42, 43, 44, **189**
  bergiana, 190, **193**, tab. **55** fig. B
  chaseana, 190, **194**
  confluens, 190, tab. **55** fig. E
  dentata, 190, **197**
    var. buchananii, 197, **198**
    var. dentata, **197,** tab. **55** fig. C
  extensa, 190, **195**
  friesii, 190, **192**
  glabrata, 207
    var. *hirsuta*, 207
  goggilodus, 198
  gongylodes, 198
  gueinziana, 190, **194**
  longicuspis, 190, **192**, tab. **55** fig. A
  madagascariensis, 190, **196,** tab. **55**
    fig. D
  palustris var. *squamigera*, 190
  pozoi, 190, **199**, tab. **55** fig. G
  prismatica, 190, **197**
  quadrangularis, 190, **195**
  spekei, 232
  squamulosa, 190
  striata, 190, **199**
  strigosa, 190, **193**
  totta, 190, **198**
  zambesiaca, 192
TODEA, 41, **46**
  africana, 46
  barbara, **46,** tab. **11**
TRICHOMANES, 41, **75**
  aerugineum, 76
  aethiopicum, 181
  borbonicum, **76,** tab. **22** fig. D
  chaerophylloides, 167
  chamaedrys, 76
  cormophyllum, 75
  cupressoides, 78
  dregei, 78
  erosum, **76**
    var. aerugineum, 76
    var. *chamaedrys*, 76
    var. erosum, **76,** tab. **22** fig. B
  frappieri, 78

  goetzei, 76
  incisum, 74
  mandiocanum, 78
  melanotrichum, 78
  palmicola, 76
  parvulum, 79
  pyxidiferum, **76**
    var. melanotrichum, **78,** tab. **22**
    fig. C
  pyxidiferum, 78
    forma *major*, 78
  rigidum, 76, **78,** tab. **22** fig. A
  sibthorpiodes, 79
  thouarsianum, 79
  tunbridgense, 80

Ugena, 57
  microphylla, 57
Urostachys, 17
  acerosus, 17
  dacrydioides, 18
  gnidioides, 18
  ophioglossoides, 18
  phlegmaria, 20
  saururus, 17
  verticillatus, 17

Vandenboschia melanotricha, 78
VITTARIA, 41, **94**
  acrostichoides, 210
  elongata, 94, **96,** tab. **29** fig. C
  gueinzii, 94
  guineensis, **96**
    var. orientalis, 94, **96**
  hildebrandtii, 96
  isoetifolia, **94,** tab. **29** fig. B
  longidentata, 94
  sarmentosa, 94
  scolpendrina, 96
  stuhlmannii, 96
  tenera, 94
  volkensii, **94,** tab. **29** fig. A
VITTARIACEAE, 92,

XIPHOPTERIS, 42, **141**
  cultrata, 142, **143**
  elastica, 143
  extensa, 144
  flabelliformis, 142, **143,** tab. **44** fig. B
  oosora, 142, **143**
  orientalis, 144
  rigescens, 143
  serrulata, 142, **144**
  villosissima, 141, **142**

FLORA ZAMBESIACA
THE GEOGRAPHICAL DIVISIONS OF THE FLORA

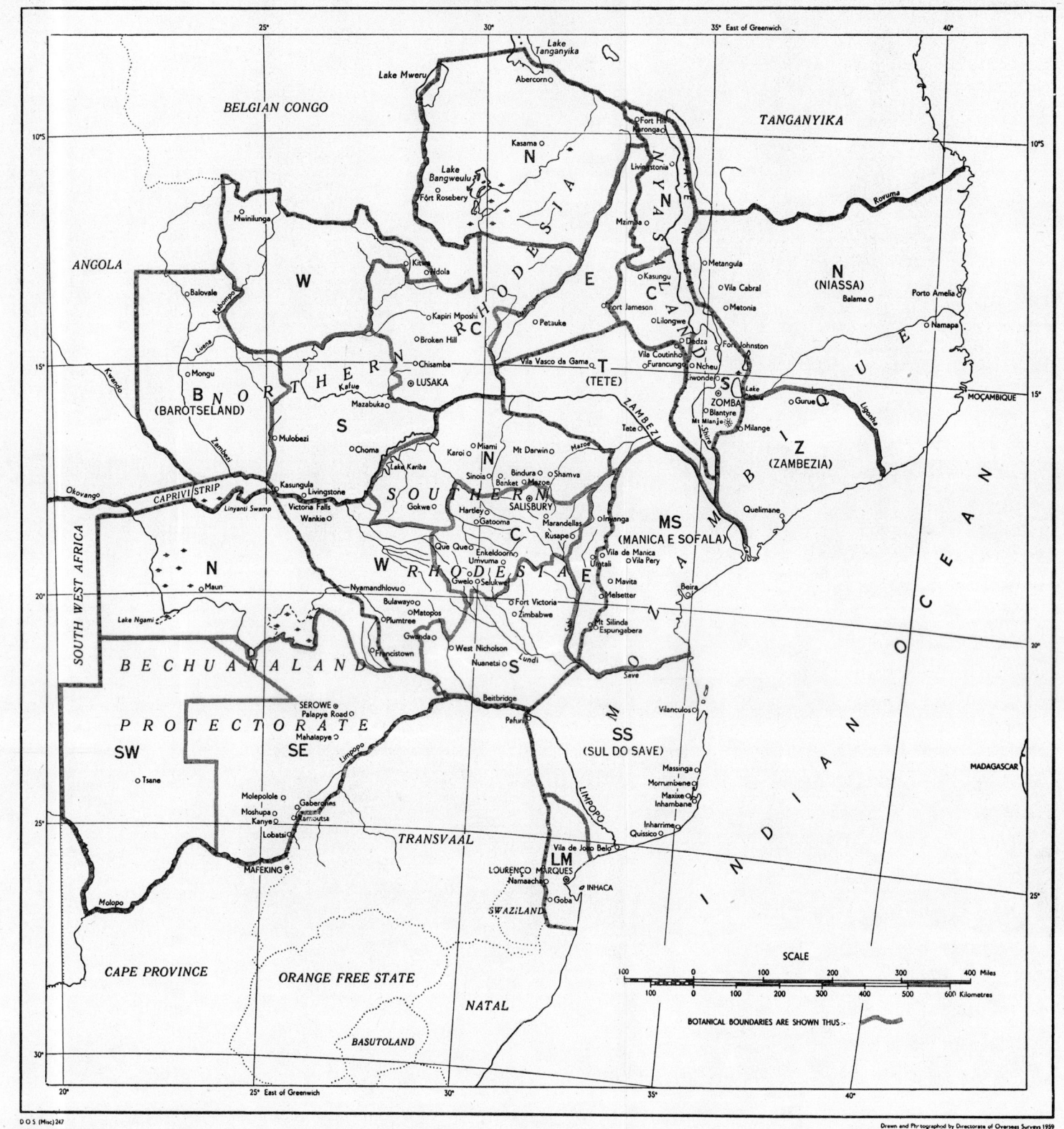

Drawn and Photographed by Directorate of Overseas Surveys 1959